# Synchronization Techniques for Digital Receivers

# Applications of Communications Theory
## Series Editor: R. W. Lucky, *Bellcore*

**Editorial Board:**
**Anthony S. Acampora,** University of Southern California
**Tingye Li,** AT&T Bell Laboratories
**William H. Tranter,** Virginia Polytechnic Institute and State University

*Recent volumes in this Series:*

BASIC CONCEPTS IN INFORMATION THEORY AND CODING:
The Adventures of Secret Agent 00111
Solomon W. Golomb, Robert E. Peile, and Robert A. Scholtz

COMMUNICATION SYSTEM DESIGN USING DSP ALGORITHMS:
With Laboratory Experiments for the TMS320C30
Steven A. Tretter

COMPUTER COMMUNICATIONS AND NETWORKS
John R. Freer

COMPUTER NETWORK ARCHITECTURES AND PROTOCOLS
Second Edition • Edited by Carl A. Sunshine

DATA COMMUNICATIONS PRINCIPLES
Richard D. Gitlin, Jeremiah F. Hayes, and Stephen B. Weinstein

FUNDAMENTALS OF DIGITAL SWITCHING
Second Edition • Edited by John C. McDonald

AN INTRODUCTION TO BROADBAND NETWORKS:
LANs, MANs, ATM, B-ISBN, and Optical Networks for Integrated
Multimedia Telecommunications
Anthony S. Acampora

AN INTRODUCTION TO PHOTONIC SWITCHING FABRICS
H. Scott Hinton

OPTICAL CHANNELS: Fibers, Clouds, Water, and the Atmosphere
Sherman Karp, Robert M. Gagliardi, Steven E. Moran, and Larry B. Stotts

SIMULATION OF COMMUNICATIONS SYSTEMS
Michel C. Jeruchim, Philip Balaban, and K. Sam Shanmugan

SYNCHRONIZATION TECHNIQUES FOR DIGITAL RECEIVERS
Umberto Mengali and Aldo N. D'Andrea

A Continuation Order Plan is available for this series. A continuation order will bring delivery of each new volume immediately upon publication. Volumes are billed only upon actual shipment. For further information please contact the publisher.

# Synchronization Techniques for Digital Receivers

Umberto Mengali and
Aldo N. D'Andrea
*University of Pisa*
*Pisa, Italy*

**PLENUM PRESS • NEW YORK AND LONDON**

Library of Congress Cataloging-in-Publication Data

On file

ISBN 0-306-45725-3

© 1997 Plenum Press, New York
A Division of Plenum Publishing Corporation
233 Spring Street, New York, N.Y. 10013

http://www.plenum.com

10 9 8 7 6 5 4 3 2 1

All rights reserved

No part of this book may be reproduced, stored in a retrieval system, or transmitted in any form or by any means, electronic, mechanical, photocopying, microfilming, recording, or otherwise, without written permission from the Publisher

Printed in the United States of America

# Preface

Synchronization is a critical function in digital communications; its failures may have catastrophic effects on the transmission system performance. Furthermore, synchronization circuits comprehend such a large part of the receiver hardware that their implementation has a substantial impact on the overall costs. For these reasons design engineers are particularly concerned with the development of new and more efficient synchronization structures. Unfortunately, the advent of digital VLSI technology has radically affected modem design rules, to a point that most analog techniques employed so far have become totally obsolete.

Although digital synchronization methods are well established by now in the literature, they only appear in the form of technical papers, often concentrating on specific performance or implementation issues. As a consequence they are hardly useful to give a unified view of an otherwise seemingly heterogeneous field. It is widely recognized that a fundamental understanding of digital synchronization can only be reached by providing the designer with a solid theoretical framework, or else he will not know where to adjust his methods when he attempts to apply them to new situations. The task of the present book is just to develop such a framework.

This is achieved by considering synchronization as a parameter estimation problem and approaching it with the techniques of estimation theory. In doing so two main goals are attained. One is to offer a coherent and systematic methodology to follow when looking for new synchronization structures. The other is to provide the designer with precise indications on the inherent performance limits of these structures.

Synchronization circuits are occasionally devised on an ad hoc basis and proven eventually by demonstration in hardware or computer simulation. Ad hoc synchronization procedures are welcome and fully acknowledged in this book. They result from application of physical insight and may lead to valuable solutions. When facing more complex problems, however, like those encountered with continuous-phase modulations, they seem of lesser efficacy and a theoretical oriented approach is indispensable.

Exercises have been inserted throughout the text as a convenient means for providing examples of application of the proposed techniques. They are not merely routine manipulations of equations. Their purpose is rather to supplement the text in various ways: (*i*) to gain familiarity with important concepts; (*ii*) to apply these concepts to practical situations; (*iii*) to fill in missing details.

The book is intended for three categories of readers. Primarily, it should be a valuable tool for design engineers in telecommunications industry. Second, it might be used as supplementary material in digital transmission courses or as a separate course in synchronization or digital modem design. As a text for a graduate-level course the book can be covered in one semester. Finally, it should be useful to researchers. On several occasions in the book we have pointed out open problems of considerable technical relevance.

The book is self-contained and any significant results are derived either in the text or in the appendices. The underlying assumptions and methods employed in the derivations are accurately outlined and the final outcomes are discussed and compared with other situations, in order to stress the physical significance. Nevertheless, as many aspects of synchronization can only be expressed in mathematical terms, the reader must have some mathematical background. In particular, a working knowledge of linear system theory, Fourier transforms, and stochastic processes is needed.

This leaves only the pleasant task of acknowledging the contribution of several people to the creation of this book. Many thanks go to our good friends and colleagues Floyd Gardner, Des Taylor, and Ruggero Reggiannini, who suggested valuable improvements and reviewed several portions of the manuscript. We would also like to express gratitude to our co-workers and students Antonio D'Amico, Alberto Ginesi, Michele Morelli, and Giorgio Vitetta, who performed many simulations, reviewed the manuscript in detail, and offered corrections and changes. There are no words to describe adequately our indebtedness to all of them.

# Contents

1. **Introduction** .................................................. 1
   1.1. What Synchronization Is About ......................... 1
   1.2. Outline of the Book .................................... 4
   *References* ................................................. 7

2. **Principles, Methods and Performance Limits** ........... 9
   2.1. Introduction ........................................... 9
   2.2. Synchronization Functions ............................. 9
      2.2.1. Timing Recovery with Baseband Systems .......... 10
      2.2.2. Degradations Due to Timing Errors .............. 12
      2.2.3. Passband PAM Systems ........................... 15
      2.2.4. Synchronization in PAM Coherent Receivers ...... 18
      2.2.5. Degradations Due to Phase Errors ............... 20
      2.2.6. Synchronization in PAM Differential Receivers .. 25
      2.2.7. Synchronization in CPM Systems ................. 28
      2.2.8. Synchronization in Simplified CPM Receivers .... 31
   2.3. Maximum Likelihood Estimation ........................ 37
      2.3.1. ML Estimation from Continuous-Time Waveforms ... 38
      2.3.2. Baseband Signaling ............................. 44
      2.3.3. ML Estimation from Sample Sequences ............ 47
      2.3.4. Baseband Signaling ............................. 52
   2.4. Performance Limits in Synchronization ................ 53
      2.4.1. True and Modified Cramer-Rao Bounds ............ 53
      2.4.2. An Alternative Approach to the Bounds .......... 57
      2.4.3. $MCRB(v)$ with PAM Modulation .................. 58
      2.4.4. $MCRB(v)$ with CPM Modulation .................. 60

|     |       |                                                              |     |
|-----|-------|--------------------------------------------------------------|-----|
|     | 2.4.5. | $MCRB(\theta)$ with PAM and CPM Modulations              | 61  |
|     | 2.4.6. | $MCRB(\tau)$ with PAM Modulation                         | 64  |
|     | 2.4.7. | $MCRB(\tau)$ with CPM Modulation                         | 67  |
| 2.5.| Key Points of the Chapter                                           | 69  |
| Appendix 2.A                                                             | 70  |
|     | 2.A.1. | Power Spectral Density for Baseband Signals              | 70  |
|     | 2.A.2. | Power Spectral Density for Bandpass Signals              | 71  |
|     | 2.A.3. | Extension to Trellis-Coded Modulations                   | 73  |
| Appendix 2.B                                                             | 73  |
| *References*                                                             | 75  |

# 3. Carrier Frequency Recovery with Linear Modulations .... 79

| 3.1. | Introduction | 79 |
|------|--------------|----|
| 3.2. | Data-Aided Frequency Estimation | 80 |
|      | 3.2.1. Maximum Likelihood Estimation | 80 |
|      | 3.2.2. Practical Frequency Estimators | 84 |
|      | 3.2.3. First Method (Kay [3]) | 86 |
|      | 3.2.4. Second Method (Fitz [4]) | 88 |
|      | 3.2.5. Third Method (Luise and Reggiannini [6]) | 89 |
|      | 3.2.6. Fourth Method (Approximate ML Estimation) | 91 |
|      | 3.2.7. Performance Comparisons | 92 |
| 3.3. | Decision-Directed Recovery with DPSK | 97 |
|      | 3.3.1. Decision-Directed Algorithms with Differential PSK | 97 |
| 3.4. | Non-Data-Aided but Clock-Aided Recovery | 100 |
|      | 3.4.1. Closed-Loop Algorithm | 100 |
|      | 3.4.2. Extension to M-ary PSK and QAM | 104 |
|      | 3.4.3. Open-Loop Algorithms | 105 |
| 3.5. | Closed-Loop Recovery with No Timing Information | 108 |
|      | 3.5.1. Likelihood Function | 108 |
|      | 3.5.2. Open-Loop Search | 115 |
|      | 3.5.3. Closed-Loop Estimator | 115 |
|      | 3.5.4. Frequency Acquisition | 120 |
|      | 3.5.5. Frequency Tracking | 124 |
|      | 3.5.6. Comparison with MCRB | 128 |
|      | 3.5.7. Other Frequency Error Detectors | 130 |
| 3.6. | Open-Loop Recovery with No Timing Information | 133 |
|      | 3.6.1. Delay-and-Multiply Method | 133 |
|      | 3.6.2. Digital Implementation | 139 |
|      | 3.6.3. Effects of Adjacent Channel Interference | 140 |

| | | |
|---|---|---|
| 3.7. | Key Points of the Chapter | 143 |
| | Appendix 3.A | 144 |
| | *References* | 145 |

## 4. Carrier Frequency Recovery with CPM Modulations — 147

- 4.1. Introduction — 147
- 4.2. Laurent Expansion — 148
- 4.3. Data-Aided Frequency Estimation — 154
  - 4.3.1. Frequency Estimation with MSK — 154
  - 4.3.2. Extension to MSK-Type Modulation — 156
- 4.4. ML-Based NDA Frequency Estimation — 157
  - 4.4.1. MSK-Type Modulation — 157
  - 4.4.2. General CPM Modulation — 161
  - 4.4.3. Loop Performance — 167
- 4.5. Delay-and-Multiply Schemes — 169
  - 4.5.1. Open-Loop Scheme — 169
  - 4.5.2. Closed-Loop Scheme — 174
- 4.6. Clock-Aided Recovery — 176
  - 4.6.1. Delay-and-Multiply Method — 176
  - 4.6.2. $2P$-Power Method with Full Response Formats — 179
- 4.7. Key Points of the Chapter — 182
- Appendix 4.A — 183
- Appendix 4.B — 185
- Appendix 4.C — 187
- *References* — 188

## 5. Carrier Phase Recovery with Linear Modulations — 189

- 5.1. Introduction — 189
- 5.2. Clock-Aided and Data-Aided Phase Recovery — 190
  - 5.2.1. ML Estimation with Non-Offset Formats — 190
  - 5.2.2. Performance with Non-Offset Formats — 192
  - 5.2.3. ML Estimation with Offset Formats — 194
  - 5.2.4. Performance with Offset Formats — 195
  - 5.2.5. Degradations Due to Frequency Errors — 198
- 5.3. Decision-Directed Phase Recovery with Non-Offset Modulation — 201
  - 5.3.1. Feedback Structures — 201

|        |        |                                                                 |     |
|--------|--------|-----------------------------------------------------------------|-----|
|        | 5.3.2. | First Approach                                                  | 202 |
|        | 5.3.3. | Second Approach                                                 | 205 |
|        | 5.3.4. | Acquisition and Tracking Characteristics                        | 206 |
|        | 5.3.5. | S-Curves for General Modulation Formats                         | 211 |
|        | 5.3.6. | Tracking Performance                                            | 213 |
|        | 5.3.7. | Effect of Frequency Errors                                      | 217 |
|        | 5.3.8. | Second-Order Tracking Loops                                     | 220 |
|        | 5.3.9. | Phase Noise                                                     | 224 |
| 5.4.   | Decision-Directed Phase Recovery with Offset Modulation         |     | 228 |
|        | 5.4.1. | Phase Estimation Loop                                           | 228 |
|        | 5.4.2. | Tracking Performance with Offset Formats                        | 232 |
|        | 5.4.3. | Effects of Phase Noise and Frequency Errors                     | 236 |
| 5.5.   | Multiple Phase-Recovery with Trellis-Coded Modulations          |     | 236 |
|        | 5.5.1. | Tentative Decisions                                             | 236 |
|        | 5.5.2. | Multiple Synchronizers                                          | 240 |
| 5.6.   | Phase Tracking with Frequency-Flat Fading                       |     | 245 |
|        | 5.6.1. | Channel Estimation Problem                                      | 245 |
|        | 5.6.2. | Pilot-Tone Assisted Modulation                                  | 251 |
|        | 5.6.3. | Pilot-Symbol Assisted Modulation                                | 252 |
|        | 5.6.4. | Per-Survivor Channel Estimation                                 | 259 |
| 5.7.   | Clock-Aided but Non-Data-Aided Phase Recovery with Non-Offset Formats | 266 |
|        | 5.7.1. | Likelihood Function                                             | 266 |
|        | 5.7.2. | High SNR                                                        | 269 |
|        | 5.7.3. | Low SNR                                                         | 271 |
|        | 5.7.4. | Feedforward Estimation with PSK                                 | 277 |
|        | 5.7.5. | Feedforward Estimation with QAM                                 | 281 |
|        | 5.7.6. | Ambiguity Resolution                                            | 282 |
|        | 5.7.7. | The Unwrapping Problem                                          | 284 |
| 5.8.   | Clock-Aided but Non-Data-Aided Phase Recovery with OQPSK        | 286 |
|        | 5.8.1. | Likelihood Function                                             | 286 |
|        | 5.8.2. | Feedback Estimation Method                                      | 288 |
|        | 5.8.3. | ML-Oriented Feedforward Method                                  | 291 |
|        | 5.8.4. | Viterbi-Like Method                                             | 295 |
| 5.9.   | Clockless Phase Recovery with PSK                               | 297 |
|        | 5.9.1. | ML-Based Feedforward Estimation                                 | 297 |
| 5.10.  | Clockless Phase Recovery with OQPSK                             | 299 |
|        | 5.10.1. | Ad Hoc Method                                                  | 299 |
| 5.11.  | Key Points of the Chapter                                       | 301 |
| *References* |                                                           | 303 |

Contents                                                                      xi

## 6. Carrier Phase Recovery with CPM Modulations .......... 307

6.1. Introduction ................................. 307
6.2. Data-Aided Phase Estimation with MSK-Type Modulation ... 308
    6.2.1. MSK-Type Receivers ...................... 308
    6.2.2. Data-Aided Phase Estimation with MSK-Type
            Modulation .............................. 311
    6.2.3. Estimator Performance with MSK .............. 312
6.3. Decision-Directed Estimation with MSK-Type Modulation ... 314
    6.3.1. Decision-Directed Estimation with MSK .......... 314
6.4. Decision-Directed Estimation with General CPM ......... 319
    6.4.1. ML Receivers for CPM .................... 319
    6.4.2. Decision-Directed Phase Estimation ............ 322
6.5. CPM Signaling over Frequency-Flat Fading Channels ...... 327
    6.5.1. ML Receiver ........................... 327
    6.5.2. Approximate ML Receiver Based on Per-Survivor-
            Processing Methods ....................... 328
    6.5.3. Improved Methods for Fast-Fading Channels ....... 333
    6.5.4. Linearly Time-Selective Channels .............. 336
6.6. Clock-Aided but Non-Data-Aided Phase Estimation ....... 339
    6.6.1. $2P$-Power Method for Full-Response Systems ...... 339
    6.6.2. ML-Oriented Phase Estimation ............... 342
6.7. Clockless Phase Estimation ....................... 346
6.8. Key Points of the Chapter ....................... 351
*References* ...................................... 351

## 7. Timing Recovery in Baseband Transmission ............ 353

7.1. Introduction ................................. 353
7.2. Synchronous Sampling .......................... 355
    7.2.1. Hybrid NCO ........................... 355
    7.2.2. Timing Adjustment for Synchronous Sampling ...... 357
7.3. Non-Synchronous Sampling ....................... 359
    7.3.1. Feedback Recovery Scheme .................. 359
    7.3.2. Piecewise Polynomial Interpolators ............. 362
    7.3.3. Timing Adjustment with Non-Synchronous Sampling .. 366
    7.3.4. Timing Adjustment with Feedforward Schemes ..... 368
7.4. Decision-Directed Timing Error Detectors .............. 371
    7.4.1. ML-Based Detectors ...................... 371
    7.4.2. S-Curve .............................. 375

7.4.3. Tracking Performance ................. 377
7.4.4. Approximate-Derivative Method .......... 383
7.4.5. Other Timing Error Detectors ............. 385
7.5. Non-Data-Aided Detectors ...................... 391
7.5.1. ML-Based Detector ..................... 391
7.5.2. The Gardner Detector .................... 393
7.5.3. Tracking Performance ................. 395
7.5.4. Self Noise Elimination with the Gardner Detector .... 396
7.6. Feedforward Estimation Schemes ................. 398
7.6.1. Non-Data-Aided ML-Based Algorithm ......... 398
7.6.2. Oerder and Meyr Algorithm ................ 402
7.7. Key Points of the Chapter ...................... 406
Appendix 7.A ................................. 407
*References* ................................... 408

## 8. Timing Recovery with Linear Modulations ............ 411

8.1. Introduction ............................. 411
8.2. Decision-Directed Joint Phase and Timing Recovery with Non-Offset Formats ........................ 412
8.2.1. ML-Based Joint Phase and Timing Estimation ...... 413
8.2.2. Remark .............................. 415
8.2.3. Ad Hoc Timing Detectors .................. 417
8.2.4. Equivalent Model of the Synchronizer .......... 418
8.2.5. Tracking Performance .................... 425
8.3. Non-Data-Aided Feedback Timing Recovery with Non-Offset Formats ............................ 428
8.3.1. ML-Oriented NDA Feedback Timing Recovery .... 428
8.3.2. The Gardner Detector and Its Performance ......... 431
8.4. Non-Data-Aided Feedforward Estimators with Non-Offset Formats ................................ 433
8.5. Timing Recovery with Frequency-Flat Fading Channels ..... 438
8.5.1. DD-CA Timing Recovery .................. 439
8.5.2. NDA Timing Recovery ................... 442
8.5.3. High SNR Approximation to the ML Estimator ...... 444
8.5.4. Modified ML Algorithm .................. 449
8.6. Decision-Directed Joint Phase and Timing Recovery with Offset Formats ............................ 452
8.6.1. ML-Based Joint Phase and Timing Estimation ...... 453
8.6.2. Other Timing Detectors ................... 459

Contents     xiii

    8.7. NDA Feedforward Joint Phase and Timing Recovery with Offset Formats .................................... 462
        8.7.1. Computation of the Likelihood Function .......... 462
        8.7.2. ML-Based Estimator ..................... 464
        8.7.3. Estimator Performance ................... 467
    8.8. Key Points of the Chapter ......................... 470
    Appendix 8.A ....................................... 470
    Appendix 8.B ....................................... 472
    Appendix 8.C ....................................... 473
    *References* ........................................ 475

## 9. Timing Recovery with CPM Modulations .............. 477

    9.1. Introduction ................................... 477
    9.2. Decision-Directed Joint Phase and Timing Recovery ....... 478
        9.2.1. ML Formulation ......................... 478
        9.2.2. Approximate Digital Differentiation ............. 483
        9.2.3. Tracking Performance and Spurious Locks ........ 485
    9.3. NDA Feedback Timing Recovery .................... 487
        9.3.1. Approximate Expression for the Likelihood Function .. 487
        9.3.2. Timing Error Detector ..................... 491
        9.3.3. Performance ........................... 493
        9.3.4. False Lock Detection ..................... 495
    9.4. Ad Hoc Feedback Schemes for MSK-Type Modulations ..... 497
    9.5. NDA Feedforward Timing Estimation ................. 502
    9.6. Ad Hoc Feedforward Schemes for MSK Modulation ....... 505
        9.6.1. MCM Scheme ......................... 505
        9.6.2. LM Scheme ........................... 509
    9.7. Key Points of the Chapter ......................... 513
    Appendix 9.A ....................................... 514
    *References* ........................................ 516

## Index ............................................ 517

# 1

# Introduction

## 1.1. What Synchronization Is About

In synchronous digital transmissions the information is conveyed by uniformly spaced pulses and the received signal is completely known except for the data symbols and a group of variables referred to as *reference parameters*. Although the ultimate task of the receiver is to produce an accurate replica of the symbol sequence with no regard to reference parameters, it is only by exploiting knowledge of the latter that the detection process can properly be performed. A few examples are sufficient to illustrate this point.

In a baseband pulse amplitude modulation (PAM) system the received waveform is first passed through a matched filter and then is sampled at the symbol rate. The optimum sampling times correspond to the maximum eye opening and are located (approximately) at the "peaks" of the signal pulses. Clearly, the locations of the pulse peaks must be accurately determined for reliable detection. A circuit that is able to predict such locations is called a *timing* (or *clock*) *synchronizer* and is a vital part of any synchronous receiver.

Coherent demodulation is used with passband digital communications when optimum error performance is of paramount importance. This means that the baseband data signal is derived making use of a local reference with the same frequency and phase as the incoming carrier. This requires accurate frequency and phase measurements insofar as phase errors introduce crosstalk between the in-phase and quadrature channels of the receiver and degrade the detection process. Circuits performing such measurements are referred to as *carrier synchronizers*.

Carrier phase information is not always needed. In applications where simplicity and robustness of implementation are more important than achieving optimum performance, differentially coherent and noncoherent demodulation are attractive alternatives to coherent detection. For example, differential demodulation of phase shift keying (PSK) signals is accomplished by computing the difference between the signal phases at two consecutive sampling times and making a decision on this difference. As the decision statistic is independent of the actual carrier phase, phase recovery is not performed. Only carrier frequency and symbol timing information is necessary.

In addition to phase, frequency, and timing, other reference parameters may be involved in the detection process. For example, this occurs with coded modulations or when the communication channel is time shared by several users on a regular basis, as happens with time division multiple access (TDMA) systems. With block coding the decoder has to know where the boundaries between codewords are. This operation is performed by *word synchronizers*. Similarly, the encoded sequence from a convolutional encoder is composed of symbol segments of fixed length and the start of each segment must be located for proper metric computation. This task is accomplished by *node synchronizers*. Finally, *frame synchronizers* are indispensable with time-shared channels to identify the boundaries between channel users and establish where the information is coming from and to where it must be routed.

All of the above examples are concerned with measuring reference parameters at the receiver, with no regard to what happens at the opposite side of the link. There are instances, however, when the transmitter assumes a positive role and, in fact, it varies the timing and frequency of its transmissions so as to meet the expectations of the receiver. This usually implies a two-way communication system, or a network, and the alignment operations are called *network synchronization*. A typical example takes place with pulse code modulation (PCM) networks where multiplexing and switching operations are performed at spatially separate nodes. Bits arriving at a given multiplexer must be available at the right time so that the assigned time slots are correctly filled and no bits are lost. Clearly, as the bits come from different nodes, it is necessary that the clocks located at those nodes, as well the local clock, all be time aligned. Another example occurs with satellite communications where many terrestrial terminals transmit signal bursts to a single satellite, trying to keep their bursts aligned in the receiver data frame. In most cases the transmitter exploits a return path from the receiver to determine the accuracy of the alignment.

From the foregoing discussion it is clear that measuring reference parameters is a vital function in data communication systems. This function is called *synchronization* and is the subject of the present book. To better define a framework for our study we think it useful to point out some limitations to the scope of our treatment and indicate distinguishing features that make the following material of particular interest for those involved in the design of modern receivers.

One basic limitation is that we shall be concerned only with timing, phase and frequency parameters. There are two basic motivations for this choice. One is the limited authors' experience with frame and network synchronization. The other is that, to a great extent, frame and network synchronization are subjects with rather specific characteristics. For example, the *marker* concept plays a fundamental role in traditional frame synchronization. A marker is a single bit or a short pattern of bits that the transmitter injects periodically into the data stream to help the receiver identify the starts of the frames. Now, the idea of using *ad hoc* means to achieve synchronization is at odds with the approach normally followed with timing and carrier recovery where it is regarded as desirable not to waste channel capacity with special signals multiplexed onto the data stream. In general, timing, phase and frequency must be directly derived from the modulated signal.

The synchronization literature is so vast as to comprise over 1000 technical papers with applications in diverse areas such as communications, telemetry, time and frequency control, and instrumentation systems. This enormous accumulation of knowledge has been incorporated and elaborated in excellent books like those by Viterbi [1], Stiffler [2], Lindsey [3], Lindsey and Simon [4], Gardner [5], Meyr and Ascheid [6], and in the ESA technical report by Gardner [7]. The present book takes advantage of all this material and develops synchronization methods for digital communications with certain features that are now indicated.

The first feature is that we focus on *digital* synchronization methods, which means that we want to recover timing, phase and carrier frequency by operating only on signal samples taken at a suitable rate. This is in contrast with the familiar *analog* methods which work on continuous-time waveforms. Although digital methods are well established in the synchronization literature by now, they are mostly in the form of technical papers, with the exceptions of report [7] and the forthcoming book by Meyr, Fechtel and Moeneclaey [8].

Digital circuits have an enormous appeal in communication technology and influence the design of all modern receivers. This is so because they do not need alignment operations, have less stringent tolerances than their analog counterparts, have low power consumption and can be integrated into small size and low cost components. Clearly, all of the above features tend to enhance performance since more complex circuitry may be used to get better functional characteristics. Also, there are some specific traits of digital circuits that directly affect the feasibility of certain synchronization algorithms. Digital memory is an important example, for it makes practicable some operations that would be complicated or even impossible in analog form.

A second feature of this study is concerned with the range of application of our results. In most synchronization books, baseband and passband PAM transmission are the dominant signaling schemes; very little space is devoted to continuous phase modulation (CPM). Of course, this lack of balance has historical and practical reasons. On the one hand, CPM techniques have become an intensive research area in the eighties, approximately with the publication of the book by Anderson, Aulin and Sundberg [8]. On the other hand, their practical application has been slowed down by implementation complexity and synchronization difficulties. Luckily, methods to reduce receiver complexity substantially and solve several synchronization problems have been proposed in the last few years. Thus, the time seems ripe for a more effective exploitation of CPM in satellite communications, digital mobile radio and low-capacity digital microwave radio systems. In this book we develop synchronization methods for both PAM and CPM.

A third feature has to do with the conceptual tools to approach synchronization problems. One possible route is to resort to heuristic reasoning. This is good as long as it works but, unfortunately, it is of limited assistance in tackling new situations such as those arising with advanced modulation schemes. On the other hand, it is widely recognized that maximum likelihood (ML) estimation techniques offer a systematic and conceptually simple guide to the solution of synchronization problems. Actually, ML methods offer two major advantages: they easily lead to appropriate circuit configurations and, under certain circumstances, provide optimum or nearly optimum performance. In this book we adopt the ML approach as our primary investigation method.

A final feature is concerned with performance evaluation. As timing, phase and frequency are continuous-valued parameters, it is natural to express synchronization accuracy in terms of *bias* and *estimation variance*. Ideally, we want zero bias and small variance, but what does "small" mean? Can other synchronizers have smaller variance? A rational answer is found in the Cramer-Rao bound (CRB), which establishes a fundamental lower limit to the variance of any unbiased estimator. As no estimator can provide lower variance, this bound can serve as a benchmark for performance evaluation purposes. Unfortunately, the CRB cannot be easily computed in many practical situations and the need arises for a more manageable performance limit. One such limit is the *modified* CRB (MCRB). In this book we consistently use the MCRB as a reference when speaking of synchronization accuracy.

Finally, a few words on prerequisites in the reader's background are useful. People involved in the design of synchronization systems need good foundations in communication theory and the underlying mathematics. Also indispensable is an adequate knowledge of digital transmission systems and modulation techniques. Textbooks like those by Benedetto, Biglieri and Castellani [10] or Proakis [11] provide excellent background material.

## 1.2. Outline of the Book

The remaining chapters are organized as follows.
*Chapter 2* lays the groundwork for further developments and is divided into three parts. The first is concerned with the effects of synchronization errors

on detection performance. Various receiver configurations and modulation formats are considered and, in each case, theoretical or simulation results are illustrated. The purpose is to establish ball-park limits on allowable synchronization errors. The second part concentrates on estimation criteria and gives a self-contained account of ML estimation methods. In particular, likelihood functions for continuous-time and discrete-time observations are derived for the additive white Gaussian noise (AWGN) channel and the concept of *wanted* and *unwanted* parameters is discussed. These parameters play a fundamental role in the computation of the MCRB. The third part gives closed-form expressions of the MCRBs for timing, phase and frequency under various modulation conditions.

*Chapter 3* investigates carrier frequency estimation with passband PAM modulation. A distinction is made between two rather different situations, depending on whether the carrier frequency offset is expected to be small or comparable with the symbol rate. Different estimation methods apply in the two cases. In particular, data-aided or decision-directed schemes can be used with small offsets whereas non-data-aided schemes are inevitable otherwise. As is intuitively clear, data-aided and decision-directed methods are much more accurate than non-data-aided ones. In fact, circuits in the first category perform close to the MCRB while the others are far from it.

*Chapter 4* concentrates on frequency estimation with CPM modulation. The same distinction between "small" and "large" frequency offsets is made as in Chapter 3. As opposed to PAM modulation, however, few methods are available for small frequency offsets and, what is worse, they are limited to binary symbols and a modulation index equal to 1/2. On the contrary, a variety of estimation schemes can be used with large frequency offsets. Their performance is far from the MCRB, however, especially with long frequency pulses. In consequence, narrow-band tracking loops are needed to achieve small estimation variances. Of course, this translates into rather long acquisition times.

*Chapter 5* is the longest and is concerned with phase estimation in PAM modulations. Its first part focuses on phase recovery for transmissions over AWGN channels. *Costas loops* are popular synchronization schemes for continuous transmissions over these channels. They are easily designed to compensate for (small) frequency offsets and have excellent tracking performance in the absence of phase noise. In any practical situation, however, some degree of phase noise is inevitable due to oscillator imperfections. The resulting tracking degradations can be limited by proper loop design. This subject is adequately addressed and criteria are provided to minimize the phase errors. The central part of the chapter considers frequency-flat fading channels. Here, the signal is affected by a multiplicative distortion (MD) which is modeled as a slowly varying Gaussian random process. As samples of the MD (taken at the symbol rate) are needed for coherent detection, the problem arises of estimating the MD random sequence from the received waveform. This is a generalization of the carrier phase recovery problem on AWGN channels where the parameter under estimation is a constant (the channel phase shift). MD estimation can be accomplished in various ways. One is to exploit known symbols multiplexed onto the data sequence. This leads to the so-called *pilot symbol assisted schemes*, which work well for fading rates up to about 1% of the symbol rate. Another solution is to perform ML joint channel and sequence estimation. This approach is effective with Doppler rates up to about 5% of the symbol rate. The last part of the chapter returns to the AWGN channel and explores open-loop phase estimation methods for applications in packet transmissions.

*Chapter 6* investigates phase recovery with CPM modulation. To some degree it has the same structure as the previous one. Its size is reduced, however, as many ideas developed earlier also apply to CPM formats. Decision-directed tracking loops are the most popular synchronization schemes for continuous transmission over the AWGN channel and are explored in the first part of the chapter. Their performance is quite close to the MCRB at signal-to-noise ratios of practical interest. With frequency-flat fading channels several approximate ML decoding schemes are available, with very good performance for fading rates up to 5% of the symbol rate. The last part of the chapter focuses on open-loop estimation methods for applications in packet transmissions.

*Chapter 7* deals with clock synchronization in baseband transmissions. The structure of the chapter reflects the fact that timing recovery consists of two basic operations: (*i*) estimation of the positions of the signal pulses relative to a local time reference; (*ii*) application of this information to the computation of symbol-spaced signal samples (*strobes*) for use in the detection process. The former is called *timing measurement*, the latter *timing adjustment*. Two approaches to timing adjustment are investigated. In one case the strobes are obtained by sampling the received signal with a clock locked to the incoming data stream (synchronous sampling). In the other, the sampling times are dictated by a free-running oscillator and the strobes are computed by interpolating between samples (non-synchronous sampling). Timing measurements are discussed in the second part of the chapter. They can be performed by either open-loop or closed-loop circuits. The former provide a direct estimate of the pulse positions relative to a local time reference. The latter compute an error signal which is proportional to the difference between the actual pulse positions and their estimates. The error signal is then exploited to update the estimates.

*Chapter 8* investigates timing recovery with passband PAM modulation. The chapter structure reflects the fact that carrier phase plays an important role in timing estimation and, in consequence, it is useful to distinguish between two scenarios. In the first one, carrier phase is estimated in conjunction with timing. This leads to joint phase and timing synchronization schemes. In the second scenario, the phase estimation problem is either postponed until after timing recovery or is not approached at all, as occurs with differential detection receivers where no phase information is needed. In these circumstances it is desirable to have timing circuits that are insensitive to phase variations. Methods to achieve this goal are analyzed. Finally, the problem of timing recovery with flat-fading is discussed. Here, the received signal is not only rotated (as happens with the AWGN channel) but is also attenuated. Thus, timing recovery is a more complex function with fading channels. Two solutions are proposed and compared. One is to use the same methods suitable for AWGN channels. The other is to employ new schemes that take the fading channel features into account.

The last topic is timing recovery with CPM signals and is covered in *Chapter 9*. The chapter is organized as the previous one except that only AWGN channels are considered. This is so because timing recovery with fading channels has not received much attention in the literature so far. Decision-directed feedback synchronizers for joint phase and timing estimation are investigated in the first part of the chapter. They have excellent tracking performance but may exhibit spurious locks, depending on the alphabet size and the frequency response of the modulator. Methods to detect and correct spurious locks are proposed. Phase information is not needed with differential detection and the problem arises of estimating timing in a phase-independent fashion. Timing algorithms that operate in this manner are investigated. They have good performance with full response systems but fail with long frequency pulses. The chapter concludes with two open-loop timing circuits for minimum shift keying (MSK) and Gaussian minimum shift keying (GMSK).

## References

[1] A.J.Viterbi, *Principles of Coherent Communication*, New York: McGraw-Hill, 1966.
[2] J.J.Stiffler, *The Theory of Synchronous Communications*, Englewood Cliffs: Prentice-Hall, 1971.
[3] W.C.Lindsey, *Synchronization Systems in Communication and Control*, Englewood Cliffs: Prentice-Hall, 1972.
[4] W.C.Lindsey and M.K.Simon, *Telecommunication System Engineering*, Englewood Cliffs: Prentice-Hall, 1973.
[5] F.M.Gardner, *Phaselock Techniques*, New York: John Wiley&Sons, 1979.
[6] H.Meyr and G.Ascheid, *Synchronization in Digital Communications*, vol. 1, New York: John Wiley&Sons, 1990.
[7] F.M.Gardner, *Demodulator Reference Recovery Techniques Suited for Digital Implementation*, European Space Agency, Final Report, ESTEC Contract No. 6847/86/NL/DG, August, 1988.
[8] H.Meyr, S.Fechtel and M.Moeneclaey, *Synchronization in Digital Communications*, vol. 2, New York: John Wiley&Sons, to be published.
[9] J.B.Anderson, T.Aulin and C-E.Sundberg, *Digital Phase Modulation*, New York: Plenum Press, 1986.

[10] S.Benedetto, E.Biglieri and V.Castellani, *Digital Transmission Theory*, Englewood Cliffs: Prentice-Hall, 1987.
[11] J.G.Proakis, *Digital Communications*, New York: McGraw-Hill, 1989.

# 2

# Principles, Methods and Performance Limits

## 2.1. Introduction

This chapter lays the groundwork for the material in the book and addresses three major themes. Section 2.2 describes synchronization functions in a digital receiver and indicates methods to pinpoint design limits on the synchronization errors. Section 2.3 is an overview of maximum likelihood parameter estimation theory, with emphasis on synchronization applications. A distinction is made between *wanted* and *unwanted* parameters, the former being those of interest in a given situation and with respect to which the maximum of a likelihood function is to be sought. The computation of likelihood functions for wanted parameters is investigated. Section 2.4 establishes limits to the performance of practical synchronizers. The most popular limit is the Cramer-Rao bound to the variance of unbiased estimators. It is argued that this limit is difficult to compute in most practical cases. A simpler limit is the *modified* Cramer-Rao bound, which is used as a benchmark in performance evaluations throughout the book.

## 2.2. Synchronization Functions

In surveying the synchronization functions we consider three signaling formats: *baseband* pulse amplitude modulation (PAM), *passband* PAM modulation (or *linear* PAM modulation) and continuous phase modulation (CPM). The discussion is kept at a conceptual level so as to highlight the synchronization aspects. The reader is assumed to be familiar with digital transmission methods at a level comparable with that of the textbooks by

Wozencraft and Jacobs [1], Benedetto and Biglieri [2] and Proakis [3].

### 2.2.1. Timing Recovery with Baseband Systems

Baseband PAM transmission is used in many commercial applications, for example in T1 carrier systems, in subscriber circuits for the integrated services digital network and in coaxial cable or fiber local area networks. In a PAM system the data stream is encoded into the amplitude values of a sequence of uniformly spaced pulses. Figure 2.1 illustrates the block diagram of a baseband PAM receiver. The incoming waveform is composed of signal plus noise:

$$r(t) = s(t) + w(t) \qquad (2.2.1)$$

The noise is a white Gaussian process with (two-sided) spectral density $N_0/2$ while the signal is constructed from time translates of a pulse $g(t)$:

$$s(t) = \sum_i c_i g(t - iT - \tau) \qquad (2.2.2)$$

In this equation $\{c_i\}$ are data symbols belonging to some $M$-ary alphabet $\{\pm 1, \pm 3, ..., \pm(M-1)\}$, $g(t)$ is the channel response, $T$ is the signaling interval, and $\tau$ represents the channel delay.

The received waveform is first filtered to remove the out-of-band noise and then is sampled at $T$-spaced instants, say $t = kT + \hat{\tau}$, $(k = 0,1,2,...)$. The samples are fed to the detector to generate estimates $\{\hat{c}_k\}$ of the transmitted data. In most practical cases the receiver filter is *matched* to $g(t)$, which means that its impulse response has the form

$$g_R(t) = g(-t + t_0) \qquad (2.2.3)$$

where $t_0$ is a delay that makes $g(-t+t_0)$ a causal function (no tails on the negative time axis). As is done in many theoretical investigations, in the sequel we set $t_0$ to zero, for this affects the filter output only by an immaterial delay. In

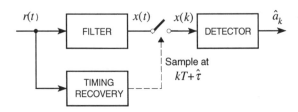

Figure 2.1. Block diagram of a baseband receiver.

other words, we write (2.2.3) as

$$g_R(t) = g(-t) \tag{2.2.4}$$

Another circumstance which is often met in practice is that the convolution $h(t) \triangleq g(t) \otimes g_R(t)$ satisfies the first Nyquist criterion

$$h(kT) = \begin{cases} 1 & \text{for } k = 0 \\ 0 & \text{for } k \neq 0 \end{cases} \tag{2.2.5}$$

In the frequency domain, the relationship $h(t) \triangleq g(t) \otimes g_R(t)$ becomes

$$H(f) = G(f)G_R(f) \tag{2.2.6}$$

and equation (2.2.4) reads

$$G_R(f) = G^*(f) \tag{2.2.7}$$

where the superscript "star" means complex conjugate. From (2.2.6)-(2.2.7) it follows that $G(f)$ and $G_R(f)$ have the same amplitude characteristic:

$$|G(f)| = |G_R(f)| = \sqrt{H(f)} \tag{2.2.8}$$

One class of Nyquist functions which is extensively used is the *raised-cosine-rolloff* characteristic

$$H(f) = \begin{cases} T & |f| \leq \dfrac{1-\alpha}{2T} \\ T\cos^2\left[\dfrac{\pi}{4\alpha}(|2fT|-1+\alpha)\right] & \dfrac{1-\alpha}{2T} \leq |f| \leq \dfrac{1+\alpha}{2T} \\ 0 & \text{otherwise} \end{cases} \tag{2.2.9}$$

where the parameter $\alpha$ is restricted to the interval $0 < \alpha \leq 1$ and is called the rolloff or excess-bandwidth factor. The inverse Fourier transform of $H(f)$ is found to be

$$h(t) = \frac{\sin(\pi t/T)}{\pi t/T} \frac{\cos(\alpha \pi t/T)}{1 - 4\alpha^2 t^2/T^2} \tag{2.2.10}$$

For further reference we note that the integral of $H(f)$ on the frequency axis equals $h(0)$. Thus, for a Nyquist function we have

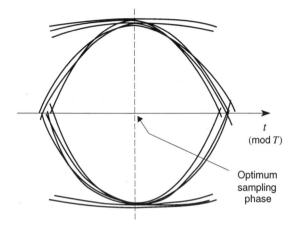

Figure 2.2. Maximum eye opening.

$$\int_{-\infty}^{\infty} H(f)df = 1 \qquad (2.2.11)$$

Substituting (2.2.9) into (2.2.8) gives a *root-raised-cosine-rolloff* function

$$G(f) = \begin{cases} \sqrt{T} & |f| \leq \dfrac{1-\alpha}{2T} \\ \sqrt{T} \cos\left[\dfrac{\pi}{4\alpha}(|2fT|-1+\alpha)\right] & \dfrac{1-\alpha}{2T} \leq |f| \leq \dfrac{1+\alpha}{2T} \\ 0 & \text{otherwise} \end{cases} \qquad (2.2.12)$$

Returning to the baseband receiver, the purpose of the timing recovery circuit (TRC) is to provide sampling instants $t = kT + \hat{\tau}$ that minimize the detector error probability. Roughly speaking, this amounts to sampling the filter output at the maximum *eye opening* (see Figure 2.2). Timing errors are unavoidable, however, and tend to degrade the detector performance, as is now illustrated.

### 2.2.2. Degradations Due to Timing Errors

Call $P(e|\hat{\tau})$ the detector error probability conditioned on a fixed sampling epoch $\hat{\tau}$. Physical reasons indicate that this is a concave-up function of $\hat{\tau}$, with

# Principles, Methods and Performance Limits   13

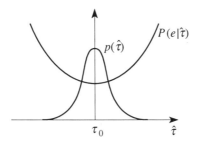

Figure 2.3. Illustrating the shape of $P(e|\hat{\tau})$ and $p(\hat{\tau})$.

a minimum at some abscissa $\hat{\tau} = \tau_0$ as shown in Figure 2.3. Thus, an ideal TRC should issue sampling pulses at $t = kT + \tau_0$. Notice that $\tau_0$ need not coincide with the channel delay $\tau$ in (2.2.2). For example, if the filter impulse response satisfies (2.2.3) and (2.2.12), then the optimum sampling epoch turns out to be $\tau_0 = \tau + t_0$. From Figure 2.1 it is clear that the sampling phase can be changed, if needed, by delaying the TRC pulses.

The fundamental feature of a *practical* TRC is that the separation between adjacent pulses is not exactly constant but varies slowly in a random manner. The variations are referred to as *timing jitter* and are a consequence of the random nature of the waveform at the TRC input. Timing jitter may be incorporated into the TRC model by describing $\hat{\tau}$ as a random variable with some probability density function $p(\hat{\tau})$.

For simplicity assume that $\hat{\tau}$ has a mean value $\tau_0$:

$$\tau_0 = \int_{-\infty}^{\infty} \hat{\tau} p(\hat{\tau}) \, d\hat{\tau} \qquad (2.2.13)$$

Note that this is not a restriction insofar as it can be satisfied by suitably delaying the TRC pulses. Then, averaging $P(e|\hat{\tau})$ over $\hat{\tau}$ gives the average error probability:

$$P(e) = \int_{-\infty}^{\infty} P(e|\hat{\tau}) p(\hat{\tau}) \, d\hat{\tau} \qquad (2.2.14)$$

Unfortunately this equation is not very useful as $P(e|\hat{\tau})$ is only known for uncoded transmissions [4]-[5]. Nevertheless, as is now indicated, the very form of (2.2.14) leads to an interesting expression of $P(e)$ involving the timing jitter variance. A more quantitative analysis is provided by Bucket and Moeneclaey in [6].

Write $P(e|\hat{\tau})$ as a truncated Taylor series in that neighborhood of $\tau_0$ where $p(\hat{\tau})$ assumes significant values (see Figure 2.3):

$$P(e|\hat{\tau}) \approx P(e|\tau_0) + \frac{1}{2} P^{(2)}(e|\tau_0)(\hat{\tau} - \tau_0)^2 \qquad (2.2.15)$$

Here, $P^{(2)}(e|\tau_0)$ is the second derivative of $P(e|\hat{\tau})$ at $\hat{\tau} = \tau_0$. Note that the term containing $(\hat{\tau} - \tau_0)$ is missing since, by assumption, $P(e|\hat{\tau})$ has a minimum at $\hat{\tau} = \tau_0$. Then, substituting into (2.2.14) yields the desired relation between $P(e)$ and the timing jitter variance $\sigma_\tau^2 \triangleq E\{(\hat{\tau} - \tau_0)^2\}$:

$$P(e) \approx P(e|\tau_0) + \frac{1}{2} P^{(2)}(e|\tau_0) \sigma_\tau^2 \qquad (2.2.16)$$

This relation indicates that $P(e)$ is degraded in proportion to $\sigma_\tau^2$. The coefficient $P^{(2)}(e|\tau_0)$ represents the jitter sensitivity of the detector. It turns out that $P^{(2)}(e|\tau_0)$ increases with the size of the symbol alphabet and decreases with the signal bandwidth. This is illustrated in the simulations in Figures 2.4-

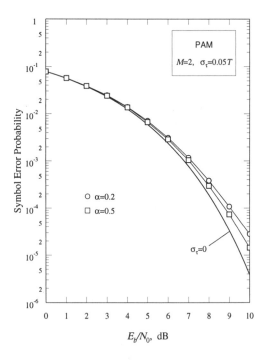

Figure 2.4. SEP degradation due to timing jitter for binary PAM modulation.

# Principles, Methods and Performance Limits

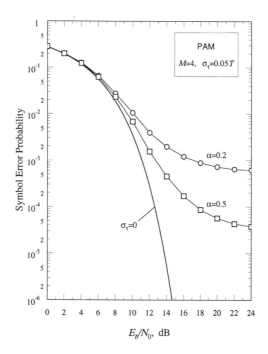

Figure 2.5. SEP degradation due to timing jitter for quaternary PAM modulation.

2.5 which show the symbol error probability (SEP) versus $E_b/N_0$ for binary ($M=2$) and quaternary ($M=4$) baseband transmission. Note that $E_b$ is the transmitted energy per bit of information and $G(f)$ and $G_R(f)$ are root-raised-cosine-rolloff functions. The parameter $\hat{\tau}$ is modeled as a zero-mean Gaussian random variable with a standard deviation of $\sigma_{\hat{\tau}}=0.05T$. The lowest curve indicates the SEP with perfect symbol timing.

An interesting feature of the upper curves in Figure 2.5 is that they exhibit an irreducible error floor as $E_b/N_0$ increases. The explanation is that timing errors generate intersymbol interference (ISI) which, in turn, produces decision errors even in the absence of noise. It should be noted that this problem is not specific to the PAM system in Figure 2.5. In fact, an error floor would eventually show up even with the case in Figure 2.4 if the signal-to-noise ratio were adequately increased.

### 2.2.3. Passband PAM Systems

Passband PAM signals are generated by *linearly modulating* baseband PAM sequences onto a sinusoidal carrier. Passband PAM signals are efficient in power and bandwidth [2]-[3] and are well suited for applications in high-speed voiceband transmission, digital microwave radio and mobile radio communications. Phase shift keying (PSK) and quadrature amplitude modulation (QAM) are two prominent members in this class. Another member of practical interest is offset quadriphase modulation (OQPSK), which is similar to conventional quadriphase PSK, except that the bit transitions on the sine and cosine carrier components are offset in time by the inverse of the bit rate. The time offset serves to limit the signal envelope variations and thereby control adjacent channel interferences in microwave radio systems employing power-efficient amplifiers [7, Ch. 4].

The mathematical model for a modulated PAM signal is

$$s_{IF}(t) = \text{Re}\{s_{CE}(t)e^{j2\pi f_c t}\} \tag{2.2.17}$$

where $f_c$ represents the carrier frequency and $s_{CE}(t)$ is the signal complex envelope relative to $f_c$. The expression for $s_{CE}(t)$ varies according to whether we consider non-offset (PSK or QAM) or offset modulation. With the former we have

$$s_{CE}(t) = \sum_i c_i g(t - iT) \tag{2.2.18}$$

where $g(t)$ is the signaling pulse and $\{c_i\}$ are information symbols. In particular, with QAM modulation $c_i$ has the form

$$c_i = a_i + jb_i \tag{2.2.19}$$

with $a_i$ and $b_i$ belonging to $\{\pm 1, \pm 3, \ldots, \pm(M-1)\}$. Vice versa, with PSK we have

$$c_i = e^{j\alpha_i} \tag{2.2.20}$$

with $\alpha_i \in \{0, 2\pi/M, \ldots, 2\pi(M-1)/M\}$. Finally, with OQPSK modulation the signal complex envelope is

$$s_{CE}(t) = \sum_i a_i g(t - iT) + j \sum_i b_i g(t - iT - T/2) \tag{2.2.21}$$

and $a_i$ and $b_i$ take values $\pm 1$.

When the signal is transmitted over a channel with a delay $\tau$, the received waveform becomes

$$r_{IF}(t) = s_{IF}(t - \tau) + w_{IF}(t) \tag{2.2.22}$$

where $w_{IF}(t)$ is the channel noise. To retrieve the transmitted information it is common practice to translate $r_{IF}(t)$ in frequency down to baseband (*demodulation*) and then operate on the resulting low-frequency waveform. It can be shown that this procedure, in addition to being convenient from an engineering point of view, does not degrade the achievable error performance of the receiver [1, Ch. 7].

As indicated in Figure 2.6, the demodulation is accomplished by multiplying $r_{IF}(t)$ by two local references $2\cos(2\pi f_{cL}t + \phi_L)$ and $-2\sin(2\pi f_{cL}t + \phi_L)$ and then feeding the products to low-pass filters to eliminate the frequency terms around $f_c + f_{cL}$. In general, the local frequency $f_{cL}$ is not exactly equal to $f_c$ and the difference $v \triangleq f_c - f_{cL}$ is referred to as *carrier frequency offset*. Assuming that the filters have a unity frequency response for the low-pass signal components and performing standard calculations [2]-[3] it can be shown that the filter outputs $r_R(t)$ and $r_I(t)$ may be represented by a single complex-valued waveform $r(t) \triangleq r_R(t) + jr_I(t)$ given by

$$r(t) = s(t) + w(t) \qquad (2.2.23)$$

with

$$s(t) \triangleq e^{j(2\pi vt + \theta)} s_{CE}(t - \tau) \qquad (2.2.24)$$

In these equations $\theta$ is a phase shift equal to $-(2\pi f_c \tau + \phi_L)$ and $w(t) = w_R(t) + jw_I(t)$ is low-pass noise. Also, $s(t)$ is given by

$$s(t) = e^{j(2\pi vt + \theta)} \sum_i c_i g(t - iT - \tau) \qquad (2.2.25)$$

with non-offset modulation and

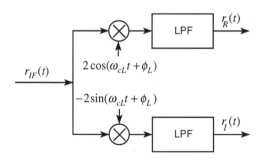

Figure 2.6. Demodulation operation.

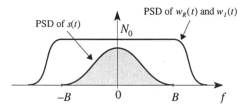

Figure 2.7. Signal and noise spectral densities.

$$s(t) = e^{j(2\pi vt + \theta)} \left\{ \sum_i a_i g(t - iT - \tau) + j \sum_i b_i g(t - iT - T/2 - \tau) \right\} \quad (2.2.26)$$

with offset modulation.

As for $w(t)$, the following remarks are useful. Suppose that the channel noise has a constant (two-sided) power spectral density (PSD) $N_0/2$ over the RF signal bandwidth. Then it can be shown that the noise components $w_R(t)$ and $w_I(t)$ are *independent* Gaussian processes with the same PSD $S_w(f) = N_0$ over the signal bandwidth $\pm B$ (see Figure 2.7). The shape of $S_w(f)$ beyond $\pm B$ is irrelevant to the detection process because further processing of $r(t)$ always involves some filtering that tends to cut off the out-of-band noise. Accordingly, in the sequel we take $S_w(f) = N_0$ over the entire frequency axis, which amounts to saying that we model $w_R(t)$ and $w_I(t)$ as *white Gaussian processes*. A more profound justification to this approach relies on the application of the *reversibility theorem* and can be found in the book by Wozencraft and Jacobs [1].

### 2.2.4. Synchronization in PAM Coherent Receivers

It is clear from (2.2.25)-(2.2.26) that the baseband signal contains unknown parameters $(v, \theta, \tau)$ in addition to the data symbols. As is now illustrated, knowledge of these parameters is vital for reliable data detection. Let us put aside timing as the subject has already been discussed in Section 2.2.2. The problem with $v$ and $\theta$ arises from the presence of the multiplicative distorsion $e^{j(2\pi vt + \theta)}$. To give an example, consider non-offset modulation and imagine what would happen if the baseband waveform $r(t)$ were matched filtered and then passed to the detector without any distortion compensation. For simplicity assume that the convolution $h(t) \triangleq g(t) \otimes g(-t)$ is Nyquist and the frequency offset $v$ is very small compared with the signal bandwidth so that the matched-filter output can be approximated as

# Principles, Methods and Performance Limits

$$x(t) = e^{j(2\pi\nu t+\theta)} \sum_i c_i h(t - iT - \tau) + n(t) \quad (2.2.27)$$

where $n(t)$ is the noise component. Then, sampling $x(t)$ at $kT + \tau$ would yield the following detector input:

$$x(k) = c_k e^{j[2\pi\nu(kT+\tau)+\theta]} + n(k) \quad (2.2.28)$$

where notations of the type $x(k) \triangleq x(kT + \tau)$ have been used. Clearly, the signal components would be rotated away from their correct positions with disabling effects on the detection process.

The above discussion points out the necessity of compensating for the distortion $e^{j(2\pi\nu t+\theta)}$ and estimating the timing epoch $\tau$. A possible method is illustrated in the block diagram of Figure 2.8. Here, the blocks indicated as frequency-, phase- and timing-recovery provide estimates $\hat{\nu}$, $\hat{\theta}$ and $\hat{\tau}$ of the synchronization parameters. It should be stressed that this receiver configuration has only illustration purposes. For example, timing can be derived from $r(t)$ (prior to frequency correction) or from the matched-filter output (after phase correction). Similarly, phase correction can be performed after matched filtering. Finally, each synchronization block may contain some kind of prefilter to hold the noise level within bounds.

The compensation for the distortion $e^{j(2\pi\nu t+\theta)}$ is performed in two steps. First, the received waveform is multiplied by $e^{-j2\pi\hat{\nu}t}$, which amounts to a counter-rotation at an angular speed $\hat{\nu}$. Next, the product $r(t)e^{-j2\pi\hat{\nu}t}$ is multiplied by $e^{-j\hat{\theta}}$. Bearing in mind equation (2.2.24), it is clear that perfect distortion suppression would require $\hat{\nu} = \nu$ and $\hat{\theta} = \theta$. In practice, $\hat{\nu}$ does not exactly coincide with $\nu$ and the task of eliminating the residual distortion $e^{j[2\pi(\nu-\hat{\nu})t+\theta]}$ is entrusted to the phase recovery circuit (PRC). This is feasible if the frequency error $\nu - \hat{\nu}$ is sufficiently small so that the angle $2\pi(\nu - \hat{\nu})t + \theta$

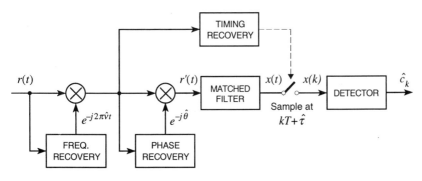

Figure 2.8. Block diagram of a coherent receiver.

is approximately constant over the measuring time $T_0$ of the PRC. In such conditions, in fact, the PRC makes periodic measurements of $2\pi(v - \hat{v})t + \theta$ (with period $T_0$) and compensates for them by counter-rotating $r(t)e^{-j2\pi\hat{v}t}$.

### 2.2.5. Degradations Due to Phase Errors

Assuming for simplicity that perfect frequency and timing estimation has been achieved, one important question is to establish the degradations in error probability resulting from inexact phase estimates. With uncoded modulation the following analytical approach may be pursued.

The first step is to model the phase error $\phi \triangleq \theta - \hat{\theta}$ as a random variable with some probability density function $p(\phi)$. A Gaussian shape for $p(\phi)$ is reasonable in most cases if the phase errors are not too large (as happens in a well designed system). So, letting $\phi_m$ and $\sigma_\phi^2$ be the mean value and the variance of $\phi$, we have

$$p(\phi) = \frac{1}{\sigma_\phi \sqrt{2\pi}} e^{-(\phi - \phi_m)^2 / 2\sigma_\phi^2} \qquad (2.2.29)$$

Note that $\phi_m$ and $\sigma_\phi^2$ depend on the operating conditions of the synchronizer and can be derived by either analytical or experimental methods.

The second step is to compute $P(e|\phi)$, the symbol error probability conditioned on a fixed value of $\phi$. This is often the most demanding task and is illustrated in Exercise 2.2.1 for quadriphase PSK (QPSK). Finally, the average error probability is evaluated by numerical integration as

$$P(e) = \int_{-\infty}^{\infty} P(e|\phi) p(\phi) d\phi \qquad (2.2.30)$$

Figure 2.9 illustrates $P(e)$ as a function of the ratio $E_s / N_0$ (energy-per-bit to noise-spectral-density) for QPSK and some values of $\sigma_\phi$. It is assumed that $\phi_m = 0$ and we have a Nyquist channel. With larger values of $\sigma_\phi$ we see that the curves exhibit a floor as $E_s / N_0$ increases. This is so because occasional large phase errors take the samples $x(k)$ away from the correct decision zone even in the absence of noise.

**Exercise 2.2.1.** Compute $P(e|\phi)$ for uncoded QPSK making the following assumptions: (*i*) transmit and receive filters are root-raised-cosine-rolloff functions; (*ii*) frequency and timing references are ideal ($\hat{v} = v$ and $\hat{\tau} = \tau$).

*Solution.* Bearing in mind that $\hat{v} = v$, from (2.2.23) and (2.2.25) we derive the following expression for the input to the matched filter (see Figure 2.8):

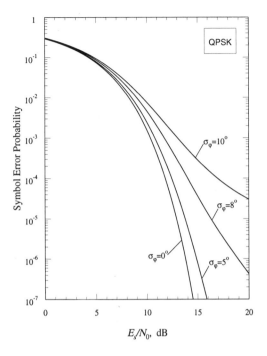

Figure 2.9. SEP degradations due to phase errors with QPSK modulation.

$$r'(t) = e^{j\phi}\sum_i c_i g(t - iT - \tau) + w'(t) \qquad (2.2.31)$$

with $\phi \triangleq \theta - \hat{\theta}$ and

$$w'(t) = w(t)e^{-j(2\pi\hat{\nu}t+\hat{\theta})} \qquad (2.2.32)$$

The filter output is the convolution of $r'(t)$ with $g(-t)$. Hence

$$x(t) = e^{j\phi}\sum_i c_i h(t - iT - \tau) + n(t) \qquad (2.2.33)$$

with $h(t) \triangleq g(t) \otimes g(-t)$ and

$$n(t) \triangleq w'(t) \otimes g(-t) \qquad (2.2.34)$$

The sampling times are $t = kT + \tau$. Thus, as $h(t)$ is Nyquist, from (2.2.33) we have

$$x(k) = c_k e^{j\phi} + n(k) \tag{2.2.35}$$

As expected, there is no intersymbol interference and the signal component is rotated by $\phi$. The QPSK symbol alphabet is $\{e^{j\pi/4}, e^{j3\pi/4}, e^{j5\pi/4}, e^{j7\pi/4}\}$, as is illustrated in Figure 2.10 where circles represent alphabet elements while dots are their rotated versions. The four quadrants are the detector decision regions. In particular, the receiver declares that $e^{j\pi/4}$ has been transmitted when $x(k)$ belongs to the first quadrant.

To get $P(e|\phi)$ we first compute the probability of correct detection, $P(c|\phi)$, and then derive $P(e|\phi)$ as

$$P(e|\phi) = 1 - P(c|\phi) \tag{2.2.36}$$

In doing so we may assume that the symbol $e^{j\pi/4}$ has been transmitted since, for symmetry, the probability of correct detection is the same with any symbol. Letting $n(k) = n_R(k) + jn_I(k)$, from (2.2.35) we have

$$x(k) = \cos(\phi + \pi/4) + n_R(k) + j[\sin(\phi + \pi/4) + n_I(k)] \tag{2.2.37}$$

and the probability that $x(k)$ belongs to the first quadrant is given by

$$P(c|\phi) = \Pr\{\cos(\phi + \pi/4) + n_R(k) \geq 0; \sin(\phi + \pi/4) + n_I(k) \geq 0\} \tag{2.2.38}$$

To proceed further we need the statistics of $n_R(k)$ and $n_I(k)$. To this end we

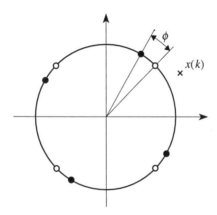

Figure 2.10. QPSK constellation.

consider the complex-valued random process $w'(t)$ in (2.2.32). It is easily seen that, since $w(t)$ is white, so is $w'(t)$, with the same PSD. It follows that $n_R(k)$ and $n_I(k)$ are independent and zero-mean Gaussian random variables with variance

$$\sigma_n^2 = N_0 \int_{-\infty}^{\infty} H(f) df \tag{2.2.39}$$

where $H(f)$ is the Fourier transform of $h(t)$. As $H(f)$ is Nyquist, the integral in (2.2.39) is unity (see (2.2.11)) and (2.2.39) reduces to

$$\sigma_n^2 = N_0 \tag{2.2.40}$$

It is desirable to express $\sigma_n^2$ as a function of the signal-to-noise ratio $E_s/N_0$. Under the previous assumptions the signal energy is $E_s = 1/2$ (see Appendix 2.A.2) and from (2.2.40) we get

$$\frac{E_s}{N_0} = \frac{1}{2\sigma_n^2} \tag{2.2.41}$$

Hence

$$\sigma_n^2 = \frac{1}{2E_s/N_0} \tag{2.2.42}$$

At this stage the probability in (2.2.38) can be written as

$$P(c|\phi) = \left[1 - Q\left(\frac{\cos(\phi + \pi/4)}{\sigma_n}\right)\right]\left[1 - Q\left(\frac{\sin(\phi + \pi/4)}{\sigma_n}\right)\right] \tag{2.2.43}$$

with

$$Q(x) \triangleq \frac{1}{\sqrt{2\pi}} \int_x^{\infty} e^{-t^2/2} dt \tag{2.2.44}$$

Finally, substituting (2.2.42)-(2.2.43) into (2.2.36) yields

$$P(e|\phi) = Q\left(\sqrt{\frac{2E_s}{N_0}}\cos(\phi + \pi/4)\right) + Q\left(\sqrt{\frac{2E_s}{N_0}}\sin(\phi + \pi/4)\right)$$
$$- Q\left(\sqrt{\frac{2E_s}{N_0}}\cos(\phi + \pi/4)\right) Q\left(\sqrt{\frac{2E_s}{N_0}}\sin(\phi + \pi/4)\right) \tag{2.2.45}$$

For example, with $E_s/N_0 = 10$ dB and $\phi = 5°$, it is found $P(e|\phi = 5°) = 2.3 \times 10^{-3}$. With a zero phase error this same probability would be achieved with $E_s/N_0 = 9.68$ dB. Thus, a phase error of $\phi = 5°$ entails a 0.3 dB loss in $E_s/N_0$.

For $\phi=0$ equation (2.2.45) reduces to the well-known formula for the symbol error probability with a perfect phase reference [2]-[3]:

$$P(e|\phi = 0) = 2Q\left(\sqrt{\frac{E_s}{N_0}}\right) - Q^2\left(\sqrt{\frac{E_s}{N_0}}\right) \qquad (2.2.46)$$

**Exercise 2.2.2.** Let $\phi$ be a random variable with zero mean and variance $\sigma_\phi^2$. Assume $\sigma_\phi^2 \ll 1$ and $E_s/N_0$ large enough so that the last term in (2.2.45) may be neglected (as its factors are much less than unity). Accordingly, $P(e|\phi)$ becomes

$$P(e|\phi) \approx Q\left(\sqrt{\frac{2E_s}{N_0}}\cos(\phi + \pi/4)\right) + Q\left(\sqrt{\frac{2E_s}{N_0}}\sin(\phi + \pi/4)\right) \qquad (2.2.47)$$

Compute the average symbol error probability $P(e)$.

*Solution.* Expanding (2.2.47) into a power series about $\phi=0$ and keeping only the terms up to the second power yields

$$P(e|\phi) \approx 2Q\left(\sqrt{\frac{E_s}{N_0}}\right) - \sqrt{\frac{E_s}{N_0}}\left(1 + \frac{E_s}{N_0}\right)Q^{(1)}\left(\sqrt{\frac{E_s}{N_0}}\right)\phi^2 \qquad (2.2.48)$$

where $Q^{(1)}(x)$ is the derivative of $Q(x)$

$$Q^{(1)}(x) = -\frac{1}{\sqrt{2\pi}}e^{-x^2/2} \qquad (2.2.49)$$

Averaging $P(e|\phi)$ with respect to $\phi$ gives the result sought:

$$P(e) \approx 2Q\left(\sqrt{\frac{E_s}{N_0}}\right) - \sqrt{\frac{E_s}{N_0}}\left(1 + \frac{E_s}{N_0}\right)Q^{(1)}\left(\sqrt{\frac{E_s}{N_0}}\right)\sigma_\phi^2 \qquad (2.2.50)$$

We see that the increase in error probability due to phase errors is proportional to $\sigma_\phi^2$. The reader should be careful when using (2.2.50) as it has been derived assuming very small phase errors. In fact, the equation gives accurate results only for $\sigma_\phi$ on the order of a few degrees.

**Exercise 2.2.3.** Consider again the QPSK communication system described in the previous exercise but assume that an ideal phase reference is now available for demodulation. Under these conditions a given error probability can be achieved at a reduced signal energy. The saving in energy represents the degradation in signal-to-noise ratio due to phase errors. Compute such a degradation as a function of $\sigma_\phi^2$.

*Solution.* Call $E_s'$ the signal energy that is needed to achieve $P(e)$ in the absence of phase errors ($\sigma_\phi^2=0$). From (2.2.50) we have

$$P(e) = 2Q\left(\sqrt{\frac{E_s'}{N_0}}\right) \qquad (2.2.51)$$

Expanding the right-hand side into a Taylor series yields

$$P(e) \approx 2Q\left(\sqrt{\frac{E_s}{N_0}}\right) + \frac{(E_s' - E_s)}{\sqrt{E_s N_0}} Q^{(1)}\left(\sqrt{\frac{E_s}{N_0}}\right) \qquad (2.2.52)$$

and comparing with (2.2.50) results in the following relation between $E_s'$ and $E_s$:

$$\frac{E_s'}{E_s} = 1 - \left(1 + \frac{E_s}{N_0}\right)\sigma_\phi^2 \qquad (2.2.53)$$

The signal-to-noise degradation is defined as

$$D \triangleq -10 \cdot \log_{10}\left(\frac{E_s'}{E_s}\right) \quad \text{dB} \qquad (2.2.54)$$

Thus, making the approximation $\log_{10}(1-\varepsilon) \approx -0.43\varepsilon$ and collecting (2.2.53)-(2.2.54) produces the desired result

$$D \approx 4.3\left(1 + \frac{E_s}{N_0}\right)\sigma_\phi^2 \qquad (2.2.55)$$

For example, with $\sigma_\phi = 5°$ and $E_s/N_0 = 10$ dB, we have $D \approx 0.36$ dB.

### 2.2.6. Synchronization in PAM Differential Receivers

Differential detection is used in applications where simplicity and robustness of implementation are more important than achieving the optimum performance of coherent receivers. Differential detection has mostly been applied to PSK modulation but its extension to QAM formats is possible and has been discussed in [8]-[9].

In an M-ary PSK differential detection system the information is represented by a sequence $\{\delta_k\}$ whose elements take values from the set $\{0, 2\pi/M, \ldots, 2\pi(M-1)/M\}$. The sequence is first differentially encoded as

$$\alpha_k = \alpha_{k-1} + \delta_k \quad \text{mod } 2\pi \quad (2.2.56)$$

and then is mapped into channel symbols $c_k = e^{j\alpha_k}$ which satisfy the recursion

$$c_k = c_{k-1} e^{j\delta_k} \quad (2.2.57)$$

At the receiver side the data $\{\delta_k\}$ are retrieved without any carrier phase knowledge, as indicated in Figure 2.11. Let us concentrate on the samples from matched filter. Paralleling the arguments leading to (2.2.28) it is found that

$$x(k) = c_k e^{j[2\pi \Delta v(kT+\tau)+\theta]} + n(k) \quad (2.2.58)$$

where $\Delta v \triangleq v - \hat{v}$ is the residual frequency error. Thus, the detector input $z(k) \triangleq x(k) x^*(k-1)$ results in

$$z(k) = c_k c_{k-1}^* e^{j2\pi \Delta vT} + N(k) \quad (2.2.59)$$

where $N(k)$ is a noise term. Alternately, bearing in mind (2.2.57) we have

$$z(k) = e^{j\delta_k} e^{j2\pi \Delta vT} + N(k) \quad (2.2.60)$$

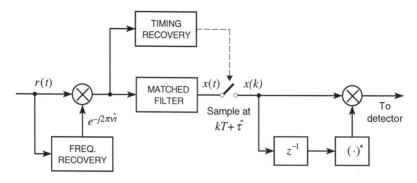

Figure 2.11. Block diagram of a differential receiver.

which has the same form as the detector input in a coherent receiver (see (2.2.35))

$$x(k) = e^{j\alpha_k}e^{j\phi} + n(k) \quad (2.2.61)$$

except that $N(k)$ has a larger variance than $n(k)$ (see [2]-[3]) and the rotation $2\pi\Delta vT$ replaces the phase error $\phi$. No trace of the carrier phase is present in (2.2.60), which means that the detector performance is phase insensitive. In fact it is only influenced by the frequency-induced rotation $2\pi\Delta vT$.

Figure 2.12 illustrates the degradations in symbol error probability due to imperfect frequency recovery. Timing is assumed ideal and the overall channel response is Nyquist. Also, the frequency error $\Delta v$ is constant and equal to a fraction of the symbol rate. The lower curve in the figure represents the performance of a coherent receiver with ideal carrier recovery. It is worth noting that the horizontal distance between the coherent curve and the lowest differential curve is about 2.3 dB. This is the minimum loss incurred by using differential rather than coherent detection. Additional losses are due to imperfect frequency compensation.

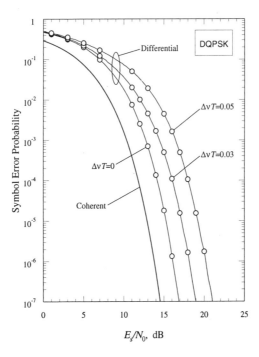

Figure 2.12. SEP degradations due to uncompensated frequency offsets.

## 2.2.7. Synchronization in CPM Systems

Continuous phase modulations (CPMs) are used in satellite communications, digital mobile radio and low-capacity digital microwave radio systems. One remarkable feature of CPM signals is that they have a constant envelope and, therefore, can be amplified without distortion by low-cost nonlinear devices operating near the saturation point. No expensive compensation for the nonlinearity is needed and no extra DC power is wasted to support a less efficient linear amplifier.

The complex envelope of a CPM signal is given by [10]

$$s_{CE}(t) = \sqrt{\frac{2E_s}{T}} e^{j\psi(t,\boldsymbol{\alpha})} \qquad (2.2.62)$$

where $E_s$ is the signal energy per symbol, $T$ is the symbol period, $\boldsymbol{\alpha} \triangleq \{\alpha_i\}$ are data symbols from the alphabet $\{\pm 1, \pm 3, .., \pm(M-1)\}$ and $\psi(t, \boldsymbol{\alpha})$ is the information-bearing phase

$$\psi(t, \boldsymbol{\alpha}) = 2\pi h \sum_i \alpha_i q(t - iT) \qquad (2.2.63)$$

The parameter $h$ is the *modulation index* and is the ratio of two relatively prime integers

$$h = \frac{K}{P} \qquad (2.2.64)$$

Also, $q(t)$ is the *phase response* of the modulator and is normalized in such a way that

$$q(t) = \begin{cases} 0 & t \leq 0 \\ 1/2 & t \geq LT \end{cases} \qquad (2.2.65)$$

Its derivative $dq(t)/dt \triangleq g(t)$ is the *frequency response* of the modulator. It is clear from (2.2.65) that the frequency response is limited to the interval $t \in (0, LT)$, where $L$ is an integer called the *correlation length*. Modulation formats with $L=1$ are said to be of *full-response* type whereas those with $L>1$ are of *partial-response* type.

By choosing different frequency responses and varying the parameters $h$ and $M$, a great variety of CPM schemes may be formed. For example, minimum shift keying (MSK) corresponds to $h=1/2$, $M=2$ and a rectangular frequency response

**Principles, Methods and Performance Limits**

$$g(t) = \begin{cases} 0 & t<0,\ t>T \\ 1/(2T) & 0 \le t \le T \end{cases} \qquad (2.2.66)$$

Alternately, Gaussian MSK (GMSK) is obtained by letting $h=1/2$, $M=2$ and taking $g(t)$ as the convolution of (2.2.66) with a Gaussian shaped pulse.

As is now explained, a CPM modulator may be viewed as a trellis encoder. For this purpose let us rearrange (2.2.63) as follows:

$$\psi(t,\boldsymbol{\alpha}) \triangleq \eta(t,C_k,\alpha_k) + \Phi_k, \qquad kT \le t < (k+1)T \qquad (2.2.67)$$

with

$$\eta(t,C_k,\alpha_k) \triangleq 2\pi h \sum_{i=k-L+1}^{k} \alpha_i q(t-iT) \qquad (2.2.68)$$

$$C_k \triangleq (\alpha_{k-L+1},\ldots,\alpha_{k-2},\alpha_{k-1}) \qquad (2.2.69)$$

$$\Phi_k \triangleq \pi h \sum_{i=0}^{k-L} \alpha_i \mod 2\pi \qquad (2.2.70)$$

In these equations the quantities $C_k$ and $\Phi_k$ are called the *correlative state* and the *phase state* of the CPM signal at the $k$-th step. From (2.2.67) it appears that $\psi(t,\boldsymbol{\alpha})$ is uniquely defined by the present symbol $\alpha_k$, the correlative state, and the phase state. Assuming independent symbols, from (2.2.69) it is clear that there are $M^{L-1}$ correlative states. Also, it can be shown [11] that $\Phi_k$ takes $P$ distinct values. It follows that $\psi(t,\boldsymbol{\alpha})$ has a total of $PM^{L-1}$ states, say $S_k \triangleq (C_k, \Phi_k)$, and the CPM modulator may be viewed as the cascade of a trellis encoder with states $\{S_k\}$ and a mapper generating the phase elements (2.2.67).

A maximum likelihood (ML) receiver takes the form of a Viterbi algorithm which searches for the most likely path in the modulator trellis. The input to the baseband receiver has the form

$$r(t) = s(t) + w(t) \qquad (2.2.71)$$

with

$$s(t) \triangleq e^{j(2\pi vt + \theta)} s_{CE}(t-\tau) \qquad (2.2.72)$$

Figure 2.13 illustrates the receiver block diagram [10]. After frequency compensation, the received waveform is fed to a bank of $M^L$ filters with impulse responses

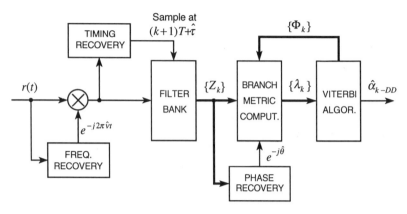

Figure 2.13. Block diagram of the ML receiver.

$$h^{(l)}(t) \triangleq \begin{cases} e^{-j\eta_l(T-t, C_0^{(l)}, \alpha_0^{(l)})} & 0 \le t \le T \\ 0 & \text{elsewhere} \end{cases} \quad (2.2.73)$$

with ($l=1,2,\ldots,M^L$). Here, $(C_0^{(l)}, \alpha_0^{(l)}) = (\alpha_{-L+1}^{(l)}, \ldots, \alpha_{-1}^{(l)}, \alpha_0^{(l)})$ is the generic realization of $(\alpha_{-L+1}, \ldots, \alpha_{-1}, \alpha_0)$ and $\eta_l(t, C_0^{(l)}, \alpha_0^{(l)})$ is expressed as

$$\eta_l(t, C_0^{(l)}, \alpha_0^{(l)}) = 2\pi h \sum_{i=-L+1}^{0} \alpha_i^{(l)} q(t - iT) \quad (2.2.74)$$

The filter outputs are sampled at $(k+1)T + \hat{\tau}$ and are used to produce the statistics

$$Z_k(C_k, \alpha_k, \hat{\tau}) \triangleq \int_{\hat{\tau}+kT}^{\hat{\tau}+(k+1)T} r(t) e^{-j2\pi \hat{v} t} e^{-j\eta(t-\tau, C_k, \alpha_k)} dt \quad (2.2.75)$$

for the branch metric computation.

The Viterbi algorithm operates on the modulator trellis as follows. There are $M$ branches stemming from a given node $\overline{S}_k = (\overline{C}_k, \overline{\Phi}_k)$, one for each possible transmitted symbol $\alpha_k$. The metric

$$\lambda_k(\overline{S}_k, \tilde{\alpha}_k) \triangleq \text{Re}\left\{ Z_k(\overline{C}_k, \tilde{\alpha}_k, \hat{\tau}) e^{-j(\hat{\theta}+\overline{\Phi}_k)} \right\} \quad (2.2.76)$$

is assigned to the branch associated with $\tilde{\alpha}_k$ and the algorithm searches for the

# Principles, Methods and Performance Limits

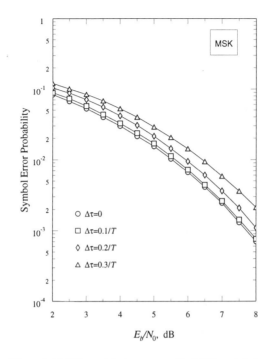

Figure 2.14. Effect of timing errors with MSK modulation

best path in the trellis. Decisions $\hat{\alpha}_{k-DD}$ are delivered with a delay $DD$ relative to the current time $k$.

Error probability degradations due to synchronization imperfections are difficult to assess analytically and, in fact, computer simulations are the only viable route. As an example, Figure 2.14 shows the sensitivity of the error probability to a constant timing offset $\Delta\tau \triangleq \tau - \hat{\tau}$. Carrier recovery is assumed ideal ($\hat{\nu} = \nu$ and $\hat{\theta} = \theta$) and the modulation is MSK. As is seen, MSK is quite tolerant of timing errors. For example, a $\Delta\tau$ as large as 20% of the symbol period produces a signal energy loss of only 0.5 dB.

## 2.2.8. Synchronization in Simplified CPM Receivers

The foregoing discussion indicates two possible obstacles to the implementation of an ML receiver. On the one hand, the modulator trellis may have a very large number of states ($PM^{L-1}$), which implies an intensive computational load for the Viterbi algorithm. On the other hand, the number of filters in the filter bank ($M^L$) may be enormous, especially with partial response and/or multilevel schemes.

Although several methods have been discovered to alleviate these difficulties (see [10], Ch. 8), most current commercial systems use suboptimum receivers. The simplest receiver employs a limiter-discriminator [12]-[14], as indicated in Figure 2.15. The input is

$$r_{IF}(t) = \text{Re}\{s_{CE}(t-\tau)e^{j2\pi f_c(t-\tau)}\} + w_{IF}(t) \qquad (2.2.77)$$

where $w_{IF}(t)$ is thermal noise (restricted in frequency to the signal bandwidth) and $s_{CE}(t)$ is the signal complex envelope:

$$s_{CE}(t) = \sqrt{\frac{2E_s}{T}} e^{j\psi(t,\alpha)} \qquad (2.2.78)$$

As the signal component from the discriminator is the derivative of the phase $\psi(t-\tau,\alpha)$, i.e.,

$$\frac{d\psi(t-\tau,\alpha)}{dt} = 2\pi h \sum_i \alpha_i g(t-iT-\tau) \qquad (2.2.79)$$

it is clear that the output from the low-pass filter is a PAM signal embedded in noise:

$$x(t) = \sum_i \alpha_i h(t-iT-\tau) + n(t) \qquad (2.2.80)$$

In this equation $h(t)$ is the filter response to $g(t)$ and $n(t)$ is a noise term. Thus, data detection can be performed by sampling $x(t)$ at the symbol rate and making decisions in a threshold detector. The samples must be taken at the maximum eye opening which occurs at some instants $kT+\tau+t_0$, with $t_0$ depending on the actual shape of $h(t)$.

Other forms of suboptimum receivers are based on differential methods or symbol-by-symbol coherent detection (as opposed to ML coherent detection considered in the previous section). For simplicity we shall concentrate on

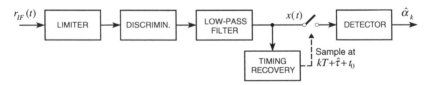

Figure 2.15. Block diagram of the limiter-discriminator receiver.

*binary* modulation with index $h=1/2$, which is referred to as *MSK-type modulation*. One feature of this format is that the exponential $e^{j\psi(t,\alpha)}$ in (2.2.62) can be approximated as a baseband PAM waveform [15] of the type

$$e^{j\psi(t,\alpha)} = \sum_i a_{0,i} h_0(t-iT) \qquad (2.2.80)$$

where the parameters $a_{0,i}$ are called *pseudo-symbols* and are related to the data $\{\alpha_0, \alpha_1, \ldots, \alpha_i\}$ by the relation

$$a_{0,i} = \exp\left(j\frac{\pi}{2}\sum_{l=0}^{i}\alpha_l\right) \qquad (2.2.81)$$

Also, the pulse $h_0(t)$ depends on the phase response $q(t)$. In particular, equation (2.2.80) holds exactly for MSK signaling, in which case $h_0(t)$ is given by

$$h_0(t) = \begin{cases} \sin\left(\dfrac{\pi t}{2T}\right) & 0 \le t \le 2T \\ 0 & \text{elsewhere} \end{cases} \qquad (2.2.82)$$

Using the above notations it is readily shown that the demodulated waveform at the receiver input takes the form

$$r(t) = s(t) + w(t) \qquad (2.2.83)$$

where

$$s(t) = e^{j(2\pi vt + \theta)} \sum_i a_{0,i} h_0(t - iT - \tau) \qquad (2.2.84)$$

and we have dropped the coefficient $\sqrt{2E_s/T}$ in the right-hand side of (2.2.84) for the sake of simplicity.

The strong similarity between (2.2.84) and the corresponding expression for linear modulations can be exploited when looking for detection algorithms. One option is to adopt a differential scheme like that in Figure 2.11. The first part of this diagram is redrawn in Figure 2.16 with two important changes. First, the low-pass filter (LPF) is no longer matched to $h_0(t)$, as one might expect. Choosing a matched filter, in fact, would result in a badly shaped pulse $h(t)$ at the LPF output and a high level of intersymbol interference. This problem has been discussed by Kawas Kaleh [16], who has provided design criteria for the LPF. The second change is in the time shift $t_0$ of the sampling times. Minimum intersymbol interference is achieved by suitably selecting $t_0$ as a function of the shape of $h(t)$.

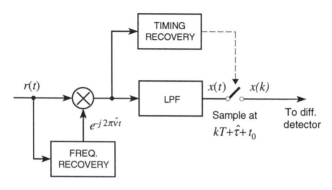

Figure 2.16. Block diagram of the differential receiver.

The differential detector operates on the samples $x(k)$ according to

$$\hat{\alpha}_k = \text{sgn}\bigl[\text{Im}\{x(k)x^*(k-1)\}\bigr] \qquad (2.2.85)$$

where sgn[$z$] equals ±1 according to whether $z$ is positive or negative. To see how this rule works, assume ideal frequency and timing recovery and neglect intersymbol interference and noise. Under these conditions $x(k)$ takes the form

$$x(k) = h(0)e^{j\theta}a_{0,k} \qquad (2.2.86)$$

Hence

$$x(k)x^*(k-1) = h^2(0)e^{j\pi\alpha_k/2} \qquad (2.2.87)$$

from which (2.2.85) is readily understood.

Symbol-by-symbol coherent detection may be used as an alternative to differential methods to improve error performance [17]-[19]. Clearly, the price to pay lies in building the phase recovery system, as indicated in Figure 2.17. In general, the LPF is not matched to $h_0(t)$ (for the same reasons given previously). One important exception occurs with MSK signaling, in which case optimum performance is achieved with an LPF impulse response that is exactly a half-sinusoidal pulse of duration $2T$ (see (2.2.82)). Design criteria for the LPF are discussed in [10], [17].

The detection process relies on the fact that the information symbols $\eta_i = \pm 1$ are differentially encoded into modulation symbols $\alpha_i = \pm 1$ as follows:

$$\alpha_k = \alpha_{k-1}\eta_k \qquad (2.2.88)$$

## Principles, Methods and Performance Limits

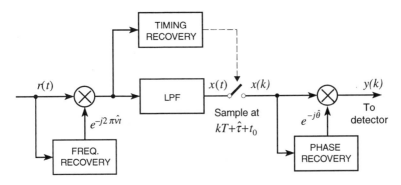

Figure 2.17. Block diagram of the coherent receiver.

Decisions are derived from the derotated samples $y(k) = x(k)e^{-j\hat{\theta}}$. Actually, the detector computes the real-valued statistic

$$z(k) \triangleq \begin{cases} \text{sgn}\{\text{Re}[y(k)]\} & k = \text{odd} \\ \text{sgn}\{\text{Im}[y(k)]\} & k = \text{even} \end{cases} \quad (2.2.89)$$

and makes decisions according to (see Exercise 2.2.4)

$$\hat{\eta}_k = -z(k)z(k-2) \quad (2.2.90)$$

Figure 2.18 shows error probability degradations due to phase errors in an MSK system. Frequency and timing recovery are ideal and the LPF is a half-sinusoidal pulse of duration $2T$. Phase errors are modeled as outcomes of a Gaussian random variable with zero mean and standard deviation $\sigma_\phi$. We see that the error curves exhibit a floor as $\sigma_\phi$ grows large. Comparing it with Figure 2.9 we see that MSK is less sensitive to phase errors than QPSK.

**Exercise 2.2.4.** Explain the decision rule (2.2.90) making the following assumptions: (*i*) synchronization is ideal; (*ii*) noise and intersymbol interference are negligible; (*iii*) pulse $h(t)$ is positive at the origin.

*Solution.* Under the assumed conditions the derotated samples $y(k)$ have the form

$$y(k) = h(0)a_{0,k} \quad (2.2.91)$$

Thus, using (2.2.81) and keeping in mind that $h(0)>0$ it is readily found that

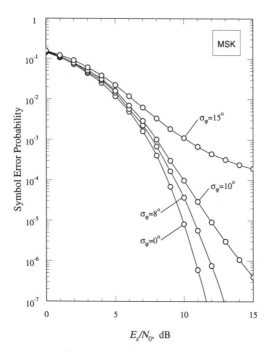

Figure 2.18. SEP degradations due to phase errors.

$$\text{sgn}\{\text{Re}[y(k)]\} = \cos\left(\frac{\pi}{2}\sum_{l=0}^{k}\alpha_l\right) \qquad k = \text{odd} \qquad (2.2.92)$$

$$\text{sgn}\{\text{Im}[y(k)]\} = \sin\left(\frac{\pi}{2}\sum_{l=0}^{k}\alpha_l\right) \qquad k = \text{even} \qquad (2.2.93)$$

On the other hand

$$\cos\left(\frac{\pi}{2}\sum_{l=0}^{k}\alpha_l\right) = \cos\left(\frac{\pi}{2}\alpha_k + \frac{\pi}{2}\sum_{l=0}^{k-1}\alpha_l\right)$$
$$= -\alpha_k \sin\left(\frac{\pi}{2}\sum_{l=0}^{k-1}\alpha_l\right) \qquad (2.2.94)$$

and, similarly,

$$\sin\left(\frac{\pi}{2}\sum_{l=0}^{k}\alpha_l\right) = \sin\left(\frac{\pi}{2}\alpha_k + \frac{\pi}{2}\sum_{l=0}^{k-1}\alpha_l\right)$$

$$= \alpha_k \cos\left(\frac{\pi}{2}\sum_{l=0}^{k-1}\alpha_l\right) \qquad (2.2.95)$$

Putting these facts together and bearing in mind (2.2.89) yields

$$\alpha_k = \begin{cases} z(k)z(k-1) & k = \text{even} \\ -z(k)z(k-1) & k = \text{odd} \end{cases} \qquad (2.2.96)$$

from which it follows that

$$\alpha_k \alpha_{k-1} = -z(k)z(k-2) \qquad (2.2.97)$$

for any integer $k$ (either even or odd). On the other hand, equation (2.2.88) may be rewritten as

$$\alpha_k \alpha_{k-1} = \eta_k \qquad (2.2.98)$$

Comparing it with (2.2.97) we conclude that (2.2.90) provides correct decisions when synchronization is ideal and both noise and intersymbol interference are negligible.

## 2.3. Maximum Likelihood Estimation

The preceding section has illustrated the synchronization functions. No details have been given on particular circuits as they will be thoroughly discussed in later chapters. For now we are only concerned with general principles and, in this context, one important question that comes to mind is whether there are general methods to derive synchronization schemes. Recent literature [20]-[22] indicates that most of the existing synchronization algorithms have been discovered through heuristic arguments or by application of ML estimation methods. As the former approach can hardly be expounded in a structured manner, in the following we concentrate on the latter and provide a short review of ML methods. Our treatment is particularly focused on the estimation of synchronization parameters. The reader interested in more general applications is referred to the textbooks by Van Trees [23] and Kay [24].

ML parameter estimation requires different mathematical tools, depending on whether the observation is a continuous-time waveform or a sample sequence. At first glance the first case looks more *natural*, as physical signals have a continuous-time support. On the other hand the second case is particularly tailored for digital receiver operations. In the sequel we first illustrate the continuous-time approach and then we extend the results to sample sequences.

### 2.3.1. ML Estimation from Continuous-Time Waveforms

For the time being we concentrate on passband signals but the discussion will later be adapted to baseband transmission. From the previous section it appears that the signal component $s(t)$ is a completely known function of time, except for a set of parameters $\boldsymbol{\gamma}$. This set may include $v$, $\theta$, $\tau$ and the data symbols, but not necessarily all these things at once. For example, if a training sequence is transmitted, the symbols are known and do not appear in $\boldsymbol{\gamma}$. Also, $\boldsymbol{\gamma}$ does not contain $\tau$ if clock information is provided to the receiver through a separate channel. Whatever the case, to stress the signal dependence on *unknown* parameters we temporarily adopt the notation $s(t, \boldsymbol{\gamma})$ in place of $s(t)$ and rewrite the baseband waveform as

$$r(t) = s(t, \boldsymbol{\gamma}) + w(t) \qquad (2.3.1)$$

Now, suppose we are allowed to observe $r(t)$ on a given interval $0 \le t \le T_0$ and are requested to provide an estimate of $\boldsymbol{\gamma}$ based on this obsevation. What can we do? The most popular approach to this problem [23]-[24] is based on the ML principle which, in rather intuitive terms, may be expressed as follows. Call $\tilde{\boldsymbol{\gamma}}$ a hypothetical (trial) value of $\boldsymbol{\gamma}$ and consider the process

$$z(t) \triangleq s(t, \tilde{\boldsymbol{\gamma}}) + w(t) \qquad (2.3.2)$$

Notice that $r(t)$ is a realization of $z(t)$ when $\tilde{\boldsymbol{\gamma}} = \boldsymbol{\gamma}$. In general, the realizations of $z(t)$ resemble $r(t)$ to various degrees, depending on how close $s(t, \tilde{\boldsymbol{\gamma}})$ is to $s(t, \boldsymbol{\gamma})$ and, ultimately, on the "distance" of $\tilde{\boldsymbol{\gamma}}$ to $\boldsymbol{\gamma}$. The ML principle suggests computing the estimate of $\boldsymbol{\gamma}$ as that $\tilde{\boldsymbol{\gamma}}$ that maximizes the "resemblance" of $r(t)$ to the realizations of $z(t)$.

A more precise formulation of the ML principle is now given in geometric terms. Denote by $\boldsymbol{z}$ and $\boldsymbol{r}$ the vector representations of $z(t)$ and $r(t)$ on a complete orthonormal basis $\{\phi_i(t)\}$ and call $p(\boldsymbol{z}|\tilde{\boldsymbol{\gamma}})$ the pdf of $\boldsymbol{z}$ for a given $\tilde{\boldsymbol{\gamma}}$. Figure 2.19 shows two functions $p(\boldsymbol{z}|\tilde{\boldsymbol{\gamma}})$ corresponding to $\tilde{\boldsymbol{\gamma}} = \tilde{\boldsymbol{\gamma}}_1$ and $\tilde{\boldsymbol{\gamma}} = \tilde{\boldsymbol{\gamma}}_2$. We see that

$$p(\boldsymbol{z} = \boldsymbol{r}|\tilde{\boldsymbol{\gamma}}_1) < p(\boldsymbol{z} = \boldsymbol{r}|\tilde{\boldsymbol{\gamma}}_2) \qquad (2.3.3)$$

# Principles, Methods and Performance Limits

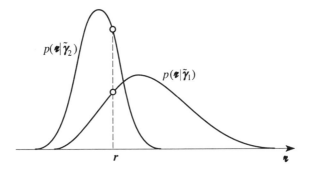

Figure 2.19. Illustrating the ML principle.

What is the physical meaning of this inequality? An answer is found bearing in mind that, when $s(t, \tilde{\gamma})$ is transmitted, the probability that $\boldsymbol{r}$ lies in a volume $d\boldsymbol{r}$ around $r$ equals $p(\boldsymbol{r} = r|\tilde{\gamma})d\boldsymbol{r}$. Thus, if many observations of $\boldsymbol{r}(t)$ are made with either $\tilde{\gamma} = \tilde{\gamma}_1$ and $\tilde{\gamma} = \tilde{\gamma}_2$, the event { $\boldsymbol{r}$ is close to $r$} will turn up more frequently with $\tilde{\gamma} = \tilde{\gamma}_2$ than with $\tilde{\gamma} = \tilde{\gamma}_1$. Accordingly, we say that $\tilde{\gamma}_2$ *is more likely* than $\tilde{\gamma}_1$ (when $r$ is observed).

Extending this idea to the case where $\tilde{\gamma}$ takes multiple values, it is readily concluded that the *most likely* value of $\gamma$ is that $\tilde{\gamma}$ where $p(\boldsymbol{r} = r|\tilde{\gamma})$ achieves a maximum. The location of the maximum is referred to as the *ML estimate* of $\gamma$ and is customarily indicated as

$$\hat{\gamma}_{ML}(r) = \arg\left\{\max_{\tilde{\gamma}}\left\{p(r|\tilde{\gamma})\right\}\right\} \qquad (2.3.4)$$

where the shorthand notation $p(r|\tilde{\gamma}) \triangleq p(\boldsymbol{r} = r|\tilde{\gamma})$ has been adopted.

In many synchronization problems the ML principle is formulated in a slightly different manner to reflect the fact that we are interested only in some components of $\gamma$, say $\lambda$. As mentioned earlier, in general $\gamma$ contains both synchronization parameters and information symbols. Now, suppose we are only interested in some synchronization parameters $\lambda$ (perhaps just one) while we do not care about the remaining (*unwanted*) components of $\gamma$, say $u$. Then, what is the ML estimate of $\lambda$?

Let $\tilde{\lambda}$ and $\tilde{u}$ be trial values of $\lambda$ and $u$ and assume that $u$ can be modeled as a random vector with a pdf $p(u)$ independent of $\lambda$. Then, application of the total probability theorem yields

$$p(r|\tilde{\lambda}) = \int_{-\infty}^{\infty} p(r|\tilde{\gamma})p(\tilde{u})d\tilde{u} \qquad (2.3.5)$$

Hence, with the same arguments used above we conclude that the ML estimate of $\lambda$ is that $\tilde{\lambda}$ that maximizes $p(r|\tilde{\lambda})$:

$$\hat{\lambda}_{ML}(r) = \arg\left\{\max_{\tilde{\lambda}}\left\{p(r|\tilde{\lambda})\right\}\right\} \qquad (2.3.6)$$

Some remarks about this procedure are of interest. In the following chapters we shall normally be concerned with the case in which $\lambda$ is a scalar, $\lambda$, which means that we shall consider the *independent estimation* of each synchronization parameter individually, rather than the *joint estimation* of all parameters simultaneously. Independent estimation is much easier to visualize and generally leads to more robust estimation schemes. Then, a problem of logical consistency arises insofar as the method leading to (2.3.6) requires that $\lambda$ be a fixed quantity of unknown value whereas the other parameters (in particular, the synchronization ones) be random. How can we get around this contradiction? There are two options: (*i*) assume that $\nu$, $\theta$ and $\tau$ are all nonrandom and consider the averaging operation (2.3.5) as an *ad hoc* approach; (*ii*) turn a blind eye to the above contradiction and take (2.3.5) as an application of the total probability theorem. To keep the flavor of ML methods in the following we adopt the latter approach.

Two drawbacks arise in the application of (2.3.5). One is that the vector representation of $r(t)$ has infinite dimensions (see Van Trees [23], Ch. 5) and, in consequence, the functions $p(r|\tilde{\gamma})$ and $p(r|\tilde{\lambda})$ are not well defined. Luckily, this obstacle can be overcome by introducing the concept of *likelihood function*, as is now explained. The other difficulty is concerned with the averaging operation in (2.3.5). In most practical cases, in fact, the integration cannot be performed analytically and one must resort to approximations which, inevitably, lead away from the true ML estimates. Here we illustrate the notion of likelihood function, while we defer approximation issues to later chapters.

Let $r$, $z$, $s(\tilde{\gamma})$ and $w$ be vector representations of $r(t)$, $z(t)$, $s(t, \tilde{\gamma})$ and $w(t)$ over a complete orthonormal basis $\{\phi_i(t)\}$ and denote by a prime their truncated versions (to $N$ components):

$$r' \triangleq (r_1, r_2, \ldots, r_N) \qquad (2.3.7)$$

$$z' \triangleq (z_1, z_2, \ldots, z_N) \qquad (2.3.8)$$

$$s'(\tilde{\gamma}) \triangleq [s_1(\tilde{\gamma}), s_2(\tilde{\gamma}), \ldots, s_N(\tilde{\gamma})] \qquad (2.3.9)$$

$$w' \triangleq (w_1, w_2, \ldots, w_N) \qquad (2.3.10)$$

**Principles, Methods and Performance Limits**

In particular, the $k$-th component of $r'$ is given by [23]

$$r_k \triangleq \int_0^{T_0} r(t)\phi_k(t)dt \qquad (2.3.11)$$

The probability density $p(\mathbf{r}'|\tilde{\boldsymbol{\gamma}})$ is computed as follows. From (2.3.2) we get

$$\mathbf{r}' = \mathbf{s}'(\tilde{\boldsymbol{\gamma}}) + \mathbf{w}' \qquad (2.3.12)$$

Next, separating $w(t)$ into real and imaginary parts

$$w(t) \triangleq w_R(t) + jw_I(t) \qquad (2.3.13)$$

and using (2.3.11) produces

$$w_k = w_{Rk} + jw_{Ik} \qquad (2.3.14)$$

with

$$w_{Rk} = \int_0^{T_0} w_R(t)\phi_k(t)dt \qquad (2.3.15)$$

$$w_{Ik} = \int_0^{T_0} w_I(t)\phi_k(t)dt \qquad (2.3.16)$$

Also, since $w_R(t)$ and $w_I(t)$ are independent Gaussian random processes with spectral density $N_0$, with the arguments of Appendix 2.B it is found that

$$\mathrm{E}\{w_{Rk}w_{In}\} = 0, \quad k,n = 1,2,\ldots,N \qquad (2.3.17)$$

$$\mathrm{E}\{w_{Rk}w_{Rn}\} = \mathrm{E}\{w_{Ik}w_{In}\} = \begin{cases} N_0 & k = n \\ 0 & \text{otherwise} \end{cases} \qquad (2.3.18)$$

Putting all these facts together and recognizing the Gaussian nature of the variables $\{w_{Rk}\}$ and $\{w_{Ik}\}$ leads to the desired result

$$p(\mathbf{r}'|\tilde{\boldsymbol{\gamma}}) = \prod_{k=1}^{N} \frac{1}{2\pi N_0} \exp\left\{-\frac{|r_k - s_k(\tilde{\boldsymbol{\gamma}})|^2}{2N_0}\right\}$$

$$= C_N \exp\left\{-\frac{1}{2N_0}\sum_{k=1}^{N}|r_k - s_k(\tilde{\boldsymbol{\gamma}})|^2\right\} \qquad (2.3.19)$$

with

$$C_N \triangleq (2\pi N_0)^{-N} \qquad (2.3.20)$$

To compute the ML estimate $\hat{\lambda}_{ML}(r)$ we would be tempted to proceed as follows: (*i*) make $N$ tend to infinity in (2.3.19) so as to get $p(\boldsymbol{z}|\tilde{\boldsymbol{\gamma}})$; (*ii*) set $\boldsymbol{z} = r$ in the limit; (*iii*) substitute the result into (2.3.5) to produce $p(r|\tilde{\boldsymbol{\lambda}})$; (*iv*) solve equation (2.3.6). Unfortunately, convergence problems prevent a direct application of this method since the sum in the second line of (2.3.19) diverges as $N$ increases (see Exercise 2.3.1). To sidestep the obstacle we shall exploit the fact that, if $f(\tilde{\boldsymbol{\lambda}})$ is an arbitrary function and we look for the location of its maximum, then we can divide $f(\tilde{\boldsymbol{\lambda}})$ by any positive quantity independent of $\tilde{\boldsymbol{\lambda}}$ without affecting the result.

To proceed, let $p(r'|\tilde{\boldsymbol{\lambda}}) \triangleq p(\boldsymbol{z}' = r'|\tilde{\boldsymbol{\lambda}})$ and keep in mind that

$$p(r'|\tilde{\boldsymbol{\lambda}}) = \int_{-\infty}^{\infty} p(r'|\tilde{\boldsymbol{\gamma}}) p(\tilde{u}) d\tilde{u} \qquad (2.3.21)$$

Based on the previous observation, the maximum of $p(r'|\tilde{\boldsymbol{\lambda}})$ can be sought by replacing $p(r'|\tilde{\boldsymbol{\gamma}})$ by

$$\Lambda(r'|\tilde{\boldsymbol{\gamma}}) \triangleq \frac{1}{B} p(r'|\tilde{\boldsymbol{\gamma}}) \qquad (2.3.22)$$

where $B$ is any positive constant independent of $\tilde{\boldsymbol{\gamma}}$. In other words, maximizing (2.3.21) is equivalent to maximizing

$$\Lambda(r'|\tilde{\boldsymbol{\lambda}}) \triangleq \int_{-\infty}^{\infty} \Lambda(r'|\tilde{\boldsymbol{\gamma}}) p(\tilde{u}) d\tilde{u} \qquad (2.3.23)$$

In particular, if we choose

$$B = C_N \exp\left\{-\frac{1}{2N_0} \sum_{k=1}^{N} |r_k|^2\right\} \qquad (2.3.24)$$

and note that

$$|r_k - s_k(\tilde{\boldsymbol{\gamma}})|^2 = |r_k|^2 + |s_k(\tilde{\boldsymbol{\gamma}})|^2 - 2\operatorname{Re}\{r_k s_k^*(\tilde{\boldsymbol{\gamma}})\} \qquad (2.3.25)$$

then equation (2.3.22) becomes

$$\Lambda(r'|\tilde{\gamma}) = \exp\left\{\frac{1}{N_0}\sum_{k=1}^{N}\text{Re}\{r_k s_k^*(\tilde{\gamma})\} - \frac{1}{2N_0}\sum_{k=1}^{N}|s_k(\tilde{\gamma})|^2\right\} \quad (2.3.26)$$

One important feature of this formula is that the argument of the exponential converges as $N$ tends to infinity [23] (as opposed to the argument in (2.3.19) which diverges).

Now, consider the location of the maximum of $\Lambda(r'|\tilde{\lambda})$

$$\hat{\lambda}(r') = \arg\left\{\max_{\tilde{\lambda}}\{\Lambda(r'|\tilde{\lambda})\}\right\} \quad (2.3.27)$$

Clearly, $\hat{\lambda}(r')$ is only an approximation to $\hat{\lambda}_{ML}(r)$ since the function $\Lambda(r'|\tilde{\lambda})$ does not incorporate all the information contained in $r(t)$. The approximation improves, however, when $N$ increases and in fact the limit of $\hat{\lambda}(r')$ as $N$ tends to infinity provides the exact value of $\hat{\lambda}_{ML}(r)$. On the other hand, we have [23]

$$\lim_{N\to\infty}\sum_{k=1}^{N}\text{Re}\{r_k s_k^*(\tilde{\gamma})\} = \int_0^{T_0}\text{Re}\{r(t)s^*(t,\tilde{\gamma})\}dt \quad (2.3.28)$$

$$\lim_{N\to\infty}\sum_{k=1}^{N}|s_k(\tilde{\gamma})|^2 = \int_0^{T_0}|s(t,\tilde{\gamma})|^2 dt \quad (2.3.29)$$

so that the limit of $\Lambda(r'|\tilde{\gamma})$ takes the form

$$\Lambda(r|\tilde{\gamma}) = \exp\left\{\frac{1}{N_0}\int_0^{T_0}\text{Re}\{r(t)s^*(t,\tilde{\gamma})\}dt - \frac{1}{2N_0}\int_0^{T_0}|s(t,\tilde{\gamma})|^2 dt\right\} \quad (2.3.30)$$

The corresponding limit of $\Lambda(r'|\tilde{\lambda})$ is obtained by averaging $\Lambda(r|\tilde{\gamma})$ over $\tilde{u}$ (see (2.3.23)), i.e.,

$$\Lambda(r|\tilde{\lambda}) \triangleq \int_{-\infty}^{\infty}\Lambda(r|\tilde{\gamma})p(\tilde{u})d\tilde{u} \quad (2.3.31)$$

and $\hat{\lambda}_{ML}(r)$ is computed as

$$\hat{\lambda}_{ML}(r) = \arg\left\{\max_{\tilde{\lambda}}\{\Lambda(r|\tilde{\lambda})\}\right\} \quad (2.3.32)$$

Functions $\Lambda(r|\tilde{\gamma})$ and $\Lambda(r|\tilde{\lambda})$ are referred to as *likelihood functions*. Comparing (2.3.31)-(2.3.32) with (2.3.5)-(2.3.6) we see that, as the dimensionality of *r* goes to infinity, the ML estimation rules still hold provided that probability densities are replaced by likelihood functions.

### 2.3.2. Baseband Signaling

The arguments leading to (2.3.30) are easily adapted to baseband signals. To do so it is sufficient to bear in mind that: (*i*) the waveforms are now real-valued; (*ii*) the spectral density of the noise equals $N_0/2$ (while, earlier, real and imaginary components of $w(t)$ had spectral density $N_0$). Skipping the details, it is found

$$\Lambda(r|\tilde{\gamma}) = \exp\left\{\frac{2}{N_0}\int_0^{T_0} r(t)s(t,\tilde{\gamma})dt - \frac{1}{N_0}\int_0^{T_0} s^2(t,\tilde{\gamma})dt\right\} \quad (2.3.33)$$

whereas (2.3.31)-(2.3.32) remain unchanged.

**Exercise 2.3.1.** In the discussion leading to (2.3.30) it has been claimed that sums of the type

$$S(N) = \sum_{k=1}^{N} |r_k|^2 \quad (2.3.34)$$

diverge as $N$ tends to infinity. Consider the simple case in which the signal component is zero and show that the expectation of $S(N)$ grows unboundedly as $N$ increases.

*Solution.* By assumption, $r_k = w_k$ and from (2.3.34) we obtain

$$E\{S(N)\} = \sum_{k=1}^{N} E\{|w_k|^2\} \quad (2.3.35)$$

On the other hand, application of (2.3.14) yields

$$|w_k|^2 = w_{Rk}^2 + w_{Ik}^2 \quad (2.3.36)$$

so that, bearing in mind (2.3.18), we conclude that

$$E\{|w_k|^2\} = 2N_0 \quad (2.3.37)$$

Substituting into (2.3.35) produces

**Principles, Methods and Performance Limits**

$$E\{S(N)\} = 2NN_0 \qquad (2.3.38)$$

which indicates that $E\{S(N)\}$ diverges as $N$ tends to infinity.

**Exercise 2.3.2.** Estimate the phase of a sinusoid $A\cos(2\pi f_0 t+\theta)$ embedded in white Gaussian noise, assuming that $A$ and $f_0$ are known.

*Solution.* Two alternative routes may be followed, depending on whether we view $A\cos(2\pi f_0 t+\theta)$ as a *baseband* or a *modulated* signal. As both views are legitimate, we arbitrarily choose the latter.

The baseband equivalent of $A\cos(2\pi f_0 t+\theta)$ is

$$s(t,\theta) = Ae^{j\theta} \qquad (2.3.39)$$

and the ML estimate of $\theta$ is obtained by maximizing the likelihood function $\Lambda(r|\tilde{\theta})$. From (2.3.30) we get

$$\Lambda(r|\tilde{\theta}) = \exp\left\{\frac{A}{N_0}\mathrm{Re}\left[e^{-j\tilde{\theta}}\int_0^{T_0} r(t)dt\right] - \frac{A^2 T_0}{2N_0}\right\} \qquad (2.3.40)$$

Clearly, maximizing $\Lambda(r|\tilde{\theta})$ is equivalent to maximizing the function

$$F(\tilde{\theta}) \triangleq \mathrm{Re}\left[e^{-j\tilde{\theta}}\int_0^{T_0} r(t)dt\right] \qquad (2.3.41)$$

or

$$F(\tilde{\theta}) = \left|\int_0^{T_0} r(t)dt\right|\cos(\psi - \tilde{\theta}) \qquad (2.3.42)$$

where $\psi$ is the argument of the integral in (2.3.41). On the other hand $F(\tilde{\theta})$ achieves a maximum for $\tilde{\theta} = \psi$. Hence, the ML estimator of $\theta$ is

$$\hat{\theta}_{ML}(r) = \arg\left\{\int_0^{T_0} r(t)dt\right\} \qquad (2.3.43)$$

Figure 2.20 illustrates a block diagram for the estimator. The left-hand side of the scheme, up to the low-pass filters, provides real and imaginary components of $r(t)$, say $r_R(t)$ and $r_I(t)$. The filters have unity gain and serve to eliminate double-frequency terms. From a functional point of view they are not necessary as their low-pass action is performed by the integrator.

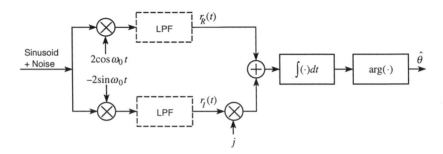

Figure 2.20. ML-estimation of the phase of a sinusoid.

**Exercise 2.3.3.** In the previous exercise the ML estimator has been computed in closed form. This is a rather exceptional case as, in general, only implicit solutions are arrived at. To give an example, consider the transmission of a real-valued pulse $g(t)$ through a channel with an unknown delay $\tau$ (see Figure 2.21). We want to estimate $\tau$ under the following circumstances:

(*i*) the noise has spectral density $N_0/2$;

(*ii*) the signal component of $r(t)$ is

$$s(t) = c_0 g(t - \tau) \qquad (2.3.44)$$

where $c_0$ is a random variable taking values $+1$ or $-1$ with the same probability;

(*iii*) the delayed pulse $g(t-\tau)$ falls entirely within the observation interval $0 \le t \le T_0$, so that the following integral is independent of $\tau$:

$$\int_0^{T_0} g^2(t - \tau) dt = E_g \qquad (2.3.45)$$

*Solution.* The delay is the only parameter of interest to us, while $c_0$ is unwanted. With the notations of Section 2.3.1 we have $\lambda = \tau$ and $u = c_0$ and the likelihood function (2.3.33) becomes

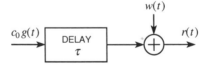

Figure 2.21. Channel with a delay $\tau$.

# Principles, Methods and Performance Limits

$$\Lambda(r|\tilde{\tau},\tilde{c}_0) = C\exp\left\{\frac{2\tilde{c}_0}{N_0}\int_0^{T_0} r(t)g(t-\tilde{\tau})dt\right\} \qquad (2.3.46)$$

where $C$ is a positive constant, independent of $\tilde{\tau}$ and $\tilde{c}_0$. Next, we average $\Lambda(r|\tilde{\tau},\tilde{c}_0)$ over $\tilde{c}_0$. As a consequence of assumption (ii), the pdf of $\tilde{c}_0$ may be written as

$$p(\tilde{c}_0) = \frac{1}{2}\delta(\tilde{c}_0 - 1) + \frac{1}{2}\delta(\tilde{c}_0 + 1) \qquad (2.3.47)$$

where $\delta(x)$ is the delta function. Thus, application of (2.3.31) yields

$$\Lambda(r|\tilde{\tau}) = C\cosh\left\{\frac{2}{N_0}\int_0^{T_0} r(t)g(t-\tilde{\tau})dt\right\} \qquad (2.3.48)$$

and the ML estimate is obtained looking for the maximum of this function.

Note that, as $\tilde{\tau}$ varies, the hyperbolic cosine achieves a maximum at the same abscissa as the absolute value of the integral (2.3.48). Hence

$$\hat{\tau}_{ML}(r) = \arg\left\{\max_{\tilde{\tau}}\left\{\left|\int_0^{T_0} r(t)g(t-\tilde{\tau})dt\right|\right\}\right\} \qquad (2.3.49)$$

Unfortunately, no explicit solution is available for (2.3.49). Approximate estimates of $\tau$ may be obtained in two ways. One is to record $r(t)$, compute the integral in (2.3.49) for different values of $\tilde{\tau}$ and look for the largest result. Alternatively, the parallel processing method indicated in Figure 2.22 may be adopted, where $\tau_1, \tau_2, ..., \tau_M$ are trial values of $\tau$.

## 2.3.3. ML Estimation from Sample Sequences

At this point we turn our attention to ML estimation when the observation consists of a sample sequence. As in Section 2.3.1, we first consider passband signaling and then we extend the results to baseband transmission. We assume that the (baseband) received waveform is first filtered in a low-pass filter $H(f)$ and then is sampled at some rate $1/T_s$, as indicated in Figure 2.23. Also, we make three major hypotheses:

(i) The filter has an ideal rectangular characteristic

$$H(f) = \begin{cases} 1 & |f| \le B_{LPF} \\ 0 & \text{elsewhere} \end{cases} \qquad (2.3.50)$$

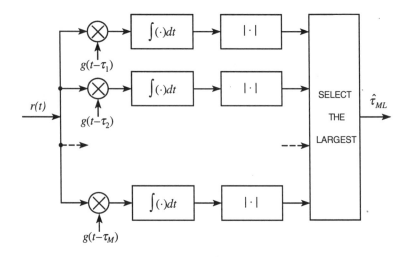

Figure 2.22. Block diagram of the delay estimator.

(*ii*) Its bandwidth is sufficiently large as to pass the signal components undistorted.

(*iii*) The sampling rate $1/T_s$ is twice the filter bandwidth

$$\frac{1}{T_s} = 2B_{LPF} \qquad (2.3.51)$$

Although mathematically convenient, the choice of a rectangular characteristic may be objectionable on practical grounds as the jumps at the band edges are physically unrealizable. One answer to this problem is that, in effect, a rectangular form is not necessary and can be readily replaced by a more realistic one with no consequences. For example, a root-raised-cosine-rolloff function (see Figure 2.24)

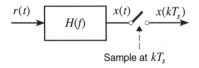

Figure 2.23. Observation in sampled form.

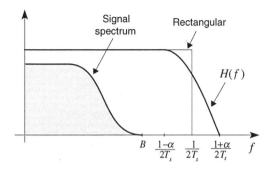

Figure 2.24. Rectangular and root-raised-cosine-rolloff characteristics.

$$H(f) = \begin{cases} 1 & |f| \le \dfrac{1-\alpha}{2T_s} \\ \cos\left[\dfrac{\pi}{4\alpha}\left(|2fT_s|-1+\alpha\right)\right] & \dfrac{1-\alpha}{2T_s} \le |f| \le \dfrac{1+\alpha}{2T_s} \\ 0 & \text{otherwise} \end{cases} \quad (2.3.52)$$

satisfies the assumption (*ii*) if the signal bandwidth $B$ is limited to

$$B < \frac{1-\alpha}{2T_s} \quad (2.3.53)$$

Furthermore, it can be shown that the samples from this filter are statistically equivalent to those from the rectangular filter if (2.3.51) holds true. In fact, they represent sufficient statistics for the estimation of the synchronization parameters [25]. For these reasons in the sequel we adopt the simple rectangular characteristic (2.3.50).

Returning to Figure 2.23, the following remarks can be made about the filter output. First, $x(t)$ has the form

$$x(t) = s(t, \boldsymbol{\gamma}) + n(t) \quad (2.3.54)$$

where $n(t)$ is complex-valued Gaussian noise with power spectral density

$$S_n(f) = \begin{cases} 2N_0 & |f| \le B_{LPF} \\ 0 & \text{elsewhere} \end{cases} \quad (2.3.55)$$

Second, the transformation $r(t) \to x(t)$ is reversible. In fact, we can go back from $x(t)$ to $r(t)$ by adding Gaussian noise to $x(t)$, say $w'(t)$, so as to make $n(t)+w'(t)$ a white process. This is achieved taking $w'(t)$ independent of $n(t)$ and with power spectral density

$$S_{w'}(f) = \begin{cases} 2N_0 & |f| > B_{LPF} \\ 0 & \text{elsewhere} \end{cases} \quad (2.3.56)$$

Third, from sampling theory it follows that $x(t)$ can be reconstructed from its samples $x \triangleq \{x(kT_s)\}$. In other words, the transformation $x(t) \to x$ is reversible.

The fourth and final point is that, as the transformations $r(t) \to x(t)$ and $x(t) \to x$ are reversible, the overall transformation $r(t) \to x$ is also reversible and anything that can be accomplished with $r(t)$ can also be accomplished with $x$ (and vice versa) without loss in performance [1].

At first glance this conclusion sounds like an invitation to put aside discrete-time methods and stick to the continuous-time estimation techniques developed earlier. Indeed, if anything that can be done with $x$ can also be done with $r(t)$ and we already know how to manage with the latter, why should we bother further? One possible answer is that this book deals with *digital* algorithms which, by definition, operate on sampled waveforms. Thus, discrete-time methods are the *natural* route to take here.

Discrete-time estimation methods may be developed by paralleling the treatment in Section 2.3.1. The major points may be summarized as follows. Call

$$x(kT_s) = s(kT_s, \gamma) + n(kT_s) \quad (2.3.57)$$

the available samples $(k = 0, 1, \ldots, L_0 - 1)$ and let $\tilde{\gamma}$ be a trial value of $\gamma$. Consider the sample sequence

$$\chi(kT_s) = s(kT_s, \tilde{\gamma}) + n(kT_s) \quad (2.3.58)$$

In particular, letting $\boldsymbol{\chi} \triangleq \{\chi(kT_s)\}$, we look for the pdf $p(\boldsymbol{\chi}|\tilde{\gamma})$. To this end we write $n(kT_s)$ in the form

$$n(kT_s) = n_R(kT_s) + jn_I(kT_s) \quad (2.3.59)$$

From (2.3.51) and (2.3.55) it can be checked that both $\{n_R(kT_s)\}$ and $\{n_I(kT_s)\}$ are white sequences satisfying the relationships

$$E\{n_{Rk}n_{Im}\} = 0 \quad \forall k, m \quad (2.3.60)$$

**Principles, Methods and Performance Limits** 51

$$E\{n_{Rk}n_{Rm}\} = E\{n_{Ik}n_{Im}\} = \begin{cases} \dfrac{N_0}{T_s} & k = m \\ 0 & \text{otherwise} \end{cases} \qquad (2.3.61)$$

Then, denoting $p(x|\tilde{\gamma}) \triangleq p(\pmb{x} = x|\tilde{\gamma})$, we have

$$p(x|\tilde{\gamma}) = \prod_{k=0}^{L_0-1} \frac{1}{2\pi N_0/T_s} \exp\left\{-\frac{|x(kT_s) - s(kT_s, \tilde{\gamma})|^2}{2N_0/T_s}\right\}$$

$$= C \exp\left\{-\frac{T_s}{2N_0} \sum_{k=0}^{L_0-1} |x(kT_s) - s(kT_s, \tilde{\gamma})|^2\right\} \qquad (2.3.62)$$

where

$$C \triangleq \left(\frac{T_s}{2\pi N_0}\right)^{L_0} \qquad (2.3.63)$$

The pdf of $x$ subject to a generic $\tilde{\lambda}$, $p(x|\tilde{\lambda})$, is obtained by averaging out the unwanted parameters from $p(x|\tilde{\gamma})$, i.e.,

$$p(x|\tilde{\lambda}) = \int_{-\infty}^{\infty} p(x|\tilde{\gamma}) p(\tilde{u}) d\tilde{u} \qquad (2.3.64)$$

Then, the ML estimate of $\lambda$ is found as that $\tilde{\lambda}$ that maximizes $p(x|\tilde{\lambda})$:

$$\hat{\lambda}_{ML}(x) = \arg\left\{\max_{\tilde{\lambda}}\left\{p(x|\tilde{\lambda})\right\}\right\} \qquad (2.3.65)$$

Note that, in contrast with the discussion in Section 2.3.1, no convergence problems arise in computing $p(x|\tilde{\gamma})$ since the summation in (2.3.62) involves only a finite number of terms. Nevertheless, scaling $p(x|\tilde{\gamma})$ is still useful to ease comparisons with estimators operating on continuous-time waveforms. Let us choose the scaling factor

$$B = C \exp\left\{-\frac{T_s}{2N_0} \sum_{k=0}^{L_0-1} |x(kT_s)|^2\right\} \qquad (2.3.66)$$

and, in analogy with (2.3.22), define the likelihood function as

$$\Lambda(x|\tilde{\gamma}) \triangleq \frac{1}{B} p(x|\tilde{\gamma}) \qquad (2.3.67)$$

Then, combining (2.3.62) and (2.3.66)-(2.3.67) yields

$$\Lambda(x|\tilde{\gamma}) = \exp\left\{\frac{T_s}{N_0} \sum_{k=0}^{L_0-1} \mathrm{Re}\left[x(kT_s)s^*(kT_s, \tilde{\gamma})\right] - \frac{T_s}{2N_0} \sum_{k=0}^{L_0-1} |s(kT_s, \tilde{\gamma})|^2 \right\} \qquad (2.3.68)$$

which is perfectly similar to (2.3.30). Finally, letting $\Lambda(x|\tilde{\lambda}) \triangleq p(x|\tilde{\lambda})/B$, from (2.3.64)-(2.3.65) we get the ML estimation equations

$$\Lambda(x|\tilde{\lambda}) \triangleq \int_{-\infty}^{\infty} \Lambda(x|\tilde{\gamma}) p(\tilde{u}) d\tilde{u} \qquad (2.3.69)$$

$$\hat{\lambda}_{ML}(x) = \arg\left\{\max_{\tilde{\lambda}}\left\{\Lambda(x|\tilde{\lambda})\right\}\right\} \qquad (2.3.70)$$

### 2.3.4. Baseband Signaling

The likelihood function $\Lambda(x|\tilde{\gamma})$ for baseband transmission is easily derived by paralleling the foregoing developments. In doing so it is sufficient to keep in mind that the waveforms are now real-valued and the noise spectral density equals $N_0/2$. Skipping the details, it is found that

$$\Lambda(x|\tilde{\gamma}) = \exp\left\{\frac{2T_s}{N_0} \sum_{k=0}^{L_0-1} x(kT_s)s(kT_s, \tilde{\gamma}) - \frac{T_s}{N_0} \sum_{k=0}^{L_0-1} s^2(kT_s, \tilde{\gamma})\right\} \qquad (2.3.71)$$

Equations (2.3.69)-(2.3.70) remain unchanged.

**Exercise 2.3.4.** A sinusoidal signal embedded in white Gaussian noise is passed through a rectangular filter with a transfer function as indicated in (2.3.50). Letting $B_{LPF}$ be the filter bandwidth and $Ae^{j\theta}$ the baseband signal component, estimate the parameter $\theta$ from the samples of the filter output taken at the rate $1/T_s = 2B_{LPF}$. Assume that the signal amplitude is known.

*Solution.* The signal component at the filter output is

$$s(t, \theta) = Ae^{j\theta} \qquad (2.3.72)$$

and the likelihood function for $\theta$ is obtained from (2.3.68) in the form

**Principles, Methods and Performance Limits**

$$\Lambda(x|\tilde{\theta}) = \exp\left\{\frac{AT_s}{N_0}\sum_{k=0}^{L_0-1}\text{Re}\left\{x(kT_s)e^{-j\tilde{\theta}}\right\} - \frac{A^2L_0T_s}{2N_0}\right\} \quad (2.3.73)$$

The maximum of $\Lambda(x|\tilde{\theta})$ is attained for that $\tilde{\theta}$ that maximizes the function

$$F(\tilde{\theta}) \triangleq \sum_{k=0}^{L_0-1}\text{Re}\left\{x(kT_s)e^{-j\tilde{\theta}}\right\} \quad (2.3.74)$$

The location of the maximum is given by

$$\hat{\theta}_{ML}(x) = \arg\left\{\sum_{k=0}^{L_0-1}x(kT_s)\right\} \quad (2.3.75)$$

This result is perfectly similar to that expressed in (2.3.43) for the continuous-time approach.

## 2.4. Performance Limits in Synchronization

In the previous section we have described ML methods for estimating synchronization parameters. At this point the question arises of assessing the ultimate accuracy that can be achieved in synchronization. Establishing bounds to this accuracy is an important goal as it provides benchmarks against which to compare the performance of actual synchronizers. Tools to approach this problem are available from parameter estimation theory in the form of Cramer-Rao bounds (CRBs) [23][24]. Other bounds are described in [26]-[29]. In the following we present a brief overview of CRBs and point out some difficulties that are encountered in their application to synchronization problems [30]-[33]. We also introduce a variant to the CRB, called the *modified* Cramer-Rao bound (MCRB), which does not exhibit such difficulties (but has some other drawbacks that will be pointed out in due time). Finally, we compute the MCRBs for various synchronization parameters and modulation formats.

### 2.4.1. True and Modified Cramer-Rao Bounds

To simplify the discussion we concentrate on the estimation of a single element of $\{\theta,\tau,\nu\}$. Such an element is denoted $\lambda$ and is viewed as a constant (not a random variable). Accordingly, the vector $u$ of unwanted parameters will contain data symbols *plus* two elements from $\{\theta,\tau,\nu\}$. In the case of baseband transmission there is only one synchronization parameter, $\tau$, and the only possible components of $u$ are data symbols.

Consider a generic estimation procedure for $\lambda$ (not necessarily ML) and let $\hat{\lambda}(r)$ be the corresponding estimate. We arbitrarily assume a continuous-time signal description but everything we shall say is also valid in a discrete-time context. Note that $\hat{\lambda}(r)$ depends on the observation $r$ and, in consequence, different observations lead to different estimates. In other words, $\hat{\lambda}(r)$ is a random variable. Its expectation may, or may not, coincide with the true value of $\lambda$. If it does (for any allowed value of $\lambda$) then we say that the estimate is *unbiased*. Being unbiased is clearly a favorable feature as, *on average,* the estimator will yield the true value of the unknown parameter. Even an unbiased estimator, however, may be unsatisfactory if the errors $\hat{\lambda}(r) - \lambda$ are widely scattered around zero. Thus, one wonders what is the minimum error dispersion that can be achieved.

An answer is given by the Cramer-Rao bound, which is a lower limit to the variance of *any* unbiased estimator. This bound is expressed as

$$\mathrm{Var}\{\hat{\lambda}(r) - \lambda\} \geq CRB(\lambda) \qquad (2.4.1)$$

where

$$CRB(\lambda) \triangleq -\frac{1}{\mathrm{E}_r\left\{\dfrac{\partial^2 \ln \Lambda(r|\lambda)}{\partial \lambda^2}\right\}}$$

$$= \frac{1}{\mathrm{E}_r\left\{\left[\dfrac{\partial \ln \Lambda(r|\lambda)}{\partial \lambda}\right]^2\right\}} \qquad (2.4.2)$$

In this equation the following notation has been used:

$$\frac{\partial \ln \Lambda(r|\lambda)}{\partial \lambda} \triangleq \left[\frac{\partial \ln \Lambda(r|\tilde{\lambda})}{\partial \tilde{\lambda}}\right]_{\tilde{\lambda}=\lambda} \qquad (2.4.3)$$

and $\mathrm{E}_r\{\cdot\}$ is the expectation of the enclosed quantity with respect to $r$. No unbiased estimator can provide smaller errors than those indicated by (2.4.2)-(2.4.3).

Unfortunately, application of this bound to practical synchronization problems is difficult due to the necessity of computing $\Lambda(r|\tilde{\lambda})$. In fact, this function is to be derived by averaging out the unwanted parameters from $\Lambda(r|\tilde{\lambda}, u)$

$$\Lambda(r|\tilde{\lambda}) = \int_{-\infty}^{\infty} \Lambda(r|\tilde{\lambda}, \tilde{u}) p(\tilde{u}) d\tilde{u} \qquad (2.4.4)$$

and this is seldom feasible, either because the integration (2.4.4) cannot be performed analytically or because the expectation in (2.4.2) poses insuperable obstacles.

As indicated in [34], a route to overcome this drawback is to resort to the *modified* Cramer-Rao bound (MCRB), which still applies to any unbiased estimator but has the following form:

$$\text{Var}\{\hat{\lambda}(r) - \lambda\} \geq MCRB(\lambda) \qquad (2.4.5)$$

with

$$MCRB(\lambda) \triangleq \frac{N_0}{E_u\left\{\int_0^{T_0} \left|\frac{\partial s(t,\lambda,u)}{\partial \lambda}\right|^2 dt\right\}} \qquad (2.4.6)$$

in the case of passband signals and

$$MCRB(\lambda) \triangleq \frac{N_0/2}{E_u\left\{\int_0^{T_0} \left|\frac{\partial s(t,\lambda,u)}{\partial \lambda}\right|^2 dt\right\}} \qquad (2.4.7)$$

with baseband signals. In (2.4.6)-(2.4.7) the notation $s(t,\lambda,u)$ is used in place of $s(t, \gamma)$ and the expectation $E_u\{\cdot\}$ is over the unwanted parameters $u$.

Two remarks about the $MCRB(\lambda)$ are useful. The first is about its calculation. This issue is addressed later but we anticipate that the $MCRB(\lambda)$ is easy to derive provided that certain assumptions on $u$ are met. The assumptions reflect our basic ignorance of the unwanted parameters and are formally expressed as follows. Denote $u$ as $u_v$, $u_\theta$ or $u_\tau$, according to whether $\lambda=v$, $\lambda=\theta$ or $\lambda=\tau$. For example, when dealing with the bound for $v$, the vector $u_v$ consists of $\theta$, $\tau$ and (possibly) the data symbols. We assume that: (*i*) the timing parameter $\tau$ in $u_v$ and $u_\theta$ is uniformly distributed over the symbol interval $(0,T)$; (*ii*) the probability density functions of $\theta$ in $u_v$ and $u_\tau$, and of $v$ in $u_\theta$ and $u_\tau$, are assigned (but need not be specified for they do not affect the final results); (*iii*) the data symbols $\{c_i\}$ are zero-mean independent random variables with

$$E\{c_i c_k^*\} = \begin{cases} C_2 & \text{for } i = k \\ 0 & \text{otherwise} \end{cases} \qquad (2.4.8)$$

In the sequel these assumptions are referred to as *standard*.

The second remark is about the relationship between $CRB(\lambda)$ and $MCRB(\lambda)$. This issue is addressed in [34], where it is shown that the former is greater than or equal to the latter, i.e.,

$$CRB(\lambda) \geq MCRB(\lambda) \qquad (2.4.9)$$

The equality holds only in two special cases: when $u$ is *perfectly known* or it is *empty* (there are no unwanted parameters).

Equation (2.4.9) indicates that $MCRB(\lambda)$ might be loose, i.e., too low even compared to the error variances of good estimators. This point is illustrated in Figure 2.25, where $\mathrm{Var}\{\hat{\lambda}(r) - \lambda\}$, $CRB(\lambda)$ and $MCRB(\lambda)$ are qualitatively drawn as a function of the signal-to-noise ratio $E_s/N_0$. Clearly, an estimator corresponding to the top line is good as its error variance is close to $CRB(\lambda)$. On the other hand, its performance looks bad as compared to $MCRB(\lambda)$. Thus, taking $MCRB(\lambda)$ as a reference may be pessimistic.

Further discussions in [34] point out that, if $MCRB(\lambda)$ is computed under the above standard assumptions while $CRB(\lambda)$ is computed taking $u$ as a *known vector*, then the two bounds coincide when the observation interval is much larger than the symbol interval. In particular, $CRB(\lambda)$ and $MCRB(\lambda)$ coincide in the following cases:

(*i*) estimation of $\theta$ when $v$, $\tau$, and data are known;

(*ii*) estimation of $\tau$ when $v$, $\theta$, and data are known;

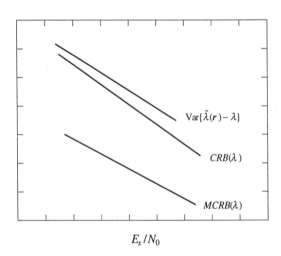

Figure 2.25. Error variance, $CRB(\lambda)$ and $MCRB(\lambda)$.

# Principles, Methods and Performance Limits

(*iii*) estimation of $v$ with M-PSK modulation, when $\tau$ and differential data are available but $\theta$ is unknown.

These results suggest that $MCRB(\lambda)$ represents a good performance reference under rather favorable operating conditions. In fact, many practical synchronizers exist that attain the bound in the circumstances (*i*)-(*iii*) above. Vice versa, $MCRB(\lambda)$ may not be a useful reference when little information is available about the unwanted parameters. For example, in the next chapter we shall see that practical frequency estimators fall short of the bound unless data symbols and a good timing reference are provided.

## 2.4.2. An Alternative Approach to the Bounds

An alternative route to establish bounds is now illustrated which provides further insight into the notion of MCRB. In the foregoing discussion the estimation of the vector $\{\theta, \tau, v\}$ has been broken into three separate operations, each concentrating on one component at a time. There is no logical necessity to proceed in this way, however. Two or even three synchronization parameters may be jointly estimated, in principle [21]. Then, one wonders whether the bounding problem becomes easier when synchronization is viewed as a single operation.

Cramer-Rao bounds in multiple parameter estimation can be derived as extensions of the corresponding scalar problem [23]-[24]. A summary of the major results is as follows. Denote $\lambda_1 \triangleq v$, $\lambda_2 \triangleq \theta$, $\lambda_3 \triangleq \tau$ and $\boldsymbol{\lambda} \triangleq (\lambda_1, \lambda_2, \lambda_3)$. Also, suppose that $\boldsymbol{\lambda}$ is deterministic (not random). Then, if $\hat{\boldsymbol{\lambda}}$ is an unbiased estimator of $\boldsymbol{\lambda}$, a bound on the variance of its $i$-th component is computed as the $(i,i)$ element of the inverse of the *Fisher information matrix* $\mathbf{I}(\boldsymbol{\lambda})$, which is defined as

$$[\mathbf{I}(\boldsymbol{\lambda})]_{i,j} \triangleq - E_r \left\{ \frac{\partial^2 \ln \Lambda(r|\boldsymbol{\lambda})}{\partial \lambda_i \partial \lambda_j} \right\} \quad (2.4.10)$$

Formally

$$\mathrm{Var}\left\{ \hat{\lambda}_i(r) - \lambda_i \right\} \geq \left[ \mathbf{I}^{-1}(\boldsymbol{\lambda}) \right]_{i,i} \quad (2.4.11)$$

Clearly, equations (2.4.10)-(2.4.11) are a generalization of the scalar CRB (2.4.2).

Let us concentrate on the computation of these bounds. A little thought reveals that we still have a problem insofar as $\Lambda(r|\tilde{\boldsymbol{\lambda}})$ is not readily available. In fact, as the data symbols $c$ are generally unknown, we should first compute $\Lambda(r|\tilde{\boldsymbol{\lambda}}, \tilde{c})$ and then pass to $\Lambda(r|\tilde{\boldsymbol{\lambda}})$ through the integral

$$\Lambda(r|\tilde{\lambda}) = \int_{-\infty}^{\infty} \Lambda(r|\tilde{\lambda},\tilde{c})p(\tilde{c})d\tilde{c} \qquad (2.4.12)$$

As we mentioned earlier, however, this route presents unsurpassable obstacles in general.

The obstacles could be overcome by assuming a known symbol sequence. Under these conditions, in fact, $p(\tilde{c})$ would become a delta function and the right-hand side of (2.4.12) would reduce to $\Lambda(r|\tilde{\lambda},c)$. How can we justify our knowing $c$ other than for reasons of mathematical convenience? An intuitive answer is that data knowledge is expected to improve estimation accuracy. Hence, taking $c$ as a known sequence should still lead to lower bounds to estimation errors.

One drawback with this approach is that the results depend on the assumed $c$. To give an example, consider a rectangular pulse sequence. Intuitively, the achievable accuracy in timing estimation depends on the number of transitions in the sequence. In particular we expect that an alternate pattern $(...,+1,-1,+1,...)$ is very suitable for clock recovery whereas an all-one pattern $(...,+1,+1,+1,...)$ is not, as it does not bear any timing information. In general, "good" sequences will provide low bounds and, vice versa, "bad" sequences will produce high bounds.

This problem can be sidestepped by considering *long* sequences. In fact, if the symbol process $\{c_i\}$ is "regular," bad and good sequences will tend to become less and less probable as the length of the sequence grows large. Asymptotically, as the length tends to infinity, the bounds should converge toward limits that depend only on the average statistics of $\{c_i\}$.

Performing long calculations it turns out that, as the observation interval grows large, these limits essentially coincide with the MCRBs discussed in the previous section. This is an interesting result as it shows that logically different routes lead to the same conclusion. It also confirms the idea that the MCRBs coincide with the true CRBs as computed assuming known data and long observation intervals.

### 2.4.3. *MCRB(v)* with PAM Modulation

We limit ourselves to non-offset PAM modulation, as offset modulation can be treated in the same manner and produces the same result. The starting point is equation (2.4.6) with $\lambda=v$. Taking $s(t,v,u)$ from (2.2.25), with simple manipulations it is found that

$$\mathrm{E}_u\left\{\int_0^{T_0}\left|\frac{\partial s(t,v,u)}{\partial v}\right|^2 dt\right\} = 4\pi^2 \int_0^{T_0} t^2\, \mathrm{E}_u\left\{|m(t)|^2\right\}dt \qquad (2.4.13)$$

# Principles, Methods and Performance Limits

with

$$m(t) \triangleq \sum_i c_i g(t - iT - \tau) \tag{2.4.14}$$

The vector $u$ is formed by $\tau$, $\theta$ and the data symbols $c \triangleq \{c_i\}$. However, as $m(t)$ is independent of $\theta$, the expectation in (2.4.13) may be limited to $c$ and $\tau$. Also, as $c$ and $\tau$ are independent, we may first compute the average over $c$ and then over $\tau$. Recalling (2.4.8) and (2.4.14) we get

$$E_c\{|m(t)|^2\} = C_2 \sum_i g^2(t - iT - \tau) \tag{2.4.15}$$

Next, application of the Poisson formula yields

$$\sum_i g^2(t - iT - \tau) = \frac{1}{T} \sum_n G_2\left(\frac{n}{T}\right) e^{j2\pi n(t-\tau)/T} \tag{2.4.16}$$

where $G_2(f)$ is the Fourier transform of $g^2(t)$. Substituting into (2.4.15) and taking $\tau$ uniformly distributed over $(0,T)$ results in

$$E_u\{|m(t)|^2\} = \frac{C_2 G_2(0)}{T}$$

$$= \frac{C_2}{T} \int_{-\infty}^{\infty} |G(f)|^2 df \tag{2.4.17}$$

In Appendix 2.A it is shown that the average signal energy per symbol is given by

$$E_s = \frac{C_2}{2} \int_{-\infty}^{\infty} |G(f)|^2 df \tag{2.4.18}$$

Thus, collecting (2.4.13) and (2.4.17)-(2.4.18), and assuming that the length of the observation interval is a multiple of $T$, say $T_0 = L_0 T$, gives

$$E_u\left\{\int_0^{T_0} \left|\frac{\partial s(t, v, u)}{\partial v}\right|^2 dt\right\} = \frac{8\pi^2 E_s}{3T}(L_0 T)^3 \tag{2.4.19}$$

Finally, inserting into (2.4.6) produces the desired result:

$$T^2 \times MCRB(v) = \frac{3}{8\pi^2 L_0^3} \frac{1}{E_s/N_0} \tag{2.4.20}$$

As is seen, $MCRB(v)$ is inversely proportional to the signal-to-noise ratio and the third power of the observation length.

In the foregoing discussion we have tacitly taken the time origin as the beginning of the observation interval. One wonders how (2.4.20) is affected choosing the interval $(t_0 \leq t \leq t_0 + T_0)$ instead. The answer is found by turning (2.4.6) into

$$MCRB(\lambda) \triangleq \frac{N_0}{E_u\left\{\int_{t_0}^{t_0+T_0} \left|\frac{\partial s(t,\lambda,u)}{\partial \lambda}\right|^2 dt\right\}} \quad (2.4.21)$$

and computing the expectation in the denominator. With the same arguments used above it is found that

$$E_u\left\{\int_{t_0}^{t_0+T_0} \left|\frac{\partial s(t,v,u)}{\partial v}\right|^2 dt\right\} = \frac{8\pi^2 E_s}{3T}\left[(t_0+T_0)^3 - t_0^3\right] \quad (2.4.22)$$

which indicates that $MCRB(v)$ does depend on $t_0$. The dependence on the beginning of the observation interval is inherent in many discussions (see [23] and [26] for example) but is generally not mentioned. Simple algebra shows that $MCRB(v)$ is maximum for $t_0 = -T_0/2$ and this maximum reads

$$T^2 \times MCRB(v) = \frac{3}{2\pi^2 L_0^3} \frac{1}{E_s/N_0} \quad (2.4.23)$$

In the sequel we take (2.4.23) as the modified CRB for frequency estimation.

### 2.4.4. $MCRB(v)$ with CPM Modulation

The modified CRB with CPM modulation is computed with the same methods of the previous section. Taking $s(t,v,u)$ as the product of $e^{j(2\pi vt+\theta)}$ by the signal complex envelope in (2.2.62) and choosing the observation interval $\pm T_0/2$ yields

$$\int_{-T_0/2}^{T_0/2} \left|\frac{\partial s(t,v,u)}{\partial v}\right|^2 dt = \frac{2\pi^2 E_s}{3T}(L_0 T)^3 \quad (2.4.24)$$

The right-hand side is independent of $u$ and, therefore, it coincides with its own expectation with respect to $u$. Thus, substituting into (2.4.6) gives

$$T^2 \times MCRB(v) = \frac{3}{2\pi^2 L_0^3} \frac{1}{E_s/N_0} \qquad (2.4.25)$$

which coincides with (2.4.23).

This result may be surprising, considering the differences between PAM and CPM modulation. To push the point further, one wonders what happens with an unmodulated carrier $A\cos[2\pi(f_0+v)t+\theta]$. The signal complex envelope is

$$s(t,v,\theta) = A e^{j(2\pi v t + \theta)} \qquad (2.4.26)$$

Hence

$$\int_{-T_0/2}^{T_0/2} \left|\frac{\partial s(t,v,\theta)}{\partial v}\right|^2 dt = \frac{\pi^2 A^2 T_0^3}{3} \qquad (2.4.27)$$

or, alternatively, bearing in mind that $T_0 = L_0 T$,

$$\int_{-T_0/2}^{T_0/2} \left|\frac{\partial s(t,v,\theta)}{\partial v}\right|^2 dt = \frac{2\pi^2 E_s}{3T}(L_0 T)^3 \qquad (2.4.28)$$

where $E_s = A^2 T/2$ is the carrier energy in $T$ seconds. Substituting into (2.4.6) produces the same bound as with PAM and CPM.

Again, this is puzzling as, intuitively, frequency measurements with unmodulated carriers should be easier than with modulated signals. The only possible explanation is that the bounds (2.4.23) and (2.4.25) are loose. This conclusion is confirmed by the observation that many practical frequency estimators are far from $MCRB(v)$. An exception occurs when both timing and data are known so that the modulation can be wiped out. Under these conditions, simple algorithms are available [35]-[37] that achieve the bound.

### 2.4.5. $MCRB(\theta)$ with PAM and CPM Modulations

We limit ourselves to non-offset PAM signaling since offset PAM and CPM modulations lead to the same result. Also, we consider the observation interval $(0, T_0)$ since it turns out that the bound is independent of where the observation begins. Taking $s(t,\theta,u)$ from (2.2.25) produces

$$E_u\left\{\int_0^{T_0} \left|\frac{\partial s(t,\theta,u)}{\partial \theta}\right|^2 dt\right\} = \int_0^{T_0} E_u\{|m(t)|^2\} dt \qquad (2.4.29)$$

where $m(t)$ is still as in (2.4.14). Here, $u$ is formed by $c$, $v$ and $\tau$. However, as $m(t)$ depends only on $c$ and $\tau$, the expectation $E_u\{|m(t)|^2\}$ is the same as that computed in Section 2.4.2. Inserting into (2.4.29) yields

$$E_u\left\{\int_0^{T_0}\left|\frac{\partial s(t,\theta,u)}{\partial \theta}\right|^2 dt\right\} = 2E_s L_0 \qquad (2.4.30)$$

and substituting into (2.4.6) gives the result sought:

$$MCRB(\theta) = \frac{1}{2L_0}\frac{1}{E_s/N_0} \qquad (2.4.31)$$

As with frequency estimation, the modified bound is inversely proportional to the signal-to-noise ratio. However, the parameter $L_0$ in the denominator is now raised to the first power, not the third, which says that phase errors are less sensitive to the observation length than frequency errors.

**Exercise 2.4.1.** Compute $MCRB(\theta)$ for an unmodulated carrier $A\cos(2\pi f_0 t + \theta)$ embedded in white Gaussian noise. Assume that the amplitude and carrier frequency are known.

*Solution.* The complex envelope of the carrier is $s(t,\theta) = Ae^{j\theta}$. Hence

$$\frac{\partial s(t,\theta)}{\partial \theta} = jAe^{j\theta} \qquad (2.4.32)$$

and

$$\left|\frac{\partial s(t,\theta)}{\partial \theta}\right| = A \qquad (2.4.33)$$

Thus, calling

$$E_s \triangleq \frac{A^2 T}{2} \qquad (2.4.34)$$

the carrier energy in $T$ seconds and substituting into (2.4.6) yields

$$MCRB(\theta) = \frac{1}{2L_0}\frac{1}{E_s/N_0} \qquad (2.4.35)$$

which coincides with (2.4.31).

**Principles, Methods and Performance Limits**

**Exercise 2.4.2.** Compute $CRB(\theta)$ for an unmodulated carrier $A\cos(2\pi f_0 t+\theta)$ embedded in white Gaussian noise. Assume again that $A$ and $f_0$ are known.

*Solution.* The likelihood function $\Lambda(r|\tilde{\theta})$ is given in (2.3.40). Thus, paralleling the arguments in Exercise 2.3.2, it is found that

$$\Lambda(r|\tilde{\theta}) = C\exp\left\{\frac{AF(\tilde{\theta})}{N_0}\right\} \quad (2.4.36)$$

where $C$ is independent of $\tilde{\theta}$ and $F(\tilde{\theta})$ is given by

$$F(\tilde{\theta}) = \cos\tilde{\theta}\int_0^{T_0} r_R(t)dt + \sin\tilde{\theta}\int_0^{T_0} r_I(t)dt \quad (2.4.37)$$

Hence

$$\frac{\partial \ln \Lambda(r|\theta)}{\partial \theta} = \frac{A\cos\theta}{N_0}\int_0^{T_0} r_I(t)dt - \frac{A\sin\theta}{N_0}\int_0^{T_0} r_R(t)dt \quad (2.4.38)$$

On the other hand, bearing in mind that

$$r_R(t) = A\cos\theta + w_R(t) \quad (2.4.39)$$

$$r_I(t) = A\sin\theta + w_I(t) \quad (2.4.40)$$

equation (2.4.38) becomes

$$\frac{\partial \ln \Lambda(r|\theta)}{\partial \theta} = \frac{A\cos\theta}{N_0}\int_0^{T_0} w_I(t)dt - \frac{A\sin\theta}{N_0}\int_0^{T_0} w_R(t)dt \quad (2.4.41)$$

Next, observe that $w_I(t)$ and $w_R(t)$ are independent processes with spectral density $N_0$. Then the two integrals in (2.4.41) are independent random variables with a common variance given by (see Appendix 2.B)

$$\sigma^2 = T_0 N_0 = L_0 T N_0 \quad (2.4.42)$$

Putting these facts together, the mean square value of (2.4.41) is found to be

$$E_r\left\{\left[\frac{\partial \ln \Lambda(r|\theta)}{\partial \theta}\right]^2\right\} = \frac{A^2 L_0 T}{N_0} \quad (2.4.43)$$

so that, substituting into (2.4.2) and defining $E_s$ as in (2.4.34) yields

$$CRB(\theta) = \frac{1}{2L_0} \frac{1}{E_s/N_0} \qquad (2.4.44)$$

which coincides with $MCRB(\theta)$ in (2.4.35). This result should be expected as there are no unwanted parameters in $A\cos(2\pi f_0 t+\theta)$ and, in consequence, CRB and MCRB coincide [34].

### 2.4.6. $MCRB(\tau)$ with PAM Modulation

As we did earlier, we limit ourselves to non-offset passband modulation since offset passband and baseband PAM may be approached in the same manner and have the same bound. Taking $s(t,\tau,\boldsymbol{u})$ from (2.2.25) yields

$$E_{\boldsymbol{u}}\left\{\int_0^{T_0} \left|\frac{\partial s(t,\tau,\boldsymbol{u})}{\partial \tau}\right|^2 dt\right\} = \int_0^{T_0} E_{\boldsymbol{u}}\{|m'(t)|^2\} dt \qquad (2.4.45)$$

with

$$m'(t) \triangleq \sum_i c_i p(t - iT - \tau) \qquad (2.4.46)$$

and $p(t) \triangleq dg(t)/dt$. It is easily checked that the Fourier transform of $p(t)$ is

$$P(f) = j2\pi f G(f) \qquad (2.4.47)$$

As $m'(t)$ is independent of $v$ and $\theta$, the expectation on the right of (2.4.45) may be limited to the data symbols. With the same arguments leading to (2.4.15)-(2.4.16) it is found that

$$E_{\boldsymbol{u}}\{|m'(t)|^2\} = \frac{C_2}{T} \sum_n P_2\left(\frac{n}{T}\right) e^{j2\pi n(t-\tau)/T} \qquad (2.4.48)$$

where $C_2 \triangleq E_c\{|c_i|^2\}$ and $P_2(f)$ is the Fourier transform of $p^2(t)$. This is related to $P(f)$ by

$$P_2(f) = \int_{-\infty}^{\infty} P(v)P(f-v) dv \qquad (2.4.49)$$

In particular, using (2.4.47) gives

$$P_2(0) = \int_{-\infty}^{\infty} [dg(t)/dt]^2 dt$$
$$= 4\pi^2 \int_{-\infty}^{\infty} f^2 |G(f)|^2 df \qquad (2.4.50)$$

Next, assuming $T_0 = L_0 T$, it can be shown that

$$\int_0^{T_0} E_u\{|m'(t)|^2\} dt = C_2 L_0 P_2(0) \qquad (2.4.51)$$

Hence, from (2.4.45) and (2.4.18) we obtain

$$E_u\left\{\int_0^{T_0} \left|\frac{\partial s(t,\tau,u)}{\partial \tau}\right|^2 dt\right\} = \frac{8\pi^2 \xi L_0 E_s}{T^2} \qquad (2.4.52)$$

where $\xi$ is an adimensional parameter:

$$\xi \triangleq T^2 \frac{\int_{-\infty}^{\infty} f^2 |G(f)|^2 df}{\int_{-\infty}^{\infty} |G(f)|^2 df} \qquad (2.4.53)$$

Finally, substituting (2.4.52) into (2.4.6) produces the desired result

$$\frac{1}{T^2} \times MCRB(\tau) = \frac{1}{8\pi^2 \xi L_0} \frac{1}{E_s/N_0} \qquad (2.4.54)$$

As with phase estimation, the modified bound is inversely proportional to the signal-to-noise ratio and the observation length. Also, it is inversely proportional to $\xi$ which may be viewed as the normalized mean square bandwidth of $G(f)$. This suggests that timing estimation should be easier with wideband signals. Physically, wideband pulses have comparatively short duration and, therefore, are better "seen" in the presence of noise.

**Exercise 2.4.3.** Compute the parameter $\xi$ assuming $G(f)$ as a root-raised-cosine function as given in (2.2.12).
*Solution.* Inserting (2.2.12) into (2.4.53) and performing straightforward manipulations, it is found that

$$\xi = \frac{1}{12} + \alpha^2 \left( \frac{1}{4} - \frac{2}{\pi^2} \right) \qquad (2.4.55)$$

As is physically intuitive, $\xi$ is an increasing function of the rolloff factor $\alpha$.

**Exercise 2.4.4.** The real-valued waveform $r(t)=g(t-\tau)+w(t)$ is observed over the infinite interval $-\infty < t < \infty$. The pulse $g(t)$ has finite energy and, in particular, it satisfies the condition

$$\lim_{t \to \pm \infty} g(t) = 0 \qquad (2.4.56)$$

The noise $w(t)$ is white, with a spectral density $N_0/2$. Compute $CRB(\tau)$ and compare the result with $MCRB(\tau)$.

*Solution.* The likelihood function $\Lambda(r|\tilde{\tau})$ is obtained from (2.3.33) letting $s(t,\tilde{\gamma}) = g(t-\tilde{\tau})$ and changing the integration interval from $(0,T_0)$ to $(-\infty,\infty)$. This results in

$$\Lambda(r|\tilde{\tau}) = \exp \left\{ \frac{2}{N_0} \int_{-\infty}^{\infty} r(t)g(t-\tilde{\tau})dt - \frac{E_s}{N_0} \right\} \qquad (2.4.57)$$

where $E_s$ is the energy of $g(t)$. Taking the logarithm yields

$$\ln \Lambda(r|\tilde{\tau}) = \frac{2}{N_0} \int_{-\infty}^{\infty} r(t)g(t-\tilde{\tau})dt - \frac{E_s}{N_0} \qquad (2.4.58)$$

To compute $CRB(\tau)$ we need the derivative $\partial \ln \Lambda(r|\tilde{\tau})/\partial \tilde{\tau}$ for $\tilde{\tau} = \tau$ (see (2.4.2)-(2.4.3)). From (2.4.58) we have

$$\frac{\partial \ln \Lambda(r|\tau)}{\partial \tau} = -\frac{2}{N_0} \int_{-\infty}^{\infty} r(t) \frac{dg(t-\tau)}{dt} dt$$

$$= -\frac{2}{N_0} \int_{-\infty}^{\infty} [g(t-\tau) + w(t)] \frac{dg(t-\tau)}{dt} dt$$

$$= -\frac{2}{N_0} \int_{-\infty}^{\infty} w(t) \frac{dg(t-\tau)}{dt} dt \qquad (2.4.59)$$

where the relationship (2.4.56) has been exploited.

The mean square value of (2.4.59) is now computed making use of the results in Appendix 2.B:

**Principles, Methods and Performance Limits**

$$E_r\left\{\left[\frac{\partial \ln \Lambda(r|\tau)}{\partial \tau}\right]^2\right\} = \frac{2}{N_0}\int_{-\infty}^{\infty}[dg(t)/dt]^2 dt \qquad (2.4.60)$$

Alternatively, because of (2.4.50), we may write

$$E_r\left\{\left[\frac{\partial \ln \Lambda(r|\tau)}{\partial \tau}\right]^2\right\} = \frac{8\pi^2}{N_0}\int_{-\infty}^{\infty} f^2|G(f)|^2 df \qquad (2.4.61)$$

Finally, substituting into (2.4.2) and rearranging yields

$$\frac{1}{T^2} \times CRB(\tau) = \frac{1}{8\pi^2 \xi}\frac{1}{E_s/N_0} \qquad (2.4.62)$$

where $\xi$ is still as defined in (2.4.53).

Comparing with (2.4.54) we see that $CRB(\tau)$ is $L_0$ times larger. It should be noted however that, in deriving (2.4.62), we have assumed a *single pulse* whereas, in (2.4.54), we considered $L_0$ adjacent pulses. Solving the present problem with a sequence of $L_0$ non-overlapped pulses leads to a result which is $L_0$ times smaller, i.e., just to (2.4.54).

### 2.4.7. $MCRB(\tau)$ with CPM Modulation

Taking $s(t,\tau,u)$ as the product of $e^{j(2\pi vt+\theta)}$ by the signal complex envelope in (2.2.62) yields

$$E_u\left\{\int_0^{T_0}\left|\frac{\partial s(t,\tau,u)}{\partial \tau}\right|^2 dt\right\} = \frac{8\pi^2 h^2 E_s}{T}\int_0^{T_0} E_u\{|m'(t)|^2\} dt \qquad (2.4.63)$$

where

$$m'(t) \triangleq \sum_i c_i g(t-iT-\tau) \qquad (2.4.64)$$

and $g(t)$ is the frequency response of the modulator.

Paralleling the argument used with (2.4.15)-(2.4.17) gives

$$E_u\{|m'(t)|^2\} = \frac{C_2}{T}\sum_i G_2\left(\frac{i}{T}\right)e^{j2\pi i(t-\tau)/T} \qquad (2.4.65)$$

where $C_2 = E\{c_i^2\}$ and $G_2(f)$ is the Fourier transform of $g^2(t)$. Note that

$$G_2(0) = \int_{-\infty}^{\infty} g^2(t)\,dt \qquad (2.4.66)$$

Next, substituting (2.4.65) into (2.4.63) and assuming $T_0 = L_0 T$ produces

$$E_u \left\{ \int_0^{T_0} \left| \frac{\partial s(t,\tau,u)}{\partial \tau} \right|^2 dt \right\} = \frac{8\pi^2 \zeta L_0 E_s}{T^2} \qquad (2.4.67)$$

where

$$\zeta \triangleq C_2 h^2 T \int_{-\infty}^{\infty} g^2(t)\,dt \qquad (2.4.68)$$

is an adimensional parameter analogous to $\xi$ in (2.4.53).

Finally, substituting into (2.4.6) yields the desired result

$$\frac{1}{T^2} \times MCRB(\tau) = \frac{1}{8\pi^2 \zeta L_0} \frac{1}{E_s/N_0} \qquad (2.4.69)$$

The similarity with (2.4.54) is striking. The two formulas differ only in the parameters $\xi$ and $\zeta$. To make a comparison let us consider the following modulations:

(*i*) QPSK with a rolloff factor $\alpha=0.5$;

(*ii*) CPM with quaternary symbols, $h=0.3$, and a frequency response

$$g(t) = \frac{1}{6T}\left[1 - \cos\frac{2\pi t}{3T}\right] \quad \text{for } 0 \le t \le 3T \qquad (2.4.70)$$

This choice is motivated by the fact that power and spectral efficiencies are almost the same in the two cases. Calculations yield $\xi=0.095$ and $\zeta=0.0375$, which indicates that timing estimation with CPM might be more difficult than with PAM.

**Exercise 2.4.5.** Spectral and power efficiencies of CPM modulations are very sensitive to the frequency response $g(t)$. In many theoretical studies two forms of $g(t)$ are adopted: rectangular (REC) and raised-cosine (RC). Thus, denoting by $L$ the duration of $g(t)$ in symbol intervals, we have either LREC or LRC pulses. Formally

$$\text{LREC}: \quad g(t) = \begin{cases} \dfrac{1}{2LT} & 0 \leq t \leq LT \\ 0 & \text{elsewhere} \end{cases} \quad (2.4.71)$$

$$\text{LRC}: \quad g(t) = \begin{cases} \dfrac{1}{2LT}\left[1 - \cos\dfrac{2\pi t}{LT}\right] & 0 \leq t \leq LT \\ 0 & \text{elsewhere} \end{cases} \quad (2.4.72)$$

Compute the $MCRB(\tau)$ with LREC and LRC pulses.

*Solution.* As a first step let us compute the energy of $g(t)$. Straightforward algebra yields

$$\int_0^{LT} g^2(t)\,dt = \frac{1}{4LT}, \quad \text{with LREC} \quad (2.4.73)$$

$$\int_0^{LT} g^2(t)\,dt = \frac{3}{8LT}, \quad \text{with LRC} \quad (2.4.74)$$

Hence, substituting into (2.4.68) and then into (2.4.69) produces

$$\frac{1}{T^2} \times MCRB(\tau) = \frac{L}{2\pi^2 C_2 h^2 L_0} \frac{1}{E_s/N_0}, \quad \text{with LREC} \quad (2.4.75)$$

$$\frac{1}{T^2} \times MCRB(\tau) = \frac{L}{3\pi^2 C_2 h^2 L_0} \frac{1}{E_s/N_0}, \quad \text{with LRC} \quad (2.4.76)$$

It is seen that $MCRB(\tau)$ increases in proportion to the length of $g(t)$, meaning that long pulses are more difficult to synchronize than short ones. Practical evidence confirms this conclusion.

## 2.5. Key Points of the Chapter

- Synchronization consists of the recovery of some reference parameters from the received signal and their application to data detection. In this book we consider three basic parameters: carrier frequency, carrier phase and timing epoch. Timing epoch is the only synchronization parameter in baseband transmission.

- Most synchronization algorithms have been discovered using either heuristic arguments or by application of maximum likelihood methods. The latter involve maximizing certain functions, referred to as likelihood functions, which depend on the observed waveform and on trial values of the synchronization parameters.

- The observation may be either a continuous-time waveform or a sample sequence. If sampling rate is properly chosen, no loss of information is incurred with sampling and the resulting ML estimates are as good as those derivable from the original continuous-time waveform. The choice between continuous-time and discrete-time methods is a question of mathematical convenience.

- Establishing bounds to parameter estimation accuracy is an important goal as it provides benchmarks for evaluating the performance of real world synchronizers. Cramer-Rao bounds indicate lower limits to estimation error variances. Unfortunately, their application to synchronization problems leads to serious mathematical difficulties. Modified Cramer-Rao bounds are much easier to employ, but are looser than true Cramer-Rao bounds.

- Modified and true Cramer-Rao bounds coincide when the data are known.

## Appendix 2.A

In this Appendix we provide a brief overview of the calculation of the power spectral density for PAM signals. The reader interested in more details is referred to the textbooks [2] and [3].

### 2.A.1. Baseband Transmission

The mathematical model for a baseband PAM signal is

$$s(t) = \sum_i c_i g(t - iT) \tag{2.A.1}$$

where $\{c_i\}$ are real-valued symbols belonging to the $M$-ary alphabet $\{\pm 1, \pm 3, \ldots, \pm(M-1)\}$. For the time being they are assumed to be equiprobable and independent so that

$$\mathrm{E}\{c_i\} = 0 \tag{2.A.2}$$

$$\mathrm{E}\{c_{i+m} c_i\} = \begin{cases} C_2 & m = 0 \\ 0 & m \neq 0 \end{cases} \tag{2.A.3}$$

**Principles, Methods and Performance Limits**

with

$$C_2 = \frac{M^2 - 1}{3} \qquad (2.A.4)$$

Under conditions (2.A.2)-(2.A.4) the power spectral density (PSD) of $s(t)$ is found to be

$$S(f) = \frac{C_2}{T}|G(f)|^2 \qquad (2.A.5)$$

where $G(f)$ is the Fourier transform of $g(t)$.

Integrating $S(f)$ yields the signal power

$$P_s = \frac{C_2}{T}\int_{-\infty}^{\infty}|G(f)|^2 df \qquad (2.A.6)$$

The signal energy per symbol equals $P_s T$ and has the expression

$$E_s = C_2 \int_{-\infty}^{\infty}|G(f)|^2 df \qquad (2.A.7)$$

In many practical cases $G(f)$ is a square-root Nyquist function

$$G(f) = \sqrt{G_{NYQ}(f)} \qquad (2.A.8)$$

with

$$\int_{-\infty}^{\infty} G_{NYQ}(f) e^{j2\pi f kT} df = \begin{cases} 1 & k = 0 \\ 0 & k \neq 0 \end{cases} \qquad (2.A.9)$$

When this happens, (2.A.7) reduces to $E_s = C_2$.

### 2.A.2. Passband Transmission

The mathematical model for a modulated signal is

$$s_{IF}(t) = \text{Re}\{s_{CE}(t)e^{j2\pi f_c t}\} \qquad (2.A.10)$$

where $s_{CE}(t)$ represents the complex envelope of $s_{IF}(t)$ with respect to the nominal carrier frequency $f_0$. The form of $s_{CE}(t)$ depends on whether modulation is non-offset or offset.

*Non-Offset Modulation*

In this case we have

$$s_{CE}(t) = \sum_i c_i g(t - iT) \qquad (2.A.11)$$

where $c_i$ are complex-valued symbols. For example, the symbol alphabet is $\{a+jb; a,b=\pm 1,\pm 3,...,\pm(M-1)\}$ with M×M-QAM and $\{e^{j\alpha}; \alpha = 0, 2\pi/M,..., 2\pi(M-1)/M\}$ with M-PSK. Again, we assume equiprobable and independent data symbols so that

$$\mathrm{E}\{c_i\} = 0 \qquad (2.A.12)$$

$$\mathrm{E}\{c_{i+m}c_i^*\} = \begin{cases} C_2 & m = 0 \\ 0 & m \neq 0 \end{cases} \qquad (2.A.13)$$

where the constant $C_2$ depends on the signal constellation. In particular

$$C_2 = \begin{cases} 2(M^2 - 1)/3 & \text{for } M \times M - QAM \\ 1 & \text{for } M - PSK \end{cases} \qquad (2.A.14)$$

*Offset Modulation*

The signal complex envelope reads

$$s_{CE}(t) = \sum_i a_i g(t - iT) + j \sum_i b_i g(t - iT - T/2) \qquad (2.A.15)$$

where $a_i$ and $b_i$ are real valued. The most common instance of offset modulation is OQPSK, where $a_i$ and $b_i$ take values ±1 independently and with the same probability. Accordingly, one has

$$\mathrm{E}\{a_{i+m}a_i\} = \mathrm{E}\{b_{i+m}b_i\} = \begin{cases} 1 & m = 0 \\ 0 & m \neq 0 \end{cases} \qquad (2.A.16)$$

Returning to (2.A.10), denote by $S_{IF}(f)$ and $S_{CE}(f)$ the power spectral densities of $s_{IF}(t)$ and $s_{CE}(t)$. It can be shown that the following relation holds (for either offset or non-offset modulations):

$$S_{IF}(f) = \frac{1}{4} S_{CE}(f - f_0) + \frac{1}{4} S_{CE}(f + f_0) \qquad (2.A.17)$$

The computation of $S_{CE}(f)$ is performed with the methods used with baseband signals and reads

$$S_{CE}(f) = \frac{C_2}{T}|G(f)|^2 \qquad (2.A.18)$$

where $C_2$ depends on the modulation format (non-offset versus offset) and the signal constellation. $C_2$ is given by (2.A.14) with non-offset modulation while it equals 2 with OQPSK.

The average power of $s_{IF}(t)$ is obtained by integrating $S_{IF}(f)$ over the frequency axis. Bearing in mind (2.A.17)-(2.A.18), this results in

$$P_s = \frac{C_2}{2T} \int_{-\infty}^{\infty} |G(f)|^2 df \qquad (2.A.19)$$

which differs from (2.A.6) by a factor 1/2. The average energy per symbol is $E_s = P_s T$ and is expressed by

$$E_s = \frac{C_2}{2} \int_{-\infty}^{\infty} |G(f)|^2 df \qquad (2.A.20)$$

When $G(f)$ is a square-root Nyquist function, the signal power becomes $P_s = C_2/2T$ and the signal energy $E_s = C_2/2$.

### 2.A.3. Extension to Trellis-Coded Modulations

The above results hold true only with zero-mean and uncorrelated symbols. In particular, conditions (2.A.12)-(2.A.13) must be satisfied with non-offset PAM signaling. It turns out, however, that these conditions are satisfied not only with uncoded modulations but also with most good trellis codes [39]-[40]. Thus, the previous spectral density formulas are generally applicable even with trellis code modulations.

## Appendix 2.B

In this Appendix we compute the mean square value of an integral of the type

$$X = \int_{-\infty}^{\infty} w(t)h(t)dt \qquad (2.B.1)$$

where $w(t)$ is a zero-mean noise process with power spectral density $S_w(f)$ and $h(t)$ is a (generally complex-valued) function with Fourier transform $H(f)$ and energy

$$E_h = \int_{-\infty}^{\infty} |H(f)|^2 df \qquad (2.B.2)$$

To begin, we write the squared amplitude of $X$ as

$$|X|^2 = \int_{-\infty}^{\infty}\int_{-\infty}^{\infty} w(t_1)w^*(t_2)h(t_1)h^*(t_2)dt_1 dt_2 \qquad (2.B.3)$$

Taking the statistical expectation yields

$$E\{|X|^2\} = \int_{-\infty}^{\infty}\int_{-\infty}^{\infty} R_w(t_1 - t_2)h(t_1)h^*(t_2)dt_1 dt_2 \qquad (2.B.4)$$

where $R_w(\tau)$ is the autocorrelation function of $w(t)$ and is related to $S_w(f)$ by

$$S_w(f) = \int_{-\infty}^{\infty} R_w(\tau)e^{-j2\pi f\tau} d\tau \qquad (2.B.5)$$

Next, letting $\tau = t_1 - t_2$ into (2.B.4) and rearranging produces

$$E\{|X|^2\} = \int_{-\infty}^{\infty}\left[\int_{-\infty}^{\infty} R_w(\tau)h(\tau + t_2)d\tau\right] h^*(t_2)dt_2 \qquad (2.B.6)$$

The internal integral in this equation can be expressed as a function of $S_w(f)$ and $H(f)$ by means of Parseval theorem:

$$\int_{-\infty}^{\infty} R_w(\tau)h(\tau + t_2)d\tau = \int_{-\infty}^{\infty} S_w(f)H(f)e^{j2\pi f t_2} df \qquad (2.B.7)$$

Then, substituting into (2.B.6) produces the desired result

$$E\{|X|^2\} = \int_{-\infty}^{\infty} S_w(f)|H(f)|^2 df \qquad (2.B.8)$$

The following special cases are of interest.

**Principles, Methods and Performance Limits**

*Case (i):* $S_w(f)$ *and* $H(f)$ *have disjoint supports.*

Suppose that $S_w(f)$ is zero where $H(f)$ takes significant values, i.e.,

$$S_w(f)H(f) = 0 \tag{2.B.9}$$

Then, (2.B.8) says that $E\{|X|^2\} = 0$. In conclusion, the random variable $X$ has zero mean and zero variance and, in consequence, is zero with probability one.

*Case (ii):* $S_w(f)$ *is a constant.*

Suppose that $w(t)$ is a white process, i.e.,

$$S_w(f) = \frac{N_0}{2} \tag{2.B.10}$$

Then, (2.B.8) yields

$$E\{|X|^2\} = \frac{E_h N_0}{2} \tag{2.B.11}$$

*Case (iii):* $S_w(f) = N_0/2$ *and* $h(t)$ *is a rectangular pulse.*

Suppose

$$h(t) = \begin{cases} 1 & \text{for } |t| \leq T_0/2 \\ 0 & \text{elsewhere} \end{cases} \tag{2.B.12}$$

Then the energy of $h(t)$ is just $T_0$ and (2.B.11) becomes

$$E\{|X|^2\} = \frac{T_0 N_0}{2} \tag{2.B.13}$$

# References

[1] J.M.Wozencraft and I.M.Jacobs, *Principles of Communication Engineering*, New York: John Wiley & Sons, 1965.
[2] S.Benedetto and E.Biglieri, *Principles of Digital Transmission with Wireless Applications*, New York: Plenum Press, 1997.
[3] J.G.Proakis, *Digital Communications*, New York: McGraw-Hill, Second Edition, 1989.
[4] E.Y.Ho and Y.S.Yeh, A New Approach to Evaluate the Error Probability in the Presence of Intersymbol Interference and Additive Gaussian Noise, *Bell Syst. Tech. J.*, **49**, 2249-2265, Nov. 1970.

[5] O.Shimbo and M.Celebiler, The Probability of Error Due to Intersymbol Interference and Gaussian Noise in Digital Communication Systems, *IEEE Trans. Commun. Technology*, **COM-19**, 113-119, April 1971.

[6] K.Bucket and M.Moeneclaey, Effect of Random Carrier Phase and Timing Errors on the Detection of Narrow-Band M-PSK and Bandlimited DS/SS, *IEEE Trans. Commun.*, **COM-43**, 1260-1263, April 1995.

[7] K.Feher, *Digital Communications: Satellite/Earth Station Engineering*, Englewood Cliffs: Prentice Hall, 1983.

[8] M.K.Simon, G.K.Huth and A.Polydoros, Differentially Coherent Detection of QASK for Frequency Hopping Systems, *IEEE Trans. Commun.*, **COM-30**, 158-164, Jan. 1982.

[9] D.Divsalar and M.K.Simon, Maximum-Likelihood Differential Detection of Uncoded and Trellis Coded Amplitude Phase Modulation over AWGN and Fading Channels: Metrics and Performance, *IEEE Trans. Commun.*, **COM-42**, 76-89, Jan. 1994.

[10] J.B.Anderson, T.Aulin and C-E.W.Sundberg, *Digital Phase Modulation*, New York: Plenum, 1986.

[11] B.Rimoldi, A Decomposition Approach to CPM, *IEEE Trans. Inform. Theory*, **IT-34**, 260-270, March 1988.

[12] R.F.Pawula, On the Theory of Error Rates for Narrow-Band Digital FM, *IEEE Trans. Commun.*, **COM-29**, 1634-1643, Nov. 1981.

[13] M.K.Simon and C.C.Wang, Differential Versus Limiter-Discriminator Detection of Narrow-Band FM, *IEEE Trans. Commun.*, **COM-31**, 1227-1234, Nov. 1983.

[14] N.A.B.Svensson and C.-E.Sundberg, Performance Evaluation of Differential and Discriminator Detection of Continuous Phase Modulation, *IEEE Trans. Vehic. Technol.*, **VT-35**, 106-116, Aug. 1986.

[15] P.A.Laurent, Exact and Approximate Construction of Digital Phase Modulations by Superpositions of Amplitude Modulated Pulses, *IEEE Trans. Commun.*, **COM-34**, 150-160, Feb. 1986.

[16] G.Kawas Kaleh, Differentially Coherent Detection of Binary Partial Response Continuous Phase Modulation with Index 0.5, *IEEE Trans. Commun.*, **COM-39**, 1335-1340, Sept. 1991.

[17] P.Galko and S.Pasupathy, Linear Receivers for Correlative Coded MSK, *IEEE Trans. Commun.*, **COM-33**, 338-347, April 1985.

[18] F. de Jager and C.B.Dekker, Tamed Frequency Modulation, a Novel Method to Achieve Spectrum Economy in Digital Transmission, *IEEE Trans. Commun.*, **COM-26**, 534-542, May 1978.

[19] K.Murota, K.Hirade, GMSK modulation for Digital Mobile Radio Telephony, *IEEE Trans. Commun.*, **COM-29**, 1044-1050, July 1981.

[20] L.E.Franks, Synchronization Subsystems: Analysis and Design, in K.Feher (ed.), *Digital Communications: Satellite/Earth Station Engineering*, Englewood Cliffs:Prentice Hall, 1983.

[21] L.E.Franks, Carrier and Bit Synchronization in Data Communication–A Tutorial Review, *IEEE Trans. Commun.*, **COM-28**, 1107-1121, Aug. 1980.

[22] F.M.Gardner, *Demodulator Reference Recovery Techniques Suited for Digital Implementation*, European Space Agency, Final Report, ESTEC Contract No. 6847/86/NL/DG, August, 1988.

[23] H.L.Van Trees, *Detection, Estimation, and Modulation: Part I*, New York: Wiley, 1968.

[24] S.M.Kay, *Fundamentals of Statistical Signal Processing: Estimation Theory*, Englewood Cliffs: Prentice-Hall, 1993.

[25] H.Meyr, M.Oerder and A.Polydoros, On Sampling Rate, Analog Prefiltering, and Sufficient Statistics for Digital Receivers, *IEEE Trans. Commun.*, **COM-41**, 3208-3214, Dec. 1994.

[26] L. P. Seidman, Performance Limitations and Error Calculations for Parameter Estimation, *Proc. IEEE*, **58**, 644-652, May 1970.

[27] J. Ziv and M. Zakai, Some Lower Bounds on Signal Parameter Estimation, *IEEE Trans. Inform. Theory*, **IT-15**, 386-391, May 1969.
[28] S. Bellini and G. Tartara, Bounds on Error in Signal Parameter Estimation, *IEEE Trans. Commun.*, **COM-22**, 340-342, March 1974.
[29] S. C. White and N. C. Beaulieu, On the Application of the Cramer-Rao and Detection Theory Bounds to Mean Square Error of Symbol Timing Recovery, *IEEE Trans. Commun.*, **COM-40**, 1635-1643, Oct. 1992.
[30] M. Moeneclaey, A Simple Lower Bound on the Linearized Performance of Practical Symbol Synchronizers, *IEEE Trans. Commun.*, **COM-31**, 1029-1032, Sept. 1983.
[31] M. Moeneclaey, A Fundamental Lower Bound to the Performance of Practical Joint Carrier and Bit Synchronizers, *IEEE Trans. Commun.*, **COM-32**, 1007-1012, Sept. 1984.
[32] M. Moeneclaey and I. Bruyland, The Joint Carrier and Symbol Synchronizability of Continuous Phase Modulated Waveforms, *Conf. Rec. ICC'86* **2**, paper 31.5.
[33] W.G.Cowley, Phase and Frequency Estimation for PSK Packets: Bounds and Algorithms, *IEEE Trans. Commun.*, **COM-44**, 26-28, Jan. 1996.
[34] A.N.D'Andrea, U.Mengali and R.Reggiannini, The Modified Cramer-Rao Bound and Its Applications to Synchronization Problems, *IEEE Trans. Commun.*, **COM-42**, No 2/3/4, 1391-1399, Feb./March/April 1994.
[35] M.P.Fitz, Planar Filtered Techniques for Burst Mode Carrier Synchronization, *Conference Record GLOBECOM'91*, paper No 12.1.
[36] M.Luise and R.Reggiannini, Carrier Frequency Recovery in All-Digital Modems for Burst-Mode Transmission, *IEEE Trans. Commun.*, **COM-43**, 1169-1178, Feb./March/April 1995.
[37] S.Kay, A Fast and Accurate Single Frequency Estimator, *IEEE Trans. Acoustics, Speech and Signal Proc.*, **37**, 1987-1990, Dec. 1989.
[38] L.E.Franks, *Signal Theory*, Prentice-Hall Inc., Englewood Cliffs, N.J., 1969.
[39] E.Biglieri, Ungerboeck Codes Do Not Shape the Signal Power Spectrum, *IEEE Trans. Inf. Theory*, **IT-32**, 595-596, July 1986.
[40] M.Moeneclaey and U.Mengali, Sufficient Conditions on Trellis Coded Modulation for Code-Independent Synchronizer Performance, *IEEE Trans. Commun.*, **COM-38**, 595-601, May 1990.

# 3

# Carrier Frequency Recovery with Linear Modulations

## 3.1. Introduction

A frequency recovery system accomplishes two basic functions: (*i*) it derives an estimate $\hat{v}$ of the carrier frequency offset; (*ii*) it compensates for this offset by counter-rotating the received waveform $r(t)$ at an angular speed $2\pi\hat{v}$. In the ensuing discussion we distinguish between two major cases [1]:

(*i*) the offset is much smaller than $1/T$,

(*ii*) the offset is on the order of the symbol rate $1/T$.

Case (*i*) occurs when a receiver is operating in steady-state conditions. In these circumstances, timing information can be recovered first, even in the presence of moderate frequency offsets, and then exploited for estimating $v$. Data symbols may or may not be available. For example, known synchronization preambles make possible data-aided operation in time-division-multiple-access (TDMA) systems. Alternately, decision-directed operation may be employed with PSK differential demodulation.

Case (*ii*) corresponds to initial frequency acquisitions in low-capacity digital radios and satellite communication systems. In these applications we can reasonably assume that data symbols, carrier phase and, perhaps, timing are all unknown. Reduction of the frequency error to a small percentage of the symbol rate is necessary before other synchronization functions can successfully begin.

This chapter is organized as follows. Frequency estimation under condition (*i*) is discussed in Sections 3.2 to 3.4. In particular, Section 3.2 deals with data-aided recovery and Section 3.3 with decision-directed recovery with differential PSK. In Section 3.4, data symbols are unknown and frequency estimation is addressed using either open-loop or closed-loop methods. Frequency recovery under condition (*ii*) is studied in Sections 3.5 and 3.6. Here, neither data nor clock information is available.

## 3.2. Data-Aided Frequency Estimation

### 3.2.1. Maximum Likelihood Estimation

In addressing maximum likelihood frequency estimation we assume that: (*i*) data symbols are known; (*ii*) timing is ideal; (*iii*) the frequency offset is a small fraction of the symbol rate. A few remarks on these restrictions are useful. In TDMA systems condition (*i*) is ensured by appending a preamble with a known pattern to the beginning of each burst. Carrier frequency offset may be reestimated on each individual burst or, alternatively, by processing several stored preambles at a time. The second procedure takes advantage of the fact that frequency varies slowly and remains constant over many bursts.

Condition (*ii*) implies that accurate timing information can be gathered even in the presence of moderate frequency errors. Actually, good clock recovery is possible even with frequency errors as large as 20% of the symbol rate.

Finally, frequency offsets in this section are in the range of a few percents of the symbol rate $1/T$, say less than 10%. For example, in a point-to-point microwave radio at 30 GHz, with typical oscillator instabilities of 5 parts per million, the combined transmit/receive oscillator instability can be as large as 300 kHz, which is less than 10% of $1/T$ only if the latter exceeds 3 Msymbols/s. Otherwise, frequency offset must be measured in two steps. A first coarse measurement (performed with the methods of Sections 3.5 and 3.6) allows one to locate the offset within a range narrower than $\pm 0.1/T$. Then, this estimate is subtracted from the carrier frequency and the residual offset is measured with the methods discussed here.

In this subsection we approach frequency estimation via ML methods. As we shall see, while the problem can be put into a simple mathematical framework, its practical solution requires some approximations.

To begin we observe that, if timing and data are known, the signal has only two unknown parameters: frequency offset and carrier phase. Based on the results of Chapter 2, the likelihood function for these parameters has the form

$$\Lambda\left(r|\tilde{v},\tilde{\theta}\right) = \exp\left\{\frac{1}{N_0}\int_0^{T_0} \mathrm{Re}\left\{r(t)\tilde{s}^*(t)\right\}dt - \frac{1}{2N_0}\int_0^{T_0}|\tilde{s}(t)|^2 dt\right\} \quad (3.2.1)$$

where

$$\tilde{s}(t) \triangleq e^{j(2\pi\tilde{v}t+\tilde{\theta})} \sum_i c_i g(t - iT - \tau) \quad (3.2.2)$$

and $\tilde{\theta}$ and $\tilde{v}$ are trial values of $\theta$ and $v$. In the following $\tilde{\theta}$ is modeled as a random variable uniformly distributed on $[0, 2\pi)$ while $\tilde{v}$ is a fixed albeit unknown quantity.

It is easily checked that the second integral in (3.2.1) is independent of $\tilde{v}$ and $\tilde{\theta}$ and, therefore, can be dropped (remember that dividing a likelihood function by anything that does not depend on the trial parameters still yields a likelihood function). Thus, maximizing $\Lambda(r|\tilde{v},\tilde{\theta})$ is equivalent to maximizing

$$\Lambda(r|\tilde{v},\tilde{\theta}) = \exp\left\{\frac{1}{N_0} \int_0^{T_0} \text{Re}\{r(t)\tilde{s}^*(t)\} dt\right\} \quad (3.2.3)$$

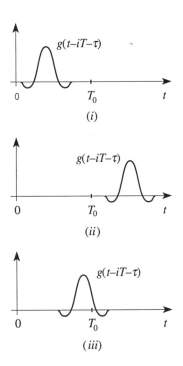

Figure 3.1. Illustrating the location of $g(t - iT - \tau)$ over the observation interval.

To proceed, let us concentrate on the integral in (3.2.3). Using (3.2.2) yields

$$\int_0^{T_0} r(t)\tilde{s}^*(t)dt = e^{-j\tilde{\theta}} \sum_i c_i^* \int_0^{T_0} r(t) e^{-j2\pi\tilde{v}t} g(t-iT-\tau)dt \qquad (3.2.4)$$

Now, bear in mind that $g(t)$ has a limited duration (say, a few symbol intervals about the origin). Thus, when computing the integrals in the right-hand side of (3.2.4), three distinct cases may occur (see Figure 3.1): (*i*) $g(t-iT-\tau)$ is totally inside the observation interval $(0,T_0)$; (*ii*) it is totally outside this interval; (*iii*) it lies across one of the extremes. In the first instance the integration can be extended to the infinite line. In the second, the integral is zero. The third situation cannot be further simplified. Nevertheless, assuming $T_0$ much greater than the duration of $g(t)$, it is realized that type-(*iii*) cases are comparatively few. Putting these facts together leads to the approximation

$$\int_0^{T_0} r(t)\tilde{s}^*(t)dt \approx e^{-j\tilde{\theta}} \sum_{k=0}^{L_0-1} c_k^* x(k) \qquad (3.2.5)$$

where $L_0 \triangleq T_0/T$ is the length of the observation interval in symbol periods (we assume $T_0$ a multiple of $T$) and $x(k)$ is the sample at $kT+\tau$ of the waveform

$$x(t) \triangleq \int_{-\infty}^{\infty} r(\xi) e^{-j2\pi\tilde{v}\xi} g(\xi-t)d\xi \qquad (3.2.6)$$

As is shown in Figure 3.2, $x(t)$ represents the response of the matched filter to the voltage $r'(t) \triangleq r(t)e^{-j2\pi\tilde{v}t}$.

Returning to (3.2.5), let us write

$$e^{-j\tilde{\theta}} \sum_{k=0}^{L_0-1} c_k^* x(k) = |X| e^{j(\psi-\tilde{\theta})} \qquad (3.2.7)$$

with

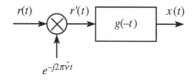

Figure 3.2. Physical interpretation of $x(t)$.

$$|X|e^{j\psi} \triangleq \sum_{k=0}^{L_0-1} c_k^* x(k) \qquad (3.2.8)$$

Then, substituting into (3.2.3) yields

$$\Lambda(r|\tilde{v},\tilde{\theta}) = \exp\left\{\frac{|X|}{N_0}\cos(\psi-\tilde{\theta})\right\} \qquad (3.2.9)$$

Elimination of $\tilde{\theta}$ from $\Lambda(r|\tilde{v},\tilde{\theta})$ is accomplished by averaging over $[0,2\pi)$. This produces

$$\Lambda(r|\tilde{v}) = I_0\left(\frac{|X|}{N_0}\right) \qquad (3.2.10)$$

where $I_0(\alpha)$ is the modified Bessel function of zero order

$$I_0(\alpha) = \frac{1}{2\pi}\int_0^{2\pi} e^{\alpha \cos z} dz \qquad (3.2.11)$$

As $I_0(\alpha)$ is an even function of $\alpha$ with an upward concavity and the argument of $I_0(\alpha)$ in (3.2.10) is positive, it is seen that $\Lambda(r|\tilde{v})$ achieves a maximum at the same location as $|X|$. From (3.2.8) we conclude that the ML estimate is obtained by maximizing

$$\Gamma(\tilde{v}) \triangleq \left|\sum_{k=0}^{L_0-1} c_k^* x(k)\right| \qquad (3.2.12)$$

Figure 3.3 illustrates the computation of $\Gamma(\tilde{v})$.

One difficulty with maximizing $\Gamma(\tilde{v})$ is apparent from the simulation results shown in Figure 3.4, where $\Gamma(\tilde{v})$ is drawn versus $\tilde{v}$ for QPSK modulation and root-raised-cosine-rolloff pulses with rolloff $\alpha=0.5$. Here, the true frequency offset $v$ equals zero, the ratio $E_s/N_0$ equals 30 dB and the

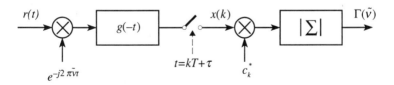

Figure 3.3. Illustrating the computation of $\Gamma(\tilde{v})$.

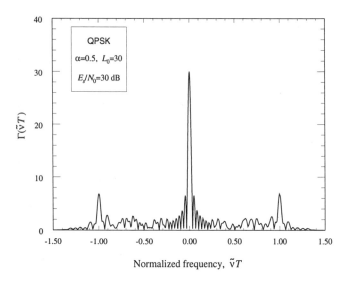

Figure 3.4. Typical shape of the $\Gamma(\tilde{v})$ function.

observation length is 30 symbols. Clearly, as there are many local maxima, the location of the global maximum must be preliminarily found.

In principle this can be done by breaking the estimation algorithm into a two-step search. The first part (*coarse search*) calculates $\Gamma(\tilde{v})$ for a set of $\tilde{v}$ values covering the uncertainty range of v and determines that $\tilde{v}$ which maximizes $\Gamma(\tilde{v})$ over this set. The second part (*fine search*) makes an interpolation between the samples of $\Gamma(\tilde{v})$ and computes the local maximum nearest to the $\tilde{v}$ that has been picked up in the first part. Note that, occasionally, $\Gamma(\tilde{v})$ will be so distorted by noise that its highest peak will occur at a distance from v. When this happens the ML algorithm makes large errors (*outliers*) [2]. In a practical situation outliers have disabling effects on the receiver performance as they result in bursts of errors. Also, outliers become more and more likely as the $E_s/N_0$ decreases. Indeed, estimation accuracy exhibits a *threshold* which is clearly visible from a graph of error variance versus $E_s/N_0$. At large $E_s/N_0$ the variance increases at a fixed rate as $E_s/N_0$ decreases. However, as $E_s/N_0$ approaches some critical value, the curve rises rapidly due to the insurgence of outliers.

### 3.2.2. Practical Frequency Estimators

The above discussion shows that ML estimation is a burdensome process and indicates that simpler methods are desirable. Here we discuss four such methods, all based on the following signal model.

Suppose that

(i) the convolution $g(t) \otimes g(-t)$ is Nyquist, i.e.,

$$\int_{-\infty}^{\infty} g(t)g(t-kT)dt = \begin{cases} 1 & k=0 \\ 0 & \text{otherwise} \end{cases} \quad (3.2.13)$$

(ii) data symbols belong to a PSK constellation

$$\{c_k = e^{j\alpha_k} : \alpha_k = 0, 2\pi/M, \ldots, 2\pi(M-1)/M\} \quad (3.2.14)$$

(iii) the uncertainty range of $v$ is small compared with the symbol rate.

Feeding

$$r(t) = e^{j(2\pi v t + \theta)} \sum_i c_i g(t - iT - \tau) + w(t) \quad (3.2.15)$$

into the matched filter $g(-t)$ produces a waveform $y(t)$ whose samples $y(k) \triangleq y(kT + \tau)$ have the form

$$y(k) = \int_{-\infty}^{\infty} r(t)g(t - kT - \tau)dt \quad (3.2.16)$$

Then, substituting (3.2.15) into (3.2.16) yields

$$y(k) = e^{j\theta} \sum_i c_i \int_{-\infty}^{\infty} e^{j2\pi v t} g(t - iT - \tau)g(t - kT - \tau)dt + n(k) \quad (3.2.17)$$

where the last term (the noise contribution) is given by

$$n(k) = \int_{-\infty}^{\infty} w(t)g(t - kT - \tau)dt \quad (3.2.18)$$

As $w(t)$ is Gaussian and has independent components with PSD $N_0$, it can be shown that $n(k)$ may be written as

$$n(k) = n_R(k) + jn_I(k) \quad (3.2.19)$$

where $n_R(k)$ and $n_I(k)$ are independent zero-mean Gaussian random variables with the same variance $N_0$.

Let us concentrate on the factor $e^{j2\pi v t}g(t-iT-\tau)$ in (3.2.17). Since $g(t-iT-\tau)$ takes significant values only over an interval $T_i$ of a few symbols around $t=iT+\tau$ and we have assumed $|v| \ll 1/T$, the exponential $e^{j2\pi v t}$ may be approximated with a constant $e^{j2\pi v(iT+\tau)}$ for $t \in T_i$ and, in consequence, we have

$$e^{j2\pi v t}g(t-iT-\tau) \approx e^{j2\pi v(iT+\tau)}g(t-iT-\tau) \qquad (3.2.20)$$

Substituting into (3.2.17) and bearing in mind (3.2.13) yields

$$y(k) = c_k e^{j[2\pi v(kT+\tau)+\theta]} + n(k) \qquad (3.2.21)$$

As is seen, the signal component in (3.2.21) depends on $v$ as well as on $\tau$, $\theta$ and the modulation $c_k$. Modulation is easily removed, however, taking advantage of the PSK property $c_k c_k^* = 1$. In fact, multiplying $y(k)$ by $c_k^*$ and letting $z(k) \triangleq y(k)c_k^*$ gives

$$z(k) = e^{j[2\pi v(kT+\tau)+\theta]} + n'(k) \qquad (3.2.22)$$

where $n'(k) \triangleq n(k)c_k^*$ has the same statistics as $n(k)$. Figure 3.5 shows the computation of $z(k)$. Equation (3.2.22) is the basis for the estimation methods discussed in the sequel. They all operate on the data set $\{z(k), k = 0, 1, \ldots, L_0 - 1\}$.

### 3.2.3. First Method (Kay [3])

Rearrange (3.2.22) as follows:

$$z(k) = \rho(k)e^{j[2\pi v(kT+\tau)+\theta+\phi(k)]} \qquad (3.2.23)$$

where $\rho(k)$ and $\phi(k)$ are implicitly defined as

$$\rho(k)e^{j\phi(k)} \triangleq 1 + n'(k)e^{-j[2\pi v(kT+\tau)+\theta]} \qquad (3.2.24)$$

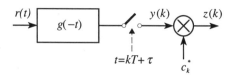

Figure 3.5. Illustrating the computation of $z(k)$.

As $E_s/N_0$ grows large it can be shown that the random variables $\{\phi(k)\}$ are approximately independent, zero-mean and Gaussian [3].

Next consider the argument of the product $z(k)z^*(k-1)$. Using (3.2.23) it is found that

$$\arg\{z(k)z^*(k-1)\} = 2\pi vT + \phi(k) - \phi(k-1) \quad (3.2.25)$$

Clearly, the quantity $\arg\{z(k)z^*(k-1)\}$ may be viewed as a noisy measurement of $2\pi vT$. Note that up to $L_0-1$ such measurements are available from the data set $\{z(k)\}$ and the question arises as to how they can be exploited to estimate $v$. The following procedure has been proposed by Kay [3].

Denote $\alpha(k) \triangleq \arg\{z(k)z^*(k-1)\}$ and let $\boldsymbol{\alpha} \triangleq \{\alpha(1), \alpha(2), \ldots, \alpha(L_0-1)\}$ be the available measurements. Also, call $\tilde{\boldsymbol{\alpha}}$ that $\boldsymbol{\alpha}$ corresponding to the trial frequency offset $\tilde{v}$. Its components have the form

$$\tilde{\alpha}(k) \triangleq 2\pi \tilde{v}T + \phi(k) - \phi(k-1), \quad k=1,2,\ldots,L_0-1 \quad (3.2.26)$$

As we mentioned earlier, the phases $\{\phi(k)\}$ are independent and approximately Gaussian at high $E_s/N_0$. Thus, $\tilde{\boldsymbol{\alpha}}$ is approximately Gaussian and its probability density function $p(\tilde{\boldsymbol{\alpha}}|\tilde{v})$ can be written in closed form. Then, maximizing $p(\tilde{\boldsymbol{\alpha}}=\boldsymbol{\alpha}|\tilde{v})$ as a function of $\tilde{v}$ yields the ML estimate of $v$ based on the observation of $\arg\{z(k)z^*(k-1)\}$, $k=1,2,\ldots,L_0-1$. It is worth stressing that this is not the same as maximizing the function $\Gamma(\tilde{v})$ in (3.2.12). Indeed, the estimator maximizing (3.2.12) observes the set $\{x(k)\}$ whereas Kay's estimator observes the sequence $\arg\{z(k)z^*(k-1)\}$.

Analysis in [3] shows that the maximum of $p(\tilde{\boldsymbol{\alpha}}=\boldsymbol{\alpha}|\tilde{v})$ is reached for

$$\hat{v} = \frac{1}{2\pi T} \sum_{k=1}^{L_0-1} \gamma(k) \arg\{z(k)z^*(k-1)\} \quad (3.2.27)$$

where $\{\gamma(k)\}$ is a smoothing function given by

$$\gamma(k) = \frac{3}{2} \frac{L_0}{L_0^2-1}\left[1 - \left(\frac{2k-L_0}{L_0}\right)^2\right], \quad k=1,2,\ldots,L_0-1 \quad (3.2.28)$$

It is also indicated in [3] that the estimator (3.2.27) is unbiased and reaches the modified Cramer-Rao bound

$$T^2 \times MCRB(v) = \frac{3}{2\pi^2 L_0^3} \frac{1}{E_s/N_0} \quad (3.2.29)$$

at high $E_s/N_0$ values.

## 3.2.4. Second Method (Fitz [4])

Fitz has proposed two frequency estimators [4]-[5]. As they have comparable performance, in the sequel we report only on that described in [4]. Call $R(m)$ the autocorrelations of $z(k)$ as obtained from the data $\{z(k)\}$, i.e.,

$$R(m) \triangleq \frac{1}{L_0 - m} \sum_{k=m}^{L_0-1} z(k)z^*(k-m), \quad 1 \leq m \leq L_0 - 1 \qquad (3.2.30)$$

Substituting (3.2.22) into this equation yields

$$R(m) = e^{j2\pi m vT} + n''(m), \quad 1 \leq m \leq L_0 - 1 \qquad (3.2.31)$$

where $n''(m)$ is a zero-mean noise term.

Next, assume that $|n''(m)|$ is small compared with unity (which is true at high $E_s/N_0$ values). At first sight one might think that the argument of $R(m)$ equals approximately $2\pi m vT$. This is not necessarily true, however, because the difference

$$e(m) \triangleq \arg\{R(m)\} - 2\pi m vT \qquad (3.2.32)$$

is occasionally large. To see why, bear in mind that the argument in (3.2.32) is taken modulo $2\pi$, which means that its values are restricted to the interval $[-\pi, \pi)$. Figure 3.6 illustrates equation (3.2.32) in two different situations. In Figure 3.6(a), $2\pi v m T$ is far from either $-\pi$ or $\pi$ and it is clear that $\arg\{R(m)\} \approx 2\pi v m T$. In Figure 3.6(b), instead, $2\pi m vT$ is close to $\pi$ and $e(m)$ may be about $-2\pi$ even with a small $n''(m)$ (an analogous situation takes place when $2\pi m vT$ is close to $-\pi$).

From the foregoing discussion it appears that the errors $e(m)$ are generally small (at high $E_s/N_0$) provided that $m$ is restricted to $1 \leq m \leq N$ and $N$ is upper bounded by

$$N < \frac{1}{2|v_{\max}|T} \qquad (3.2.33)$$

where $\pm v_{\max}$ denotes the uncertainty range of $v$. Then, summing (3.2.32) for $1 \leq m \leq N$ and dividing by $N$ yields

$$\frac{1}{N}\sum_{m=1}^{N} e(m) = \frac{1}{N}\sum_{m=1}^{N} \arg\{R(m)\} - \pi(N+1)vT \qquad (3.2.34)$$

The left-hand side is now small as its terms tend to compensate each other.

## Carrier Frequency Recovery with Linear Modulations

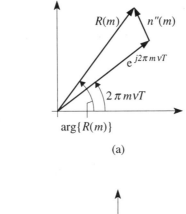

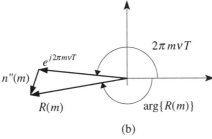

Figure 3.6. Illustrating equation (3.2.32).

Thus, setting to zero the righthand side and solving for ν produces the Fitz estimation formula

$$\hat{\nu} = \frac{1}{\pi N(N+1)T} \sum_{m=1}^{N} \arg\{R(m)\} \qquad (3.2.35)$$

The Fitz estimator is found to be unbiased in the range $|\nu| \leq 1/(2NT)$ and achieves the $MCRB(\nu)$ provided that $N$ equals $L_0/2$. Its estimation accuracy degrades as $N$ decreases but, at the same time, the computational load gets lighter and the estimation range wider, as is seen from (3.2.33). Thus, there is a tradeoff between estimation range, on the one hand, and accuracy and computational simplicity on the other.

### 3.2.5. Third Method (Luise and Reggiannini [6])

The approach adopted by Luise and Reggiannini (L&R) in [6] starts again from (3.2.31) but follows a different route. As a first step, the index $m$ is restricted to $1 \leq m \leq N$, where $N$ is a design parameter less than $L_0-1$. Thus,

$$R(m) = e^{j2\pi mvT} + n''(m), \qquad 1 \le m \le N \tag{3.2.36}$$

Next, the average of (3.2.36) is computed so as to smooth out the noise. This yields

$$\frac{1}{N}\sum_{m=1}^{N} R(m) = \frac{1}{N}\sum_{m=1}^{N} e^{j2\pi mvT} + \frac{1}{N}\sum_{m=1}^{N} n''(m) \tag{3.2.37}$$

which reduces to

$$\sum_{m=1}^{N} e^{j2\pi mvT} \approx \sum_{m=1}^{N} R(m) \tag{3.2.38}$$

presuming that the last term in (3.2.37) is negligible.

Equation (3.2.38) is now solved for $v$. To this end consider the identity

$$\sum_{m=1}^{N} e^{j2\pi mvT} = \frac{\sin \pi NvT}{\sin \pi vT} e^{j\pi(N+1)vT} \tag{3.2.39}$$

and observe that the ratio $\sin(\pi NvT)/\sin(\pi vT)$ is positive for

$$|v| \le \frac{1}{NT} \tag{3.2.40}$$

Thus, taking the argument of both sides of (3.2.39) gives

$$v = \frac{1}{\pi(N+1)T} \arg\left\{\sum_{m=1}^{N} e^{j2\pi mvT}\right\} \tag{3.2.41}$$

and using (3.2.38) leads to the L&R formula

$$\hat{v} = \frac{1}{\pi(N+1)T} \arg\left\{\sum_{m=1}^{N} R(m)\right\} \tag{3.2.42}$$

Figure 3.7 illustrates a block diagram for the L&R estimator. A look-up table is employed to compute the argument of the sum.

Analysis and computer simulations reported in [6] indicate that the L&R estimator is unbiased in the range (3.2.40). It is also shown that it achieves the $MCRB(v)$ at $E_s/N_0$ values as low as 0 dB for $N \approx L_0/2$. As happens with the Fitz estimator, the accuracy degrades rather slowly as $N$ decreases while the estimation range gets wider (see (3.2.40)).

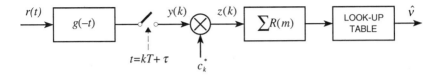

Figure 3.7. Block diagram of the L&R frequency estimator.

### 3.2.6. Fourth Method (Approximate ML Estimation)

The last method relies on an approximation to the function $\Gamma(\tilde{v})$ in (3.2.12), which is valid when $v$ is confined in a small interval around the origin. To see how this comes about let us compute $x(k) \triangleq x(kT+\tau)$ from (3.2.6):

$$x(k) = \int_{-\infty}^{\infty} r(t) e^{-j2\pi \tilde{v} t} g(t - kT - \tau) dt \qquad (3.2.43)$$

Then, assuming $\tilde{v}$ sufficiently small and reasoning as with (3.2.20) yields

$$e^{j2\pi v t} g(t - iT - \tau) \approx e^{j2\pi v(iT+\tau)} g(t - iT - \tau) \qquad (3.2.44)$$

from which we get

$$x(k) \approx e^{-j2\pi \tilde{v}(kT+\tau)} \int_{-\infty}^{\infty} r(t) g(t - kT - \tau) dt$$

$$= y(k) e^{-j2\pi \tilde{v}(kT+\tau)} \qquad (3.2.45)$$

where $y(k)$ is the matched-filter output at $kT+\tau$ (see Figure 3.7). Substituting this equation into (3.2.12) and letting $z(k) \triangleq y(k) c_k^*$ produces the desired approximation

$$\Gamma(\tilde{v}) \approx \left| \sum_{k=1}^{L_0} z(k) e^{-j2\pi \tilde{v} kT} \right| \qquad (3.2.46)$$

At first sight it seems we are still bogged down by the same difficulties encountered with (3.2.12). It should be noted, however, that the variable $\tilde{v}$ in (3.2.12) is hidden in the samples $x(k)$ whereas it appears explicitly in (3.2.46). This means that while the samples $\{x(k)\}$ must be recomputed for every $\tilde{v}$, the $\{z(k)\}$ need only be computed once. In addition, the coarse search for the maximum of (3.2.46) can be made rapid by means of fast Fourier transform (FFT) techniques [2].

It is interesting to compare the complexity of this (approximate) ML estimator with that of the Fitz or L&R methods (which are approximately equivalent for a given $N$). Consider (3.2.42), for example. Here, the bulk of the operations goes into computing the correlations $\{R(m)\}$. Actually, from (3.2.30) and (3.2.42) we see that a total of $N(2L_0-N-1)/2$ complex multiplications is required. As for the ML estimator, most of the computational effort goes into computing $\Gamma(\tilde{v})$ with a sufficiently high resolution. It would seem sufficient to take $\Gamma(\tilde{v})$ at the points $\tilde{v}=n/L_0$, $n=0,1,\ldots,L_0-1$. It turns out, however, that to achieve the MCRB even at low $E_s/N_0$ the number of points must be increased by a factor $M$ (the so-called *zero-stuffing factor*) equal to 2 or 4, depending on $L_0$ (see [2]). In general, $M=4$ is needed when $L_0$ is small (say, less than 64) while $M=2$ is sufficient with a larger $L_0$. Using an FFT of size $ML_0$ results in a complexity on the order of $(1/2)ML_0 \times \log_2(ML_0)$ multiplications.

As we mentioned earlier, the L&R algorithm achieves the MCRB for $N=L_0/2$. For $L_0=128$, this amounts to 6112 multiplications. With the ML algorithm, instead, we have 1024 multiplications. Thus, the latter is simpler. Now, suppose we decrease $N$ from 64 to 8. The complexity of the L&R algorithm reduces to 988, less than with the ML algorithm. At the same time performance deteriorates, however. Simulations run with QPSK and 50% rolloff show a degradation of about 6 dB in terms of $E_s/N_0$. Thus, the L&R algorithm (as well as Fitz's) provides a trade-off between accuracy and implementation complexity.

### 3.2.7. Performance Comparisons

Comparisons between the above estimators are not straightforward as their performance depends on several parameters such as: (*i*) signal-to-noise ratio $E_s/N_0$; (*ii*) observation length $L_0$; (*iii*) parameter $N$ (in Fitz and L&R estimators). In the following we report on simulation results obtained with $L_0=32$ and $N=16$. As mentioned earlier, this choice of $N$ provides the minimum error variance with the Fitz and L&R estimators. Also, the zero-stuffing factor $M$ in the ML estimator is set to 4. A QPSK scheme is assumed, with an overall raised-cosine-rolloff response and a rolloff factor $\alpha=0.5$.

Figure 3.8 illustrates the (normalized) average estimates, $E\{\hat{v}T\}$, versus $vT$ at $E_s/N_0=2$ dB for all the above algorithms, except the ML (which is found to be unbiased over the full range ±15% of the symbol rate). From the figure it appears that the Kay algorithm has a (very small) bias only toward the extremes of the interval $|vT| \leq 0.15$. The L&R and Fitz algorithms have a much narrower operating range.

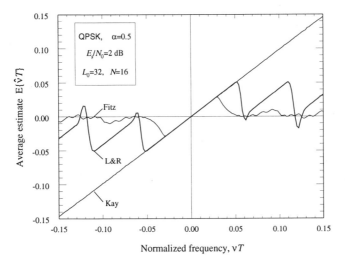

Figure 3.8. Expected value of $\hat{\nu}T$ versus $\nu T$ for QPSK at $E_s/N_0 = 2$ dB.

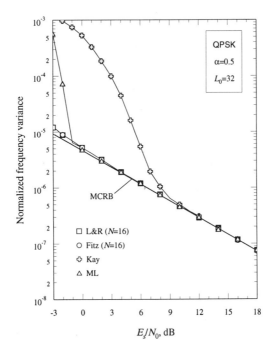

Figure 3.9. Estimation variance versus $E_s/N_0$ for QPSK.

Figure 3.9 shows estimation variance versus $E_s/N_0$. It appears that the Kay method departs from the $MCRB(v)$ at moderate $E_s/N_0$ values, say below 8.5 dB, whereas the other three stay close to $MCRB(v)$ up to 0 dB.

**Exercise 3.2.1.** The simulation results in Figure 3.9 indicate that the Kay algorithm has a much worse threshold than the other methods. Provide an intuitive interpretation for this fact.

*Solution.* In writing (3.2.25) we have overlooked the fact that the arg-function takes values in the interval $[-\pi, \pi)$ whereas the right-hand side may be greater than $\pi$ in absolute value. A modulo $2\pi$ operation is needed, which amounts to changing (3.2.25) into

$$\arg\{z(k)z^*(k-1)\} = 2\pi vT + \phi(k) - \phi(k-1) - 2\pi m(k) \qquad (3.2.47)$$

where $m(k)$ is an integer. Note that $m(k)$ is normally zero at high signal-to-noise ratio (SNR) since the noise-induced phase errors $\phi(k)$ are small and we have assumed $2\pi vT \ll 1$. As the SNR decreases, however, the chances that $m(k)$ be nonzero increase. Clearly, an event $m(k) \neq 0$ makes $\arg\{z(k)z^*(k-1)\}$ a bad estimate of $2\pi vT$. Accordingly, the Kay method is bound to fail as SNR decreases.

It might be argued that this same drawback is also inherent in the Fitz and L&R algorithms although it does not produce visible degradations. The answer is that Kay operates on $z(k)z^*(k-1)$ by *first* applying the hard arctan nonlinearity and *then* smoothing the results (see (3.2.27)) whereas Fitz and L&R apply smoothing *before* the nonlinearity (see (3.2.34) and (3.2.37)).

**Exercise 3.2.2.** In a TDMA transmission system each burst consists of a preamble of 20 known symbols plus a segment of $M=400$ data symbols. The modulation is uncoded QPSK and the signal-to-noise ratio is $E_s/N_0 = 10$ dB. Timing recovery is ideal. The carrier phase measurements (independently derived from each preamble) are so accurate that negligible phase errors can be assumed *at the start* of each data segment. Due to uncompensated frequency errors, however, the carrier phase changes across a data segment and achieves the value

$$\Delta\phi = 2\pi M(v - \hat{v})T \qquad (3.2.48)$$

at the end of the segment. Clearly, a non-negligible $\Delta\phi$ degrades the symbol error rate (SER). Indicate the maximum error $|v - \hat{v}|$ that corresponds to a SER degradation of about 0.3 dB at the end of the data segment. Also, suppose that frequency estimates are derived with the methods described in the previous sections. Each estimate is based on an observation of $L_0$ symbols and is possibly longer than a single preamble (this is made possible by putting together several consecutive preambles). Compute the total observation length $L_0$ that is needed to achieve a frequency estimation accuracy consistent with an SER degradation of 0.3 dB. How many preambles should be collected to achieve this accuracy?

*Solution.* In Exercise 2.2.3 of Chapter 2 it has been found that an SER degradation of 0.36 dB corresponds to phase errors of about 5° at $E_s/N_0 = 10$ dB. Substituting this result into (3.2.48) yields

$$|v - \hat{v}| \approx \frac{3.5 \cdot 10^{-5}}{T} \qquad (3.2.49)$$

All of the frequency estimation methods discussed earlier achieve the MCRB at $E_s/N_0 = 10$ dB. The value of the bound is

$$MCRB(v) = \frac{0.3}{2\pi^2 L_0^3 T^2} \qquad (3.2.50)$$

The standard deviation of the estimates is the square root of (3.2.50). Equating this quantity to the error in (3.2.49) and solving for $L_0$ results in

$$L_0 \approx 230 \text{ symbols} \qquad (3.2.51)$$

As each preamble has a length of 20 symbols, about 12 preambles are needed to produce a frequency estimate.

**Exercise 3.2.3.** For $N=1$ the L&R estimator takes the form

$$\hat{v} = \frac{1}{2\pi T} \arg\left\{ \sum_{k=1}^{L_0-1} z(k) z^*(k-1) \right\} \qquad (3.2.52)$$

Compute its variance as $E_s/N_0$ grows large.
*Solution.* Start from equation (3.2.22) written in the form

$$z(k) = e^{j[2\pi v(kT+\tau)+\theta]}[1 + V(k)] \qquad (3.2.53)$$

with

$$V(k) \triangleq n'(k) e^{-j[2\pi v(kT+\tau)+\theta]} \qquad (3.2.54)$$

Recalling that $\{n'(k)\}$ are independent and Gaussian random variables, it is easily seen that the sequence $\{V(k)\}$ is statistically equivalent to $\{n'(k)\}$. In particular, real and imaginary components of $V(k)$ have the same variance $N_0$.

Next, using (3.2.53) we get

$$z(k)z^*(k-1) = e^{j2\pi vT}\left[1 + V(k) + V^*(k-1) + V(k)V^*(k-1)\right] \quad (3.2.55)$$

When $E_s/N_0$ is large, $V(k)$ and $V^*(k-1)$ have small amplitudes relative to unity and the product $V(k)V^*(k-1)$ can be neglected in (3.2.55). Also, only the imaginary part of $V(k) + V^*(k-1)$ may be kept as the real part is small in comparison with unity. Thus, letting

$$V(k) \triangleq V_R(k) + jV_I(k) \quad (3.2.56)$$

we obtain

$$z(k)z^*(k-1) \approx e^{j2\pi vT}\left[1 + jV_I(k) - jV_I(k-1)\right] \quad (3.2.57)$$

Hence

$$\sum_{k=1}^{L_0-1} z(k)z^*(k-1) \approx (L_0-1)e^{j2\pi vT}\left[1 + j\frac{V_I(L_0-1) - V_I(0)}{L_0-1}\right] \quad (3.2.58)$$

from which, taking the argument and inserting into (3.2.52) produces

$$\hat{v} \approx v + \frac{V_I(L_0-1) - V_I(0)}{2\pi T(L_0-1)} \quad (3.2.59)$$

The variance of $\hat{v}$ equals the mean square value of the last term in (3.2.59). Formally

$$\text{Var}\{\hat{v}\} = \frac{2N_0}{4\pi^2 T^2 (L_0-1)^2} \quad (3.2.60)$$

To express this result in terms of $E_s/N_0$ we note that the signal energy equals 1/2. This is easily checked from the results of Appendix 2.A.2 of Chapter 2, bearing in mind that the channel is Nyquist and the signal constellation is circular with a unity radius. In summary

$$N_0 = \frac{1}{2E_s/N_0} \quad (3.2.61)$$

and (3.2.60) becomes

$$T^2 \times \text{Var}\{\hat{v}\} = \frac{1}{4\pi^2 (L_0-1)^2} \frac{1}{E_s/N_0} \quad (3.2.62)$$

Comparing with $MCRB(v)$ in (3.2.29) it is seen that the variance (3.2.62) is inversely proportional to $L_0^2$ while the $MCRB(v)$ is inversely proportional to $L_0^3$.

## 3.3. Decision-Directed Recovery with DPSK

### 3.3.1. Decision-Directed Algorithms with Differential PSK

In the foregoing discussion the data have been taken from a preamble. Alternatively, they can be derived from the detector (assuming that decisions are sufficiently accurate). In the latter case we speak of decision-directed (DD) methods rather than data-aided (DA) methods. One obvious question is whether reliable decisions can be obtained even in the presence of a carrier frequency offset. In coherent detection systems this is not the case since frequency recovery is a prerequisite to phase recovery and, ultimately, to correct decisions. For PSK with differential detection, vice versa, reliable differential decisions are possible even in the presence of (moderate) frequency offsets. This suggests that DD frequency estimation might be feasible.

To investigate this point consider again the samples $y(k)$ from the matched filter output and let $D(k)$ be the detector decision corresponding to $y(k)y^*(k-1)$ (see Figure 3.10). Also, assume reliable decisions so that

$$D(k) \approx c_k c_{k-1}^* \qquad (3.3.1)$$

For $m=1$, equation (3.2.30) reads

$$R(1) = \frac{1}{L_0 - 1} \sum_{k=1}^{L_0 - 1} z(k) z^*(k-1) \qquad (3.3.2)$$

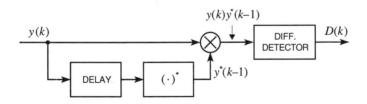

Figure 3.10. Differential detection with PSK modulation.

where $z(k) \triangleq y(k)c_k^*$. On the other hand, we have

$$z(k)z^*(k-1) = y(k)y^*(k-1)(c_k c_{k-1}^*)^*$$
$$\approx y(k)y^*(k-1)D^*(k) \qquad (3.3.3)$$

Hence

$$R(1) = \frac{1}{L_0 - 1} \sum_{k=1}^{L_0 - 1} y(k)y^*(k-1)D^*(k) \qquad (3.3.4)$$

Substituting this result into either Fitz formula (3.2.35) or L&R formula (3.2.42) with $N=1$ yields the following DD frequency estimator:

$$\hat{v} = \frac{1}{2\pi T} \arg \left\{ \sum_{k=1}^{L_0 - 1} y(k)y^*(k-1)D^*(k) \right\} \qquad (3.3.5)$$

This algorithm is reminiscent of the *rotational frequency detector* described in [7] and is similar (but not identical) to the method proposed in [8].

Figure 3.11 shows simulation results for the variance of the normalized estimates $\hat{v}T$ as obtained from (3.3.5) with QPSK modulation. The overall channel response $g(t) \otimes g(-t)$ is Nyquist with a 50% rolloff factor and the observation interval is of 100 symbols. The true offset value is either zero or 5% of the symbol rate. We see that in both cases the threshold is rather high, on the order of 15-20 dB.

**Exercise 3.3.1.** The assumption made earlier about reliable differential decisions is valid only with a limited frequency offset. Provide a ball-park value for the maximum $v$ that is consistent with the hypothesis of correct decisions. Assume QPSK modulation.

*Solution.* Using (3.2.21), the input to the differential detector may be written as

$$y(k)y^*(k-1) \approx c_k c_{k-1}^* e^{j2\pi vT} + n'(k) \qquad (3.3.6)$$

where $n'(k)$ is a noise term contributed by Signal×Noise and Noise×Noise interactions in the product $y(k)y^*(k-1)$. Denoting $c_k = e^{j\alpha_k}$ as the generic QPSK symbol, the information is transmitted through the differences $\alpha_k - \alpha_{k-1}$. The detector tries to establish the value of $\alpha_k - \alpha_{k-1}$ in the set $\{m\pi/2, m=0,1,2,3\}$ or, which is the same, the value of the integer $m$ corresponding to $\alpha_k - \alpha_{k-1}$. To achieve this goal it chooses that $m$ which minimizes the distance of $m\pi/2$ to $\phi(k) \triangleq \arg\{y(k)y^*(k-1)\}$, i.e.,

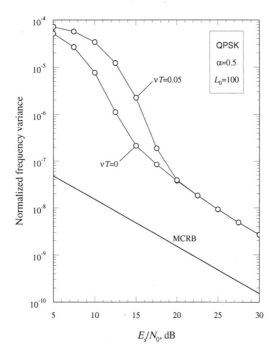

Figure 3.11. Normalized estimation variance.

$$\hat{m} = \arg\left\{\min_{m}\left\{|\phi(k) - m\pi/2|\right\}\right\} \qquad (3.3.7)$$

For the sake of argument, suppose $\alpha_k - \alpha_{k-1} = 0$. Then, in the absence of noise and with $\nu=0$, from (3.3.6) it is seen that $\phi(k) = 0$. Correspondingly, (3.3.7) yields $\hat{m} = 0$, which is correct. Suppose instead $\nu \neq 0$ (but still a negligible noise level). Then, $\phi(k) = 2\pi\nu T$ and the detector decision depends on $\nu$. Inspection of (3.3.7) reveals that correct decisions still occur provided that $2\pi\nu T$ is less than $\pi/4$ in absolute value.

In the presence of noise the situation is more complex as decision errors may occur anyway (with certain probabilities). From the foregoing discussion, however, it is clear that the error probability increases as $\nu$ departs from zero and the increase may be significant unless $|2\pi\nu T|$ is a small fraction of $\pi/4$, say 10-20%. Taking this as a reasonable figure, it is concluded that reliable differential decisions require $\nu$ values limited to a few percents of the symbol rate.

## 3.4. Non-Data-Aided but Clock-Aided Recovery

### 3.4.1. Closed-Loop Algorithm

In this section we concentrate on frequency estimation for PAM signaling with coherent detection. In doing so we still assume that timing is ideal and the Nyquist condition is satisfied. However, since some (moderate) frequency errors are involved, we do not expect that symbol decisions are correct. A closed-loop estimation scheme is described first, while open-loop methods are treated in Section 3.4.2.

For simplicity we start with QPSK. Extensions to general PSK and QAM constellations are discussed later. Our aim is to illustrate a frequency recovery scheme proposed in [9] and represented in the block diagram of Figure 3.12. Here, samples from the matched filter are multiplied by $\hat{c}_k^*$, the complex-conjugate decisions from the detector, and are fed to an error generator. The purpose of the generator is to give an indication of the difference between $v$ and its current estimate $\hat{v}(t)$ provided by the voltage controlled oscillator (VCO). The error signal is filtered (to smooth out the noise) and used to steer the VCO frequency toward $v$. Note that the VCO in the figure is the *baseband* equivalent for the real VCO. In particular, the oscillating frequency $\hat{v}(t)$ in the former equals the difference between oscillating frequency and free running frequency in the latter. The offset $v$ is tracked under the action of the signal $u(k)$ provided by the loop filter. As we shall see, when $\hat{v}(t)$ is less than $v$, $u(k)$ has a positive DC component and the VCO is forced to speed up. Similarly, when $\hat{v}(t)$ is greater than $v$, the VCO is forced to slow down.

The heart of the scheme is the error generator whose operation is now described. Noise is neglected for simplicity and the Nyquist condition is assumed. Accordingly, the received signal is modeled as

$$r(t) = e^{j(2\pi vt + \theta)} \sum_i c_i g(t - iT - \tau) \qquad (3.4.1)$$

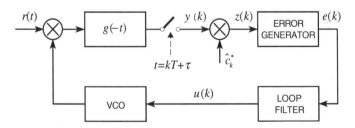

Figure 3.12. Generation of the frequency error.

# Carrier Frequency Recovery with Linear Modulations

In all practical cases the VCO frequency varies slowly and may be considered a constant over several symbol periods. Thus, for several transmission intervals around the generic sampling time $kT+\tau$ the VCO output is well approximated by an exponential $e^{-j2\pi\hat{v}t}$ ($\hat{v}$=constant) and the matched-filter input becomes

$$r(t)e^{-j2\pi\hat{v}t} = e^{j(2\pi f_d T+\theta)}\sum_i c_i g(t-iT-\tau) \qquad (3.4.2)$$

with $f_d \triangleq v - \hat{v}$. Then, with the reasoning leading to (3.2.21) it is easily concluded that the $k$-th sample from the matched filter has the form

$$y(k) = c_k e^{j\phi(k)} \qquad (3.4.3)$$

with

$$\phi(k) \triangleq 2\pi f_d(kT+\tau) + \theta \qquad (3.4.4)$$

Alternately, as $c_k$ has unit amplitude, we may write

$$y(k) = e^{j\psi(k)} \qquad (3.4.5)$$

with

$$\psi(k) = \arg\{c_k\} + \phi(k) \qquad (3.4.6)$$

Next, we turn our attention to the decision rule. The detector makes the decision $\hat{c}_k = e^{j\hat{m}\pi/2}$, where $\hat{m}$ is that integer that minimizes the difference between $\psi(k)$ and $m\pi/2$ in absolute value:

$$\hat{m} = \arg\left\{\min_m\{|\psi(k) - m\pi/2|\}\right\} \qquad (3.4.7)$$

This decision rule is illustrated in Figure 3.13, where circles represent QPSK constellation points. Calling $z(k) = y(k)\hat{c}_k^*$ the input to the error generator, from this figure it is seen that the argument of $z(k)$ is always in the range $\pm\pi/4$. Formally

$$\arg\{z(k)\} = [\psi(k) - \hat{m}\pi/2]_{-\pi/4}^{\pi/4} \qquad (3.4.8)$$

where $[x]_{-\varphi}^{\varphi}$ means "$x$ reduced to the interval $[-\varphi,\varphi)$." Thus, substituting (3.4.4)-(3.4.6) into (3.4.8) and bearing in mind that both $\arg\{c_k\}$ and $\hat{m}\pi/2$ are multiples of $\pi/2$, it is concluded that

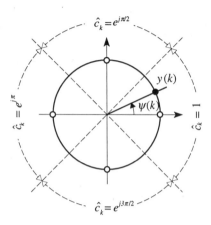

Figure 3.13. Decision rule for QPSK.

$$\arg\{z(k)\} = \left[2\pi f_d(kT+\tau) + \theta\right]_{-\pi/4}^{\pi/4} \tag{3.4.9}$$

Figure 3.14(a) and (b) illustrate $\arg\{z(k)\}$ versus time for $f_d > 0$ and $f_d < 0$ respectively. We see that $\arg\{z(k)\}$ varies in a saw-tooth fashion, with increasing or decreasing ramps depending on the sign of $f_d$. As is explained soon, this is a crucial point to understand the operation of the error generator.

The purpose of the error generator is to provide a signal that, on average, has the same sign as $f_d$. The method proposed in [9] is to relate the error signal $e(k)$ to the argument of $z(k)$ as follows:

$$e(k) \triangleq \begin{cases} \arg\{z(k)\} & \text{if } |\arg\{z(k)\}| < \alpha \\ e(k-1) & \text{otherwise} \end{cases} \tag{3.4.10}$$

where $\alpha$ is a positive parameter less than $\pi/4$. The rationale behind this rule is apparent from inspection of Figure 3.15(a)-(b) which illustrates $e(k)$ versus time for the same $\arg\{z(k)\}$ values indicated in Figure 3.14. These figures show that the average of $e(k)$, $S(f_d)$, is positive for $f_d > 0$ and negative in the opposite case.

The exact dependence of $S(f_d)$ on the frequency error $f_d$ is difficult to establish analytically. Figure 3.16 illustrates the shape of $S(f_d)$ as obtained by simulation for two values of the signal-to-noise ratio. The overall channel response $g(t) \otimes g(-t)$ is Nyquist with a 35% rolloff factor. The parameter $\alpha$ in (3.4.10) is chosen equal to $\pi/8$. As is seen, the range where $S(f_d)$ takes significant values is on the order of ±10% of the symbol rate and represents the

# Carrier Frequency Recovery with Linear Modulations

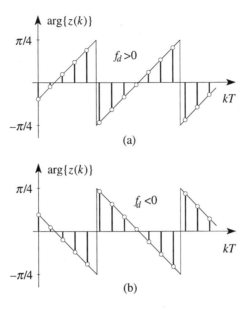

Figure 3.14. Function arg{$z(k)$} versus time.

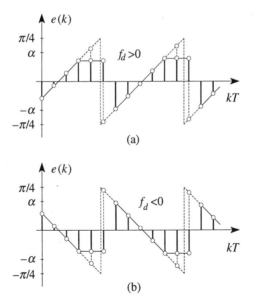

Figure 3.15. Explaining the operation of the error generator.

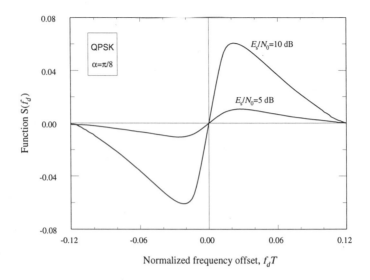

Figure 3.16. Function $S(f_d)$ for QPSK.

*acquisition range* of the loop. The name comes from the fact that an error signal with a nonzero DC component is needed to move the VCO frequency, which can happen only if $S(f_d) \neq 0$. Further discussion on this concept is deferred to Section 3.5.4.

### 3.4.2. Extension to M-ary PSK and QAM

The above ideas are easily extended to M-ary PSK. In this case equation (3.4.9) becomes

$$\arg\{z(k)\} = \left[2\pi f_d(kT+\tau)+\theta\right]_{-\pi/M}^{\pi/M} \tag{3.4.11}$$

Correspondingly, the error signal is still as indicated in (3.4.10) except that $\alpha$ is less than $\pi/M$.

The case of QAM modulation is trickier as a consequence of the more complex signal constellation. A possible solution for 16-QAM is as follows. From Figure 3.17 we see that a 16-QAM constellation is formed by two QPSK sub-constellations with a total of 8 points. The remaining points lie in an annular region between the indicated circles. The idea is to distinguish between points from the sub-constellations and those from the annular region. When the former are transmitted the signal error is computed as indicated for QPSK; otherwise it is left unchanged. This is done by turning (3.4.10) into

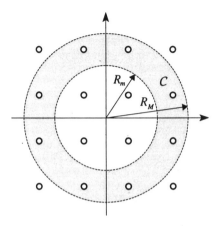

Figure 3.17. QAM constellation.

$$e(k) \triangleq \begin{cases} \arg\{z(k)\} & \text{if } |\arg\{z(k)\}| < \alpha \text{ and } z(k) \notin C \\ e(k-1) & \text{otherwise} \end{cases} \quad (3.4.12)$$

where $C$ is the annular region, i.e.,

$$C \triangleq \{R_m < |z(k)| < R_M\} \quad (3.4.13)$$

Curves of $S(f_d)$ versus $f_d$ are provided in [9] for 16-QAM modulation. They take significant values in a range of about ±3% of the symbol rate. Frequency detectors of this type are currently incorporated in many modems for digital microwave radios.

### 3.4.3. Open-Loop Algorithms

The frequency recovery methods illustrated above have a closed-loop structure and are suitable for continuous mode transmission. Open-loop schemes are more appealing for burst mode applications because of their shorter estimation times. Here we describe one such scheme, which has been proposed in [10] for QPSK modulation but can be extended to $M$-ary PSK.

As in Section 3.2, we assume that $v$ is a small fraction of the symbol rate. Accordingly, the matched-filter output is approximately

$$y(k) = c_k e^{j[2\pi v(kT+\tau)+\theta]} + n(k) \quad (3.4.14)$$

Let us first concentrate on a QPSK constellation $\{e^{jm\pi/2}, m = 0,1,2,3\}$. It is easily checked that $c_k^4 = 1$. Thus, raising $y(k)$ to the fourth power yields

$$y^4(k) = e^{j[8\pi v(kT+\tau)+4\theta]} + n'(k) \tag{3.4.15}$$

where $n'(k)$ is a noise term resulting from Signal×Noise and Noise×Noise interactions. We see that the modulation has been removed from $y^4(k)$. Next, multiplying $y^4(k)$ by $[y^*(k-1)]^4$ yields

$$\left[y(k)y^*(k-1)\right]^4 = e^{j8\pi vT} + n''(k) \tag{3.4.16}$$

where, again, $n''(k)$ comes from Signal×Noise and Noise×Noise products. Equation (3.4.16) indicates that $[y(k)y^*(k-1)]^4$ is an estimate of $e^{j8\pi vT}$. The estimation accuracy can be improved by smoothing out the noise as follows:

$$\frac{1}{L_0-1}\sum_{k=1}^{L_0-1}\left[y(k)y^*(k-1)\right]^4 = e^{j8\pi vT} + \frac{1}{L_0-1}\sum_{k=1}^{L_0-1}n''(k) \tag{3.4.17}$$

Finally, presuming that the last term in (3.4.17) is small in amplitude as compared with unity and taking the argument of both sides yields

$$\hat{v} = \frac{1}{8\pi T}\arg\left\{\sum_{k=1}^{L_0-1}\left[y(k)y^*(k-1)\right]^4\right\} \tag{3.4.18}$$

Figure 3.18 illustrates a block diagram for the algorithm (3.4.18). It should be noted that, as $\arg\{\cdot\}$ takes values in the range $\pm\pi$, the estimates vary between $\pm 1/(8T)$. This is in keeping with the simulation results of Figure 3.19 wherein the average of $\hat{v}$ (normalized to the symbol rate) is drawn versus the true frequency offset for two values of the signal-to-noise ratio. The channel response is Nyquist with a 50% rolloff factor. As is seen, $E\{\hat{v}T\}$ is proportional to $vT$ in the range $|vT| < 1/8$. Figure 3.20 shows the estimation variance as a

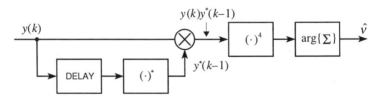

Figure 3.18. Block diagram of the open-loop estimator.

# Carrier Frequency Recovery with Linear Modulations

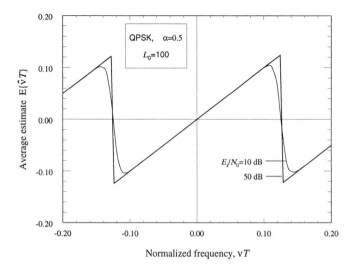

Figure 3.19. Normalized average estimate versus $\nu T$.

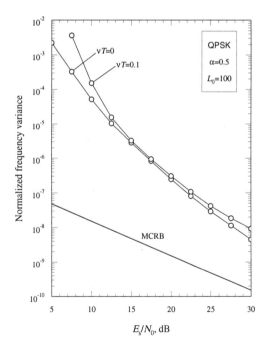

Figure 3.20. Normalized frequency variance as a function of $E_s/N_0$.

function of $E_s/N_0$ when the true offset is either zero or 10% of the symbol rate. The MCRB is also indicated as a reference. Comparing with Figure 3.11 we see that the degradations as $vT$ departs from zero are now quite limited.

The above concepts can be extended to $M$-ary PSK modulation in a straightforward manner. In particular, modulation removal requires raising $y(k)$ to the $M$-th power. This produces

$$y^M(k) = e^{j[2M\pi v(kT+\tau)+M\theta]} + n'(k) \qquad (3.4.19)$$

Then, reasoning as before yields

$$\hat{v} = \frac{1}{2M\pi T}\arg\left\{\sum_{k=1}^{L_0-1}\left[y(k)y^*(k-1)\right]^M\right\} \qquad (3.4.20)$$

Note that the estimation range is now reduced to (approximately) $\pm 1/(2MT)$.

## 3.5. Closed-Loop Recovery with No Timing Information

### 3.5.1. Likelihood Function

So far a very small frequency offset has been assumed. Henceforth we take the opposite viewpoint and allow $v$ to achieve values on the order of the symbol rate. A first consequence of this change in perspective is that previous assumptions of an ideal clock and, even more, of known data symbols are no longer tenable. With a large frequency offset, in fact, data symbols are not available and timing information is totally lacking, meaning that our best guess of the timing phase is anything in the interval $(0,T)$. Thus, significant changes must be made on the signal model, as is now indicated.

Let us start from the complex envelope of the incoming waveform

$$r(t) = s(t) + w(t) \qquad (3.5.1)$$

The signal has two different expressions, according to whether *non-offset* or *offset* modulation is considered. For now we concentrate on the former but the results are subsequently extended to the latter. For non-offset modulation we have

$$s(t) = e^{j(2\pi vt+\theta)}\sum_i c_i g(t-iT-\tau) \qquad (3.5.2)$$

# Carrier Frequency Recovery with Linear Modulations

The parameters involved in (3.5.2) are modeled as follows:

- $v$ is an unknown constant, in the range $\pm 1/T$;
- $\theta$ is a random variable uniformly distributed over $[0, 2\pi)$;
- $\tau$ is a random variable uniformly distributed over $[0, T)$;
- $\{c_i\}$ are zero-mean independent random variables with the following second-order moments:

$$E\{c_i c_k^*\} = \begin{cases} C_2 & \text{for } i = k \\ 0 & \text{elsewhere} \end{cases} \tag{3.5.3}$$

- $\theta$, $\tau$, and $\{c_i\}$ are independent of each other.

Consider the likelihood function

$$\Lambda(r|\tilde{v},\tilde{\theta},\tilde{\tau},\tilde{c}) = \exp\left\{\frac{1}{N_0}\int_0^{T_0} \text{Re}[r(t)\tilde{s}^*(t)]dt - \frac{1}{2N_0}\int_0^{T_0} |\tilde{s}(t)|^2 dt\right\} \tag{3.5.4}$$

where $\tilde{s}(t)$ is the trial signal

$$\tilde{s}(t) \triangleq e^{j(2\pi\tilde{v}t+\tilde{\theta})} \sum_i \tilde{c}_i g(t - iT - \tilde{\tau}) \tag{3.5.5}$$

To compute the marginal likelihood function $\Lambda(r|\tilde{v})$ we must average $\Lambda(r|\tilde{v},\tilde{\theta},\tilde{\tau},\tilde{c})$ with respect to $\tilde{\theta}$, $\tilde{\tau}$ and $\tilde{c}$. Unfortunately, this operation is difficult and we are compelled to make approximations. We assume a low SNR, such that the expansion of the exponential in (3.5.4) into a power series can be truncated to the quadratic term. Letting

$$X_{rs} \triangleq \int_0^{T_0} \text{Re}[r(t)\tilde{s}^*(t)]dt \tag{3.5.6}$$

$$X_{ss} \triangleq \int_0^{T_0} |\tilde{s}(t)|^2 dt \tag{3.5.7}$$

this amounts to writing (3.5.4) as

$$\Lambda(r|\tilde{v},\tilde{\theta},\tilde{\tau},\tilde{c}) \approx 1 + \frac{1}{2N_0}(2X_{rs} - X_{ss}) + \frac{1}{8N_0^2}(2X_{rs} - X_{ss})^2 \tag{3.5.8}$$

It is worth noting that this assumption is not valid in many practical cases and, in fact, its true *raison d'être* is that it leads to a mathematically convenient formula involving only quadratic nonlinearities. As we shall see repeatedly in this book, mathematics with quadratic nonlinearities can often be carried through to closed-form solutions with reasonable efforts. Higher nonlinearities would generally prevent any useful conclusion.

The right-hand side of (3.5.8) contains several terms that must be averaged with respect to the unwanted parameters $\tilde{u} \triangleq \{\tilde{\theta}, \tilde{\tau}, \tilde{c}\}$. This operation is straightforward but tedious and is skipped. Nevertheless, it turns out that $X_{ss}$, $X_{rs}$, $X_{ss}^2$ and the product $X_{ss}X_{rs}$ all have expectations independent of $\tilde{v}$. Thus, they can be ignored as they do not affect the maximization. In conclusion, letting $\Lambda(r|\tilde{v})$ be the expectation of $\Lambda(r|\tilde{v}, \tilde{\theta}, \tilde{\tau}, \tilde{c})$ with respect to $\tilde{u}$, we have

$$\Lambda(r|\tilde{v}) = A_1 \, \mathrm{E}_{\tilde{u}}\left\{X_{rs}^2\right\} + A_2 \tag{3.5.9}$$

where $A_1$ and $A_2$ are constants independent of $\tilde{v}$ (in particular, $A_1$ is positive). Furthermore, as we are not interested in the actual value of the maximum of $\Lambda(r|\tilde{v})$ but in its location, the values of $A_1$ and $A_2$ are immaterial and we may concentrate on maximizing

$$\Lambda'(r|\tilde{v}) \triangleq \mathrm{E}_{\tilde{u}}\left\{X_{rs}^2\right\} \tag{3.5.10}$$

As a first step in this direction consider the integral in (3.5.6). Using (3.5.5) and the same arguments leading to (3.2.5) it is found that

$$\int_0^{T_0} r(t)\tilde{s}^*(t)dt \approx e^{-j\tilde{\theta}} \sum_{i=0}^{L_0-1} \tilde{c}_i^* x(iT + \tilde{\tau}) \tag{3.5.11}$$

where $L_0 \triangleq T_0/T$ is the length of the observation interval in symbol periods and $x(iT + \tilde{\tau})$ is the sample of

$$x(t) \triangleq \int_{-\infty}^{\infty} r(\xi) e^{-j2\pi\tilde{v}\xi} g(\xi - t) d\xi \tag{3.5.12}$$

at $t = iT + \tilde{\tau}$. As illustrated in Figure 3.21, $x(t)$ is the response of the matched filter to $r'(t) \triangleq r(t)e^{-j2\pi\tilde{v}t}$.

Next, collecting (3.5.6)-(3.5.11) produces

$$X_{rs} = \frac{1}{2}e^{-j\tilde{\theta}}\sum_{i=0}^{L_0-1}\tilde{c}_i^* x(iT+\tilde{\tau}) + \frac{1}{2}e^{j\tilde{\theta}}\sum_{i=0}^{L_0-1}\tilde{c}_i x^*(iT+\tilde{\tau}) \tag{3.5.13}$$

# Carrier Frequency Recovery with Linear Modulations

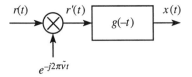

Figure 3.21. Physical interpretation of $x(t)$.

from which we get

$$X_{rs}^2 = \frac{1}{2} \sum_{i=0}^{L_0-1} \sum_{k=0}^{L_0-1} \tilde{c}_i^* \tilde{c}_k x(iT+\tilde{\tau}) x^*(kT+\tilde{\tau})$$

$$+ \frac{1}{4} e^{-j2\tilde{\theta}} \sum_{i=0}^{L_0-1} \sum_{k=0}^{L_0-1} \tilde{c}_i^* \tilde{c}_k^* x(iT+\tilde{\tau}) x(kT+\tilde{\tau})$$

$$+ \frac{1}{4} e^{j2\tilde{\theta}} \sum_{i=0}^{L_0-1} \sum_{k=0}^{L_0-1} \tilde{c}_i \tilde{c}_k x^*(iT+\tilde{\tau}) x^*(kT+\tilde{\tau}) \quad (3.5.14)$$

We now perform the expectation of $X_{rs}^2$ indicated in (3.5.10). Clearly, the last two terms in (3.5.14) give a zero contribution as the exponential $e^{\pm j2\tilde{\theta}}$ has zero mean with respect to $\tilde{\theta}$. Thus, bearing in mind (3.5.3) we obtain

$$E_{\tilde{\theta},\tilde{c}}\{X_{rs}^2\} = \frac{C_2}{2} \sum_{i=0}^{L_0-1} |x(iT+\tilde{\tau})|^2 \quad (3.5.15)$$

Also, averaging with respect to $\tilde{\tau}$ (which is uniformly distributed over $[0,T)$) yields

$$\Lambda'(r|\tilde{v}) = \frac{C_2}{2T} \sum_{i=0}^{L_0-1} \int_0^T |x(iT+\tilde{\tau})|^2 d\tilde{\tau} \quad (3.5.16)$$

As a final step we observe that the sum in (3.5.16) equals the integral of $|x(t)|^2$ over the observation interval. Thus, discarding the immaterial factor $C_2/2T$, we are led to maximizing

$$\Lambda''(r|\tilde{v}) \triangleq \int_0^{T_0} |x(t)|^2 dt \quad (3.5.17)$$

The integral in (3.5.17) represents the energy of the matched-filter output. In the following some methods to find where the maximum energy occurs are considered. Before proceeding, however, we offer a physical interpretation of (3.5.17).

The ratio of the integral (3.5.17) to $T_0$ gives the average power of $x(t)$. When $T_0/T$ grows large, such a power tends to the *statistical* power of $x(t)$. This is the sum of two terms, $P_S$ and $P_N$, associated with signal and noise respectively. The noise power $P_N$ is contributed by the voltage $w'(t) \triangleq w(t)\exp\{-j2\pi\tilde{v}t\}$ at the matched-filter input (see Figure 3.21). As $w(t)$ is white, so is $w'(t)$, which implies that $P_N$ is independent of $\tilde{v}$. Thus, maximizing $P_S + P_N$ amounts to maximizing $P_S$.

The quantity $P_S$ can be computed from the power spectral density of the signal component in $r(t)$, which reads (see Appendix 2.A.2 to Chapter 2)

$$S(f) = \frac{C_2}{T}|G(f - \Delta v)|^2 \tag{3.5.18}$$

with $\Delta v \triangleq v - \tilde{v}$. Hence (see Figure 3.22)

$$P_s = \int_{-\infty}^{\infty} S(f)|G(f)|^2 df$$

$$= \frac{C_2}{T}\int_{-\infty}^{\infty} |G(f - \Delta v)|^2 |G(f)|^2 df \tag{3.5.19}$$

Application of the Schwartz inequality [11, p. 395] to (3.5.19) indicates that $P_S$ reaches a maximum for $\Delta v = 0$. It is concluded that the integral in (3.5.17) has a maximum for $\tilde{v} = v$.

The above argument suggests a potential drawback when operating with frequency selective channels. In fact, a condition for the maximum of (3.5.17) to occur at $\tilde{v} = v$ is that the channel has a flat frequency response. To see why, assume that the channel transfer function is $C(f)$. Then the spectrum of the signal component at the matched filter input becomes

$$S(f) = \frac{C_2}{T}|G(f - \Delta v)|^2 |C(f + \tilde{v})|^2 \tag{3.5.20}$$

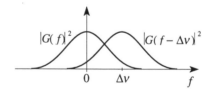

Figure 3.22. Functions $|G(f)|^2$ and $|G(f - \Delta v)|^2$.

## Carrier Frequency Recovery with Linear Modulations

and the signal power at the filter output results in

$$P_s = \frac{C_2}{T} \int_{-\infty}^{\infty} |G(f - \Delta v)|^2 |C(f + \tilde{v})|^2 |G(f)|^2 df \quad (3.5.21)$$

From this equation it is seen that $P_S$ need not be maximum at $\Delta v=0$ and, in consequence, maximizing $P_S$ may result in an estimation error. How large this error is depends both on the channel transfer function and the shape of the signal spectrum. In general, small errors are incurred with limited excess bandwidth factors. In fact, Exercise 3.5.1 shows that there would be no errors with an ideal rectangular spectrum.

So far we have considered non-offset PAM modulation. The case of offset modulation may be treated in a similar manner and leads again to formula (3.5.17). Thus, even with offset modulation, an approximate ML estimate is obtained by feeding the matched filter with $r(t)\exp\{-j2\pi\tilde{v}t\}$ and adjusting the demodulating frequency $\tilde{v}$ so as to maximize the output power.

**Exercise 3.5.1.** Assuming a $G(f)$ with a rectangular shape

$$G(f) = \begin{cases} \sqrt{T} & \text{for } |f| \leq \frac{1}{2T} \\ 0 & \text{elsewhere} \end{cases} \quad (3.5.22)$$

show that the maximum of (3.5.21) occurs for $\tilde{v} = v$ for any $C(f)$.

*Solution.* For simplicity we solve the problem letting $v=0$ but the discussion is readily extended to non-zero offsets. For $v=0$ equation (3.5.21) becomes

$$P_s = \frac{C_2}{T} \int_{-\infty}^{\infty} |G(f + \tilde{v})|^2 |C(f + \tilde{v})|^2 |G(f)|^2 df \quad (3.5.23)$$

or, making the change of variable $f_1 = f + \tilde{v}$,

$$P_s = \frac{C_2}{T} \int_{-\infty}^{\infty} |G(f_1)|^2 |C(f_1)|^2 |G(f_1 - \tilde{v})|^2 df_1 \quad (3.5.24)$$

Since $G(f)$ is rectangular in the range $\pm 1/(2T)$, the product $G(f_1)G(f_1 - \tilde{v})$ is zero if $|\tilde{v}|$ exceeds $1/T$. Hence, the right hand side of (3.5.24) is zero for $|\tilde{v}| > 1/T$. Vice versa, if $\tilde{v}$ is confined in the interval $\pm 1/T$, from (3.5.24) we obtain (see Figure 3.23)

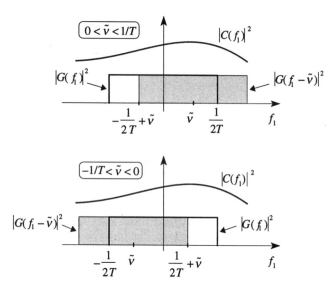

Figure 3.23. Illustrating the computation of $P_S$.

$$P_s = C_2 T \int_{-1/(2T)+\tilde{v}}^{1/(2T)} |C(f_1)|^2 df_1 \quad \text{for } 0 < \tilde{v} < 1/T \quad (3.5.25)$$

$$P_s = C_2 T \int_{-1/(2T)}^{1/(2T)+\tilde{v}} |C(f_1)|^2 df_1 \quad \text{for } -1/T < \tilde{v} < 0 \quad (3.5.26)$$

$$P_s = C_2 T \int_{-1/(2T)}^{1/(2T)} |C(f_1)|^2 df_1 \quad \text{for } \tilde{v} = 0 \quad (3.5.27)$$

To see where $P_S$ achieves its maximum, we compute the derivative of $P_S$ with respect to $\tilde{v}$. With simple manipulations it is found that

$$\frac{dP_s}{d\tilde{v}} = -C_2 T \left| C\left(-\frac{1}{2T} + \tilde{v}\right) \right|^2 \quad \text{for } 0 < \tilde{v} < \frac{1}{T} \quad (3.5.28)$$

$$\frac{dP_s}{d\tilde{v}} = C_2 T \left| C\left(\frac{1}{2T} + \tilde{v}\right) \right|^2 \quad \text{for } -\frac{1}{T} < \tilde{v} < 0 \quad (3.5.29)$$

which indicate that $dP_s/d\tilde{v}$ is positive for $\tilde{v} < 0$ and negative for $\tilde{v} > 0$. This means that $P_S$ decreases anyway as $\tilde{v}$ departs from zero and achieves its maximum just at $\tilde{v} = 0$. No estimation error is incurred, whatever the shape of the channel transfer function.

### 3.5.2. Open-Loop Search

Returning to equation (3.5.17), a first method to maximize the integral is to divide the range of $\tilde{v}$ into small intervals with midpoints $v_k = v_0 + k\Delta v$, $k=0, 1, 2,\ldots,N-1$, and proceed as follows:

(*i*) take a record of $r(t)$ over $(0,T_0)$;

(*ii*) for each $v_k$ compute $r(t)e^{-j2\pi v_k t}$ and the energy $E_k$ of $x(t)$, as indicated in Figure 3.24;

(*iii*) take the greatest energy, $E^{(\max)}$, and approximate $v$ with the corresponding $v_k$.

The unavoidable quantization error involved in this procedure can be reduced by interpolating between the $E_k$ values closest to $E^{(\max)}$ and looking for the location where the interpolating curve is maximum. Clearly, the method is computationally simple but time consuming.

An alternative approach is to use a parallel structure, as indicated in Figure 3.25. Here, the estimation time is just $T_0$, which entails a reduction by a factor $N$ with respect to the serial processing. The system complexity is $N$ times larger, however.

### 3.5.3. Closed-Loop Estimator

A third approach described in [12]-[14] is to employ a closed loop structure wherein an error signal is generated (at multiples of the symbol period) which is proportional to the difference between $v$ and its current estimate $\hat{v}$. This signal is then used to improve the estimate in a recursive fashion. Again, the algorithm applies to both non-offset and offset PAM modulation.

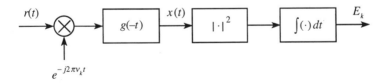

Figure 3.24. Arrangement to measure $E_k$.

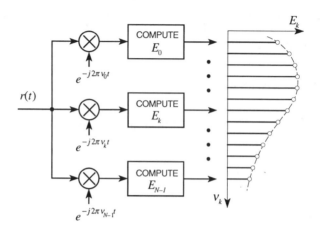

Figure 3.25. Parallel processing for $r(t)$.

To explain the procedure we start with the derivative of (3.5.17) with respect to $\tilde{v}$, say $d\Lambda''(r|\tilde{v})/d\tilde{v}$. Clearly, maximizing $\Lambda''(r|\tilde{v})$ amounts to solving the equation $d\Lambda''(r|\tilde{v})/d\tilde{v}=0$. As we shall see, this can be done in a manner that is suitable for digital implementation. From (3.5.17) we have

$$\frac{d\Lambda''(r|\tilde{v})}{d\tilde{v}} = 2\int_0^{T_0} \text{Re}\left\{x(t)\frac{\partial x^*(t)}{\partial \tilde{v}}\right\}dt \tag{3.5.30}$$

Also, from (3.5.12) we obtain after some manipulations

$$\frac{\partial x(t)}{\partial \tilde{v}} = jy(t) - j2\pi t x(t) \tag{3.5.31}$$

where $y(t)$ is defined as

$$y(t) \triangleq \int_{-\infty}^{\infty} r(\xi)e^{-j2\pi\tilde{v}\xi}2\pi(t-\xi)g(\xi-t)d\xi \tag{3.5.32}$$

Substituting into (3.5.30) results in

$$\frac{d\Lambda''(r|\tilde{v})}{d\tilde{v}} = 2\int_0^{T_0} \text{Im}\left\{x(t)y^*(t)\right\}dt \tag{3.5.33}$$

where $\text{Im}\{z\}$ means "imaginary part of $z$."

# Carrier Frequency Recovery with Linear Modulations

A physical interpretation of $y(t)$ is illustrated in Figure 3.26. It is seen that $y(t)$ is the response of the filter $h(t) \triangleq 2\pi t g(-t)$ to $r(t)e^{-j2\pi \tilde{v}t}$. Denoting by $G(f)$ the Fourier transform of $g(t)$, it is easily shown that the Fourier transform of $h(t)$ is proportional to the conjugate derivative of $G(f)$:

$$H(f) = j\frac{dG^*(f)}{df} \qquad (3.5.34)$$

For this reason $2\pi t g(-t)$ is often referred to as the *derivative matched filter*.

Next we turn our attention to (3.5.33). Voltages $x(t)$ and $y(t)$ are outputs from low-pass filters (matched filter and derivative matched filter). As they have no components beyond $|f| > 1/T$ (we assume $G(f)$ to be bandlimited to $\pm 1/T$), the integrand $\text{Im}\{x(t)y^*(t)\}$ is bandlimited to $|f| \leq 2/T$ and the integral in (3.5.33) can be computed through the samples of the integrand taken at twice the symbol rate. Formally,

$$\frac{d\Lambda''(r|\tilde{v})}{d\tilde{v}} \approx 2T_s \sum_{k=0}^{2L_0-1} \text{Im}\{x(kT_s)y^*(kT_s)\} \qquad (3.5.35)$$

where $L_0 = T_0/T$ is the observation length in symbol periods and $T_s \triangleq T/2$.

To solve the equation

$$\frac{d\Lambda''(r|\tilde{v})}{d\tilde{v}} = 0 \qquad (3.5.36)$$

we resort to a recursive procedure in which the parameter $\tilde{v}$ is replaced by a time-varying function $\hat{v}(kT_s)$ and the summation (3.5.35) is used as an error signal to steer $\hat{v}(kT_s)$ toward $v$. In practice the idea is implemented as follows.

Assume that a mechanism to produce $\hat{v}(kT_s)$ has already been devised and take

$$u(kT_s) = \sum_{i=k-N+1}^{k} \text{Im}\{x(iT_s)y^*(iT_s)\} \qquad (3.5.37)$$

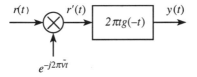

Figure 3.26. Physical interpretation for $y(t)$.

as an error signal, where $N$ is some integer parameter. In this equation $x(iT_s)$ and $y(iT_s)$ are computed letting $\tilde{v} = \hat{v}(iT_s)$ in (3.5.12) and (3.5.32). The problem is to choose the sequence $\hat{v}(iT_s)$, $i=1, 2,...$, so as to force $u(kT_s)$ to zero. Intuitively, $u(kT_s)$ will tend to vanish when $\hat{v}(kT_s)$ (and its $N$ prior values) approaches $v$. Vice versa, a non-zero $u(kT_s)$ will indicate that $\hat{v}(kT_s)$ is still far from the desired value and a better estimate must be sought. In this case a reasonable move is to change $\hat{v}(kT_s)$ in proportion to $u(kT_s)$, i.e.,

$$\hat{v}[(k+1)T_s] = \hat{v}(kT_s) + \gamma u(kT_s) \tag{3.5.38}$$

where $\gamma$ is a suitable constant (*step-size*).

It should be noted that (3.5.38) represents a digital integrator and, as such, has a low-pass action on $u(kT_s)$. On the other hand, $u(kT_s)$ is a smoothed version of $\text{Im}\{x(kT_s)y^*(kT_s)\}$. Thus, there are two low-pass transformations in cascade, $\text{Im}\{x(kT_s)y^*(kT_s)\} \Rightarrow u(kT_s)$ and $u(kT_s) \Rightarrow \hat{v}(kT_s)$, and the former can be suppressed for the sake of simplicity. This is accomplished by replacing $u(kT_s)$ by $\text{Im}\{x(kT_s)y^*(kT_s)\}$ in (3.5.38), which results in

$$\hat{v}[(k+1)T_s] = \hat{v}(kT_s) + \gamma \text{Im}\{x(kT_s)y^*(kT_s)\} \tag{3.5.39}$$

Frequency estimates in (3.5.39) are updated at twice the symbol rate ($1/T_s=2/T$). A symbol-rate updating is preferable, however, to ease the computing load. To do so we replace $\text{Im}\{x(kT_s)y^*(kT_s)\}$ by its average over one symbol interval. In other words, defining

$$e(kT) \triangleq \frac{1}{2}\text{Im}\{x(2kT_s)y^*(2kT_s)\} + \frac{1}{2}\text{Im}\{x[(2k+1)T_s]y^*[(2k+1)T_s]\} \tag{3.5.40}$$

we update $\hat{v}(kT)$ according to

$$\hat{v}[(k+1)T] = \hat{v}(kT) + \gamma e(kT) \tag{3.5.41}$$

Figure 3.27 illustrates the block diagram of a frequency recovery circuit based on equations (3.5.40)-(3.5.41). Here, the blocks MF and DMF represent the matched filter and the derivative matched filter. The digital integration (3.5.41) is performed by the loop filter whereas the VCO generates an exponential $e^{-j\phi(t)}$, with $\phi(t)$ given by

$$\frac{d\phi(t)}{dt} = 2\pi\hat{v}(kT) \qquad \text{for } kT \le t < (k+1)T \tag{3.5.42}$$

The above equations establish an algorithm for solving (3.5.36). Other methods exist that achieve similar results. An interesting option is to keep only

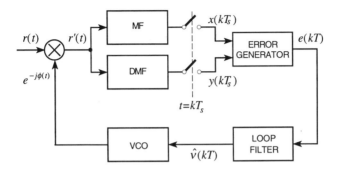

Figure 3.27. Block diagram for the frequency estimator.

one term in the right-hand side of (3.5.40). This turns out to be useful for further savings in computing load. The interested reader is referred to [13]-[14] for an in-depth discussion on the pros and cons of such an approximation.

So far phase rotations and filtering have been described as continuous-time (analog) operations. In a practical implementation $r(t)$ is first sampled and all further processing is done digitally. If the sampling rate is sufficiently high, analog and digital models are equivalent.

The digital counterpart of the scheme in Figure 3.27 is illustrated in Figure 3.28. The received waveform is first fed to an anti-aliasing filter (not shown in the figure) and then is sampled at some rate $1/T_N \triangleq N/T$. The filter bandwidth $B$ must be large enough to pass the signal components undistorted and the oversampling factor $N$ must be greater than $2BT$ to avoid aliasing. In these conditions no loss of information is incurred with sampling. The samples $r(nT_N)$ are counter-rotated by $\phi(nT_N)$ (as is done with $r(t)$ on a continuous-time basis) and are fed to the MF and DMF. Filter outputs are decimated to $1/T_s$ before entering the error generator.

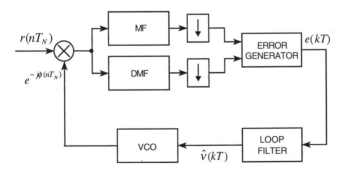

Figure 3.28. Digital implementation of the algorithm.

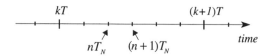

Figure 3.29. Partition of the $k$-th symbol interval.

The sequence $\phi(nT_N)$ is derived as follows. Dividing the interval $kT \leq t < (k+1)T$ into $N$ sub-intervals of length $T_N = T/N$ (see Figure 3.29) and integrating (3.5.42) over the generic sub-interval, say $nT_N \leq t < (n+1)T_N$, yields

$$\phi[(n+1)T_N] = \phi(nT_N) + 2\pi \hat{v}(kT)T/N \qquad (3.5.43)$$

This equation involves two indexes: a sample index $n$ and a symbol index $k$. From Figure 3.29, it appears they are related by

$$k = \text{int}\left(\frac{n}{N}\right) \qquad (3.5.44)$$

where int($z$) means "the largest integer not exceeding $z$." In practice, computing $\phi(nT_N)$ through (3.5.43) may not be easy as the phase may grow large, causing overflows in the computing unit. Overflows are avoided by taking $\phi$ modulo $2\pi$, i.e.,

$$\phi[(n+1)T_N] = \phi(nT_N) + 2\pi \hat{v}(kT)T/N \qquad \text{mod } 2\pi \qquad (3.5.45)$$

The performance of the above algorithms is qualitatively discussed in the next subsection. No quantitative details are provided as they involve lengthy calculations. The interested reader is referred to [1] and [12] for an in-depth discussion on their acquisition characteristics and to [13]-[14] for performance assessments. Some simulation results are shown later.

### 3.5.4. Frequency Acquisition

To understand the operation of the loop in Figure 3.27 it is expedient to disconnect the VCO from the mixer and drive the latter at a fixed frequency $\hat{v}$. Under these conditions the resulting error signal has an average $\text{E}\{e(kT)|\hat{v}\}$ that depends on the frequency difference $f_d \triangleq v - \hat{v}$, i.e.,

$$S(f_d) \triangleq \text{E}\{e(kT)|\hat{v}\} \qquad (3.5.46)$$

In practice, function $S(f_d)$ looks like an "S" (rotated by 90°) and is usually dubbed the "S-curve." Figure 3.30 shows $S(f_d)$ for QPSK modulation, as

# Carrier Frequency Recovery with Linear Modulations

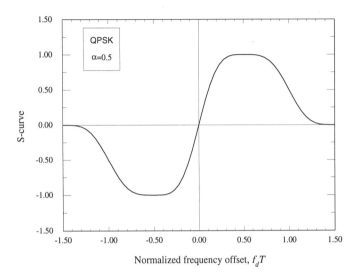

Figure 3.30. S-curve for QPSK modulation.

obtained taking a root-raised-cosine-rolloff channel with $\alpha=0.5$ and random data (significant deviations from the "regular" shape illustrated in the figure are possible for particular data patterns [1], [12]). It appears that $S(f_d)$ is zero at the origin and extends over the range $f_d = \pm 1.5/T$. As is now explained, the loop will eventually lock on the incoming carrier frequency provided that $f_d$ is within this range.

To see how this comes about let us return to Figure 3.27. The VCO instantaneous frequency equals $\hat{v}(kT)$ and the DC component in the signal error is $S[v - \hat{v}(kT)]$. Thus, $e(kT)$ is the sum of $S[v - \hat{v}(kT)]$ plus some zero-mean disturbance $n(kT)$, which accounts for the thermal noise and data pattern

$$e(kT) = S[v - \hat{v}(kT)] + n(kT) \qquad (3.5.47)$$

Collecting (3.5.41) and (3.5.47) yields

$$\hat{v}[(k+1)T] = \hat{v}(kT) + \gamma S[v - \hat{v}(kT)] + \gamma n(kT) \qquad (3.5.48)$$

which suggests the loop-equivalent model in Figure 3.31.

A quantitative analysis of this circuit is difficult since $n(kT)$ depends in a complex way on the data and thermal noise. Some insight into the loop behavior may be gathered by ignoring $n(kT)$. Under these conditions, (3.5.48) becomes an autonomous equation and its solution is found with methods that

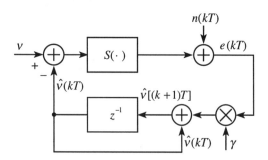

Figure 3.31. Equivalent circuit for the frequency loop.

are now summarized (see [15] for further details).

Briefly, let $f_d(kT) \triangleq v - \hat{v}(kT)$ be the frequency error at the $k$-th step and rewrite (3.5.48) in the form

$$f_d[(k+1)T] = f_d(kT) - \gamma S[f_d(kT)] \qquad (3.5.49)$$

Figure 3.32(a) illustrates the shape of $S(f_d)$. Next, draw the curves $y_1(f_d) \triangleq f_d$ and $y_2(f_d) \triangleq f_d - \gamma S(f_d)$ on the same reference system as indicated in Figure 3.32(b). Start from $P_0$ and move vertically to meet $y_2(f_d)$ at $P_1$. It is clear from (3.5.49) that the ordinate of $P_1$ is $f_d(0) - \gamma S[f_d(0)] = f_d(T)$. Next, move horizontally to meet the line $y_1(f_d)$. As $y_1(f_d)$ is the bisector of the coordinate axes, the abscissa of $P_2$ equals the ordinate of $P_1$, $f_d(T)$. At this stage we look for $f_d(2T)$. To this end, move vertically from $P_2$ to meet $y_2(f_d)$ at $P_3$. From (3.5.49) it is seen that the ordinate of $P_3$ is $f_d(2T)$. Next, move horizontally up to $P_4$ on the straight line $y_1(f_d)$ ... and so on.

It is clear from the figure that the trajectory $P_0$, $P_1$, $P_2$ ..., etc. converges to the origin provided that the initial frequency error $f_d(0)$ is within the range where $S(f_d)$ is nonzero. This range is referred to as the loop *acquisition range*. In [12] it is found that this range equals $\pm 2B_s$ ($B_s$ is the signal bandwidth) and is independent of the modulation format (either offset or non-offset).

Examination of Figure 3.32(b) reveals that the number of iterations required to achieve the origin depends on the vertical distance between $y_1(f_d)$ and $y_2(f_d)$. The larger the distance, the quicker the convergence. As the difference $y_1(f_d) - y_2(f_d)$ equals $\gamma S(f_d)$, it follows that the acquisition process grows faster as $\gamma$ increases. As we shall see, however, increasing $\gamma$ deteriorates the loop tracking performance. Thus, acquisition rapidity and tracking accuracy are contrasting goals and some trade-off is needed to meet a satisfactory balance between them. We shall return to this subject later, after discussing loop tracking performance.

# Carrier Frequency Recovery with Linear Modulations

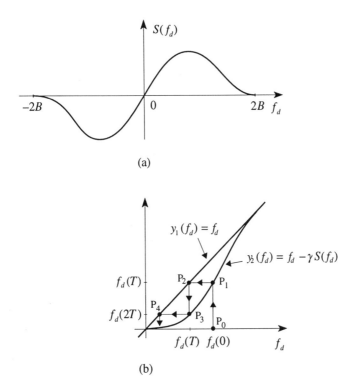

Figure 3.32. Graphical solution for (3.5.49).

**Exercise 3.5.2.** Solve (3.5.49) under the assumption that frequency errors are so small that $S(f_d)$ can be approximated by $Af_d$, where $A$ is the slope of the $S$-curve at the origin.

*Solution.* From (3.5.49) we get in succession

$$f_d(T) = f_d(0)(1-\gamma A)$$
$$f_d(2T) = f_d(T)(1-\gamma A) = f_d(0)(1-\gamma A)^2$$
$$f_d(3T) = f_d(2T)(1-\gamma A) = f_d(0)(1-\gamma A)^3$$
$$\vdots$$
$$f_d(kT) = f_d(0)(1-\gamma A)^k \qquad (3.5.50)$$

which is the solution sought.

Letting

$$\xi \triangleq \ln\left(\frac{1}{1-\gamma A}\right) \quad (3.5.51)$$

equation (3.5.50) may also be written as

$$f_d(kT) = f_d(0)e^{-\xi k} \quad (3.5.52)$$

### 3.5.5. Frequency Tracking

The tracking performance of the loop in Figure 3.27 is now assessed with the methods indicated in [13]. The major steps in the analysis may be summarized as follows. In the steady state the errors $f_d(kT) = v - \hat{v}(kT)$ are so small that the approximation $S(f_d) \approx Af_d$ can be made. In consequence (3.5.48) reduces to

$$f_d[(k+1)T] = (1-\gamma A)f_d(kT) - \gamma n(kT) \quad (3.5.53)$$

Note that, as $n(kT)$ is zero mean, so is $f_d(kT)$. Also, application of Z-transform methods shows that $f_d(kT)$ may be viewed as the response to $n(kT)$ of a digital filter with transfer function

$$\mathcal{H}(z) = -\frac{\gamma}{z-(1-\gamma A)} \quad (3.5.54)$$

or, which is the same, with impulse response

$$h(kT) = \begin{cases} -\gamma(1-\gamma A)^{k-1} & k \geq 1 \\ 0 & k < 1 \end{cases} \quad (3.5.55)$$

Accordingly, (3.5.53) becomes

$$f_d(kT) = \sum_i n(iT)h[(k-i)T] \quad (3.5.56)$$

From this equation the frequency error variance of $f_d(kT)$ is computed as

$$\sigma^2 = \sum_{m=-\infty}^{\infty} R_n(m)\eta(m) \quad (3.5.57)$$

where $R_n(m) \triangleq \mathrm{E}\{n[(k+m)T]n(kT)\}$ is the noise autocorrelation function and $\eta(mT)$ is the convolution of $h(k)$ with $h(-k)$

## Carrier Frequency Recovery with Linear Modulations

$$\eta(mT) = \sum_{i=-\infty}^{\infty} h(iT)h[(i-m)T] \qquad (3.5.58)$$

On the other hand, collecting (3.5.55) and (3.5.58), after some manipulations it is found that

$$\eta(mT) = \frac{\gamma}{A(2-\gamma A)}(1-\gamma A)^{|m|} \qquad (3.5.59)$$

Hence, substituting into (3.5.57) yields

$$\sigma^2 = \frac{\gamma}{A(2-\gamma A)} \sum_{m=-\infty}^{\infty} R_n(m)(1-\gamma A)^{|m|} \qquad (3.5.60)$$

In particular, with uncorrelated noise, (3.5.60) reduces to

$$\sigma^2 = \frac{\gamma}{A(2-\gamma A)} R_n(0) \qquad (3.5.61)$$

An alternative method to compute $\sigma^2$ involves spectral analysis techiques. Let $S_n(f)$ be the power spectral density of $n(kT)$

$$S_n(f) = T \sum_{m=-\infty}^{\infty} R_n(m)e^{-j2\pi mfT} \qquad (3.5.62)$$

and denote by $H(f)$ the right-hand side of (3.5.54) for $z = e^{j2\pi fT}$, i.e.,

$$H(f) \triangleq -\frac{\gamma}{e^{j2\pi fT} - (1-\gamma A)} \qquad (3.5.63)$$

Then, the error variance $\sigma^2$ is given by [11, p. 332]

$$\sigma^2 = \int_{-1/(2T)}^{1/(2T)} S_n(f)|H(f)|^2 df \qquad (3.5.64)$$

In some practical cases $S_n(f)$ is nearly flat over the interval $\pm B_L$ around the origin, where $H(f)$ takes significant values and (3.5.64) becomes

$$\sigma^2 = S_n(0) \int_{-1/(2T)}^{1/(2T)} |H(f)|^2 df \qquad (3.5.65)$$

The parameter $B_L$ is referred to as the noise equivalent bandwidth of the loop and is defined as

$$B_L \triangleq \frac{1}{2|H(0)|^2} \int_{-1/(2T)}^{1/(2T)} |H(f)|^2 df \qquad (3.5.66)$$

Using (3.5.63) it is found that

$$B_L T = \frac{\gamma A}{2(2-\gamma A)}$$

$$\approx \frac{\gamma A}{4} \qquad (3.5.67)$$

since $\gamma A$ is usually much less than unity.

In conclusion, collecting (3.5.65)-(3.5.67) yields

$$\sigma^2 = \frac{S_n(0)}{A^2} 2B_L \qquad (3.5.68)$$

which establishes a proportionality between error variance and loop bandwidth. This equation indicates that $\sigma^2$ can be made as small as desired by reducing $B_L$ or, which is the same, the step size $\gamma$. As we have pointed out earlier, however, decreasing $B_L$ may result in acquisitions that are too long. Thus, a trade-off is needed between acquisition length and tracking performance.

An interesting question is whether a relation can be established between acquisition time $T_{acq}$ and loop bandwidth $B_L$. The answer is not simple because $T_{acq}$ is not a fixed quantity that can be computed as a function of the loop parameters (as happens with $B_L$). Indeed, examination of the loop equivalent model in Figure 3.31 indicates that $T_{acq}$ is a random variable whose outcomes depend on the noise level and the initial error $f_d(0)$.

An approximate relationship between $T_{acq}$ and $B_L$ could be obtained if $f_d(0)$ were sufficiently small to allow a linear analysis and noise were negligible. For example, in Exercise 3.5.3 it is shown that the time needed for the frequency error to pass from $0.1/T$ to $0.001/T$ (the latter value being in the range of practical values when the loop is in steady-state conditions) is approximately

$$T_{acq} \approx \frac{1.15}{B_L} \qquad (3.5.69)$$

Unfortunately, the assumption of linear operations is normally not valid as initial frequency errors may be large. When this happens, the acquisition is longer than predicted by the linear analysis. This may be visualized by ignoring

# Carrier Frequency Recovery with Linear Modulations

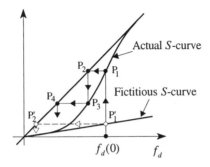

Figure 3.33. Actual versus fictitious acquisition.

noise and drawing trajectories like those in Figure 3.32(b) for two different S-curves: the actual $S(f_d)$ and a fictitious curve, $S_{fic}(f_d)$, which is just a straight line with the same slope as $S(f_d)$ at the origin, i.e.,

$$S_{fic}(f_d) = A f_d \qquad (3.5.70)$$

Intuitively, $S_{fic}(f_d)$ tells us how things would go under linear operations. From Figure 3.33 it appears that real acquisitions may be much longer than fictitious ones and, accordingly, $T_{acq}$ may be quite longer than (3.5.69). Considering these facts, the following rule-of-thumb formula is sometimes adopted as a rough estimation of $T_{acq}$:

$$T_{acq} \approx \frac{\eta}{B_L} \qquad (3.5.71)$$

with $\eta$ varying from 1.5 to 2.5.

**Exercise 3.5.3.** With reference to the equivalent model in Figure 3.31 assume $S(f_d) \approx A f_d$ and neglect $n(kT)$. Compute the time needed to pass from an initial error $f_d(0)$ to a fraction of this error, say $f_d(0)/N$.

*Solution.* Denote by $kT$ the time wherein $f_d(kT)$ attains the value $f_d(0)/N$. Application of (3.5.52) yields

$$k = \frac{1}{\xi} \ln N \qquad (3.5.72)$$

where $\xi$ is given in (3.5.51). For $\gamma A \ll 1$, equation (3.5.51) gives $\xi \approx \gamma A$ or, taking (3.5.67) into account,

$$\xi \approx 4 B_L T \qquad (3.5.73)$$

Substituting into (3.5.72) yields the "acquisition time"

$$T_{acq} \approx \frac{1}{4B_L} \ln N \qquad (3.5.74)$$

**Exercise 3.5.4.** Assume that a linear amplifier with gain $K>1$ is put in front of the frequency loop in Figure 3.27. How does this affect the frequency error variance?

*Solution.* Looking at Figure 3.27 it is clear that $x$ and $y$ are both multiplied by $K$ and, in consequence, $e(kT)$ is multiplied by $K^2$. It follows that:

(*i*) the slope of the $S$-curve increases by a factor $K^2$

(*ii*) the loop bandwidth increases by a factor $K^2$ (see (3.5.67))

(*iii*) the noise $n(kT)$ increases by a factor $K^2$

(*iv*) the power spectral density of $n(kT)$ increases by a factor $K^4$.

Putting all these facts together it is seen from (3.5.68) that the error variance becomes $K^2$ times larger. This emphasizes the need to keep the amplifier gain constant (by means of an automatic gain control) to prevent changes in the loop operating conditions.

### 3.5.6. Comparison with MCRB

The modified Cramer-Rao bound for carrier frequency estimation is given by

$$T^2 \times MCRB(v) = \frac{3}{2\pi^2 L_0^3} \frac{1}{E_s/N_0} \qquad (3.5.75)$$

In comparing this bound with (3.5.68) a difficulty arises in that $MCRB(v)$ has been derived for estimators operating over finite-length observations whereas the scheme in Figure 3.27 observes *all the past* up to the current time, as is readily recognized from the presence of an integrator in the loop. Even if the past is not uniformly weighted (its recent part counts more), it is not obvious how an infinitely long "weighted" observation compares with a time-limited "uniformly weighted" observation.

To address this problem we transform the original scheme (OS) in Figure 3.27 into an equivalent scheme (ES) with the same estimation errors (in mean square sense) but with a finite observation length. This length will be taken as the *equivalent observation length* for the OS.

# Carrier Frequency Recovery with Linear Modulations

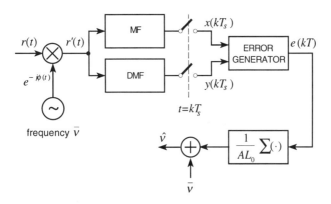

Figure 3.34. Block diagram for the equivalent estimator.

A block diagram for the ES is shown in Figure 3.34. Here, an oscillator operating at a fixed frequency $\bar{v}$ is used in place of the VCO. The oscillator output is given by $e^{-j\phi(t)}$, where

$$\frac{d\phi(t)}{dt} = 2\pi\bar{v} \quad \text{for } 0 \leq t \leq L_0 T \tag{3.5.76}$$

and an estimate of $v$ is provided according to the rule

$$\hat{v} = \bar{v} + \frac{1}{AL_0}\sum_{k=1}^{L_0} e(kT) \tag{3.5.77}$$

where $e(kT)$ is the error generator output and $A$ is the slope of the S-curve at the origin. As indicated in (3.5.47), $e(kT)$ is the sum of a DC component, $S(v-\bar{v})$, plus some zero-mean noise $n(kT)$:

$$e(kT) = S(v-\bar{v}) + n(kT) \tag{3.5.78}$$

To understand the ES operation, assume that $\bar{v}$ is close to $v$ so that

$$S(v-\bar{v}) \approx A(v-\bar{v}) \tag{3.5.79}$$

Then, substituting (3.5.78)-(3.5.79) into (3.5.77) results in

$$\hat{v} = v + \frac{1}{AL_0}\sum_{k=1}^{L_0} n(kT) \tag{3.5.80}$$

which says that, on average, the estimator gives the correct $v$.

The variance of the estimation error $f_d \triangleq v - \hat{v}$ is computed assuming the noise samples to be uncorrelated, i.e.,

$$E\{n[(k+m)T]n(kT)\} = \begin{cases} S_n(0)/T & \text{for } m = 0 \\ 0 & \text{elsewhere} \end{cases} \quad (3.5.81)$$

Under these conditions from (3.5.80) it is easily found that

$$\sigma^2 = \frac{S_n(0)}{A^2 L_0 T} \quad (3.5.82)$$

Comparing (3.5.82) with (3.5.68) it is concluded that OS and ES have the same tracking errors provided that $L_0$ is related to the OS noise equivalent bandwidth by

$$L_0 = \frac{1}{2B_L T} \quad (3.5.83)$$

Having established an equivalent length for the OS observations, it is interesting to compare the OS tracking performance with $MCRB(v)$. Figure 3.35 shows the simulated error variance for the ML-based tracking loop discussed in Section 3.5.3 (see Figure 3.28). The modulation format is QPSK and the overall channel response is Nyquist with an excess bandwidth factor $\alpha = 0.5$. Also, the anti-aliasing filter is an 8-th order Butterworth type with a $-3$ dB bandwidth of $1/T$ and the oversampling factor is $N = 2$. A loop bandwidth of $B_L T = 5 \cdot 10^{-3}$ is used, which corresponds to an observation of $L_0 = 100$ symbols. We see that the estimator variance falls short of the MCRB by orders of magnitude and exhibits a floor as the SNR increases. This is a manifestation of the so-called *self noise*, which means that the signal error has a considerable thermal-noise-independent component contributed by Signal×Signal interactions.

### 3.5.7. Other Frequency Error Detectors

Error generators of the type in (3.5.40) are referred to as frequency error detectors (FEDs) as they measure frequency offsets from a locally generated reference. In Section 3.5.3 an FED has been derived from maximum likelihood methods. Other types proposed in the literature have been discovered by *ad hoc* reasoning. Interestingly enough, they are close or equivalent to the ML-based FED. In the following we briefly report on two such FEDs: the *quadricorrelator* [1], [16]-[20] and the *dual filter detector* [1], [21].

# Carrier Frequency Recovery with Linear Modulations

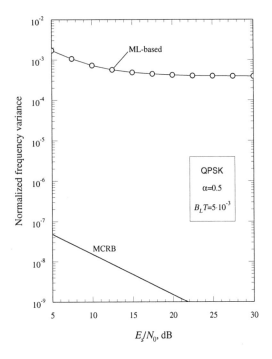

Figure 3.35. Normalized error variance for the ML-based estimator.

Figure 3.36 illustrates the block diagram of a quadricorrelator. The error signal $e(t)$ has the form

$$e(t) = \text{Im}\{x(t)y^*(t)\} \qquad (3.5.84)$$

where $x(t)$ and $y(t)$ are obtained by low-pass filtering $r'(t)$ in $h_1(t)$ and $h_2(t)$. In particular, taking $h_1(t)=g(-t)$ and $h_2(t)=2\pi t g(-t)$ makes (3.5.84) the continuous-time version of the FED in (3.5.39).

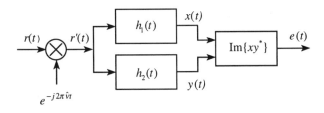

Figure 3.36. Block diagram of a quadricorrelator.

An interesting question is how $h_1(t)$ and $h_2(t)$ should be chosen for optimum performance and how such a performance compares with that of an ML-based FED. Notice that the ML-based FED is not necessarily optimum since a number of approximations have been made in its derivation. The first issue is addressed in [19]-[20], where it is shown that self noise can be entirely eliminated by suitably designing $h_1(t)$ and $h_2(t)$. Without entering into details, Figure 3.37 shows simulations for the estimation error variance with an optimized quadricorrelator. The operating conditions are the same as in Figure 3.35 and the filters $h_1(t)$ and $h_2(t)$ are designed according to [20]. We see that the quadricorrelator error variance has no floor, which means that self noise has been deleted. The quadricorrelator performance is much better than the ML-based detector's, even though its distance from the MCRB is still huge. The explanation is that it uses very limited information about the signal. This is in contrast with the algorithms in Section 3.2, which are data-aided and clock aided and, in fact, come close to or even attain the MCRB.

The block diagram of a dual filter detector (DFD) is depicted in Figure 3.38. It is formed by two parallel branches, each comprising a band-pass filter

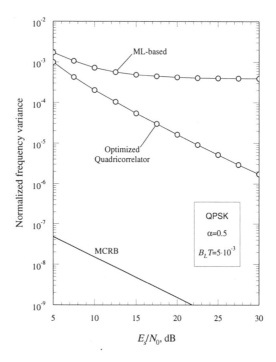

Figure 3.37. Normalized error variance for the optimized quadricorrelator.

# Carrier Frequency Recovery with Linear Modulations

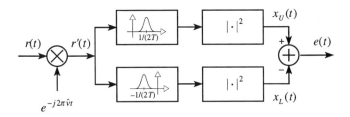

Figure 3.38. Block diagram of a dual filter detector.

in cascade with a square-law nonlinearity. Filters are centered at $\pm 1/(2T)$. The DFD operation is easily understood from Figure 3.39, which shows the spectrum of $r'(t)$, $S(f)$, along with the frequency responses of the filters. The DC components in $x_U(t)$ and $x_L(t)$ represent the signal powers from the upper and lower filter, respectively. When the difference $f_d \triangleq v - \hat{v}$ is zero, $S(f)$ is centered about the origin and the above powers are equal. Thus, the output $e(t)$ has no DC component. On the contrary, if $f_d$ is nonzero, a power unbalance arises that contributes to the DC part of $e(t)$ in proportion to $f_d$.

At first glance there seems to be no connection between DFDs and quadricorrelators. It can be shown, vice versa, that they can be designed so as to be equivalent [19]. This means that choosing between DFDs and quadricorrelators is only a question of practical implementation.

## 3.6. Open-Loop Recovery with No Timing Information

### 3.6.1. Delay-and-Multiply Method

The acquisition time of closed-loop schemes depends on the loop bandwidth in a way that is roughly expressed in (3.5.71). Bearing in mind (3.5.83),

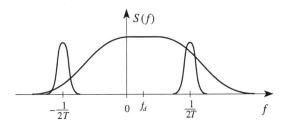

Figure 3.39. Explaining the DFD operation.

it follows that the acquisition time may be related to the "equivalent" observation length of the loop by

$$T_{acq} \approx 2\eta L_0 T \qquad (3.6.1)$$

As $\eta$ is in the range from 1.5 to 2.5, equation (3.6.1) says that a closed-loop estimator needs from three to five "equivalent" observation intervals to release its estimate. On the contrary, an open-loop estimator does it in just $L_0 T$ seconds. In the following we discuss an open-loop estimator having an acquisition range of about $\pm 1/T$ and mean square errors comparable with those of the closed-loop methods. Because of its shorter acquisition time this new scheme is more suited for burst-mode transmission.

The proposed method is based on the delay-and-multiply arrangement depicted in Figure 3.40. For simplicity we assume that the low-pass filter (LPF) has a rectangular characteristic. Also, its bandwidth $B_{LPF}$ is sufficiently large to pass the signal components undistorted. Although a rectangular characteristic is not physically realizable, the ensuing discussion can be readily adapted to practical situations. As is now explained, the estimation scheme is based on the statistics of the voltage

$$z(t) = x(t)x^*(t - \Delta T) \qquad (3.6.2)$$

where the value of delay $\Delta T$ is a design parameter.

Suppose the modulation is non-offset and write the incoming signal as

$$s(t) = e^{j(2\pi v t + \theta)} \sum_i c_i g(t - iT - \tau) \qquad (3.6.3)$$

Since the LPF does not distort $s(t)$, its output may be expressed as

$$x(t) = e^{j(2\pi v t + \theta)} \sum_i c_i g(t - iT - \tau) + n(t) \qquad (3.6.4)$$

where the noise $n(t)$ has the same power spectral density as $w(t)$ for $|f| \leq B_{LPF}$ and is zero elsewhere. Then, inserting into (3.6.2) results in

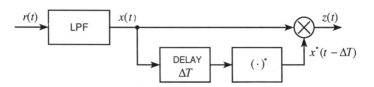

Figure 3.40. Delay-and-multiply scheme.

## Carrier Frequency Recovery with Linear Modulations

$$z(t) = e^{j2\pi v \Delta T} \sum_i \sum_k c_i c_k^* g(t - iT - \tau) g(t - kT - \tau - \Delta T)$$
$$+ s(t) n^*(t - \Delta T) + n(t) s^*(t - \Delta T) + n(t) n^*(t - \Delta T) \qquad (3.6.5)$$

With zero-mean uncorrelated symbols, the expected value of $z(t)$ over data and thermal noise is found to be

$$\mathrm{E}\{z(t)\} = C_2 e^{j2\pi v \Delta T} A(t - \tau) + R_n(\Delta T) \qquad (3.6.6)$$

where $C_2 \triangleq \mathrm{E}\{|c_i|^2\}$, the function $A(t)$ is defined as

$$A(t) \triangleq \sum_i g(t - iT) g(t - iT - \Delta T) \qquad (3.6.7)$$

and $R_n(\xi)$ is the autocorrelation of $n(t)$:

$$R_n(\xi) = 4 N_0 B_{LPF} \frac{\sin 2\pi B_{LPF} \xi}{2\pi B_{LPF} \xi} \qquad (3.6.8)$$

Clearly, $A(t)$ is a periodic function of period $T$. Thus the voltage $z(t)$ may be seen as the sum of a periodic component, $\mathrm{E}\{z(t)\}$, plus a zero-mean random process $N(t)$:

$$z(t) = \mathrm{E}\{z(t)\} + N(t) \qquad (3.6.9)$$

Integrating (3.6.9) and bearing in mind (3.6.6) yields

$$\frac{1}{T_0} \int_0^{T_0} z(t) dt = C_2 A_0 e^{j2\pi v \Delta T} + R_n(\Delta T) + X \qquad (3.6.10)$$

where $A_0$ is the DC component of $A(t)$ and $X$ is the time average of $N(t)$:

$$X \triangleq \frac{1}{T_0} \int_0^{T_0} N(t) dt \qquad (3.6.11)$$

At this point we note that: (*i*) $A_0$ is positive for moderate values of $\Delta T$ (see Exercise 3.6.1); (*ii*) $X$ is a zero-mean random variable; (*iii*) $R_n(\Delta T)$ vanishes for

$$\Delta T = \frac{k}{2 B_{LPF}}, \quad k = 1, 2, \ldots \qquad (3.6.12)$$

Thus, assuming that $\Delta T$ satisfies (3.6.12), equation (3.6.10) reduces to

$$\frac{1}{T_0}\int_0^{T_0} z(t)dt = C_2 A_0 e^{j2\pi\nu\Delta T} + X \tag{3.6.13}$$

so that, taking the arguments of both sides under the presumption of a small $X$ yields the frequency offset estimator

$$\hat{\nu} = \frac{1}{2\pi\Delta T}\arg\left\{\int_0^{T_0} z(t)dt\right\} \tag{3.6.14}$$

The performance analysis of this algorithm involves lengthy calculations and is not addressed here. The interested reader is referred to [22] for an in-depth discussion. The estimation range, instead, can be computed as follows.

Rewrite (3.6.13) in the form

$$\frac{1}{T_0}\int_0^{T_0} z(t)dt = C_2 A_0\left(1 + X_I + jX_Q\right)e^{j2\pi\nu\Delta T} \tag{3.6.15}$$

with

$$X_I + jX_Q \triangleq \frac{X}{C_2 A_0}e^{-j2\pi\nu\Delta T} \tag{3.6.16}$$

and note that, as $X$ is zero mean, so are $X_I$ and $X_Q$. Also, if $X_I$ and $X_Q$ are small compared with unity (as happens in all practical cases) and $A_0$ is positive, then substituting (3.6.15) into (3.6.14) yields

$$\hat{\nu} \approx \frac{1}{2\pi\Delta T}\arg\left\{(1 + jX_Q)e^{j2\pi\nu\Delta T}\right\} \tag{3.6.17}$$

It is easily seen that $\arg\{(1 + jX_Q)e^{j2\pi\nu\Delta T}\}$ is approximately equal to $2\pi\nu\Delta T + X_Q$, provided that $2\pi\nu\Delta T$ is not close to the extremes of the interval $\pm\pi$. Hence

$$\hat{\nu} \approx \nu + \frac{X_Q}{2\pi\Delta T} \tag{3.6.18}$$

which indicates that $\hat{\nu}$ is unbiased since $X_Q$ has zero mean.

It should be noted that this conclusion is no longer true if $2\pi\nu\Delta T$ is close to either $\pi$ or $-\pi$. To see this point, suppose that $2\pi\nu\Delta T$ is slightly less than $\pi$ or (which is the same) $\nu$ is slightly less than $1/(2\Delta T)$. Then, as illustrated in Figure 3.41, even a small $X_Q$ can make $\arg\{(1 + jX_Q)e^{j2\pi\nu\Delta T}\}$ overcome $\pi$ and reach $-\pi$. When this happens, (3.6.14) gives an estimate which is near $-1/(2\Delta T)$ rather

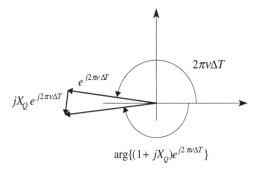

Figure 3.41. Determining the estimation range of the algorithm.

than $1/(2\Delta T)$. A similar shortcoming occurs when $v$ is near $-1/(2\Delta T)$ (with the estimator saying $\hat{v} \approx 1/(2\Delta T)$ instead of $\hat{v} \approx -1/(2\Delta T)$). In summary, consistent frequency estimates are obtained only if $v$ is well within $\pm 1/(2\Delta T)$.

**Exercise 3.6.1.** Compute the parameter $A_0$ in (3.6.10) as a function of the delay $\Delta T$ and $G(f)$, the Fourier transform of $g(t)$. Specify the result when $G(f)$ is a root-raised-cosine-rolloff function with rolloff $\alpha$.

*Solution.* As $A(t)$ is periodic of period $T$, its time average reads

$$A_0 = \frac{1}{T}\int_0^T A(t)dt \qquad (3.6.19)$$

or, using (3.6.7),

$$A_0 = \frac{1}{T}\sum_i \int_0^T g(t-iT)g(t-iT-\Delta T)dt \qquad (3.6.20)$$

Making the change of variable $t_1 = t - iT$ yields

$$A_0 = \frac{1}{T}\sum_i \int_{-iT}^{-iT+T} g(t_1)g(t_1 - \Delta T)dt_1 \qquad (3.6.21)$$

or

$$A_0 = \frac{1}{T}\int_{-\infty}^{\infty} g(t)g(t - \Delta T)dt \qquad (3.6.22)$$

since the sum of the integrals may be written as a single integral on the entire line. Finally, application of the Parseval formula produces the desired result

$$A_0 = \frac{1}{T}\int_{-\infty}^{\infty}|G(f)|^2 e^{j2\pi f \Delta T}df \qquad (3.6.23)$$

which shows that $A_0$ is the inverse Fourier transform of $|G(f)|^2/T$ as computed at $t = \Delta T$.

If $G(f)$ is a root-raised-cosine-rolloff function, $|G(f)|^2$ is Nyquist and its inverse Fourier transform is expressed by

$$g(t) = \frac{\sin(\pi t/T)}{\pi t/T}\frac{\cos(\alpha \pi t/T)}{1-4\alpha^2 t^2/T^2} \qquad (3.6.24)$$

Then, from (3.6.23) we get

$$A_0 = \frac{1}{T}\frac{\sin(\pi \Delta T/T)}{\pi \Delta T/T}\frac{\cos(\alpha\pi\Delta T/T)}{1-4\alpha^2(\Delta T/T)^2} \qquad (3.6.25)$$

It should be noted that $A_0$ vanishes when $\Delta T$ is a multiple of the symbol period. Thus, $\Delta T$ values close to multiples of $T$ should be avoided since the estimation accuracy deteriorates as $A_0$ becomes small.

**Exercise 3.6.2.** In Section 3.6.1 a non-offset PAM modulation has been assumed. Show that the estimator (3.6.14) can also be used with OQPSK signaling.

*Solution.* The OQPSK signal model is

$$s(t) = e^{j(2\pi vt+\theta)}\left\{\sum_i [a_i g(t-iT-\tau) + jb_i g(t-iT-T/2-\tau)]\right\} \qquad (3.6.26)$$

where $a_i$ and $b_i$ take independently the values $\pm 1$ with the same probability. Paralleling the passages leading to (3.6.5) yields

$$\begin{aligned}z(t) = &\; e^{j2\pi v\Delta T}\sum_i\sum_k a_i a_k g(t-iT-\tau)g(t-kT-\Delta T-\tau)\\
&+ e^{j2\pi v\Delta T}\sum_i\sum_k b_i b_k g(t-iT-T/2-\tau)g(t-kT-T/2-\Delta T-\tau)\\
&- je^{j2\pi v\Delta T}\sum_i\sum_k a_i b_k g(t-iT-\tau)g(t-kT-T/2-\Delta T-\tau)\\
&+ je^{j2\pi v\Delta T}\sum_i\sum_k b_i a_k g(t-iT-T/2-\tau)g(t-kT-\Delta T-\tau)\\
&+ s(t)n^*(t-\Delta T) + n(t)s^*(t-\Delta T) + n(t)n^*(t-\Delta T)\end{aligned} \qquad (3.6.27)$$

Let us compute the expectation of $z(t)$. By assumption

$$E\{a_i a_k\} = E\{b_i b_k\} = \begin{cases} 1 & \text{for } i = k \\ 0 & \text{otherwise} \end{cases} \quad (3.6.28)$$

$$E\{a_i b_k\} = 0 \quad \forall i, k \quad (3.6.29)$$

Hence, from (3.6.27) it is found that

$$E\{z(t)\} = e^{j2\pi v \Delta T} A(t - \tau) + R_n(\Delta T) \quad (3.6.30)$$

with

$$A(t) \triangleq \sum_i [g(t - iT)g(t - iT - \Delta T) \\ + g(t - iT - T/2)g(t - iT - T/2 - \Delta T)] \quad (3.6.31)$$

Clearly, the function $A(t)$ has the same features as in (3.6.7) (it is periodic of period $T$ and has nonzero DC). Then, following the lines of Section 3.6.1, it is concluded that the estimator (3.6.14) applies also to OQPSK signaling.

### 3.6.2. Digital Implementation

The digital implementation of the delay-and-multiply scheme in Figure 3.40 proceeds as follows. Start with the LPF bandwidth and observe that: (*i*) $B_{LPF}$ must be large enough to pass $s(t)$ undistorted even when the frequency offset is at its maximum, say $\pm v_{max}$; (*ii*) the signal bandwidth equals $(1+\alpha)/2T$, where $\alpha$ is the rolloff factor. Thus, for $v_{max}$ on the order of $1/T$ and $\alpha$ about 0.5, one needs an LPF bandwidth of approximately $2/T$. In the sequel we take $B_{LPF}=2/T$.

Next, let us concentrate on the integral in (3.6.14). As $x(t)$ is bandlimited within $\pm 2/T$, it follows that $z(t) = x(t)x^*(t - \Delta T)$ is bandlimited within $\pm 4/T$. Then, if $T_0$ is much larger than $T$, it can be shown (see Appendix 3.A) that the integral can be computed from the samples of $z(t)$ taken at a rate $R=4/T$. Formally

$$\int_0^{T_0} z(t) dt \approx \frac{T}{4} \sum_{k=0}^{4L_0 - 1} x(kT/4 + t_0) x^*(kT/4 + t_0 - \Delta T) \quad (3.6.32)$$

where $L_0 = T_0/T$ and $t_0$ is an arbitrary sampling phase. Actually, $R$ need not be exactly $4/T$. Small deviations from this value are equivalent to periodically

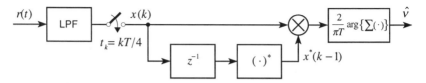

Figure 3.42. Block diagram for the digital estimator.

sweeping $t_0$ between 0 and $T/4$. If the sweeping process is so slow that $t_0$ keeps approximately constant over $T_0$ seconds, then (3.6.32) remains valid.

Setting $t_0=0$ for simplicity and substituting (3.6.32) into (3.6.14) yields the desired estimation algorithm

$$\hat{v} = \frac{1}{2\pi\Delta T}\arg\left\{\sum_{k=0}^{4L_0-1} x(kT/4)x^*(kT/4 - \Delta T)\right\} \quad (3.6.33)$$

In general, the summation in (3.6.33) involves $8L_0$ samples of $x(t)$. This number may be halved, however, if $\Delta T$ equals a multiple of $T/4$ (note that this choice is consistent with condition (3.6.12) because $B_{LPF}=2/T$). In particular, choosing $\Delta T=T/4$ results in

$$\hat{v} = \frac{2}{\pi T}\arg\left\{\sum_{k=0}^{4L_0-1} x(kT/4)x^*[(k-1)T/4]\right\} \quad (3.6.34)$$

Figure 3.42 shows a block diagram for the algorithm (3.6.34). Here, $x(k)$ stands for $x(kT/4)$ and $z^{-1}$ represents a $T/4$ delay. As mentioned earlier, performance analysis of this estimator is complex and is pursued in [22]. Figure 3.43 illustrates simulation results for the estimation variance with QPSK signaling. The pulse $g(t)$ corresponds to a root-raised-cosine-rolloff filter with 50% of excess bandwidth. The curve has been drawn for $v = 0$ but the same results are obtained for any $v$ in the range $\pm 1/T$. Comparing with Figure 3.35 it is seen that the accuracy of the two estimators is virtually identical.

### 3.6.3. Effects of Adjacent Channel Interference

A weakness of the delay-and-multiply estimator (and of all the non-data-aided (NDA) frequency estimators, in general) is the sensitivity to adjacent channel interference. To illustrate this point let us return to the analog model in Figure 3.40 and suppose that the received waveform is the sum of the desired signal at carrier frequency $f_0$ plus an interfering PAM signal at frequency $f_0+F$. As we did earlier, we assume that the desired signal is passed through the LPF

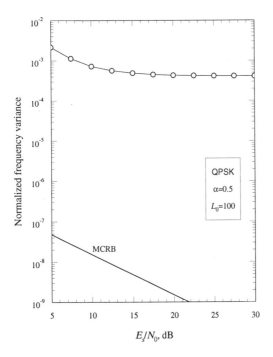

Figure 3.43. Error variance of the delay-and-multiply estimator.

undistorted. The interfering signal, instead, is distorted but not totally eliminated. Under these conditions the LPF output reads

$$x(t) = e^{j(2\pi\nu t+\theta)}\sum_i c_i g(t-iT-\tau)$$
$$+e^{j[2\pi(\nu+F)t+\theta']}\sum_i c'_i g'(t-iT-\tau')+n(t) \quad (3.6.35)$$

where $\{c'_i\}$ and $\{c_i\}$ are independent data sequences with the same statistics and $g'(t)$ is some complex-valued function, whose energy depends on the overlap of the interfering signal spectrum and the LPF response.

With the arguments of Section 3.6.1 it is seen that the expected value of the multiplier output is now

$$E\{z(t)\} = C_2 e^{j2\pi\nu\Delta T}A(t-\tau) + C_2 e^{j2\pi(\nu+F)\Delta T}A'(t-\tau') + R_n(\Delta T) \quad (3.6.36)$$

with

$$A'(t) \triangleq \sum_i g'(t-iT)g'^*(t-iT-\Delta T) \tag{3.6.37}$$

Correspondingly, the time average of $z(t)$ over the observation interval becomes

$$\frac{1}{T_0}\int_0^{T_0} z(t)dt \approx C_2 e^{j2\pi\nu\Delta T}\left(A_0 + A_0' e^{j2\pi F\Delta T}\right) + R_n(\Delta T) + X' \tag{3.6.38}$$

where $A_0'$ represents the DC component of $A'(t)$ and $X'$ is a zero-mean random variable. Note that $A_0 + A_0' e^{j2\pi F\Delta T}$ is a complex number, in general.

From this equation the effect of adjacent channel interference is apparent. Even ignoring $X'$ and assuming $R_n(\Delta T)=0$, the argument of the right-hand side is no longer proportional to $\nu$, as happens with (3.6.13), and the estimates (3.6.14) are biased.

An obvious question is whether a similar problem arises with the closed-loop scheme in Figure 3.27. Clearly, when the VCO frequency is near $\nu$, the low-pass action of the MF and DMF filters will tend to attenuate the interfering signal components and, in consequence, their effects on the loop operation. A total elimination is not possible, however, unless the interfering signal has no spectral overlap into the above filters.

**Exercise 3.6.3.** Suppose that the LPF in Figure 3.40 has a rectangular transfer function. Assuming that the desired and interfering signals are unmodulated carriers, $s(t) = X_M e^{j2\pi\nu t}$ and $s'(t) = \rho X_M e^{j2\pi(\nu+F)t}$ with $0 \le \rho < 1$, compute the estimation errors of (3.6.14) in the absence of thermal noise.

*Solution.* Suppose that the interfering signal passes through the LPF (otherwise there would be no interference). Then, the LPF output reads

$$x(t) = X_M e^{j2\pi\nu t} + \rho X_M e^{j2\pi(\nu+F)t} \tag{3.6.39}$$

and the voltage $z(t) \triangleq x(t)x^*(t-\Delta T)$ is easily found to be

$$z(t) = X_M^2 e^{j2\pi\nu\Delta T}\left(1 + \rho^2 e^{j2\pi F\Delta T} + \rho e^{j2\pi F\Delta T}e^{-j2\pi Ft} + \rho e^{j2\pi Ft}\right) \tag{3.6.40}$$

Substituting into (3.6.14) yields (with an observation time much longer than $1/F$)

$$\hat{\nu} \approx \frac{1}{2\pi\Delta T}\arg\left\{e^{j2\pi\nu\Delta T}\left(1 + \rho^2 e^{j2\pi F\Delta T}\right)\right\} \tag{3.6.41}$$

For $\rho^2 \ll 1$ and $\nu$ not too close to $\pm 1/(2\Delta T)$, from this equation it is found that

$$\hat{v} \approx v + \rho^2 \frac{\sin(2\pi F \Delta T)}{2\pi \Delta T} \qquad (3.6.42)$$

As expected, the estimation error decreases with $\rho$.

## 3.7. Key Points of the Chapter

- Data-aided ML frequency estimation is time-consuming and involves a two-step search routine. The first step requires calculating the maximum of $\Gamma(\tilde{v})$ over a set of $\tilde{v}$ values covering the range of interest. The second part locates the maximum of $\Gamma(\tilde{v})$ nearest to the point picked up in the first step.

- Because of its implementation complexity, true data-aided ML frequency estimation is not used, in general, and simpler methods are resorted to. In this context the algorithms by Kay [3], Fitz [4]-[5], Luise and Reggiannini (L&R) [6] and the approximate ML estimator in Section 3.2.6 are remarkable. The Kay algorithm has a rather high threshold, however, and is unsuitable with many coded modulations. The Fitz and L&R methods have comparable complexity and achieve the Cramer-Rao bound up to about zero dB. Their computational load may be reduced at the expense of some performance degradation. The approximate ML estimator is as efficient as the Fitz and L&R methods or better and, depending on the operating conditions, may be easier to implement.

- Decision-directed frequency estimation is simple to implement with differential PSK, as is illustrated in Section 3. The basic idea is that modulation can be removed from signal samples making use of differential decisions. Once this is done, frequency offset can be estimated by measuring phase rotations between consecutive samples.

- Modulation removal with M-ary PSK can also be obtained by raising the signal samples to the M-th power.

- This trick does not work with QAM modulation. More complex algorithms are needed to cope with constellations that do not exhibit the PSK rotational symmetry. One possibility is indicated in Section 3.4.

- Estimation of large frequency offsets can be accomplished with either closed-loop or open-loop (feedforward) schemes. A variety of frequency difference detectors exist for use in frequency loops. Delay-and-multiply methods appear simpler to implement and achieve comparable performance.

## Appendix 3.A

Let $x(t)$ be a realization of a random process with a power spectral density $S(f)$ restricted to $|f| < B$, i.e.,

$$S(f) = 0 \quad \text{for} \quad |f| \geq B \tag{3.A.1}$$

In this Appendix we argue that the integral of $x(t)$ over an interval much longer than $1/B$ can be expressed as a function of the samples of $x(t)$ taken at a sampling rate

$$\frac{1}{T_s} = B \tag{3.A.2}$$

Formally, we maintain that

$$\int_0^{T_0} x(t)dt \approx T_s \sum_{k=1}^{L_0} x(kT_s + t_0) \tag{3.A.3}$$

where $L_0 \triangleq T_0/T_s$ and the sampling phase $t_0$ is arbitrarily chosen within $0 \leq t_0 \leq T_s$.

The proof may be broken into three steps. First, define

$$x_{T_0}(t) \triangleq \begin{cases} x(t) & 0 \leq t \leq T_0 \\ 0 & \text{elsewhere} \end{cases} \tag{3.A.4}$$

and call $X_{T_0}(f)$ the Fourier transform of $x_{T_0}(t)$. As the integral of $x(t)$ over $(0, T_0)$ equals the integral of $x_{T_0}(t)$ over the entire line, and as the latter equals $X_{T_0}(0)$, we have

$$\int_0^{T_0} x(t)dt = X_{T_0}(0) \tag{3.A.5}$$

Second, define the *periodogram* of $x(t)$ over $0 \leq t \leq T_0$ as

$$S(f, T_0) \triangleq \frac{|X_{T_0}(f)|^2}{T_0} \tag{3.A.6}$$

It can be shown [11, Ch. 13] that, when $T_0$ is large compared with $1/B$, the expectation of $S(f, T_0)$ tends to the power spectral density of $x(t)$:

$$\lim_{T_0 \to \infty} E\{S(f,T_0)\} = S(f) \tag{3.A.7}$$

Thus, recalling (3.A.1)-(3.A.2), from (3.A.6)-(3.A.7) it is recognized that

$$X_{T_0}(f) \approx 0 \quad \text{for} \quad |f| \geq 1/T_s \tag{3.A.8}$$

provided that $T_0$ is sufficiently large.

Third, application of the Poisson sum formula [11, p. 395] yields

$$T_s \sum_{k=-\infty}^{\infty} x_{T_0}(kT_s + t_0) = \sum_{k=-\infty}^{\infty} X_{T_0}(k/T_s) e^{j2\pi k t_0/T_s} \tag{3.A.9}$$

from which, bearing in mind (3.A.8), we obtain

$$T_s \sum_{k=-\infty}^{\infty} x_{T_0}(kT_s + t_0) = X_{T_0}(0) \tag{3.A.10}$$

At this point (3.A.3) follows easily by combining (3.A.5) and (3.A.10) and taking the definition of $x_{T_0}(t)$ into account.

## References

[1] F.M.Gardner, *Demodulator Reference Recovery Techniques Suited for Digital Implementation*, European Space Agency, Final Report, ESTEC Contract No 6847/86/NL/DG, August, 1988.
[2] D.C.Rife and R.R.Boorstyn, Single-Tone Parameter Estimation from Discrete-Time Observations, *IEEE Trans. Inform. Theory*, **IT-20**, 591-598, Sept. 1974.
[3] S.Kay, A Fast and Accurate Single Frequency Estimator, *IEEE Trans. Acoust. Speech, Signal Processing*, **ASSP-37**, 1987-1990, Dec. 1989.
[4] M.P.Fitz, Planar Filtered Techniques for Burst Mode Carrier Synchronization, Conf. Rec. GLOBECOM'91, Phoenix, Arizona, Dec. 2-5, 1991, paper 12.1.
[5] M.P.Fitz, Further Results in the Fast Estimation of a Single Frequency, *IEEE Trans. Commun.*, **COM-42**, 862-864, March 1994.
[6] M.Luise and R.Reggiannini, Carrier Frequency Recovery in All-Digital Modems for Burst-Mode Transmissions, *IEEE Trans. Commun.*, **COM-43**, 1169-1178, Feb./March/April 1995.
[7] D.G.Messerschmitt, Frequency Detectors for PLL Acquisition in Timing and Carrier Recovery, *IEEE Trans. Commun.*, **COM-27**, 1288-1295, Sept. 1979.
[8] F.Classen, H.Meyr and P.Sehier, Maximum Likelihood Open Loop Carrier Synchronizer for Digital Radio, Conf. Rec. ICC'93, Geneva, May 23-26, 1993, pp. 493-497.
[9] H.Sari and S.Moridi, New Phase and Frequency Detectors for Carrier Recovery in PSK and QAM Systems, *IEEE Trans. Commun.*, **COM-36**, 1035-1043, Sept. 1988.
[10] J.C.–I. Chuang and N.R.Sollenberger, Burst Coherent Demodulation with Combined Symbol Timing, Frequency Offset Estimation, and Diversity Selection, *IEEE Trans. Commun.*, **COM–39**, 1157-1164, July 1991.

[11] A.Papoulis, *Probability, Random Variables, and Stochastic Processes*, New York: McGraw-Hill, 1991.
[12] F.M.Gardner, Frequency Detectors for Digital Demodulators Via Maximum Likelihood Derivation, ESA Final Report: Part II, ESTEC contract No 8022/88/NL/DG, March 1990.
[13] A.N.D'Andrea and U.Mengali, Noise Performance of Two Frequency-Error Detectors Derived from Maximum Likelihood Estimation Methods, *IEEE Trans. Commun.*, **COM-42**, 793-802, Feb./March/April 1994.
[14] G.Karam, F.Daffara and H.Sari, Simplified Versions of the Maximum-Likelihood Frequency Detector, Conf. Rec. GLOBECOM'92, Orlando, FL, Dec. 6-9, 1992, paper 11.02.
[15] A.N.D'Andrea and F.Russo, First-Order DPLL's: A Survey of a Peculiar Methodology and Some New Applications, *Alta Frequenza*, 495-506, Dec. 1983.
[16] C.R.Cahn, Improving Frequency Acquisition of a Costas Loop, *IEEE Trans. Commun.*, **COM-25**, 1453-1459, Dec. 1977.
[17] F.D.Natali, AFC Tracking Algorithms, *IEEE Trans. Commun.*, **COM-32**, 935-947, Aug. 1984.
[18] F.M.Gardner, Properties of Frequency Difference Detectors, *IEEE Trans. Commun.*, **COM-33**, 131-138, Feb. 1985.
[19] A.N.D'Andrea and U.Mengali, Performance of a Quadricorrelator Driven by Modulated Signals, *IEEE Trans. Commun.*, **COM-38**, 1952-1957, Nov. 1990.
[20] _____, Design of Quadricorrelators for Automatic Frequency Control Systems, *IEEE Trans. Commun.*, **COM-41**, 988-997, June 1993.
[21] T.Alberty and V.Hespelt, A New Pattern Jitter Free Frequency Error Detector, *IEEE Trans. Commun.*, **COM-37**, 159-163, Feb. 1989.
[22] F.Classen and H.Meyr, Two Frequency Estimation Schemes Operating Independently of Timing Information, Conf. Rec. GLOBECOM'93, Houston, TX, Nov. 29-Dec. 2, 1993, pp. 1996-2000.

# 4

# Carrier Frequency Recovery with CPM Modulations

## 4.1. Introduction

Continuous phase modulation (CPM) encompasses a class of signaling schemes that conserve and reduce signal energy and bandwidth at the same time. Furthermore, the signals in this class have a constant envelope and therefore are very attractive in radio channels employing low-cost non-linear power amplifiers. Notwithstanding these favorable features, current CPM applications are still limited to a few simple modulation schemes (basically, MSK and its generalizations) because of implementation complexity and synchronization problems [1]. Research efforts are under way and advances in these areas are expected in the near future.

We have seen in Chapter 3 that a number of algorithms are available to accomplish carrier frequency recovery with PAM modulation under diverse operating conditions. Unfortunately, the state of the art with CPM is less developed. Studies in this area are quite recent and results are still limited. Most of the material in this chapter is concerned with the estimation of "large" frequency offsets, on the order of the symbol rate. As pointed out in Chapter 3, this generally implies that frequency recovery must be performed without exploiting data and timing information. There are two exceptions, however, which are of interest in burst transmission applications. The first one is discussed in Section 4.3, where we address data-aided and clock-aided frequency recovery for MSK. This is an interesting issue not only because MSK is so popular but also in view of possible extensions of the same ideas to MSK-type modulation. The other is considered in Section 4.6 in the context of clock-aided (but non-data-aided) frequency estimation. The motivation for discussing clock-aided methods is that, in the presence of moderate frequency offsets, timing information can be gathered first (in a non-data-aided fashion) and then exploited for frequency recovery.

The chapter is organized as follows. The next section summarizes basic notations for MSK-type signals and gives an overview of the so-called *Laurent expansion*. As we shall see, this is a useful mathematical tool that provides good insight into the notion of MSK-type modulation and forms the basis for later discussions. Section 4.3 deals with data-aided and clock-aided frequency recovery with MSK-type signals. Non-data-aided and non-clock-aided methods are investigated in Section 4.4. The same approach, but with general multilevel CPM, is discussed in Section 4.5. Clock-aided (but not data-aided) recovery is treated in Section 4.6.

## 4.2. Laurent Expansion

In this section we concentrate on a subset of CPM formats denoted MSK-type modulations. Here, the information symbols are binary, the modulation index is $h = 1/2$, and the signal complex envelope has the form [1]-[2]

$$s_{CE}(t) = \sqrt{\frac{2E_s}{T}} e^{j\psi(t,\alpha)} \qquad (4.2.1)$$

with

$$\psi(t,\alpha) \triangleq \pi \sum_i \alpha_i q(t - iT) \qquad (4.2.2)$$

In these equations, $E_s$ represents the energy per symbol, $T$ is the signaling interval, $\alpha \triangleq \{\alpha_i\}$ are independent data symbols taking values $\pm 1$ with same probability, and $q(t)$ is the *phase response* of the modulator, which is related to the *frequency response*, $g(t)$, by the relationship

$$q(t) = \int_{-\infty}^{t} g(\tau) d\tau \qquad (4.2.3)$$

The frequency response is time-limited to the interval $(0, LT)$ and satisfies the following conditions:

$$\int_0^{LT} g(t) dt = \frac{1}{2} \qquad (4.2.4)$$

$$g(t) = g(LT - t) \qquad (4.2.5)$$

Condition (4.2.4) implies a scaling on $g(t)$ whereas (4.2.5) means that $g(t)$ is symmetric around the instant $t=LT/2$.

In many theoretical studies $g(t)$ is given one of the following shapes: rectangular (REC), raised-cosine (RC) and Gaussian-MSK (GMSK). A rectangular frequency pulse of length $L$ is denoted (LREC). For example, 1REC pulses are used with MSK. Similarly, (LRC) means RC of length $L$. Formally,

$$\text{LREC}: \quad g(t) = \begin{cases} \dfrac{1}{2LT} & 0 \leq t \leq LT \\ 0 & \text{elsewhere} \end{cases} \qquad (4.2.6)$$

$$\text{LRC}: \quad g(t) = \begin{cases} \dfrac{1}{2LT}\left[1-\cos\dfrac{2\pi t}{LT}\right] & 0 \leq t \leq LT \\ 0 & \text{elsewhere} \end{cases} \qquad (4.2.7)$$

$$\text{GMSK}: \quad g(t) = \frac{1}{2T}\left\{Q\left[\frac{2\pi B}{\sqrt{\ln 2}}\left(t-\frac{(L+1)T}{2}\right)\right]\right.$$
$$\left. - Q\left[\frac{2\pi B}{\sqrt{\ln 2}}\left(t-\frac{(L-1)T}{2}\right)\right]\right\} \qquad (4.2.8)$$

with

$$Q[x] \triangleq \frac{1}{\sqrt{2\pi}} \int_{x}^{\infty} e^{-t^2/2} dt \qquad (4.2.9)$$

The parameter $B$ in (4.2.8) represents the −3 dB bandwidth of the Gaussian pulse-shaping filter prior to the modulator and $g(t)$ is the response (delayed by $(L+1)T/2$ seconds) of this filter to a 1REC pulse. In particular, the pan-European digital cellular mobile radio system adopts a bandwidth $BT = 0.3$. The parameter $L$ in (4.2.8) must be chosen sufficiently large so that $g(t)$ is approximately limited to the interval $(0,LT)$. For example, $L=4$ is adequate with $BT=0.3$.

From (4.2.1)-(4.2.2) it appears that the signal depends in a nonlinear manner on the data $\{\alpha_i\}$. In many theoretical investigations this is a drawback as it considerably complicates the analysis. It has been shown by P.A. Laurent [3] however that a *binary* CPM signal with an arbitrary modulation index may be written as a superposition of a few time functions that look like linearly modulated PAM waveforms. This is the so-called Laurent expansion which is now overviewed for a modulation index of 1/2 (MSK-type signaling).

As indicated in [3], the exponential function in (4.2.1) may be expressed as the superposition of $M = 2^{L-1}$ PAM waveforms

$$e^{j\psi(t,\alpha)} = \sum_{m=0}^{M-1}\sum_{i} a_{m,i} h_m(t-iT) \qquad (4.2.10)$$

where $h_m(t)$ is given by

$$h_m(t) = c(t-LT)\prod_{l=1}^{L-1} c\big(t-LT+lT+\gamma_{m,l}LT\big) \qquad (4.2.11)$$

In this equation the pulse $c(t)$ is defined as

$$c(t) \triangleq \begin{cases} \cos[\pi q(t)] & 0 \le t \le LT \\ c(-t) & -LT \le t < 0 \\ 0 & \text{elsewhere} \end{cases} \qquad (4.2.12)$$

and the coefficient $\gamma_{m,l}$ is the $l$-th digit (0 or 1) in the binary representation of the integer $m$, i.e.,

$$m = \sum_{l=1}^{L-1} \gamma_{m,l} 2^{l-1} \qquad (4.2.13)$$

The coefficients $a_{m,i}$ look like data symbols and will referred to as *pseudo-symbols* in the sequel. It turns out that they are related to the information symbols $\alpha_i$ by the relationship

$$a_{m,i} = \exp\left(j\frac{\pi}{2}\alpha_i\right)\exp\left(j\frac{\pi}{2}\sum_{l=1}^{L-1}\bar{\gamma}_{m,l}\alpha_{i-l}\right)\exp\left(j\frac{\pi}{2}\sum_{l=-\infty}^{i-L}\alpha_l\right) \qquad (4.2.14)$$

with $\bar{\gamma}_{m,l} \triangleq 1 - \gamma_{m,l}$.

The following remarks are of interest:

(*i*) The right-hand side in (4.2.10) is an *exact* representation of the exponential. It may be intriguing that a limited number of PAM waveforms add up to a unity amplitude time function but this is precisely the meaning of Laurent expansion. The interested reader may want to look at Figures 5-7 in Laurent's paper [3] for a pictorial explanation of (4.2.10).

(*ii*) With *full response* systems ($L=1$), the integer $M$ is unity and the Laurent expansion has a single PAM component. With *partial response* schemes ($L>1$), vice versa, $M$ may be large and, in consequence, (4.2.10) may be

# Carrier Frequency Recovery with CPM Modulations

awkward to handle. Fortunately, in most cases of practical interest the signal power is mostly concentrated in the first component, i.e., the one corresponding to the zero-order pulse

$$h_0(t) = \prod_{l=1}^{L} c(t - lT) \tag{4.2.15}$$

When this happens, the Laurent expansion reduces to

$$e^{j\psi(t,\alpha)} \approx \sum_i a_{0,i} h_0(t - iT) \tag{4.2.16}$$

(*iii*) Moments of the pseudo-symbols $a_{m,i}$ are needed in later developments. It can been shown that the first-order moments are zero and the second-order ones are given by

$$E\{a_{m,i} a_{n,k}^*\} = \begin{cases} 1 & m = n \text{ and } i = k \\ 0 & \text{otherwise} \end{cases} \tag{4.2.17}$$

**Exercise 4.2.1.** Apply Laurent expansion to MSK signaling.
*Solution.* With MSK modulation the parameter $L$ is unity and the Laurent expansion has just one component (recall that $M = 2^{L-1}$). Hence,

$$e^{j\psi(t,\alpha)} = \sum_i a_{0,i} h_0(t - iT) \tag{4.2.18}$$

This means that an MSK signal can be *exactly* represented as a PAM waveform. Indeed it can be viewed as an OQPSK signal with half-cycle sine-shaped pulses.

To see how this comes about, bear in mind that MSK uses 1REC frequency pulses. Accordingly, the phase response $q(t)$ has the form

$$q(t) = \begin{cases} t/(2T) & 0 \leq t \leq T \\ 0 & t < 0 \\ 1/2 & t > T \end{cases} \tag{4.2.19}$$

and (4.2.12) yields

$$c(t) = \begin{cases} \cos\left(\dfrac{\pi t}{2T}\right) & |t| \leq T \\ 0 & \text{elsewhere} \end{cases} \tag{4.2.20}$$

Correspondingly (4.2.15) becomes

$$h_0(t) = \begin{cases} \sin\left(\dfrac{\pi t}{2T}\right) & 0 \le t \le 2T \\ 0 & \text{elsewhere} \end{cases} \qquad (4.2.21)$$

which is a half-cycle sine function of length $2T$.

Next, let us concentrate on the pseudo-symbols $a_{0,i}$. They are derived from (4.2.14) letting $L=1$ and $m=0$. Assuming $i>0$, we have

$$\begin{aligned} a_{0,i} &= \exp\left(j\frac{\pi}{2}\sum_{l=-\infty}^{i}\alpha_l\right) \\ &= e^{j\phi}\exp\left(j\frac{\pi}{2}\sum_{l=1}^{i}\alpha_l\right) \end{aligned} \qquad (4.2.22)$$

with

$$e^{j\phi} \triangleq \exp\left(j\frac{\pi}{2}\sum_{l=-\infty}^{0}\alpha_l\right) \qquad (4.2.23)$$

It is easily seen that $\phi$ takes the values $\{0, \pi/2, \pi, 3\pi/2\}$, depending on the data pattern.

The summation in the second line of (4.2.22) is either even or odd, depending on the index $i$. As a consequence we have

$$\exp\left(j\frac{\pi}{2}\sum_{l=1}^{i}\alpha_l\right) = \begin{cases} \cos\left(\dfrac{\pi}{2}\sum_{l=1}^{i}\alpha_l\right) & i = \text{even} \\ j\sin\left(\dfrac{\pi}{2}\sum_{l=1}^{i}\alpha_l\right) & i = \text{odd} \end{cases} \qquad (4.2.24)$$

Thus, letting

$$a_{2i} \triangleq \cos\left(\frac{\pi}{2}\sum_{l=1}^{2i}\alpha_l\right) \qquad (4.2.25)$$

$$a_{2i+1} \triangleq \sin\left(\frac{\pi}{2}\sum_{l=1}^{2i+1}\alpha_l\right) \qquad (4.2.26)$$

and substituting into (4.2.18) results in

$$e^{j\psi(t,\alpha)} = e^{j\phi}\left[\sum_{i=1}^{\infty} a_{2i} h_0(t-2iT) + j\sum_{i=1}^{\infty} a_{2i-1} h_0[t-(2i-1)T]\right] \quad (4.2.27)$$

which is the traditional OQPSK representation for MSK signals.

**Exercise 4.2.2.** Compute $h_0(t)$ for 2REC pulses.
*Solution.* With 2REC pulses the phase response $q(t)$ has the form

$$q(t) = \begin{cases} t/(4T) & 0 \le t \le 2T \\ 0 & t < 0 \\ 1/2 & t > 2T \end{cases} \quad (4.2.28)$$

Substituting into (4.2.12) yields

$$c(t) = \begin{cases} \cos\left(\dfrac{\pi t}{4T}\right) & |t| \le 2T \\ 0 & \text{elsewhere} \end{cases} \quad (4.2.29)$$

Finally, using (4.2.15) after some further manipulations produces

$$h_0(t) = \begin{cases} \dfrac{1}{2\sqrt{2}}\left\{1 + \sqrt{2}\sin\left[\dfrac{\pi}{2T}\left(t - \dfrac{T}{2}\right)\right]\right\} & 0 \le t \le 3T \\ 0 & \text{elsewhere} \end{cases} \quad (4.2.30)$$

**Exercise 4.2.3.** Show that MSK-type signals can be expressed in an approximate manner as OQPSK waveforms.
*Solution.* As mentioned earlier, in many practical cases the PAM waveform with index $m=0$ in (4.2.10) contains most of the signal power. Thus, keeping only this waveform in the Laurent expansion yields (4.2.16). This equation is formally identical to (4.2.18) except that the pulse $h_0(t)$ in the latter is a half-cycle sine function whereas, in (4.2.16), it has a more general form that can be computed from (4.2.15). As for the pseudo-symbols $a_{0,i}$, it is recognized from (4.2.14) that they are the same as with MSK (since $\bar{\gamma}_{0,l} = 0$) and, therefore, an MSK-type signal has (approximately) the OQPSK structure indicated in (4.2.27).

## 4.3. Data-Aided Frequency Estimation

### 4.3.1. Frequency Estimation with MSK

In this section we investigate data-aided and clock-aided carrier frequency estimation with MSK signaling. The problem is similar to that discussed in Chapter 3 with PSK modulation and, in fact, we shall adopt the same approach here.

As a start, collecting (4.2.1) and (4.2.18) yields

$$s_{CE}(t) = \sqrt{\frac{2E_s}{T}} \sum_i a_{0,i} h_0(t - iT) \qquad (4.3.1)$$

The pulse $h_0(t)$ is a half-cycle sinusoid, as indicated in (4.2.21), and the pseudo-symbols $a_{0,i}$ are given in (4.2.22). It is worth stressing that they must be thought of as known quantities since so are the data $\{\alpha_i\}$.

The signal component in the demodulated waveform is obtained from (4.3.1) by introducing an exponential factor $e^{j(2\pi v t + \theta)}$ (to account for phase/frequency errors in the demodulation process) and delaying the pulses by $\tau$. This leads to the following expression for the received waveform:

$$r(t) = s(t) + w(t), \qquad (4.3.2)$$

where $w(t)$ is white Gaussian noise and $s(t)$ has the form

$$s(t) = \sqrt{\frac{2E_s}{T}} e^{j(2\pi v t + \theta)} \sum_i a_{0,i} h_0(t - iT - \tau) \qquad (4.3.3)$$

Our task is to derive an estimate of $v$ from the observation of $r(t)$. In doing so the parameters $\{a_{0,i}\}$ and $\tau$ are viewed as known quantities while $\theta$ is unknown and can be anywhere in the interval $[0, 2\pi)$. The operations to perform on $r(t)$ are illustrated in Figure 4.1. The block LPF is a low-pass filter with a bandwidth sufficiently large to pass the signal components undistorted.

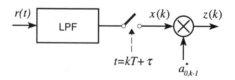

Figure 4.1. Received waveform processing.

# Carrier Frequency Recovery with CPM Modulations

Thus, the filter output $x(t)$ is formed by (4.3.3) plus some low-pass noise $n(t)$. Samples of $x(t)$ are taken at $t=kT+\tau$ and are denoted $x(k)$. Formally,

$$x(k) = \sqrt{\frac{2E_s}{T}} e^{j[2\pi v(kT+\tau)+\theta]} \sum_i a_{0,i} h_0[(k-i)T] + n(k) \qquad (4.3.4)$$

This formula can be simplified bearing in mind the form of $h_0(t)$ in (4.2.21). In fact, since

$$h_0(kT) = \begin{cases} 1 & \text{for} \quad k=1 \\ 0 & \text{otherwise} \end{cases} \qquad (4.3.5)$$

equation (4.3.4) becomes

$$x(k) = \sqrt{\frac{2E_s}{T}} e^{j[2\pi v(kT+\tau)+\theta]} a_{0,k-1} + n(k) \qquad (4.3.6)$$

Clearly, $x(k)$ depends on the modulation format through the pseudo-symbols. On the other hand, from (4.2.22) it is seen that $|a_{0,k-1}|^2 = 1$. Hence, the modulation can be wiped out by multiplying $x(k)$ by $a_{0,k-1}^*$ and this produces

$$z(k) = \sqrt{\frac{2E_s}{T}} e^{j[2\pi v(kT+\tau)+\theta]} + n'(k) \qquad (4.3.7)$$

where $n'(k) \triangleq n(k) a_{0,k-1}^*$ is a noise term. Note that $n'(k)$ has the same variance as $n(k)$.

The right-hand side of (4.3.7) represents a discrete-time sinusoid embedded in noise and our aim is to estimate the sinusoid's frequency. As this problem has already been discussed in Chapter 3 in connection with PSK modulation, the same solutions (for example, either the Fitz or Luise and Reggiannini algorithms) can be adopted.

An interesting question arises about the accuracy of these algorithms in the present circumstances. In particular, recalling that they achieve the modified Cramer-Rao bound with PSK, we wonder whether this same performance is obtained with MSK. Intuitively, it should not be so since no matched filtering is used here. In fact, matching the LPF response to $h_0(t)$ would result in a significant amount of intersymbol interference and, eventually, in large performance losses. A more quantitative answer is now given under some restrictive conditions.

Assume an ideal LPF with a rectangular characteristic over $|f| \leq 1/T$. With most MSK-type modulations this LPF passes the signal components with only minor distortions. In addition, it has the nice property that the noise samples at its output are independent (recall that sampling is performed at the symbol rate). The sample noise variance is readily found to be $E\{|n(k)|^2\} = 4N_0/T$. Then, dividing both sides of (4.3.7) by $\sqrt{2E_s/T}$ and letting $z'(k) \triangleq z(k)/\sqrt{2E_s/T}$ produces

$$z'(k) = e^{j[2\pi v(kT+\tau)+\theta]} + n''(k) \qquad (4.3.8)$$

where $n''(k)$ are independent and Gaussian random variables with variance $2N_0/E_s$. This equation is formally identical to (3.2.22) in Chapter 3, except that the variance of the noise term is doubled. Thus, the estimation accuracy is degraded by 3 dB with respect to PSK.

### 4.3.2. Extension to MSK-Type Modulation

In many practical cases the signal power of an MSK-type signal is concentrated in the first Laurent component. When this happens the representation (4.3.1) is still approximately valid, although the shape of $h_0(t)$ is no longer a half-cycle sinusoid as with MSK. Nevertheless, $h_0(t)$ may be transformed into a Nyquist pulse by proper equalization. Now, suppose that the equalized pulse $\tilde{h}_0(t)$ satisfies the condition (see Figure 4.2)

$$\tilde{h}_0(kT + t_0) \approx \begin{cases} 1 & \text{for } k = K \\ 0 & \text{elsewhere} \end{cases} \qquad (4.3.9)$$

for some value of $t_0$. Then, absorbing the equalization operations into the LPF in Figure 4.1, the signal component at the filter output is still as indicated in (4.3.3) (perhaps, with a different initial phase)

$$s(t) \approx \sqrt{\frac{2E_s}{T}} e^{j(2\pi vt+\theta')} \sum_i a_{0,i} \tilde{h}_0(t - iT - \tau) \qquad (4.3.10)$$

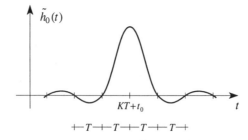

Figure 4.2. Equalized pulse $\tilde{h}_0(t)$.

and the same procedure described for MSK can be adopted to estimate $v$. In fact, sampling the LPF output at $t=kT+t_0+\tau$ yields

$$x(k) = \sqrt{\frac{2E_s}{T}}\, e^{j[2\pi v(kT+t_0+\tau)+\theta']} a_{0,k-K} + n(k) \qquad (4.3.11)$$

which has the same form as (4.3.6). Modulation is eliminated from $x(k)$ by multiplying by $a_{0,k-K}^*$ and this produces a result essentially identical to (4.3.7). All further steps remain the same. A frequency estimation method based on these ideas has been proposed in [4]-[5].

## 4.4. ML-Based NDA Frequency Estimation

### 4.4.1. MSK-Type Modulation

Symbols and timing information have been exploited for frequency estimation in the previous discussion. The underlying idea is that symbols can be taken from a known preamble, while timing can be established prior to frequency offset compensation. Henceforth we concentrate on frequency estimation methods that dispense with data symbols and, in general, with timing information as well. In particular, in this section we report on algorithms based on ML methods whereas, in Section 4.5, we describe *ad hoc* techniques. As we have seen in Chapter 2, ML methods can be applied either to continuous-time waveforms or sampled versions thereof. The continuous-time approach is now adopted with MSK-type modulations. The sample-based method will be used in Section 4.4.2 for general multilevel CPM signaling.

Non-data-aided and non-clock-aided frequency estimation with MSK-type signaling can be formulated within the same framework described in Section 3.5.1 of Chapter 3 in connection with PAM modulation. Therefore, we do not need to reiterate previous developments and may limit ourselves to drawing conclusions directly from that chapter. To see how this comes about, let us compare MSK-type and QPSK complex envelopes:

$$\text{MSK-type}: \quad s_{CE}(t) \approx \sqrt{\frac{2E_s}{T}} \sum_i a_{0,i} h_0(t-iT) \qquad (4.4.1)$$

$$\text{QPSK}: \quad s_{CE}(t) = \sum_i c_i g(t-iT) \qquad (4.4.2)$$

Note that the Laurent expansion has been limited to the first term in (4.4.1). We see that, apart from an immaterial factor $\sqrt{2E_s/T}$ and the substitution

$h_0(t) \to g(t)$, the differences in the envelopes are limited to the symbol statistics. Although $a_{0,i}$ and $c_i$ belong to the same alphabet { $e^{j\pi m/2}$, $m$=0,1,2,3}, the former are intrinsically coded (as is clear from (4.2.14)) whereas the $c_i$ are independent (assuming uncoded QPSK). The crucial point is to see whether this has any bearing on the derivation of ML-based estimators.

To address this issue recall from Section 3.5 of Chapter 3 that (assuming low signal-to-noise ratios and random data) the ML estimate $\hat{v}_{ML}$ is that $\tilde{v}$ that maximizes the energy of the matched-filter output $x(t)$ when the input is fed by $r(t)e^{-j2\pi\tilde{v}t}$. Actually, the energy of $x(t)$ depends on the data symbols only through the symbol correlations. Higher-order moments do not matter. On the other hand, as the $c_i$ are uncorrelated, so are the $a_{0,i}$ (see (4.2.17)), and this implies that the difference in symbol statistics with QPSK and MSK-type is immaterial as far as the energy of $x(t)$ is concerned. We conclude that the methods discussed in Section 3.5 are still valid in the present context.

In particular the closed-loop scheme discussed in Section 3.5 can be employed. Figure 4.3 illustrates such a scheme. Here, MF is matched to $h_0(t)$ while DMF is matched to $-2\pi t h_0(t)$. Sampling is performed at twice the symbol rate and the error signal $e(kT)$ is computed according to the formula

$$e(kT) = \frac{1}{2}\text{Im}\{x(kT)y^*(kT)\} + \frac{1}{2}\text{Im}\{x(kT+T/2)y^*(kT+T/2)\} \quad (4.4.3)$$

Finally, the current offset estimate $\hat{v}(kT)$ is updated at symbol rate as follows:

$$\hat{v}[(k+1)T] = \hat{v}(kT) + \gamma e(kT), \quad (4.4.4)$$

where $\gamma$ is the step size.

Figures 4.4-4.5 show S-curves for the frequency error detector (4.4.3) as obtained with LREC and GMSK pulses (see [6] for detailed calculations).

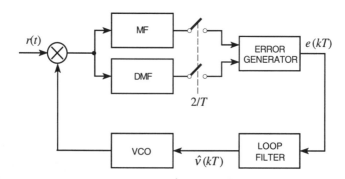

Figure 4.3. Block diagram of the closed-loop frequency estimator.

# Carrier Frequency Recovery with CPM Modulations

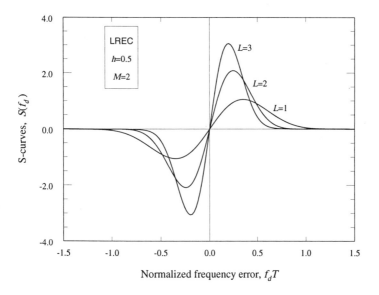

Figure 4.4. S-curves with LREC pulses.

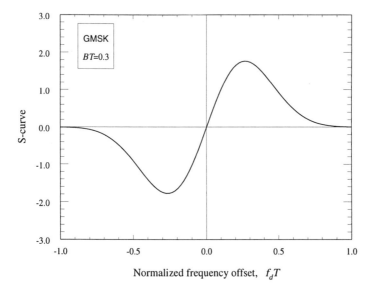

Figure 4.5. S-curves with GMSK pulses.

Here, $f_d$ is the difference $v - \hat{v}$ and, in consequence, the variable $f_d T$ represents the frequency error normalized to the symbol rate. From the width of the curves we see that the loop has an acquisition range of about 50-70% of $1/T$.

Digital implementation, acquisition and tracking performance of the above scheme can be addressed with the methods of Section 3.5 and are reported in [6]. Figure 4.6 shows the simulated error variance of the synchronizer with GMSK modulation ($BT=0.3$) and a loop bandwidth $B_L T = 5 \cdot 10^{-3}$. We see that the curve is approximately horizontal, which means an overwhelming predominance of self noise. It is interesting to notice that the same performance is obtained with a similar loop operating with OQPSK signals [7]. Of course, this is not surprising as GMSK is an approximate form of linear offset modulation.

An important question is whether it is possible to reduce self noise by exploiting the methods proposed by Alberty and Hespelt [8] and D'Andrea and Mengali [9] in the context of linear modulations. Considering the similarity between MSK-type and OQPSK modulations it is tempting to think that this is the case. Unfortunately, things might not be so simple. In fact the procedure discussed in [9] can be easily adapted to eliminate the self noise contribution

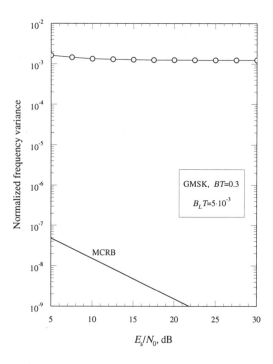

Figure 4.6. Tracking error variance of the synchronizer with GMSK.

from the *first* (most important) Laurent component. It is not clear, however, to what extent the interference from other components would add to the overall noise. This subject deserves in-depth examination and is left as an item for further research.

### 4.4.2. General CPM Modulation

In the preceding subsection the Laurent expansion has been exploited to transform MSK-type signals into sums of linearly modulated waveforms. In this way the ML estimation problem has been approached with the same methods previously developed for PAM modulation. In dealing with *multilevel* CPM we follow a different route based on discrete-time methods. To this end the received waveform $r(t)$ is passed through a rectangular anti-aliasing filter (AAF) and then is sampled at a rate $1/T_s$, as indicated in Figure 4.7. The filter bandwidth $B_{LPF}$ is assumed large enough not to distort the signal components and the sampling rate equals $2B_{LPF}$.

The samples from the filter have the form

$$x(kT_s) = s(kT_s) + n(kT_s) \tag{4.4.5}$$

where $n(kT_s)$ represents noise and $s(kT_s)$ is a sample of

$$s(t) = e^{j(2\pi vt + \theta)} \sqrt{\frac{2E_s}{T}} e^{j\psi(t-\tau,\alpha)} \tag{4.4.6}$$

with

$$\psi(t,\alpha) = 2\pi h \sum_i \alpha_i q(t - iT) \tag{4.4.7}$$

The filter output is observed over the interval $(0, T_0)$ and we assume that the ratio $T_0/T$ is an integer, $L_0$. Also, we take $T_s$ a sub-multiple $N$ of the symbol period, i.e., we set $T_s = T/N$. Then, denoting by

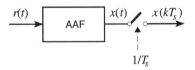

Figure 4.7. Filtering and sampling operations.

$$x = \{x(0), x(T_s), x(2T_s), \ldots, x[(NL_0 - 1)T_s]\} \qquad (4.4.8)$$

the samples of $x(t)$, the likelihood function $\Lambda(x|\tilde{\alpha}, \tilde{v}, \tilde{\theta}, \tilde{\tau})$ may be written as

$$\Lambda(x|\tilde{\alpha}, \tilde{v}, \tilde{\theta}, \tilde{\tau}) = \exp\left\{\frac{T_s}{N_0}\text{Re}\left\{\sum_{k=0}^{NL_0-1} x(kT_s)\tilde{s}^*(kT_s)\right\} - \frac{T_s}{2N_0}\sum_{k=0}^{NL_0-1}|\tilde{s}(kT_s)|^2\right\} \qquad (4.4.9)$$

with

$$\tilde{s}(t) \triangleq e^{j(2\pi\tilde{v}t+\tilde{\theta})}\sqrt{\frac{2E_s}{T}}e^{j\psi(t-\tilde{\tau},\tilde{\alpha})} \qquad (4.4.10)$$

As $\tilde{s}(t)$ has a constant envelope, the second summation in (4.4.9) is independent of the unknown signal parameters and can be dropped for simplicity. Accordingly, we have

$$\Lambda(x|\tilde{\alpha}, \tilde{v}, \tilde{\theta}, \tilde{\tau}) = \exp\left\{\frac{T_s}{N_0}\text{Re}\left\{\sum_{k=0}^{NL_0-1} x(kT_s)\tilde{s}^*(kT_s)\right\}\right\} \qquad (4.4.11)$$

and eventually, using (4.4.10),

$$\Lambda(x|\tilde{\alpha}, \tilde{v}, \tilde{\theta}, \tilde{\tau}) = \exp\left\{C \times \text{Re}\left\{e^{-j\tilde{\theta}}\sum_{k=0}^{NL_0-1} x(kT_s)e^{-j2\pi\tilde{v}kT_s}e^{-j\psi(kT_s-\tilde{\tau},\tilde{\alpha})}\right\}\right\} \qquad (4.4.12)$$

with

$$C \triangleq \frac{T_s}{N_0}\sqrt{\frac{2E_s}{T}} \qquad (4.4.13)$$

The derivation of the frequency estimator may be divided into two steps: first we compute the average of $\Lambda(x|\tilde{\alpha}, \tilde{v}, \tilde{\theta}, \tilde{\tau})$ with respect to $(\tilde{\alpha}, \tilde{\theta}, \tilde{\tau})$ so as to obtain $\Lambda(x|\tilde{v})$, the marginal likelihood function. Then, we propose an algorithm to locate the maximum of $\Lambda(x|\tilde{v})$. In doing so, the parameters $\tilde{\theta}$ and $\tilde{\tau}$ are taken uniformly distributed over $[0,2\pi)$ and $[0,T)$, respectively, and the symbols $\alpha_i$ are modelled as independent and equally likely random variables belonging to the alphabet $\{\pm 1, \pm 3, \ldots, \pm(M-1)\}$.

To begin, let us define

$$X \triangleq \sum_{k=0}^{NL_0-1} x(kT_s)e^{-j2\pi\tilde{v}kT_s}e^{-j\psi(kT_s-\tilde{\tau},\tilde{\alpha})} \qquad (4.4.14)$$

and put $X = |X|e^{j\phi_x}$. Also, observe that $X$ is a function of $\tilde{\alpha}$, $\tilde{v}$ and $\tilde{\tau}$ but not of $\tilde{\theta}$. Then, inserting (4.4.14) into (4.4.12) produces

$$\Lambda(x|\tilde{\alpha},\tilde{v},\tilde{\theta},\tilde{\tau}) = \exp\{C|X|\cos(\phi_x - \tilde{\theta})\} \quad (4.4.15)$$

Also, averaging with respect to $\tilde{\theta}$ yields

$$\Lambda(x|\tilde{\alpha},\tilde{v},\tilde{\tau}) = I_0(C|X|) \quad (4.4.16)$$

where

$$I_0(\xi) \triangleq \frac{1}{2\pi}\int_0^{2\pi} e^{\xi\cos\alpha}d\alpha \quad (4.4.17)$$

is the zero-order modified Bessel function.

Next, we perform the expectation of $\Lambda(x|\tilde{\alpha},\tilde{v},\tilde{\tau})$ over $\tilde{\alpha}$ and $\tilde{\tau}$. In doing so we assume that the signal-to-noise ratio is sufficiently low so that the power series expansion of $I_0(C|X|)$ can be truncated to the quadratic term, i.e.,

$$I_0(C|X|) \approx 1 + \frac{C^2}{4}|X|^2 \quad (4.4.18)$$

Averaging (4.4.18) yields

$$\Lambda(x|\tilde{v}) = 1 + \frac{C^2}{4}\mathrm{E}_{\tilde{\alpha},\tilde{\tau}}\{|X|^2\} \quad (4.4.19)$$

Clearly, maximizing $\Lambda(x|\tilde{v})$ amounts to maximizing

$$\Gamma(\tilde{v}) \triangleq \mathrm{E}_{\tilde{\alpha},\tilde{\tau}}\{|X|^2\} \quad (4.4.20)$$

In Appendix 4.A it is shown that $\Gamma(\tilde{v})$ may be written as

$$\Gamma(\tilde{v}) = \sum_{k_1=0}^{NL_0-1}\sum_{k_2=0}^{NL_0-1} x(k_1T_s)x^*(k_2T_s)e^{-j2\pi(k_1-k_2)\tilde{v}T_s}H[(k_2-k_1)T_s] \quad (4.4.21)$$

where $H(kT_s)$ is a real-valued function defined as

$$H(kT_s) \triangleq \frac{1}{T}\int_0^T \prod_{i=-\infty}^{\infty}\left\{\frac{1}{M}\frac{\sin[2\pi hMp(t-iT,kT_s)]}{\sin[2\pi hp(t-iT,kT_s)]}\right\}dt \quad (4.4.22)$$

In this equation $M$ is the size of the symbol alphabet and $p(t,kT_s)$ is related to the phase response of the modulator by

$$p(t,kT_s) \triangleq q(t) - q(t - kT_s) \quad (4.4.23)$$

Useful remarks for the numerical calculation of the integrand in (4.4.22) are given in Appendix 4.C.

Next, we look for the value of $\tilde{v}$ that maximizes $\Gamma(\tilde{v})$ or, equivalently, that makes the derivative $d\Gamma(\tilde{v})/d\tilde{v}$ vanish. The expression of $d\Gamma(\tilde{v})/d\tilde{v}$ is readily derived from (4.4.21) and reads

$$\frac{d\Gamma(\tilde{v})}{d\tilde{v}} = j2\pi T_s \sum_{k_1=0}^{NL_0-1} \sum_{k_2=0}^{NL_0-1} y(k_1 T_s) y^*(k_2 T_s) h[(k_2 - k_1)T_s] \quad (4.4.24)$$

where $y(kT_s)$ is a rotated version of $x(kT_s)$

$$y(kT_s) \triangleq x(kT_s) e^{-j2\pi k \tilde{v} T_s} \quad (4.4.25)$$

and $h(kT_s)$ is related to $H(kT_s)$ by

$$h(kT_s) \triangleq kH(kT_s) \quad (4.4.26)$$

Figures 4.8-4.9 show $h(kT_s)$ for 1REC frequency pulses and a sampling interval $T_s = T/4$. In particular, a binary alphabet and a modulation index of 0.5 is considered in Figure 4.8, whereas Figure 4.9 corresponds to an octal alphabet and a modulation index of 0.125. It appears that $h(kT_s)$ takes significant values only for $|kT_s| \leq 2T$ in both cases.

Equation (4.4.24) is now put in a form that leads to a practical solution for the equation $d\Gamma(\tilde{v})/d\tilde{v} = 0$. As a first step note that, since $\Gamma(\tilde{v})$ is real-valued (see (4.4.20)), so is $d\Gamma(\tilde{v})/d\tilde{v}$ and, in consequence, the double summation in (4.4.24) is a purely imaginary quantity. Hence

$$\frac{d\Gamma(\tilde{v})}{d\tilde{v}} = -2\pi T_s \operatorname{Im}\left\{ \sum_{k_1=0}^{NL_0-1} \sum_{k_2=0}^{NL_0-1} y(k_1 T_s) y^*(k_2 T_s) h[(k_2 - k_1)T_s] \right\} \quad (4.4.27)$$

Second, assume an observation interval much longer than the duration of $h(kT_s)$. Then the summation with respect to $k_1$ in (4.4.27) can be extended from $-\infty$ to $\infty$, i.e.,

$$\sum_{k_1=0}^{NL_0-1} y(k_1 T_s) h[(k_2 - k_1)T_s] \approx \sum_{k_1=-\infty}^{\infty} y(k_1 T_s) h[(k_2 - k_1)T_s] \quad (4.4.28)$$

# Carrier Frequency Recovery with CPM Modulations

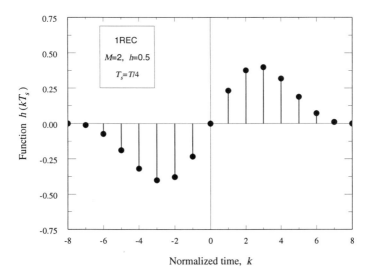

Figure 4.8. Function $h(kT_s)$ with 1REC pulses and a binary alphabet.

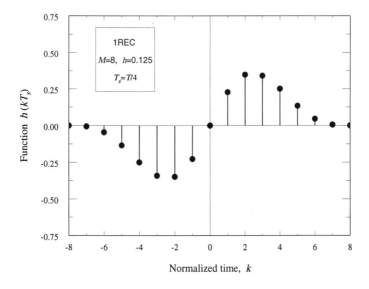

Figure 4.9. Function $h(kT_s)$ with 1REC pulses and octal alphabet.

so that, denoting

$$z(kT_s) \triangleq T_s \sum_{k_1=-\infty}^{\infty} y(k_1 T_s) h[(k-k_1)T_s] \qquad (4.4.29)$$

and substituting into (4.4.27), after some algebra we obtain

$$\frac{d\Gamma(\tilde{v})}{d\tilde{v}} \approx 2\pi \sum_{k=0}^{NL_0-1} \text{Im}\{y(kT_s) z^*(kT_s)\} \qquad (4.4.30)$$

Third, the quantity $z(kT_s)$ may be viewed as the output of a filter $h(kT_s)$ driven by $y(kT_s)$. The filter is non-causal but can be made causal by suitably delaying $h(kT_s)$ by some sampling intervals, say $D$. For example, $D=2N$ is an adequate delay in the cases indicated in Figures 4.8-4.9. The output of the causal filter is then $w(kT_s) = y(kT_s) \otimes h[(k-D)T_s]$ and (4.4.30) may be rewritten in terms of $w(kT_s)$ and $y(kT_s)$ as

$$\frac{d\Gamma(\tilde{v})}{d\tilde{v}} \approx 2\pi \sum_{k=D}^{NL_0+D-1} \text{Im}\{y[(k-D)T_s] w^*(kT_s)\} \qquad (4.4.31)$$

An algorithm to solve the equation $d\Gamma(\tilde{v})/d\tilde{v} = 0$ is now within reach. The basic idea is to exploit the sum of some consecutive terms in (4.4.31) as an error signal to drive $d\Gamma(\tilde{v})/d\tilde{v}$ toward zero. The application of this idea is discussed in Appendix 3C and leads to the loop indicated in Figure 4.10. In this diagram two time indexes are used, the *symbol index n* and the *sample index k*. They are related by

$$n = \text{int}\left(\frac{k}{N}\right) \qquad (4.4.32)$$

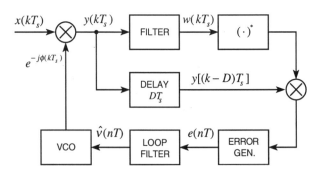

Figure 4.10. Block diagram of the frequency loop.

# Carrier Frequency Recovery with CPM Modulations

where int($z$) means "the largest integer not exceeding $z$." In essence, $n$ tells us the symbol interval corresponding to the $k$-th sample.

The filter generating $w(kT_s)$ has impulse response $h[(k-D)T_s]$ and the error signal is given by

$$e(nT) = \sum_{k=nN}^{(n+1)N-1} \text{Im}\left\{y[(k-D)T_s]w^*(kT_s)\right\} \qquad (4.4.33)$$

Also, the frequency estimates are updated (at symbol rate) according to the formula

$$\hat{v}[(n+1)T] = \hat{v}(nT) + \gamma\, e(nT) \qquad (4.4.34)$$

where $\gamma$ is a step-size parameter. Finally, the VCO computes the phases

$$\phi[(k+1)T_s] = \phi(kT_s) + 2\pi T_s \hat{v}(nT) \qquad \text{mod } 2\pi \qquad (4.4.35)$$

and produces the mapping $\phi(kT_s) \to e^{-j\phi(kT_s)}$.

## 4.4.3. Loop Performance

The performance of the frequency loop in Figure 4.10 can be assessed with the methods indicated in Section 3.5 of Chapter 3. As the calculations are exceedingly long, in the sequel we limit ourselves to some comments on numerical results drawn from [10].

Acquisition capability is established by the S-curve of the error generator, which is the expectation of the error signal for a fixed frequency estimate $\hat{v}(nT) = \hat{v}$. This expectation turns out to depend on the difference $f_d \triangleq v - \hat{v}$ between the true frequency offset and its estimate $\hat{v}$ and is denoted $S(f_d)$. As discussed in Chapter 3, the width of $S(f_d)$ establishes the acquisition range of the loop.

S-curves with LREC pulses, binary symbols and a modulation index of 0.5 have been computed with an oversampling factor $N=4$ and a loop delay $D=8$ (corresponding to two symbol intervals). They are virtually identical to those for the frequency loop in Section 4.4.1 (see Figure 4.4). This is not surprising as the approach followed there exploits the same quadratic approximation to the likelihood function. The novelty is that the present results apply to any modulation index (not just 0.5, as in Section 4.4.1) and to general $M$-ary alphabets. For example, Figure 4.11 illustrates the S-curve for 1REC, quaternary modulation and modulation index $h=0.25$ (again, an oversampling factor $N=4$ and a loop delay $D=8$ have been used). We see that the curve is rather wide, which means that the loop has large acquisition ranges (on the

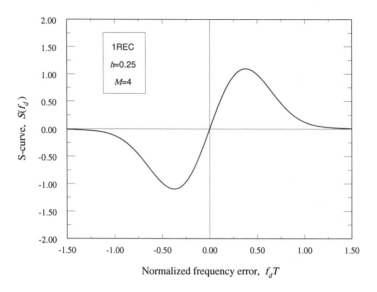

Figure 4.11. S-curve with 1REC pulses, $M=4$ and $h=0.25$.

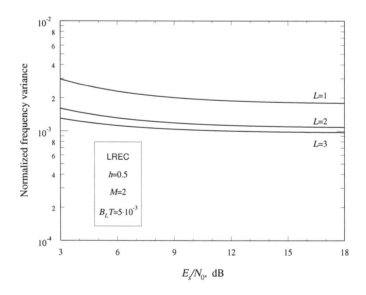

Figure 4.12. Tracking error variance with LREC pulses and binary modulation.

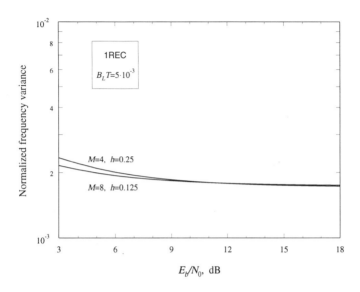

Figure 4.13. Tracking error variance with 1REC pulses and quaternary/octal modulation.

order of the signal bandwidth).

Numerical calculations reported in [10] indicate that, with a binary alphabet and a modulation index of 0.5, the loop tracking accuracy is essentially the same as that with the scheme in Figure 4.3. Again, this is intuitively clear in view of the equivalence between the discrete-time approach adopted here and the treatment in Section 4.1 based on the Laurent expansion. Figure 4.12 shows the tracking error variance (normalized to the squared symbol rate) with LREC pulses, binary alphabet and $h=0.5$. The loop noise bandwidth is $B_L T = 5 \cdot 10^{-3}$.

Figure 4.13 provides analogous results for multilevel modulation. Here, the horizontal axis represents the ratio of the energy per bit $E_b$ to the noise spectral density ($E_b$ is related to the energy per symbol by $E_b = E_s / \log_2 M$). As in Figure 4.12, self noise appears as the prevailing disturbance.

## 4.5. Delay-and-Multiply Schemes

### 4.5.1. Open-Loop Scheme

The frequency loops in the previous sections have been derived with ML arguments. In the following we address the same problem with *ad hoc* methods. In particular, we discuss *delay-and-multiply* schemes, wherein the frequency offset is estimated from products of signal samples. These schemes operate in a non-data-aided and non-clock-aided fashion and have either an open-loop or a closed-loop topology.

Let us start with the open-loop scheme indicated in Figure 4.14. The sample sequence $\{x(kT_s)\}$ is obtained from the received waveform through filtering and sampling operations as shown in Figure 4.7. The same assumptions as in Section 4.4.2 are made here and, in particular, a rectangular filter with a bandwidth of twice the sampling rate is used. In these conditions the noise samples are uncorrelated. No restrictions are imposed on the modulation format and, in particular, the modulation index is arbitrary.

Carrier frequency offset is estimated through the formula

$$\hat{v} = \frac{1}{2\pi DT_s} \arg\left\{ \sum_{k=0}^{NL_0-1} z(kT_s) \right\} \quad (4.5.1)$$

where $L_0$ is the observation length and $z(kT_s)$ is given by

$$z(kT_s) = x(kT_s) x^*[(k-D)T_s] \quad (4.5.2)$$

Equation (4.5.1) has the following interpretation. Let us split $x(kT_s)$ into signal and noise

$$x(kT_s) = s(kT_s) + n(kT_s) \quad (4.5.3)$$

with

$$s(kT_s) = e^{j(2\pi v kT_s + \theta)} \sqrt{\frac{2E_s}{T}} e^{j\psi(kT_s - \tau, \alpha)} \quad (4.5.4)$$

Now, substituting into (4.5.2) yields

$$z(kT_s) = \frac{2E_s}{T} e^{j2\pi vDT_s} e^{j\{\psi(kT_s - \tau, \alpha) - \psi[(k-D)T_s - \tau, \alpha]\}} + N(kT_s) \quad (4.5.5)$$

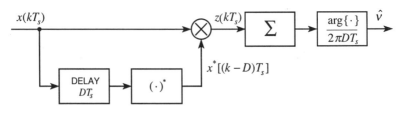

Figure 4.14. Open-loop delay-and-multiply frequency estimator.

# Carrier Frequency Recovery with CPM Modulations

where $N(kT_s)$ is a zero-mean noise term. Finally, averaging produces

$$E\{z(kT_s)\} = \frac{2E_s}{T} A(kT_s - \tau) e^{j2\pi v DT_s} \qquad (4.5.6)$$

where $A(t)$ is defined as

$$A(t) = E\left\{e^{j[\psi(t,\alpha) - \psi(t - DT_s, \alpha)]}\right\} \qquad (4.5.7)$$

Lengthy calculations in Appendix 4.B show that $A(t)$ may be expressed as

$$A(t) = \prod_{i=-\infty}^{\infty} \left\{ \frac{1}{M} \frac{\sin[2\pi h M p(t - iT, DT_s)]}{\sin[2\pi h p(t - iT, DT_s)]} \right\} \qquad (4.5.8)$$

with $p(t, DT_s) \triangleq q(t) - q(t - DT_s)$. Clearly, $A(t)$ is a real-valued function. It is also periodic of period $T$, as is seen from the fact that (4.5.8) does not change if $t$ is replaced by $t+T$. Figure 4.15 illustrates the shape of $A(t)$ for MSK modulation.

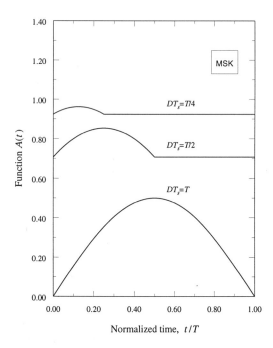

Figure 4.15. Shape of $A(t)$ with MSK.

Returning to (4.5.6) and summing over $0 \le k \le NL_0 - 1$ yields

$$\mathrm{E}\left\{\sum_{k=0}^{NL_0-1} z(kT_s)\right\} = \frac{2E_s}{T} NL_0 \overline{A} e^{j2\pi vDT_s} \qquad (4.5.9)$$

where $\overline{A}$ is the sample average of $A(t)$

$$\overline{A} \triangleq \frac{1}{N}\sum_{k=0}^{N-1} A(kT_s - \tau) \qquad (4.5.10)$$

At this point the following interpretation of (4.5.1) can be given. In all practical cases $\overline{A}$ is positive. Then, taking the arguments of the two sides of (4.5.9) and solving for $v$ results in

$$v = \frac{1}{2\pi DT_s}\arg\left\{\mathrm{E}\left\{\sum_{k=0}^{NL_0-1} z(kT_s)\right\}\right\} \qquad (4.5.11)$$

This equation says that $v$ could be computed *exactly* if the expectation of the sum were available. As this is not the case, we replace the expectation by the sum itself and this results in equation (4.5.1).

It is worth noting that, as the arg-function takes values in the range $\pm\pi$, the estimates (4.5.1) vary between $\pm 1/(2DT_s)$. For example, with $DT_s = T/2$ the estimation range equals $\pm 1/T$. In practice, useful operation is limited to a narrower range, as indicated in Figure 4.16 which shows simulation results for $\mathrm{E}\{\hat{v}T\}$ versus $vT$ with 1REC pulses, $M=4$ and a modulation index of 0.5. Here, the delay $DT_s$ equals $T/3$, the anti-alias filter bandwidth equals $3/T$, and the oversampling factor is set to 6. As is seen, as $v$ approaches $\pm 1.5/T$, the average $\mathrm{E}\{\hat{v}T\}$ deviates from the true value and approaches zero.

The estimation accuracy of (4.5.1) is investigated in [12] where formulas are given for the error variance as a function of the modulation parameters and the signal-to-noise ratio. It turns out that the results are insensitive to the sampling phase when the oversampling factor is sufficiently large (say 4 or 6, depending on the signal bandwidth). Comparisons with the ML-based schemes in Section 4.4 give mixed results, depending on the signal alphabet and the modulation index. In any case the error variance curves exhibit a high level of self noise. An example is given in Figure 4.17 in the case of MSK signaling. The oversampling factor is $N=4$ and two values of the delay $DT_s$ are used. We see that self noise is overwhelming.

# Carrier Frequency Recovery with CPM Modulations

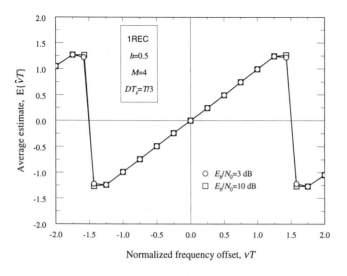

Figure 4.16. Expectation of $\hat{v}$ versus $v$ with a quaternary alphabet.

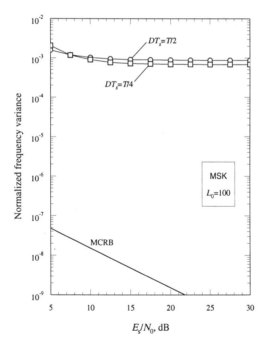

Figure 4.17. Error variance with MSK.

## 4.5.2. Closed-Loop Scheme

The delay-and-multiply method can also be employed in a closed-loop configuration. To see how this comes about consider Figure 4.18. Denoting

$$x'(kT_s) \triangleq x(kT_s)e^{-j2\pi\hat{v}kT_s} \qquad (4.5.12)$$

and

$$z'(kT_s) \triangleq x'(kT_s){x'}^*\big[(k-D)T_s\big] \qquad (4.5.13)$$

let us concentrate on the quantity

$$Q(n) \triangleq \frac{1}{2\pi DT_s}\arg\left\{\sum_{k=nN}^{(n+1)N-1} z'(kT_s)\right\} \qquad (4.5.14)$$

From (4.5.3)-(4.5.4) and (4.5.12) it is seen that $x'(kT_s)$ and $x(kT_s)$ have the same signal component, except that the frequency offset is now $v - \hat{v}$, not $v$. Then, $Q(n)$ may be viewed as an estimate of $v - \hat{v}$ and can be used as an error signal to steer $\hat{v}$ toward $v$. This operation is performed as indicated in Figure 4.19. Here, $n$ represents the symbol index and is related to the sample index $k$ by $k = \text{int}(n/N)$. The error generator produces

$$e(nT) \triangleq \frac{1}{2\pi DT_s}\arg\left\{\sum_{k=nN}^{(n+1)N-1} z'(kT_s)\right\} \qquad (4.5.15)$$

which serves to update the VCO frequency

$$\hat{v}[(n+1)T] = \hat{v}(nT) + \gamma e(nT) \qquad (4.5.16)$$

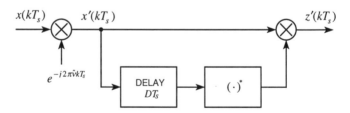

Figure 4.18. Explaining the delay-and-multiply concept for closed loops.

# Carrier Frequency Recovery with CPM Modulations

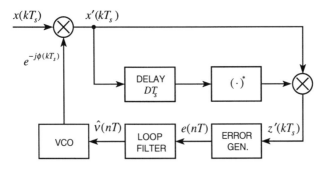

Figure 4.19. Frequency estimation loop.

Finally, the phase $\phi(kT_s)$ is updated according to

$$\phi[(k+1)T_s] = \phi(kT_s) + 2\pi T_s \hat{v}(nT) \quad \text{mod } 2\pi \tag{4.5.17}$$

Interestingly enough, a simpler error signal can be adopted in place of (4.5.15). In fact the arg-function takes very small values when $\hat{v}(nT)$ is close to $v$ (as happens in the tracking mode) and the following approximation can be made:

$$\arg\{\Sigma\} \approx \frac{\text{Im}\{\Sigma\}}{\text{Re}\{\Sigma\}} \tag{4.5.18}$$

Hence, replacing $\text{Re}\{\Sigma\}$ with its average value and substituting into (4.5.15) yields the simpler error signal

$$e(nT) = \sum_{k=nN}^{(n+1)N-1} \text{Im}\{z'(kT_s)\} \tag{4.5.19}$$

where an immaterial factor has been dropped, for it can be absorbed into the step-size parameter.

There is some degree of resemblance between (4.5.19) and the error signal (4.4.33) in Section 4.4.2. In fact it can be checked that the frequency loop in Figure 4.10 is transformed into that in Figure 4.19 (with the error generator (4.5.19)) by setting $D = 0$ and replacing the filter by a delay $DT_s$.

**Exercise 4.5.1.** Compute the S-curve of the frequency error detector (4.5.19).

*Solution.* Let us open the loop in Figure 4.19 and drive the leftmost multiplier by the exponential $e^{-j[2\pi \hat{v}kT_s + \phi(0)]}$. In these conditions the voltage $x'(kT_s)$ becomes

$$x'(kT_s) = x(kT_s)e^{-j[2\pi\hat{v}kT_s + \phi(0)]} \qquad (4.5.20)$$

Since

$$x(kT_s) = s(kT_s) + n(kT_s) \qquad (4.5.21)$$

we have

$$x'(kT_s) = s(kT_s)e^{-j[2\pi\hat{v}kT_s + \phi(0)]} + n'(kT_s) \qquad (4.5.22)$$

where $n'(kT_s)$ is a noise term equivalent to $n(kT_s)$.

Next, compare (4.5.22) with the expression of $x(kT_s)$ resulting from (4.5.3)-(4.5.4). It can be checked that there are only two differences: the signal component in (4.5.22) has a frequency offset equal to $v - \hat{v}$ instead of $v$ and an initial phase equal to $\theta - \phi(0)$ instead of $\theta$. It follows that the expectation

$$\mathrm{E}\left\{\sum_{k=nN}^{(n+1)N-1} z'(kT_s)\right\} \qquad (4.5.23)$$

can be computed by paralleling the arguments leading to (4.5.9). As a result we get

$$\mathrm{E}\left\{\sum_{k=nN}^{(n+1)N-1} z'(kT_s)\right\} = \frac{2E_s}{T} N\overline{A} e^{j2\pi(v-\hat{v})DT_s} \qquad (4.5.24)$$

Hence, taking the expectation of (4.5.19) and using (4.5.24) yields the result sought

$$S(f_d) = \frac{2E_s}{T} N\overline{A} \sin(2\pi f_d DT_s) \qquad (4.5.25)$$

with $f_d \triangleq v - \hat{v}$. As is seen, the S-curve has a sinusoidal shape of period $1/(DT_s)$.

The definition of $\overline{A}$ in (4.5.10) might suggest that the amplitude of the S-curve depends on the sampling phase. Computer simulations indicate however that this is not the case as long as the oversampling factor is sufficiently large.

## 4.6. Clock-Aided Recovery

### 4.6.1. Delay-and-Multiply Method

In discussing the estimation schemes in Sections 4.4 and 4.5 we have assumed that no timing information is available. Actually, the performance of

# Carrier Frequency Recovery with CPM Modulations

the resulting schemes is largely insensitive to the sampling phase (but it is plagued by a great deal of self noise). Of course, the insensitivity to timing is an appealing feature as clock information is generally unavailable in the presence of large frequency offsets. With small frequency errors, however, timing information can be derived first and then exploited for frequency estimation. In these conditions the question arises as to whether delay-and-multiply methods can be rearranged so as to exploit timing information, perhaps to alleviate the self noise problem. This subject is now addressed concentrating first on MSK modulation.

Return to the scheme in Figure 4.14 and assume that: (*i*) sampling is performed at the instants $kT + \tau$; (*ii*) the delay $DT_s$ in the lower branch equals $T$; (*iii*) the anti-aliasing filter (not visible in the figure) introduces negligible signal distortions. In summary, let us model the samples from the filter as

$$x(kT + \tau) = s(kT + \tau) + n(kT + \tau) \tag{4.6.1}$$

where

$$s(kT + \tau) = e^{j[2\pi v(kT+\tau)+\theta]}\sqrt{\frac{2E_s}{T}} e^{j\psi(kT,\alpha)} \tag{4.6.2}$$

In these conditions the following frequency estimator has been proposed for MSK in [13]:

$$\hat{v} = -\frac{1}{4\pi T}\arg\left\{\sum_{k=0}^{L_0-1} z(kT+\tau)\right\} \tag{4.6.3}$$

with

$$z(kT+\tau) \triangleq x^2(kT+\tau)\left[x^2(kT-T+\tau)\right]^* \tag{4.6.4}$$

An interesting feature of this estimator is that it has no self noise, which amounts to saying that the estimates tend to the true parameter value as the SNR grows large. To prove this claim let $n(kT + \tau) = 0$ in (4.6.1). Then, (4.6.4) becomes

$$z(kT+\tau) = \left(\frac{2E_s}{T}\right)^2 e^{j4\pi vT} e^{j2[\psi(kT,\alpha)-\psi(kT-T,\alpha)]} \tag{4.6.5}$$

On the other hand, with MSK modulation it is readily shown that

$$2[\psi(t,\alpha) - \psi(t-T,\alpha)] = 2\pi \sum_i \alpha_i p(t-iT,T) \tag{4.6.6}$$

with

$$p(t,T) \triangleq \begin{cases} t/(2T) & 0 \le t \le T \\ 1 - t/(2T) & T < t \le 2T \\ 0 & \text{elsewhere} \end{cases} \quad (4.6.7)$$

Hence

$$2[\psi(kT,\alpha) - \psi(kT-T,\alpha)] = \pm \pi \quad (4.6.8)$$

and the last exponential in (4.6.5) equals −1. Then, substituting into (4.6.3) gives $\hat{v} = v$, which proves the claim.

Figure 4.20 illustrates the estimation error variance as obtained by simulation with MSK and GMSK formats. The AAF filter is an 8th-order Butterworth type of bandwidth $1/T$. Comparing with Figure 4.17 it is seen that self noise is practically eliminated with MSK. The residual floor in the variance curve is caused by signal distortions in the AAF (whose bandwidth $1/T$ is not sufficiently large to pass the MSK signal). Self noise with GMSK is still high, however. Actually, it turns out that with GMSK the present clock-aided method is not much better than the non-clock-aided scheme [13].

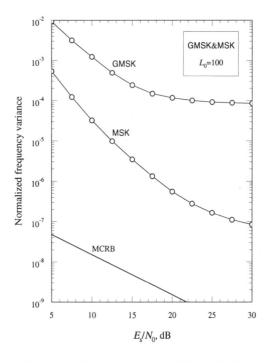

Figure 4.20. Error variance with MSK and GMSK.

### 4.6.2. 2P-Power Method with Full Response Formats

Another way to exploit timing information is to remove the modulation from the signal samples. Return to the mathematical expression for the signal

$$s(t) = e^{j(2\pi vt+\theta)}\sqrt{\frac{2E_s}{T}}\,e^{j\psi(t-\tau,\boldsymbol{\alpha})} \qquad (4.6.9)$$

with

$$\psi(t,\boldsymbol{\alpha}) = 2\pi h \sum_i \alpha_i q(t-iT) \qquad (4.6.10)$$

and consider *full response* formats. This means that $q(t)=0$ for $t<0$ and $q(t)=1/2$ for $t>T$. Assuming that transmission starts at $t=0$ (so that $\alpha_i=0$ for $i<0$) from (4.6.10) we get

$$\psi(kT,\boldsymbol{\alpha}) = \pi h \sum_{i=0}^{k-1}\alpha_i \qquad (4.6.11)$$

In all practical cases the modulation index is a rational number, say

$$h = \frac{K}{P} \qquad (4.6.12)$$

where $K$ and $P$ are integers with no factors in common. For example, $K$ is unity and $P=2$ with MSK signaling. Thus, equation (4.6.11) takes the form

$$\psi(kT,\boldsymbol{\alpha}) = \pi \frac{K}{P}\sum_{i=0}^{k-1}\alpha_i \qquad (4.6.13)$$

from which it is easily realized that

$$[2P\psi(kT,\boldsymbol{\alpha})]_{\bmod 2\pi} \in \{2\pi p,\ 0 \le p \le (P-1)\} \qquad (4.6.14)$$

This equation says that $2P\psi(kT,\boldsymbol{\alpha})$ is a multiple of $2\pi$ and, in consequence, $e^{j2P\psi(kT,\boldsymbol{\alpha})}$ is unity. Thus, sampling $s(t)$ at $t = kT+\tau$ and raising to the 2P-th power yields

$$s^{2P}(kT+\tau) = \left(\frac{2E_s}{T}\right)^P e^{j(4\pi kPvT+\phi)} \qquad (4.6.15)$$

with $\phi \triangleq 4\pi Pv\tau + 2P\theta$. As is seen, $s^{2P}(kT+\tau)$ is a discrete-time sinewave at frequency $2Pv$.

In practice $s(t)$ is embedded in thermal noise and only the sum $r(t) = s(t) + w(t)$ is available. As an approximation, however, $r(t)$ can be filtered (to eliminate as much noise as possible) and used in place of $s(t)$. Denoting by $x(t)$ the filtered version of $r(t)$ (see Figure 4.21), and $n(t)$ the residual noise, we have

$$x(kT + \tau) = s(kT + \tau) + n(kT + \tau) \qquad (4.6.16)$$

from which, letting $z(k) \triangleq x^{2P}(kT + \tau)$, it is easily found that

$$z(k) = \left(\frac{2E_s}{T}\right)^P e^{j(4\pi kPvT + \phi)} + N(kT + \tau) \qquad (4.6.17)$$

where $N(kT + \tau)$ results from the products Signal×Noise and Noise×Noise in the binomial expansion of $(s + n)^{2P}$.

The sequence $\{z(k)\}$ represents measurements of a sinewave embedded in noise, and our task is to estimate the sinewave's frequency. This problem has already been discussed in Chapter 3 in connection with PSK modulation and can be solved with the same methods proposed there. For example, either the Fitz or Luise and Reggiannini algorithms can be employed.

Figures 4.22-4.23 yield simulation results corresponding to MSK and an observation interval of $L_0=100$ symbols. Timing is assumed ideal and the low-pass filter prior to the $2P$-th power nonlinearity has a rectangular transfer function of bandwidth $B_{LPF}T=1.2/T$. Also, the samples $\{z(k)\}$ are fed to a Luise and Reggiannini frequency estimator that computes

$$\hat{v} = \frac{1}{2\pi P(N+1)T} \arg\left\{\sum_{m=1}^{N} R(m)\right\} \qquad (4.6.18)$$

with

$$R(m) \triangleq \frac{1}{L_0 - m} \sum_{k=m}^{L_0-1} z(k) z^*(k-m) \qquad (4.6.19)$$

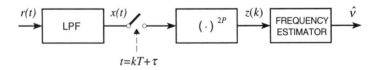

Figure 4.21. Block diagram of the $2P$-power frequency estimator.

# Carrier Frequency Recovery with CPM Modulations 181

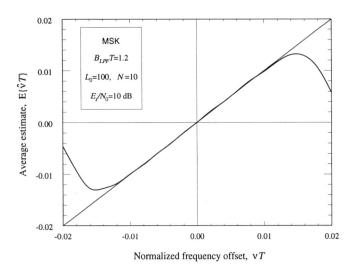

Figure 4.22. Average value of $\hat{\nu}T$ as a function of the frequency offset.

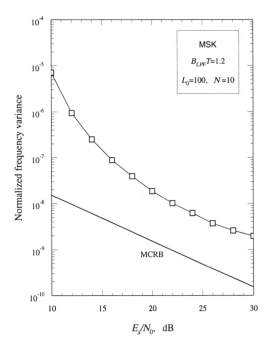

Figure 4.23. Normalized variance of the estimates.

The parameter $N$ in (4.6.18) equals 10. Figure 4.22 shows the expected value of the estimate (normalized to the symbol rate) versus the true frequency offset. It appears that the estimates are practically unbiased for frequency offsets within ±1% of the symbol rate. In Figure 4.23 the variance of the estimates (normalized to the symbol rate) is shown. We see that at low SNR the slope of the curve is steeper than for MCRB. Clearly this is a consequence of the noise enhancement due to the 4-power nonlinearity. A negligible self noise level is observed at high SNR. Comparing with Figure 4.20 it appears that the $2P$-power is superior though it is more complex to implement.

## 4.7. Key Points of the Chapter

- Data-aided and clock-aided frequency estimation with MSK-type formats can be approached with the same methods adopted with linear modulations. This is so because MSK-type signals can be approximated as OQPSK waveforms (Laurent expansion). Estimation accuracy is about 3 dB from the modified Cramer-Rao bound with true MSK. With general MSK-type modulations, instead, performance is worse, due to approximations in truncating the Laurent expansion.

- A variety of algorithms are available to estimate large carrier frequency offsets. They are non-data-aided and, in general, non-clock-aided. Some of them are derived from maximum likelihood methods and have a closed-loop configuration. Their acquisition range is on the order of the signal bandwidth. Accuracy is far from the modified Cramer-Rao bound, however. With binary modulations, in particular, it is comparable with that of analogous estimation algorithms for OQPSK.

- Alternatively, *ad hoc* estimation schemes can be used. They are of the delay-and-multiply type and are simple to implement. Their topology may be either open- or closed-loop. Their estimation range is as large as with the ML-based schemes. Accuracy may be either better or worse than with ML-based methods, depending on the specific modulation format.

- Clock information can be exploited to improve frequency estimation accuracy. This subject has not received much attention in the literature but a few results concerning MSK clearly indicate that the idea is promising and can provide significant improvements.

## Appendix 4.A

In this appendix we compute the following expectation:

$$\Gamma(\tilde{v}) = \mathrm{E}_{\tilde{\alpha},\tilde{\tau}}\{|X|^2\} \tag{4.A.1}$$

where $X$ is defined as

$$X = \sum_{k=0}^{NL_0-1} x(kT_s) e^{-j2\pi\tilde{v}kT_s} e^{-j\psi(kT_s-\tilde{\tau},\tilde{\alpha})} \tag{4.A.2}$$

As a first step let us rewrite (4.A.1) in the form

$$\Gamma(\tilde{v}) = \mathrm{E}_{\tilde{\tau}}\{\mathrm{E}_{\tilde{\alpha}}\{|X|^2\}\} \tag{4.A.3}$$

meaning that $|X|^2$ is first averaged over the symbols and then over the timing epoch. Noting that

$$|X|^2 = \sum_{k_1=0}^{NL_0-1}\sum_{k_2=0}^{NL_0-1} x(k_1T_s) x^*(k_2T_s) e^{-j2\pi\tilde{v}(k_1-k_2)T_s} e^{j[\psi(k_2T_s-\tilde{\tau},\tilde{\alpha})-\psi(k_1T_s-\tilde{\tau},\tilde{\alpha})]} \tag{4.A.4}$$

we have

$$\mathrm{E}_{\tilde{\alpha}}\{|X|^2\} = \sum_{k_1=0}^{NL_0-1}\sum_{k_2=0}^{NL_0-1} x(k_1T_s) x^*(k_2T_s) e^{-j2\pi\tilde{v}(k_1-k_2)T_s}$$
$$\times \mathrm{E}_{\tilde{\alpha}}\left\{e^{j[\psi(k_2T_s-\tilde{\tau},\tilde{\alpha})-\psi(k_1T_s-\tilde{\tau},\tilde{\alpha})]}\right\} \tag{4.A.5}$$

The expectation $\mathrm{E}_{\tilde{\alpha}}\left\{e^{j[\psi(k_2T_s-\tilde{\tau},\tilde{\alpha})-\psi(k_1T_s-\tilde{\tau},\tilde{\alpha})]}\right\}$ is computed in Appendix 4.B and reads

$$\mathrm{E}_{\tilde{\alpha}}\left\{e^{j[\psi(k_2T_s-\tilde{\tau},\tilde{\alpha})-\psi(k_1T_s-\tilde{\tau},\tilde{\alpha})]}\right\}$$
$$= \prod_{i=-\infty}^{\infty}\left[\frac{1}{M}\frac{\sin\{2\pi hMp[k_2T_s-\tilde{\tau}-iT,(k_2-k_1)T_s]\}}{\sin\{2\pi hp[k_2T_s-\tilde{\tau}-iT,(k_2-k_1)T_s]\}}\right] \tag{4.A.6}$$

where $p(t,\Delta T)$ is related to the phase response $q(t)$ of the modulator by

$$p(t, \Delta T) \triangleq q(t) - q(t - \Delta T) \tag{4.A.7}$$

and the generic factor

$$\frac{1}{M} \frac{\sin\{2\pi h M p[k_2 T_s - \tilde{\tau} - iT, (k_2 - k_1)T_s]\}}{\sin\{2\pi h p[k_2 T_s - \tilde{\tau} - iT, (k_2 - k_1)T_s]\}} \tag{4.A.8}$$

is replaced by unity when $p[k_2 T_s - \tilde{\tau} - iT, (k_2 - k_1)T_s] = 0$ (this is because (4.A.8) tends to unity as $p[k_2 T_s - \tilde{\tau} - iT, (k_2 - k_1)T_s]$ goes to zero).

Next, we note that the right-hand side in (4.A.6) is a periodic function of $\tilde{\tau}$ with period $T$ (in fact, the right-hand side does not change if $\tilde{\tau}$ is changed to $\tilde{\tau} + T$). Also, averaging $E_{\tilde{a}}\{|X|^2\}$ with respect to $\tilde{\tau}$ yields $\Gamma(\tilde{\nu})$, as indicated in (4.A.3). Thus, bearing in mind that $\tilde{\tau}$ is uniformly distributed between 0 and $T$, from (4.A.5)-(4.A.6) we get

$$\Gamma(\tilde{\nu}) = \sum_{k_1=0}^{NL_0-1} \sum_{k_2=0}^{NL_0-1} x(k_1 T_s) x^*(k_2 T_s) e^{-j2\pi\tilde{\nu}(k_1-k_2)T_s} H[(k_2 - k_1)T_s] \tag{4.A.9}$$

with

$$H[(k_2 - k_1)T_s] \triangleq \frac{1}{T} \int_0^T \prod_{i=-\infty}^{\infty} \left[ \frac{1}{M} \frac{\sin\{2\pi h M p[k_2 T_s - \tilde{\tau} - iT, (k_2 - k_1)T_s]\}}{\sin\{2\pi h p[k_2 T_s - \tilde{\tau} - iT, (k_2 - k_1)T_s]\}} \right] d\tilde{\tau}$$

$$= \frac{1}{T} \int_0^T \prod_{i=-\infty}^{\infty} \left[ \frac{1}{M} \frac{\sin\{2\pi h M p[t - iT, (k_2 - k_1)T_s]\}}{\sin\{2\pi h p[t - iT, (k_2 - k_1)T_s]\}} \right] dt \tag{4.A.10}$$

The second line follows from the periodicity of the integrand with respect to $\tilde{\tau}$.

Assuming that the symmetry condition

$$q(t) = \frac{1}{2} - q(LT - t) \tag{4.A.11}$$

is satisfied, it can be shown that $H(kT_s)$ is an even function. In fact, substituting (4.A.11) into (4.A.7) yields

$$p(t, -\Delta T) = -p(LT - t, \Delta T) \tag{4.A.12}$$

and inserting this result into (4.A.10) produces

# Carrier Frequency Recovery with CPM Modulations

$$H(-kT_s) = \frac{1}{T}\int_0^T \prod_{i=-\infty}^{\infty}\left[\frac{1}{M}\frac{\sin\{2\pi hMp(t-iT,-kT_s)\}}{\sin\{2\pi hp(t-iT,-kT_s)\}}\right]dt$$

$$= \frac{1}{T}\int_0^T \prod_{i=-\infty}^{\infty}\left[\frac{1}{M}\frac{\sin\{2\pi hMp(LT-t+iT,kT_s)\}}{\sin\{2\pi hp(LT-t+iT,kT_s)\}}\right]dt$$

$$= H(kT_s) \qquad (4.A.13)$$

which proves the claim.

## Appendix 4.B

In this appendix we compute the expectation $E_{\tilde{\alpha}}\{e^{j[\psi(t_2,\tilde{\alpha})-\psi(t_1,\tilde{\alpha})]}\}$, where

$$\psi(t,\tilde{\alpha}) = 2\pi h \sum_{i=-\infty}^{\infty} \tilde{\alpha}_i q(t-iT) \qquad (4.B.1)$$

Using the definition (4.A.7) we have

$$E_{\tilde{\alpha}}\{e^{j[\psi(t_2,\tilde{\alpha})-\psi(t_1,\tilde{\alpha})]}\} = E_{\tilde{\alpha}}\left\{\exp\left[j2\pi h\sum_{i=-\infty}^{\infty}\tilde{\alpha}_i p(t_2-iT,t_2-t_1)\right]\right\}$$

$$= E_{\tilde{\alpha}}\left\{\prod_{i=-\infty}^{\infty}\exp[j2\pi h\tilde{\alpha}_i p(t_2-iT,t_2-t_1)]\right\} \qquad (4.B.2)$$

As the symbols are independent, each factor in the second line can be averaged separately. Hence,

$$E_{\tilde{\alpha}}\{e^{j[\psi(t_2,\tilde{\alpha})-\psi(t_1,\tilde{\alpha})]}\} = \prod_{i=-\infty}^{\infty} E_{\tilde{\alpha}_i}\{e^{j2\pi h\tilde{\alpha}_i p(t_2-iT,t_2-t_1)}\} \qquad (4.B.3)$$

On the other hand, as $\alpha_i \in \{\pm 1, \pm 3, \ldots \pm(M-1)\}$, the factors on the right-hand side in (4.B.3) take the form

$$E_{\tilde{\alpha}_i}\{e^{j2\pi h\tilde{\alpha}_i p(t_2-iT,t_2-t_1)}\} = \frac{1}{M}\sum_{m=1,3,\ldots}^{M-1} e^{j2\pi hmp(t_2-iT,t_2-t_1)}$$

$$+ \frac{1}{M}\sum_{m=1,3,\ldots}^{M-1} e^{-j2\pi hmp(t_2-iT,t_2-t_1)} \qquad (4.B.4)$$

Also, it is easily proved that

$$\sum_{m=1,3,\ldots}^{M-1} e^{j2\pi hmp(t_2-iT,t_2-t_1)} = \sum_{n=0}^{M/2-1} e^{j2\pi h(2n+1)p(t_2-iT,t_2-t_1)}$$

$$= \frac{\sin[\pi hMp(t_2-iT,t_2-t_1)]}{\sin[2\pi hp(t_2-iT,t_2-t_1)]} e^{j\pi hMp(t_2-iT,t_2-t_1)} \quad (4.B.5)$$

Hence

$$E_{\tilde{\alpha}_i}\left\{e^{j2\pi h\tilde{\alpha}_i p(t_2-iT,t_2-t_1)}\right\} = \frac{1}{M}\frac{\sin[2\pi hMp(t_2-iT,t_2-t_1)]}{\sin[2\pi hp(t_2-iT,t_2-t_1)]} \quad (4.B.6)$$

and inserting this into (4.B.3) yields the final result

$$E_{\tilde{\alpha}}\left\{e^{j[\psi(t_2,\tilde{\alpha})-\psi(t_1,\tilde{\alpha})]}\right\} = \prod_{i=-\infty}^{\infty}\left[\frac{1}{M}\frac{\sin[2\pi hMp(t_2-iT,t_2-t_1)]}{\sin[2\pi hp(t_2-iT,t_2-t_1)]}\right] \quad (4.B.7)$$

In deriving (4.B.6) it has been implicitly assumed that $p(t_2-iT,t_2-t_1)$ is nonzero. When $p(t_2-iT,t_2-t_1)$ equals zero the exponential $e^{j2\pi h\tilde{\alpha}_i p(t_2-iT,t_2-t_1)}$ equals one and we have

$$E_{\tilde{\alpha}_i}\left\{e^{j2\pi h\tilde{\alpha}_i p(t_2-iT,t_2-t_1)}\right\} = 1 \quad (4.B.8)$$

This equation is consistent with (4.B.6) provided that the right-hand side of (4.B.6) is computed as the limit for $p \to 0$. It is concluded that equation (4.B.7) is valid in general provided that the factors with $p(t_2-iT,t_2-t_1)=0$ are set to unity.

When computing the right-hand side of (4.B.7) the question arises of recognizing the non-unity factors. In other words, we want to establish the set $\mathcal{J}$ such that $p(t_2-iT,t_2-t_1) \neq 0$ for $i \in \mathcal{J}$. From (4.B.2) it is clear that $p(t_2-iT,0)$ is zero anyway, which means that $\mathcal{J}$ is empty for $t_1=t_2$. The case $t_1 \neq t_2$ can be handled bearing in mind (4.B.2) and the very form of the phase response $q(t)$:

$$q(t) = \begin{cases} 0 & t \leq 0 \\ 1/2 & t \geq LT \end{cases} \quad (4.B.9)$$

In this way it is readily concluded that

$$\mathcal{J} = \left\{\frac{\min(t_1,t_2)}{T} - L < i < \frac{\max(t_1,t_2)}{T}\right\} \quad (4.B.10)$$

# Appendix 4.C

In this appendix we illustrate the steps leading to the block diagram in Figure 4.10. In doing so we set $D=0$ throughout in the formulas in Section 3.4.2, for this greatly simplifies the discussion without affecting the conclusions. The starting point is the derivative of $\Gamma(\tilde{v})$ (see equation (4.4.31) in the text)

$$\frac{d\Gamma(\tilde{v})}{d\tilde{v}} \approx 2\pi \sum_{k=0}^{NL_0-1} \text{Im}\{y(kT_s)w^*(kT_s)\} \tag{4.C.1}$$

where

$$y(kT_s) \triangleq x(kT_s)e^{-j2\pi k \tilde{v} T_s} \tag{4.C.2}$$

Also, $x(kT_s)$ is the output from the anti-aliasing filter and $w(kT_s)$ is the response to $y(kT_s)$ of the filter $h(kT_s)$:

$$w(kT_s) = y(kT_s) \otimes h(kT_s) \tag{4.C.3}$$

Our task is to compute that $\tilde{v}$ where $d\Gamma(\tilde{v})/d\tilde{v}$ vanishes. The basic idea is to exploit the sum of some consecutive terms in (4.C.1) as an error signal to drive $d\Gamma(\tilde{v})/d\tilde{v}$ toward zero.

For convenience we take $N$ terms (as many as the samples in a symbol period) and introduce a *symbol index n*, which is related to the *sample index k* by

$$n = \text{int}\left(\frac{k}{N}\right) \tag{4.C.4}$$

where $\text{int}(z)$ means "the largest integer not exceeding $z$." In essence, $n$ gives the symbol interval corresponding to the $k$-th sample. Then, the frequency estimates are updated according to

$$\hat{v}[(n+1)T] = \hat{v}(nT) + \gamma e(nT) \tag{4.C.5}$$

where $\gamma$ is a step-size parameter and $e(nT)$ is the error signal:

$$e(nT) = \sum_{k=nN}^{(n+1)N-1} \text{Im}\{y(kT_s)w^*(kT_s)\} \tag{4.C.6}$$

Let us concentrate on the computation of the samples $y(kT_s)$ appearing in (4.C.6) (once the $y(kT_s)$ are known, the corresponding $w(kT_s)$ are derived from (4.C.3)). To this end consider the piecewise varying function $\phi(t)$ such that

$$\frac{d\phi(t)}{dt} = 2\pi\hat{v}(nT) \quad \text{for} \quad nT \leq t < (n+1)T \tag{4.C.7}$$

Bearing in mind (4.C.2), it is recognized that the following approximation holds true:

$$y(kT_s) \approx x(kT_s)e^{-j\phi(kT_s)} \qquad (4.C.8)$$

Thus, to compute $y(kT_s)$ we need $\phi(kT_s)$ which, in turn, can be derived by integrating (4.C.7) over $kT_s \leq t \leq (k+1)T_s$:

$$\phi[(k+1)T_s] = \phi(kT_s) + 2\pi T_s \hat{v}(nT) \qquad \text{mod } 2\pi \qquad (4.C.9)$$

To summarize, suppose that $\hat{v}(nT)$ is known. Then, equation (4.C.9) gives $\{\phi(kT_s)\}$ and (4.C.8) gives $\{y(kT_s)\}$. Next, equation (4.C.3) yields $\{w(kT_s)\}$ and substituting into (4.C.6) produces $e(nT)$. Finally, the new estimate $\hat{v}[(n+1)T]$ is derived from (4.C.5). These steps are illustrated in the block diagram in Figure 4.10.

## References

[1] J.B.Anderson and C-E.Sundberg, Advances in Constant Envelope Coded Modulation, *IEEE Communications Magazine*, **29**, No 12, 36-45, Dec. 1991.
[2] J.B.Anderson, T.Aulin and C.-E.Sundberg, *Digital Phase Modulation*, New York: Plenum, 1986.
[3] P.A.Laurent, Exact and Approximate Construction of Digital Phase Modulations by Superposition of Amplitude Modulated Pulses, *IEEE Trans. Commun.*, **COM-34**, 150-160, Feb. 1986.
[4] M.Luise and R.Reggiannini, An Efficient Frequency Recovery Scheme for GSM Receivers, Conf. Rec. Commun. Theory Mini-Conference, Orlando, FL, Dec. 6-9, 1992, pp. 36-40.
[5] M.Luise and R.Reggiannini, Carrier Frequency Recovery in All-Digital Modems for Burst-Mode Transmissions, *IEEE Trans. Commun.*, **COM-43**, 1169-1178, March 1995.
[6] A.N.D'Andrea, A.Ginesi and U.Mengali, Frequency Detectors for CPM Signals, *IEEE Trans. Commun.*, **COM-43**, 1829-1837, March 1995.
[7] A.N.D'Andrea and U.Mengali, Noise Performance of Two Frequency-Error Detectors Derived from Maximum Likelihood Estimation Methods, *IEEE Trans. Commun.*, **COM-42**, 793-802, Feb./March/April 1994.
[8] T.Alberty and V.Hespelt, A New Pattern Jitter Free Frequency Error Detector, *IEEE Trans. Commun.*, **COM-37**, 159-163, Feb. 1989.
[9] A.N.D'Andrea and U.Mengali, Design of Quadricorrelators for Automatic Frequency Control Systems, *IEEE Trans. Commun.*, **COM-41**, 988-997, June 1993.
[10] S.Bravo, Carrier Frequency Acquisition in CPM modulations, Laurea Thesis, Department of Information Engineering, University of Pisa, June 1995.
[11] J.M.Wozencraft and I.M.Jacobs, *Principles of Communication Engineering*, New York: John Wiley & Sons, 1965.
[12] A.N.D'Andrea, A.Ginesi and U.Mengali, Digital Carrier Frequency Estimation for Multilevel CPM Signals, Conf. Rec. ICC'95, Seattle, WA, June 18-22, 1995, pp. 1041-1045.
[13] R.Mehlan, Yong-En Chen and H.Meyr, A Fully Digital Feedforward MSK Demodulator with Joint Frequency Offset and Symbol Timing Estimation for Burst Mode Mobile Radio, *IEEE Trans. Veh. Tech.*, **VT-42**, 434-443, Nov. 1993.

# 5

# Carrier Phase Recovery with Linear Modulations

## 5.1. Introduction

In this chapter we investigate algorithms for carrier phase estimation with linear modulations. As with frequency recovery, ML estimation methods will play a central role in our study. We shall see that various approximations to the ML formulation are possible, leading to different estimation methods. Thus, a rather disparate set of synchronization schemes is anticipated. This is also a consequence of the many scenarios that can be thought of, depending on the specific modulation format and the availability of data/clock information. In this regard the following categories may be envisaged:

(*i*) *Modulation format*:
- Modulation may be either offset or non-offset.

(*ii*) *Additional knowledge*:
- Clock information may be available or not.
- Information symbols may be known or not. When they are, they may come either from a known preamble (*data-aided* schemes) or from the detector output (*decision-directed* schemes)

(*iii*) *Estimator topology*:
- Estimators may be either open loop or closed loop.

Another distinction arises from the presence of carrier frequency offsets. For the sake of simplicity phase estimation is usually approached assuming that frequency recovery has already been accomplished. This is in keeping with the fact that most phase estimators can cope with moderate residual frequency errors. Frequency errors are not always that moderate, however. When this happens, it is necessary to endow the phase synchronizer with extra frequency acquisition capabilities.

The chapter is organized as follows. In the next few sections we address data-aided and clock-aided phase estimation. Next, we gradually loosen these assumptions and eventually discuss phase recovery with no data and no clock information.

## 5.2. Clock-Aided and Data-Aided Phase Recovery

### 5.2.1. ML Estimation with Non-Offset Formats

Start with the complex envelope of the received waveform

$$r(t) = s(t) + w(t) \quad (5.2.1)$$

where

$$s(t) = e^{j(2\pi vt + \theta)} \sum_i c_i g(t - iT - \tau) \quad (5.2.2)$$

and $w(t)$ is thermal noise. Its real and imaginary components are independent and each have a power spectral density $N_0$. The parameter $v$ represents the frequency offset, $\theta$ is the carrier phase we want to estimate, $\tau$ is the timing phase, $\{c_i\}$ are information symbols, $T$ is the symbol period and $g(t)$ is the signaling pulse shape.

The phase $\theta$ is an unknown constant, taking values in the range $\pm \pi$. All the other parameters, $v$, $\tau$ and $\{c_i\}$, are assumed to be perfectly known to the receiver. As mentioned earlier, knowledge of the data symbols may come from a known preamble. Carrier frequency and symbol epoch may either be estimated in advance and independently of the carrier phase or, as happens in some burst mode transmissions, can be accurately tracked between bursts.

To estimate $\theta$ with ML methods we need the likelihood function $\Lambda(r|\tilde{\theta})$. This has been derived in Chapter 2 and is expressed by

$$\Lambda(r|\tilde{\theta}) = \exp\left\{\frac{1}{N_0} \int_0^{T_0} \text{Re}\{r(t)\tilde{s}^*(t)\}dt - \frac{1}{2N_0} \int_0^{T_0} |\tilde{s}(t)|^2 dt\right\} \quad (5.2.3)$$

where $0 \leq t \leq T_0$ is the observation interval and $\tilde{s}(t)$ is the trial signal

## Carrier Phase Recovery with Linear Modulations

$$\tilde{s}(t) \triangleq e^{j(2\pi vt + \tilde{\theta})} \sum_i c_i g(t - iT - \tau) \tag{5.2.4}$$

Note that $\tilde{\theta}$ is the only parameter bearing a "tilde". This reflects the fact that the other parameters are all known. It is also worth observing that $|\tilde{s}(t)|$ is independent of $\tilde{\theta}$. Hence, taking the logarithm of (5.2.3) yields (within some constants)

$$\ln \Lambda(r|\tilde{\theta}) = \text{Re}\left\{ \int_0^{T_0} r(t)\tilde{s}^*(t)\,dt \right\} \tag{5.2.5}$$

Using (5.2.4) we get

$$\int_0^{T_0} r(t)\tilde{s}^*(t)\,dt = e^{-j\tilde{\theta}} \sum_i c_i^* \int_0^{T_0} \left[ r(t) e^{-j2\pi vt} \right] g(t - iT - \tau)\,dt \tag{5.2.6}$$

Hence, reasoning as in Section 3.2.1, it is easily found that the integral in (5.2.5) may be expressed as

$$\int_0^{T_0} r(t)\tilde{s}^*(t)\,dt \approx e^{-j\tilde{\theta}} \sum_{k=0}^{L_0-1} c_k^* x(k) \tag{5.2.7}$$

where $L_0 \triangleq T_0/T$ is the length of the observation interval in symbol periods and $x(k)$ represents the sample at $t = kT + \tau$ of the convolution

$$x(t) \triangleq \left[ r(t) e^{-j2\pi vt} \right] \otimes g(-t) \tag{5.2.8}$$

In summary, collecting (5.2.5) and (5.2.7) yields

$$\ln \Lambda(r|\tilde{\theta}) = \text{Re}\left\{ e^{-j\tilde{\theta}} \sum_{k=0}^{L_0-1} c_k^* x(k) \right\} \tag{5.2.9}$$

and the maximum of $\Lambda(r|\tilde{\theta})$ is achieved for

$$\hat{\theta} = \arg\left\{ \sum_{k=0}^{L_0-1} c_k^* x(k) \right\} \tag{5.2.10}$$

The block diagram of the ML phase estimator is illustrated in Figure 5.1.

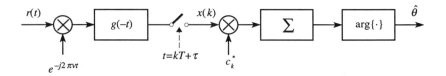

Figure 5.1. Block diagram of the ML phase estimator.

### 5.2.2. Performance with Non-Offset Formats

We maintain that algorithm (5.2.10) achieves the Cramer-Rao bound $CRB(\theta)$ when the ratio $E_s/N_0$ is sufficiently large and the convolution $g(t) \otimes g(-t) \triangleq h(t)$ is Nyquist, i.e.,

$$h(kT) = \begin{cases} 1 & k = 0 \\ 0 & k \neq 0 \end{cases} \quad (5.2.11)$$

To prove our claim recall from Chapter 2 that, if there are no unwanted parameters (as happens in the present case), the $CRB(\theta)$ coincides with the modified bound $MCRB(\theta)$. So, we only need to show that the error variance of the estimator (5.2.10) equals the modified bound

$$MCRB(\theta) = \frac{1}{2L_0} \frac{1}{E_s/N_0} \quad (5.2.12)$$

To proceed, consider the matched-filter output

$$x(t) = e^{j\theta} \sum_i c_i h(t - iT - \tau) + n(t) \quad (5.2.13)$$

with

$$n(t) \triangleq \left[ w(t) e^{-j2\pi vt} \right] \otimes g(-t) \quad (5.2.14)$$

It is easily checked that $n(t)$ has independent real and imaginary components, each with variance $N_0$. Also, the signal energy $E_s$ equals $C_2/2$, where $C_2$ is the expectation of $|c_i|^2$ (see Appendix 2.A). Thus, we have

$$N_0 = \frac{C_2}{2E_s/N_0} \quad (5.2.15)$$

Now observe that the samples of $x(t)$ at $kT+\tau$ are given by

$$x(k) = e^{j\theta} c_k + n(k) \tag{5.2.16}$$

from which, multiplying both sides by $c_k^*$ yields

$$c_k^* x(k) = e^{j\theta}\left[|c_k|^2 + c_k^* n'(k)\right] \tag{5.2.17}$$

with $n'(k) \triangleq n(k)e^{-j\theta}$. Note that $n'(k)$ has the same statistics as $n(k)$ and, in particular, its real and imaginary components are zero mean and independent random variables of variance $N_0$. Substituting into (5.2.10) and rearranging yields

$$\hat{\theta} = \arg\{e^{j\theta}(1 + N_R + jN_I)\} \tag{5.2.18}$$

with

$$N_R + jN_I \triangleq \frac{\sum_{k=0}^{L_0-1} c_k^* n'(k)}{\sum_{k=0}^{L_0-1} |c_k|^2} \tag{5.2.19}$$

From this equation it is easily seen that $N_R$ and $N_I$ both have zero mean. Thus, assuming that they are small compared with unity, (5.2.18) reduces to

$$\hat{\theta} \approx \theta + N_I \tag{5.2.20}$$

which says that the estimate of $\theta$ is unbiased and the estimation variance is given by

$$\text{Var}\{\hat{\theta} - \theta\} = \text{E}\{N_I^2\} \tag{5.2.21}$$

Next, write

$$\text{Var}\{\hat{\theta} - \theta\} = \text{E}\{\text{E}\{N_I^2 | c\}\} \tag{5.2.22}$$

meaning that we first take the expectation with respect to noise (while keeping $c \triangleq (c_0, c_1, \ldots, c_{L_0-1})$ fixed) and then we average the result over the symbols. Solving (5.2.19) for $N_I$ and substituting into (5.2.22) it is found, after some algebra, that

$$\text{Var}\{\hat{\theta} - \theta\} = \frac{N_0 C_2}{L_0} \text{E}\left\{\frac{1}{\overline{C_2^2}}\right\} \tag{5.2.23}$$

where $C_2$ is the expectation of $|c_k|^2$ and $\overline{C}_2$ is the aritmetic mean

$$\overline{C}_2 \triangleq \frac{1}{L_0} \sum_{k=0}^{L_0-1} |c_k|^2 \qquad (5.2.24)$$

Bearing in mind (5.2.15), equation (5.2.23) may be put in the alternative form

$$\operatorname{Var}\{\hat{\theta}-\theta\} = \frac{1}{2L_0} \frac{C_2^2}{E_s/N_0} \operatorname{E}\left\{\frac{1}{\overline{C}_2^2}\right\} \qquad (5.2.25)$$

Now, as $L_0$ increases, the mean $\overline{C}_2$ approaches $C_2$ and (5.2.25) becomes

$$\operatorname{Var}\{\hat{\theta}-\theta\} = \frac{1}{2L_0} \frac{1}{E_s/N_0} \qquad (5.2.26)$$

which shows that the Cramer-Rao bound is achieved. For example, an error standard deviation of 4.8° is obtained with only $E_s/N_0$ =6 dB and $L_0$=18.

### 5.2.3. ML Estimation with Offset Formats

We now address ML phase estimation with offset signals. The signal model is

$$s(t) = e^{j(2\pi\nu t + \theta)} \left\{ \sum_i a_i g(t-iT-\tau) + j \sum_i b_i g(t-iT-T/2-\tau) \right\} \qquad (5.2.27)$$

where $a_i$ and $b_i$ are real-valued information symbols. In particular, with OQPSK modulation they take the values ±1. As the derivation of the ML estimator follows the same lines described earlier, we limit ourselves to highlighting the major steps.

Paralleling the passages from (5.2.3) to (5.2.9) produces

$$\ln \Lambda(r|\tilde{\theta}) = \operatorname{Re}\left\{ e^{-j\tilde{\theta}} \left[ \sum_{k=0}^{L_0-1} a_k x(k) - j \sum_{k=0}^{L_0-1} b_k x(k+1/2) \right] \right\} \qquad (5.2.28)$$

where $x(t)$ is the matched-filter output and $x(k)$ and $x(k+1/2)$ are its samples taken at $t=kT+\tau$ and $t=kT+T/2+\tau$, respectively. Maximizing (5.2.28) as a function of $\tilde{\theta}$ gives the desired estimate

# Carrier Phase Recovery with Linear Modulations

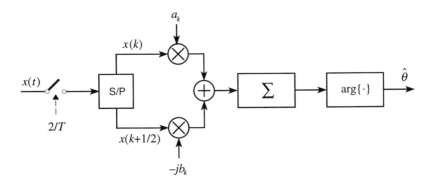

Figure 5.2. Block diagram of the ML estimator.

$$\hat{\theta} = \arg\left\{ \sum_{k=0}^{L_0-1} a_k x(k) - j \sum_{k=0}^{L_0-1} b_k x(k+1/2) \right\} \quad (5.2.29)$$

Figure 5.2 illustrates the block diagram of the estimator. As is seen, the samples from the matched filter are now taken at the rate $2/T$ and a serial-to-parallel converter (S/P) separates the sequences $\{x(k)\}$ and $\{x(k+1/2)\}$.

### 5.2.4. Performance with Offset Formats

The performance of estimator (5.2.29) is now assessed with the same methods adopted earlier. In doing so we still assume that the Nyquist condition holds and the modulation is uncoded.

The matched-filter output has the form

$$x(t) = e^{j\theta}\left\{ \sum_i a_i h(t-iT-\tau) + j \sum_i b_i h(t-iT-T/2-\tau) \right\} + n(t) \quad (5.2.30)$$

Again, real and imaginary components of the noise have zero mean and variance $\sigma_n^2 = N_0$. Thus, as the signal energy is unity (see Appendix 2.A), we have

$$N_0 = \frac{1}{E_s/N_0} \quad (5.2.31)$$

and, correspondingly,

$$\sigma_n^2 = \frac{1}{E_s/N_0} \qquad (5.2.32)$$

Sampling $x(t)$ at $kT+\tau$ and $kT+T/2+\tau$ and bearing in mind that $h(t)$ is Nyquist yields

$$x(k) = e^{j\theta}\left[a_k + j\sum_i b_i h(k-i-1/2)\right] + n(k) \qquad (5.2.33)$$

$$x(k+1/2) = e^{j\theta}\left[jb_k + \sum_i a_i h(k-i+1/2)\right] + n(k+1/2) \qquad (5.2.34)$$

Then, substituting into (5.2.29) and rearranging produces

$$\hat{\theta} = \arg\{e^{j\theta}(1 + jS + N_R + jN_I)\} \qquad (5.2.35)$$

with

$$S \triangleq \frac{1}{2L_0} \sum_{k=0}^{L_0-1} \left[\sum_i a_k b_i h(k-i-1/2) - \sum_i b_k a_i h(k-i+1/2)\right] \qquad (5.2.36)$$

$$N_R + jN_I \triangleq \frac{1}{2L_0} \sum_{k=0}^{L_0-1} [n'(k) + n''(k+1/2)] \qquad (5.2.37)$$

$$n'(k) \triangleq n(k)a_k e^{-j\theta} \qquad (5.2.38)$$

$$n''(k+1/2) \triangleq -jn(k+1/2)b_k e^{-j\theta} \qquad (5.2.39)$$

It may be checked that $N_R$ and $N_I$ have zero mean and the same variance

$$\sigma_{N_R}^2 = \sigma_{N_I}^2 = \frac{1}{2L_0}\frac{1}{E_s/N_0} \qquad (5.2.40)$$

Also, straightforward manipulations show that the random variable $S$ in (5.2.36) has zero mean and variance

$$\sigma_S^2 = \frac{C}{2L_0} \qquad (5.2.41)$$

with

$$C \triangleq \sum_{k=-\infty}^{\infty} h^2(k-1/2) - \frac{1}{L_0} \sum_{k_1=1}^{L_0} \sum_{k_2=1}^{L_0} h(k_2 - k_1 + 1/2) h(k_1 - k_2 - 1/2) \quad (5.2.42)$$

Next, assume that $E_s/N_0$ and $L_0$ are sufficiently large so that $S$, $N_R$ and $N_I$ are all small compared to unity. Then, (5.2.35) reduces to

$$\hat{\theta} \approx \theta + S + N_I \quad (5.2.43)$$

from which the variance of the estimation errors is seen to be $\sigma_S^2 + \sigma_{N_I}^2$, i.e.,

$$\mathrm{Var}\{\hat{\theta} - \theta\} = \frac{1}{2L_0} \frac{1}{E_s/N_0} + \frac{C}{2L_0} \quad (5.2.44)$$

This equation indicates that the estimator (5.2.29) does not achieve the Cramer-Rao bound. Note that the second term $C/(2L_0)$ in (5.2.44) is independent of the thermal noise level and, in fact, it is contributed by interactions between signal components (self noise). Figure 5.3 compares qualitatively (5.2.44) with the Cramer-Rao bound.

In practice (5.2.44) may be quite close to the bound. For example, assume $L_0=18$ and suppose that the Fourier transform of $h(t)$ has a raised-cosine-rolloff shape with $\alpha=0.5$. From (5.2.42) it is found that $C=0.0225$. Thus, for $E_s/N_0 =$ 6 dB equation (5.2.44) yields a phase error standard deviation of about 5° (instead of 4.8°, which corresponds to the bound).

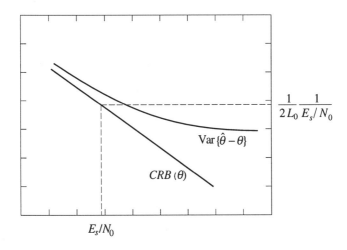

Figure 5.3. Comparison with CRB.

### 5.2.5. Degradations Due to Frequency Errors

To assess the effects of frequency errors on phase recovery it is useful to bear in mind that: (*i*) $\theta$ represents the phase of the demodulated carrier $e^{j(2\pi vt+\theta)}$ *at the time origin*; (*ii*) in the previous discussion the time origin has been chosen as the beginning of the observation interval. To investigate phase estimation in more general terms it is expedient to shift the time origin to a point internal to the observation interval. In particular we choose the center of the observation interval at a generic time $t=mT$, as indicated in Figure 5.4. This will allow us to assess the dependence of the phase estimates on the parameter $mT$.

For the sake of simplicity let us take an observation interval with an odd number of symbol intervals. Then, proceeding as in Section 5.2.1, the ML estimate of $\theta$ with non-offset modulation is found to be

$$\hat{\theta} = \arg\left\{\frac{1}{L_0} \sum_{k=m-(L_0-1)/2}^{m+(L_0-1)/2} c_k^* x(k)\right\} \quad (5.2.45)$$

where $x(t)$ is still as defined in (5.2.8). Note that an immaterial factor $1/L_0$ has been inserted in (5.2.45) for convenience. Our aim is to assess the performance of this estimator in the presence of a frequency error and, in particular, to compute the first- and second-order moments of $\hat{\theta}$. In doing so we make the following simplifying assumptions:

(*i*) the modulation is PSK;

(*ii*) the difference $f_d$ between $v$ and its estimate is much less than $1/T$;

(*iii*) the Nyquist condition is satisfied.

To account for frequency errors, we suppose that the demodulation operation indicated in Figure 5.1 is performed with a reference frequency $\hat{v}$ other than $v$. Then the filter output $x(t)$ becomes

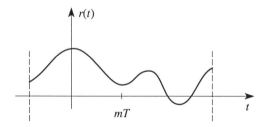

Figure 5.4. Observation interval centered around $t=mT$.

$$x(t) = \left[r(t)e^{-j2\pi\hat{v}t}\right] \otimes g(-t) \qquad (5.2.46)$$

Substituting $r(t)=s(t)+w(t)$ into (5.2.46) and using (5.2.2) yields (instead of (5.2.13))

$$x(t) = e^{j\theta}\sum_i c_i\left[e^{j2\pi f_d t}g(t-iT-\tau)\right] \otimes g(-t) + n(t) \qquad (5.2.47)$$

As $g(t-iT-\tau)$ takes significant values only over a few symbol periods around $iT+\tau$ and $|f_d T|$ is much less than unity (by assumption), the exponential $e^{j2\pi f_d t}$ in (5.2.47) can be approximated as $e^{j2\pi i f_d T}$. Thus, bearing in mind the relation $g(t) \otimes g(-t) = h(t)$, we have

$$x(t) = e^{j\theta}\sum_i c_i e^{j2\pi i f_d T} h(t-iT-\tau) + n(t) \qquad (5.2.48)$$

Sampling at $kT+\tau$ produces

$$x(k) = c_k e^{j\theta} e^{j2\pi k f_d T} + n(k) \qquad (5.2.49)$$

from which we get (recall that $|c_k|^2 = 1$)

$$c_k^* x(k) = e^{j\theta}\left[e^{j2\pi k f_d T} + n'(k)\right] \qquad (5.2.50)$$

with $n'(k) \triangleq n(k) c_k^* e^{-j\theta}$. Finally, substituting into (5.2.45) yields

$$\hat{\theta} = \arg\left\{e^{j\theta}\left[\frac{1}{L_0}\sum_{k=m-(L_0-1)/2}^{m+(L_0-1)/2} e^{j2\pi k f_d T} + N_R + jN_I\right]\right\} \qquad (5.2.51)$$

where the complex number $N_R+jN_I$ is defined as

$$N_R + jN_I \triangleq \frac{1}{L_0}\sum_{k=m-(L_0-1)/2}^{m+(L_0-1)/2} n'(k) \qquad (5.2.52)$$

Simple manipulations show that the summation in (5.2.51) may be put in the form

$$\frac{1}{L_0}\sum_{k=m-(L_0-1)/2}^{m+(L_0-1)/2} e^{j2\pi k f_d T} = \rho(f_d T)e^{j2\pi m f_d T} \qquad (5.2.53)$$

with

$$\rho(f_d T) \triangleq \frac{\sin(\pi L_0 f_d T)}{L_0 \sin(\pi f_d T)} \tag{5.2.54}$$

Note that $\rho(0)=1$. Furthermore, in a small interval around the origin, $\rho(f_d T)$ decreases as $f_d T$ departs from zero.

Collecting (5.2.51) and (5.2.53) and rearranging yields

$$\hat{\theta} = \arg\left\{e^{j(2\pi m f_d T + \theta)}(1 + V_R + jV_I)\right\} \tag{5.2.55}$$

where $V_R$ and $V_I$ are related to $N_R$ and $N_I$ by

$$\begin{cases} V_R \triangleq N_R \cdot \dfrac{e^{-j2\pi m f_d T}}{\rho(f_d T)} \\ V_I \triangleq N_I \cdot \dfrac{e^{-j2\pi m f_d T}}{\rho(f_d T)} \end{cases} \tag{5.2.56}$$

It is easily checked that $V_R$ and $V_I$ have zero mean and variance

$$\sigma_{V_R}^2 = \sigma_{V_I}^2 = \frac{1}{2L_0 \rho^2(f_d T)} \frac{1}{E_s/N_0} \tag{5.2.57}$$

The performance of the estimator is readily assessed at high signal-to-noise ratio. In these conditions in fact $V_R$ and $V_I$ are small compared with unity, and (5.2.55) reduces to

$$\hat{\theta} \approx \theta + 2\pi m f_d T + V_I \tag{5.2.58}$$

from which it is clear that the average estimate is $\theta + 2\pi m f_d T$ and coincides with the phase of $e^{j(2\pi f_d t + \theta)}$ *at the center* of the observation interval. Thus, if this same estimate is used for demodulation over the entire interval, a position-dependent bias is incurred. The estimation variance is given by

$$\text{Var}\{\hat{\theta} - \theta\} = \frac{1}{2L_0 \rho^2(f_d T)} \frac{1}{E_s/N_0} \tag{5.2.59}$$

and is larger than the $CRB(\theta)$ for $f_d \neq 0$ (since $\rho(f_d T)<1$).

As an example consider again the case discussed in Section 5.2.2, i.e., non-offset modulation with $E_s/N_0 = 6$ dB and $L_0=18$. We have seen that, for zero frequency error, the estimates have a mean $\theta$ and a standard deviation of $4.8°$. With $f_d=0.001/T$, the mean increases by $3.2°$ while the standard deviation remains essentially the same.

## 5.3. Decision-Directed Phase Recovery with Non-Offset Modulation

### 5.3.1. Feedback Structures

From the preceding discussion it appears that ML data-aided phase estimation methods lead to *open-loop* (*feedforward*) schemes. As is now explained, *closed-loop* (*feedback*) structures are unavoidable when detector decisions are exploited in place of true data.

Assume that carrier frequency and timing are ideal. To get an estimate of $\theta$ it would seem sufficient to replace the known symbols by their estimates $\hat{c}_k$ in (5.2.10). This produces

$$\hat{\theta} = \arg\left\{\sum_{k=0}^{L_0-1} \hat{c}_k^* x(k)\right\} \tag{5.3.1}$$

Unfortunately a closer look at the problem reveals that estimator (5.3.1) does not work. In fact, as indicated in (5.2.16), the signal constellation at the detector input is rotated by an angle $\theta$ from its correct position (see Figure 5.5) and, in consequence, the quality of the detector decisions strongly depends on the amount of rotation. As an example consider QPSK modulation and $\theta=\pi/4$. Also, assume Nyquist pulses so that

$$x(k) = c_k e^{j\pi/4} + n(k) \tag{5.3.2}$$

In these conditions 50% of the decisions are wrong on average. In fact, as $c_k e^{j\pi/4}$ lies on the borderline between the decision zones for $c_k$ and $c_k e^{j\pi/2}$, the decision $\hat{c}_k$ will be either $c_k$ or $c_k e^{j\pi/2}$ with the same probability. Formally, $\hat{c}_k$

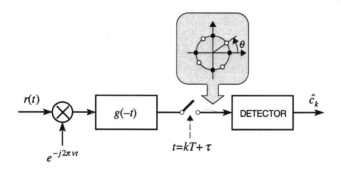

Figure 5.5. Phase rotation of the signal constellation.

may be modelled as

$$\hat{c}_k = c_k u(k) + c_k e^{j\pi/2}[1 - u(k)] \tag{5.3.3}$$

where $u(k)$ is a random variable taking values 1 or 0 with probability 1/2. Substituting into the sum in (5.3.1) yields (recall that $|c_k|^2 = 1$)

$$\sum_{k=0}^{L_0-1} \hat{c}_k^* x(k) = e^{j\pi/4} \sum_{k=0}^{L_0-1} u(k) + e^{-j\pi/4} \sum_{k=0}^{L_0-1} [1 - u(k)] \tag{5.3.4}$$

where the noise term $n(k)$ in (5.3.2) has been neglected for simplicity. As the expected value of (5.3.4) equals $L_0(e^{j\pi/4}/2 + e^{-j\pi/4}/2) = L_0/\sqrt{2}$, the mean value of the argument of (5.3.4) is zero and not $\pi/4$, as should be.

Clearly, the stumbling block is the uncompensated phase rotation at the detector input and the only way to sidestep the obstacle is to compensate for the rotation before entering the detector. In the sequel we investigate two approaches to this problem.

### 5.3.2. First Approach

The first approach is based on a proper adjustment of (5.3.1). At any time $kT+\tau$ a phase estimate $\hat{\theta}(k)$ is computed on the basis of the last $L_0$ symbols as follows:

$$\hat{\theta}(k) = \arg\left\{ \sum_{l=k-L_0}^{k-1} \hat{c}_l^* x(l) \right\} \tag{5.3.5}$$

Next, $x(k)$ is counter-rotated by $\hat{\theta}(k)$ and fed to the detector, as indicated in Figure 5.6. In this manner the residual rotation at the detector input is reduced to $\theta - \hat{\theta}(k)$, which is hopefully less than $\theta$. Note that, as $\hat{c}_k$ depends on $\theta - \hat{\theta}(k)$, the estimator (5.3.5) has been turned into a feedback scheme. It is the feedback mechanism that will eventually drive the system toward a steady-state regime, as is now intuitively explained.

Suppose that initially $\hat{\theta}(k)$ and $\theta$ are far from each other. Then, most detector decisions will be incorrect and $\hat{\theta}(k)$ will fluctuate in a random manner. Note that, with large values of $L_0$, $\hat{\theta}(k)$ is slowly varying in time and, in fact, it will be almost constant over several signaling intervals. Sooner or later, however, $\hat{\theta}(k)$ will approach $\theta$ and the detector decisions will improve. Then, better phase estimates will be produced and this will further enhance the decision quality. Eventually, a steady-state regime will be reached wherein $\hat{\theta}(k)$ will steadily fluctuate around $\theta$.

Unfortunately, such a regime is not unique and, in fact, $\hat{\theta}(k)$ may settle around $\theta$ plus multiples of $\alpha$, the symmetry angle of the signal constellation. In particular, $\alpha$ equals $2\pi/M$ with $M$-ary PSK and $\pi/2$ with QAM modulation. This estimation ambiguity occurs with any NDA or DD estimation algorithm and is a consequence of the rotational symmetry of the signaling scheme. A possible explanation is as follows.

Return to (5.2.16) and assume for simplicity that the noise is negligible and the modulation is QPSK. Then, the matched-filter output becomes

$$x(k) = c_k e^{j\theta} \tag{5.3.6}$$

where $c_k$ belongs to the alphabet $\{1, e^{j\pi/2}, e^{j\pi}, e^{j3\pi/2}\}$. We maintain that $\hat{\theta}(k) = \theta + m\pi/2$ ($m = 0, 1, 2, 3$) are all equilibrium points for the loop in Figure 5.6. To prove our claim let us temporarily disconnect the phase estimator from the multiplier and drive the latter with some external voltage $e^{-j\bar{\theta}}$, with $\bar{\theta} = \theta + m\pi/2$ ($m = 0, 1, 2, 3$). As we shall see, in these conditions the estimate (5.3.5) equals $\bar{\theta}$. Therefore, the loop operation will not be affected by re-establishing the connection between the phase estimator and the multiplier and, in consequence, $\hat{\theta}(k)$ will keep its previous value (admittedly, the argument is incomplete as it does not prove that $\hat{\theta}(k) = \theta + m\pi/2$ is a *stable point*. Indeed, stability must be assessed through other reasoning which is developed later).

To proceed, consider the input to the decision device. Since $\bar{\theta} = \theta + m\pi/2$, we have

$$y(k) = x(k) e^{-j\theta} e^{-jm\pi/2} \tag{5.3.7}$$

or, as a consequence of (5.3.6),

$$y(k) = c_k e^{-jm\pi/2} \tag{5.3.8}$$

Correspondingly, the detector produces the decisions

$$\hat{c}_k = c_k e^{-jm\pi/2} \tag{5.3.9}$$

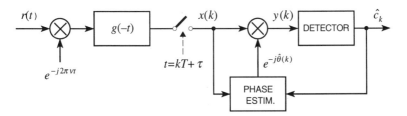

Figure 5.6. Phase recovery scheme.

so that

$$\sum_{l=k-L_0}^{k-1} \hat{c}_l^* x(l) = L_0 e^{j(m\pi/2 + \theta)} \qquad (5.3.10)$$

which proves that the right-hand side of (5.3.5) equals $\theta + m\pi/2$.

Another way to look at the ambiguity problem is to recognize that the final task of the receiver is to estimate the data sequence $\{c_k\}$. The difficulty in doing so is that there are several combinations of $\theta$ values and symbol sequences that correspond to the same observation $\{x(k)\}$. This is easily seen by rewriting (5.3.6) as

$$x(k) = c_k' e^{j\theta'} \qquad (5.3.11)$$

with

$$\theta' \triangleq \theta + m\pi/2 \qquad (5.3.12)$$

$$c_k' \triangleq c_k e^{-jm\pi/2} \qquad (5.3.13)$$

As $c_k'$ is an element of the symbol alphabet, any $\{c_k'\}$ is a legitimate sequence and there is no way to distinguish among different sequences unless additional information is provided within the message (for example, by inserting a unique word).

It should be pointed out that the preceding discussion does not apply to coded modulations. In the presence of coding, in fact, a rotated sequence may not be a legitimate sequence and the above argument fails. This indicates that phase ambiguities with coded systems are fewer than with uncoded systems.

An important issue is how to cope with phase ambiguities in the detection process. A common method with uncoded modulation (especially with continuous transmissions) is to resort to differential encoding at the transmitter and *differentially coherent* decoding at the receiver. Consider QPSK signaling, for simplicity, and let $\{\eta_k\}$ be the information sequence, with $\eta_k \in \{0, \pi/2, \pi, 3\pi/2\}$. Mapping $\{\eta_k\}$ into channel symbols $\{e^{j\eta_k}\}$ and demodulating the received sequence with an ambiguous phase reference would result in decisions of the type $\hat{c}_k = e^{j(\eta_k - m\pi/2)}$ from which the recovery of $\{\eta_k\}$ would be impossible. Suppose instead that $\{\eta_k\}$ is first differentially encoded into a new sequence $\{\delta_k\}$ according to the rule

$$\delta_k = \delta_{k-1} + \eta_k \quad \mod 2\pi \qquad (5.3.14)$$

or, equivalently,

$$\eta_k = \delta_k - \delta_{k-1} \quad \text{mod } 2\pi \tag{5.3.15}$$

Next, the phases $\{\delta_k\}$ are mapped into channel symbols $c_k = e^{j\delta_k}$. In these conditions the detector decisions are $\hat{c}_k = e^{j(\delta_k - m\pi/2)}$. Hence, computing the arguments $\arg\{\hat{c}_k \hat{c}_{k-1}^*\}$ yields

$$\begin{aligned}
\arg\{\hat{c}_k \hat{c}_{k-1}^*\} &= (\delta_k - m\pi/2) - (\delta_{k-1} - m\pi/2) \quad \text{mod } 2\pi \\
&= \delta_k - \delta_{k-1} \quad \text{mod } 2\pi \\
&= \eta_k
\end{aligned} \tag{5.3.16}$$

which says that the information can be recovered in spite of phase ambiguities.

With coded modulations the problem is often solved by resorting to rotationally invariant codes. With such codes phase rotations by multiples of the constellation's symmetry angle do not affect the detection process as they still produce legitimate sequences with the same information. In some applications the coding scheme is not rotationally invariant, however, and the ambiguity problem is approached by means of special algorithms [1]-[4] that can tell whether the input sequence is correctly rotated.

The phase recovery system in Figure 5.6 has been discussed in [5]-[9]. Analysis and computer simulations reported in [8]-[9] indicate that, with QPSK and negligible carrier frequency offsets, its tracking performance is close to the modified Cramer-Rao bound at SNR values of practical interest. Unfortunately, significant degradations take place in the presence of frequency errors and there seems to be no way to adjust the scheme so as to sidestep this drawback. By contrast, the problem can be easily solved with the second approach we discuss next.

### 5.3.3. Second Approach

The basic idea is a recursive method to compute the zero of the derivative of the log-likelihood function (5.2.9). Taking the derivative of (5.2.9) with respect to $\tilde{\theta}$ and rearranging yields

$$\frac{d}{d\tilde{\theta}} \ln \Lambda(r|\tilde{\theta}) = \sum_{k=0}^{L_0-1} \text{Im}\left\{c_k^* x(k) e^{-j\tilde{\theta}}\right\} \tag{5.3.17}$$

A procedure to make the sum vanish is as follows. First, $c_k$ is replaced with the decision $\hat{c}_k$ from the detector. Second, the generic term in the sum is computed setting $\tilde{\theta}$ equal to the current estimate $\hat{\theta}(k)$. Third, the result is used as an error signal to improve the phase estimate. Formally, the following recursion is generated:

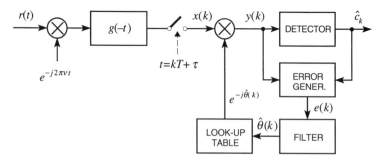

Figure 5.7. Costas phase recovery loop.

$$\hat{\theta}(k+1) = \hat{\theta}(k) + \gamma e(k) \quad (5.3.18)$$

where

$$e(k) \triangleq \text{Im}\left\{\hat{c}_k^* x(k) e^{-j\hat{\theta}(k)}\right\} \quad (5.3.19)$$

and $\gamma$ is a step-size parameter.

Figure 5.7 illustrates a block diagram for this algorithm. The error generator computes $e(k)$ and the loop filter operates according to (5.3.18). Finally, a look-up table produces the map $\hat{\theta}(k) \to e^{-j\hat{\theta}(k)}$. This scheme is very popular for continuous-transmission applications and may be viewed as a generalization of the well-known Costas detector [10]-[11]. In the sequel it is referred to as a *Costas loop*. Its characteristics have been analyzed in many papers in the past several years (see [12]-[22] and references therein). Its acquisition capabilities and tracking features are discussed in the next section.

### 5.3.4. Acquisition and Tracking Characteristics

A key tool to investigate phase acquisition is the S-curve of the phase error generator. This is the expectation of the error signal $e(k)$, conditioned on a fixed value of the difference $\phi \triangleq \theta - \hat{\theta}$, i.e.,

$$S(\phi) \triangleq \text{E}\{e(k)|\phi\} \quad (5.3.20)$$

Experimentally $S(\phi)$ is obtained by opening the loop and measuring the time average of the error signal as indicated in Figure 5.8.

A mathematical model for the Costas loop is found by paralleling the arguments used in Chapter 3 in connection with carrier frequency recovery.

# Carrier Phase Recovery with Linear Modulations

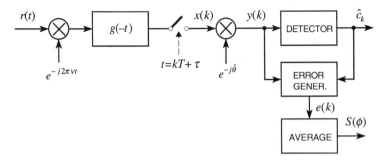

Figure 5.8. Measuring the S-curve of the detector.

Bearing in mind that the output of the look-up table in Figure 5.7 is $e^{-j\hat{\theta}(k)}$ and that $\hat{\theta}(k)$ varies slowly in time, the error signal may be viewed as the sum of a "local" average, $S[\theta - \hat{\theta}(k)]$, plus a noise term $N(k)$ such that

$$e(k) = S[\theta - \hat{\theta}(k)] + N(k) \qquad (5.3.21)$$

Thus, substituting into (5.3.18) yields

$$\hat{\theta}(k+1) = \hat{\theta}(k) + \gamma S[\theta - \hat{\theta}(k)] + \gamma N(k) \qquad (5.3.22)$$

as illustrated in Figure 5.9. Here the digital integrator (5.3.18) is drawn in a dashed rectangle.

The tracking behavior of the loop is investigated by first looking for the equilibrium points of the autonomous equation

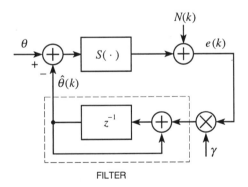

Figure 5.9. Equivalent model for the tracking loop.

$$\hat{\theta}(k+1) = \hat{\theta}(k) + \gamma S[\theta - \hat{\theta}(k)] \tag{5.3.23}$$

and then computing the noise-induced fluctuations around the operating point. A necessary condition for equilibrium (i.e., $\hat{\theta}(k) = \hat{\theta}_{eq} = \text{constant}$) is a null in the S-curve

$$S(\theta - \hat{\theta}_{eq}) = 0 \tag{5.3.24}$$

It is easily realized from (5.3.23), however, that $\hat{\theta}_{eq}$ is a *stable* solution only if $S(\phi)$ has a positive slope at $\phi = \theta - \hat{\theta}_{eq}$. Negative slopes correspond to unstable solutions, as is pictorially indicated in Figure 5.10.

Intuitively we expect that $\hat{\theta}(k) = \theta$ is a stable solution. As is now explained it is not unique, however. For the sake of argument consider uncoded QPSK and assume that the Nyquist condition is satisfied. Then, the matched-filter output becomes

$$x(k) = c_k e^{j\theta} + n(k) \tag{5.3.25}$$

and the detector input in Figure 5.7 is found to be

$$y(k) = c_k e^{j\phi} + n'(k) \tag{5.3.26}$$

with $n'(k) \triangleq n(k)e^{-j\hat{\theta}}$. At high SNR the detector decisions are

$$\hat{c}_k = c_k e^{jm(\phi)\pi/2} \tag{5.3.27}$$

where $m(\phi)$ is that integer such that

$$\left|\phi - m(\phi)\frac{\pi}{2}\right| < \frac{\pi}{4} \tag{5.3.28}$$

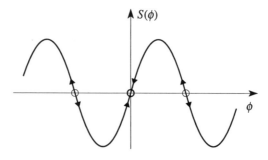

Figure 5.10. Stable and unstable equilibrium points.

# Carrier Phase Recovery with Linear Modulations

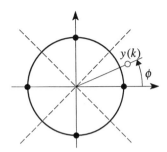

Figure 5.11. Illustrating equation (5.3.28).

For example, for $\phi \in (-\pi/4, \pi/4)$ (see Figure 5.11) equation (5.3.28) yields $m(\phi) = 0$ and (5.3.27) says that the decisions are correct. Vice versa, for $\phi \in (\pi/4, 3\pi/4)$ we have $m(\phi) = 1$ and (5.3.27) gives the incorrect results $\hat{c}_k = c_k e^{j\pi/2}$.

With these results at hand the S-curve can be easily computed. In fact, substituting (5.3.25) and (5.3.27) into the error signal (5.3.19) and bearing in mind that $|c_k|^2 = 1$ yields after some manipulations

$$e(k) \approx \sin[\phi - m(\phi)\pi/2] + \mathrm{Im}\left\{c_k^* n(k) e^{-j[\hat{\theta} + m(\phi)\pi/2]}\right\} \quad (5.3.29)$$

The second term has zero mean while the first is a constant (for a fixed $\phi$). Hence, taking the expectation we get the S-curve

$$S(\phi) \approx \sin[\phi - m(\phi)\pi/2] \quad (5.3.30)$$

which is represented in the solid line in Figure 5.12(a).

Thermal noise plays an important role in the decision process, especially when $y(k)$ is close to the decision boundaries, i.e., when $\phi$ is about a multiple of $\pi/4$ (see Figure 5.11). In practice, the thermal noise rounds off the discontinuities at these multiples and the actual S-curve takes the form indicated in Figure 5.12(b).

The following remarks are of interest.

- Figure 5.12 indicates that $\phi = 0$ (i.e., $\hat{\theta}_{eq} = \theta$) is a stable point but it is not unique. There are three more points in the range $(0, 2\pi]$, at a distance of $\pi/2$ from each other. This corresponds to the same four-fold ambiguity we have observed in connection with the recovery scheme in Figure 5.6.

- A problem that may limit the application of Costas loops is the so-called *hangup* phenomenon, which occasionally occurs during phase acquisitions.

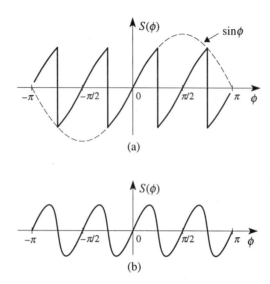

Figure 5.12. S-curves for uncoded QPSK: (a) infinite SNR; (b) finite SNR.

Assume that the initial value of $\hat{\theta}(k)$ is close to a reverse-slope null. The system will tend to move to a stable point but, in doing so, its evolution will be governed by noise since the steering force $S[\theta - \hat{\theta}(k)]$ is small. As a result, $\hat{\theta}(k)$ may dwell about its initial position for a while, thus making the acquisition long. Long acquisitions are hardly allowed in certain applications as, for example, in burst transmission. In-depth investigations on hangup phenomena in phase-lock loops are reported in [23]-[25].

- Another drawback is the *cycle slipping*, a phenomenon that manifests itself in any tracking scheme (be it closed loop or open loop) in the form of short synchronization failures. Briefly, assume that the loop is in the tracking mode and $\hat{\theta}(k)$ is fluctuating around the true carrier phase $\theta$. Fluctuations are usually small. Occasionally, however, thermal noise and self noise concur to produce a large phase deviation so that $\hat{\theta}(k)$ will be attracted toward some nearby equilibrium point. When this happens we say that a *phase slip* has occurred. A slip implies a net change of $\hat{\theta}(k)$ by multiples of $\pi/2$ in the case of Figure 5.12 and, in consequence, a burst of errors in the data detection process. Cycle slips must be rare events in a well-designed loop.

- Many efforts have been invested in theoretical and experimental studies on cycle slipping. References [26]-[28] provide formulas for the mean-time between slips, $T_s$, in continuous-time loops. A link between continuous- and discrete-time loops is established in [29]. The following approximate expression for $T_s$ is given in [29] for small $E_s/N_0$ values:

$$2B_L T_s \approx \frac{\pi}{2} \exp\left(\frac{p^2}{2\pi^2 \sigma^2}\right) \qquad (5.3.31)$$

where $B_L$ is the loop noise bandwidth and $\sigma^2$ the phase error variance in the steady state (see next section). Also, $p$ is the period of the S-curve (i.e., $S(\phi + p) = S(\phi)$). For example, for QPSK we have $p = \pi/2$, as indicated in Figure 5.12.

### 5.3.5. S-Curves for General Modulation Formats

S-curves with modulation formats other than QPSK are difficult to compute. Interesting results are reported in [29] for quadrature shift keying (QSK) and general $M$-ary PSK signal constellations. As is intuitively clear, $M$-ary PSK gives rise to S-curves with a $2\pi/M$ periodicity, whereas rectangular or "cross" shaped constellations produce a $\pi/2$ periodicity. Whatever the case, S-curves with complex signal constellations may exhibit spurious locks that cannot be resolved by differential encoding (see Figure 5.13). Changes in the error signal form (5.3.19) are needed to prevent such spurious points [21]-[22].

So far only uncoded systems have been considered. With trellis-coded (TC) modulations, S-curves are usually derived by simulation since analytical methods are too complex. Figure 5.14 provides a qualitative idea of S-curves for three TC-PSK systems. In particular, Figure 5.14(a) and 5.14(b) corresponds to an eight-state and a four-state Ungerboeck TC-8PSK code [30], respectively; while Figure 5.14(c) corresponds to an eight-state TC-16PSK proposed by Pietrobon et al. [31].

As is seen, $S(\phi)$ extends over a range of $\pm \pi/4$ around the origin in Figure 5.14(a). In Figure 5.14(b) it has a similar shape around the origin but there is a replica around $\pm \pi$. This corresponds to a two-fold phase ambiguity which, however, has no consequences since the code is invariant to phase shifts of $\pi$. Finally, in Figure 5.14(c), $S(\phi)$ is periodic with period $\pi/8$. Again, this ambiguity is immaterial as the code is invariant to rotations by multiples of $\pi/8$.

Figure 5.13. S-curve with spurious stable points.

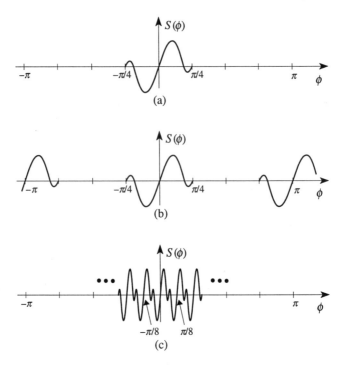

Figure 5.14. S-curves for a few TC-PSK systems.

S-curves of the type in Figure 5.14 are very bad for they have extensive dead zones and spurious equilibria. If the initial phase error $\phi$ lies in a dead zone there will be no steering force to lead the loop toward a nearby lock condition and the system will perform a long random walk before locking.

**Exercise 5.3.1.** Compute the S-curve for the phase detector of a Costas loop assuming that: (i) the modulation is binary PSK (BPSK) or, equivalently, the symbols are $c_k = \pm 1$; (ii) the Nyquist condition is satisfied; (iii) $E_s/N_0 \gg 1$.

*Solution.* Paralleling the arguments leading to (5.3.29) it is easily seen that the error signal becomes

$$e(k) \approx \sin[\phi - m(\phi)\pi] + \text{Im}\left\{c_k^* n(k) e^{-j[\hat{\theta} + m(\phi)\pi]}\right\} \quad (5.3.32)$$

where $m(\phi)$ takes the values $\pm 1$ so as to make $|\phi - m(\phi)\pi|$ less than $\pi/2$. Thus, taking the expectation of (5.3.32) and letting $C_2 \triangleq E\{|c_k|^2\}$ yields the desired result

$$S(\phi) \approx \sin[\phi - m(\phi)\pi] \qquad (5.3.33)$$

**Exercise 5.3.2.** Solve the problem in Exercise 5.3.1 for an $M$-ary PSK constellation $c_k = e^{j\alpha_k}$, $\alpha_k \in \{0, 2\pi/M, \ldots, 2\pi(M-1)/M\}$.

*Solution.* With the same arguments used in the previous exercise it is found that

$$S(\phi) \approx \sin[\phi - m(\phi) 2\pi/M] \qquad (5.3.34)$$

where $m(\phi)$ is that integer satisfying the inequality

$$\left|\phi - m(\phi)\frac{2\pi}{M}\right| < \frac{\pi}{M} \qquad (5.3.35)$$

From (5.3.34)-(5.3.35) it is recognized that the detector exhibits an $M$-fold ambiguity.

### 5.3.6. Tracking Performance

The tracking performance of the loop in Figure 5.9 can be investigated with the methods of Chapter 3, which are now summarized for completeness. Further details may be found in [20] and [21]. To proceed, assume that steady-state conditions have already been achieved and $\hat{\theta}(k)$ fluctuates around $\theta$. In all practical cases the fluctuations are so small that the following approximation holds true:

$$S[\theta - \hat{\theta}(k)] \approx A[\theta - \hat{\theta}(k)] \qquad (5.3.36)$$

where $A$ is the slope of the S-curve at the origin. Denoting $\phi(k) \triangleq \theta - \hat{\theta}(k)$ the phase error and inserting into (5.3.22) yields

$$\phi(k+1) = (1 - \gamma A)\phi(k) - \gamma N(k) \qquad (5.3.37)$$

In this equation $\phi(k)$ may be viewed as the response to $N(k)$ of a filter with transfer function

$$\mathcal{H}_N(z) = -\frac{\gamma}{z - (1 - \gamma A)} \qquad (5.3.38)$$

The phase error variance $\sigma^2$ is now expressed either as a function of the autocorrelation

$$R_N(m) = E\{N(k+m)N(k)\} \qquad (5.3.39)$$

or in terms of the power spectral density

$$S_N(f) = T \sum_{m=-\infty}^{\infty} R_N(m) e^{-j2\pi mfT} \qquad (5.3.40)$$

The former procedure results in

$$\sigma^2 = \frac{\gamma}{A(2-\gamma A)} \sum_{m=-\infty}^{\infty} R_N(m)(1-\gamma A)^{|m|} \qquad (5.3.41)$$

while the latter yields

$$\sigma^2 = \int_{-1/2T}^{1/2T} S_N(f) |H_N(f)|^2 df \qquad (5.3.42)$$

where $H_N(f)$ is the right-hand side of (5.3.38) for $z = e^{j2\pi fT}$, i.e.,

$$H_N(f) = -\frac{\gamma}{e^{j2\pi fT} - (1-\gamma A)} \qquad (5.3.43)$$

In some practical cases the noise spectral density $S_N(f)$ is nearly flat over the frequency range where $H_N(f)$ takes significant values and (5.3.42) reduces to

$$\sigma^2 \approx \frac{S_N(0)}{A^2} 2B_L \qquad (5.3.44)$$

where $B_L$ represents the noise equivalent bandwidth of the loop

$$B_L T = \frac{\gamma A}{2(2-\gamma A)} \qquad (5.3.45)$$

This formula can be further approximated as

$$B_L T \approx \frac{\gamma A}{4} \qquad (5.3.46)$$

since $\gamma A$ is usually small as compared with unity.

**Exercise 5.3.3.** Compute the phase error variance of a Costas loop with QPSK signaling assuming that: (*i*) the Nyquist condition is satisfied; (*ii*) the

## Carrier Phase Recovery with Linear Modulations

signal shaping is evenly apportioned between transmit and receive filters (i.e., the receive filter is matched to the transmitted pulse); (*iii*) the ratio $E_s/N_0$ is large. Compare the result with the Cramer-Rao bound

$$CRB(\theta) = \frac{1}{2L_0} \frac{1}{E_s/N_0} \qquad (5.3.47)$$

*Solution.* Looking at Figure 5.9 it is apparent that for $\hat{\theta}(k) \approx \theta$ (as happens in steady-state conditions) the noise term $N(k)$ is approximately equal to $e(k)$. Thus, setting $\hat{c}_k \approx c_k$ and $\hat{\theta}(k) \approx \theta$ into (5.3.19) yields

$$N(k) \approx \text{Im}\{c_k^* x(k) e^{-j\theta}\} \qquad (5.3.48)$$

or, making use of (5.3.25),

$$N(k) \approx \text{Im}\{c_k^* n(k) e^{-j\theta}\} \qquad (5.3.49)$$

On the other hand, as $|c_k^* e^{-j\theta}| = 1$ and $\{n(k)\}$ is a white Gaussian sequence, it follows that $\{c_k^* n(k) e^{-j\theta}\}$ is equivalent to $\{n(k)\}$, which implies that $\{N(k)\}$ are independent random variables with zero mean and variance $N_0$. Thus, the power spectral density of $\{N(k)\}$ is constant:

$$S_N(f) = T N_0 \qquad (5.3.50)$$

This result may be put in a more convenient form bearing in mind that, because of the assumptions (*i*)-(*ii*), the signal energy equals 1/2 (see Appendix 2.A). Hence

$$\frac{E_s}{N_0} = \frac{1}{2N_0} \qquad (5.3.51)$$

from which we have $N_0 = 1/(2E_s/N_0)$. Thus, (5.3.50) becomes

$$S_N(f) = \frac{T}{2E_s/N_0} \qquad (5.3.52)$$

Next, we concentrate on the expression of the phase error variance in (5.3.44). From (5.3.30) it is seen that the slope of the S-curve at the origin is unity. Hence, taking (5.3.52) into account gives the desired result

$$\sigma^2 = B_L T \frac{1}{E_s/N_0} \qquad (5.3.53)$$

In comparing (5.3.53) with the Cramer-Rao bound we face the same conceptual difficulty that has been encountered in Chapter 3 when dealing with closed-loop frequency estimators. Costas estimators make observations over the infinite past with a weighting procedure that tends to emphasize recent data against old ones. By contrast, the Cramer-Rao bound is a lower limit to the variance of estimators which make observations of finite length. A solution to this difficulty is found with the same methods adopted in Chapter 3. In essence, the closed-loop scheme is transformed into an equivalent open-loop scheme with the same estimation errors and the observation length of the latter is taken as the *equivalent* observation length for the former. This length turns out to be (in symbol interval)

$$L_{eq} = \frac{1}{2B_L T} \qquad (5.3.54)$$

Thus, substituting into (5.3.53) and setting $L_{eq}=L_0$ yields

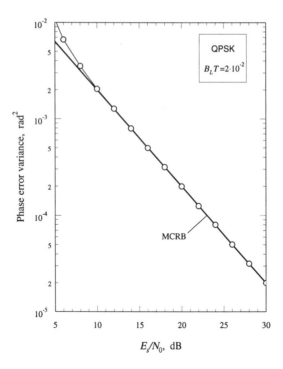

Figure 5.15. Tracking error variance of a Costas loop with QPSK.

# Carrier Phase Recovery with Linear Modulations

$$\sigma^2 = \frac{1}{2L_0} \frac{1}{E_s/N_0} \quad (5.3.55)$$

which coincides with the $CRB(\theta)$.

It is worth noting that this coincidence holds only asymptotically, as $E_s/N_0$ tends to infinity. In fact (5.3.53) is not correct at moderate $E_s/N_0$ values as a consequence of decision errors and/or loop nonlinearities. Figure 5.15 shows simulation results for the tracking variance of a Costas loop with QPSK modulation. The loop bandwidth is 2% of the symbol rate and the channel response is Nyquist with a 50% rolloff. We see that the variance stays close to the MCRB for $E_s/N_0 \geq 10$ dB.

## 5.3.7. Effect of Frequency Errors

In this section we investigate the effects of frequency errors on the tracking performance of Costas loops. To simplify the analysis we assume that frequency errors are small compared with the symbol rate.

The block diagram of the loop is shown in Figure 5.16. It coincides with the scheme in Figure 5.7, except that the leftmost mixer is driven at a reference frequency $\hat{v}$ which is generally different from $v$. Denoting by $f_d \triangleq v - \hat{v}$ the error and assuming $|f_d T| \ll 1$, it can be shown that the samples from the matched filter are now expressed as

$$x(k) = \sum_i c_i e^{j\theta(i)} h[(k-i)T] + n(k) \quad (5.3.56)$$

with $h(t) \triangleq g(t) \otimes g(-t)$ and

$$\theta(i) = \theta(i-1) + 2\pi f_d T \quad (5.3.57)$$

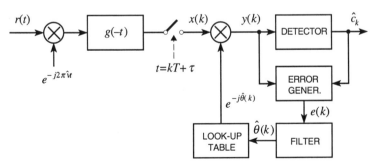

Figure 5.16. Block diagram of a Costas tracking system.

The error signal is still as in (5.3.19) and is repeated here for convenience:

$$e(k) = \text{Im}\{\hat{c}_k^* x(k) e^{-j\hat{\theta}(k)}\} \tag{5.3.58}$$

Our task is to derive the equivalent model for the loop. To this end we first concentrate on the S-curve, which is the expectation of $e(k)$ conditioned on a constant phase error $\phi(k) = \theta(k) - \hat{\theta}(k)$, i.e.,

$$S(\phi) = \text{E}\{e(k)|\phi(k) = \phi\} \tag{5.3.59}$$

An important question is how this curve is affected by frequency errors. We maintain that the dependence is negligible as a consequence of the assumption $|f_d T| \ll 1$. To prove our claim it is sufficient to show that the quantity $x(k)e^{-j\hat{\theta}(k)}$ is independent of $f_d$. Indeed, if this is true, the detector decisions are also independent of $f_d$ (see Figure 5.16) and, in consequence, $e(k)$ is not affected by the frequency error.

To proceed, from (5.3.56) we get

$$x(k)e^{-j\hat{\theta}(k)} = \sum_i c_i e^{j[\theta(i) - \hat{\theta}(k)]} h[(k-i)T] + n'(k) \tag{5.3.60}$$

where $n'(k)$ is a noise term statistically equivalent to $n(k)$. Also, from (5.3.57) it is readily shown that

$$\theta(i) = \theta(k) + 2\pi(i-k)f_d T \tag{5.3.61}$$

Next, bearing in mind that $\theta(k) - \hat{\theta}(k) = \phi$, equation (5.3.60) becomes

$$x(k)e^{-j\hat{\theta}(k)} = e^{j\phi} \sum_i c_i e^{-j2\pi(k-i)f_d T} h[(k-i)T] + n'(k) \tag{5.3.62}$$

Now, as $h[(k-i)T]$ takes significant values only for $|k-i|$ on the order of a few units, the exponential in (5.3.62) is approximately unity (since $|f_d T| \ll 1$) and this renders $x(k)e^{-j\hat{\theta}(k)}$ independent of $f_d$, as anticipated.

At this stage the derivation of the loop equivalent model is straightforward. Denoting by

$$N(k) \triangleq e(k) - S[\theta(k) - \hat{\theta}(k)] \tag{5.3.63}$$

the equivalent noise and bearing in mind that the loop filter is still governed by (5.3.18) we have

## Carrier Phase Recovery with Linear Modulations

$$\hat{\theta}(k+1) = \hat{\theta}(k) + \gamma S\left[\theta(k) - \hat{\theta}(k)\right] + \gamma N(k) \qquad (5.3.64)$$

which corresponds to the block diagram in Figure 5.9 where $\theta$ is no longer a constant but varies in time according to (5.3.57).

Equation (5.3.64) can be expressed in terms of the phase error $\phi(k)$ as follows. By definition

$$\hat{\theta}(k) = \theta(k) - \phi(k) \qquad (5.3.65)$$

hence, taking (5.3.57) into account, yields

$$\begin{aligned}\hat{\theta}(k+1) &= \theta(k+1) - \phi(k+1) \\ &= \theta(k) + 2\pi f_d T - \phi(k+1)\end{aligned} \qquad (5.3.66)$$

and collecting (5.3.64)-(5.3.66) produces the desired result

$$\phi(k+1) = \phi(k) + 2\pi f_d T - \gamma S[\phi(k)] - \gamma N(k) \qquad (5.3.67)$$

In the steady state $\phi(k)$ oscillates around some average value $\phi_{ss}$ which is a stable solution of the autonomous system

$$\phi(k+1) = \phi(k) + 2\pi f_d T - \gamma S[\phi(k)] \qquad (5.3.68)$$

From this equation it appears that $\phi_{ss}$ must be a root of the equation (see also Figure 5.17)

$$S(\phi_{ss}) = \frac{2\pi f_d T}{\gamma} \qquad (5.3.69)$$

The slope of $S(\phi)$ must be positive at $\phi = \phi_{ss}$ in order that this root be stable. In particular, if $\phi_{ss}$ is small so that $S(\phi_{ss}) \approx A\phi_{ss}$, then from (5.3.69) we get

$$\phi_{ss} = \frac{2\pi f_d T}{\gamma A} \qquad (5.3.70)$$

or, introducing the loop bandwidth $B_L T \approx \gamma A / 4$,

$$\phi_{ss} = \frac{\pi f_d}{2 B_L} \qquad (5.3.71)$$

In conclusion, frequency errors cause static phase shifts. These may be sufficiently large as to degrade the error probability. For example, with a loop

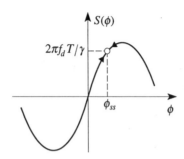

Figure 5.17. Illustrating the stable solutions.

bandwidth $B_L=10^{-3}/T$, even a frequency error as small as $10^{-4}/T$ causes a static error of 9° which may be disabling in many applications. Remedies to this drawback are discussed in the next section.

Before doing so let us return to Figure 5.17. Clearly, there is a maximum value of $|f_d|$ beyond which equation (5.3.69) has no solutions, which means that the loop cannot achieve a steady-state condition. This critical value is readily found to be

$$f_d^{(\max)} = \frac{2B_L}{\pi} \cdot \frac{S_{\max}}{A} \qquad (5.3.72)$$

where $S_{\max}$ is the maximum amplitude of the S-curve. The interval $\pm f_d^{(\max)}$ within which a locking condition is eventually found is called the *locking range*. The locking range is of the order of the loop bandwidth.

### 5.3.8. Second-Order Tracking Loops

The standard method to cope with frequency errors is to turn the original first-order loop filter into a second-order filter containing a further integrator. A widely used scheme is indicated in Figure 5.18. By inspection it is seen that the governing equations are

$$\hat{\theta}(k+1) = \hat{\theta}(k) + \xi(k) \qquad (5.3.73)$$

$$\xi(k) = \xi(k-1) + \gamma(1+\rho)e(k) - \gamma e(k-1) \qquad (5.3.74)$$

where $\rho$ is a positive constant. Clearly, the filter becomes first-order for $\rho=0$.

The influence of the second integrator on the static phase error may be assessed as follows. First, rewrite (5.3.73) in terms of $\phi(k)$. Using (5.3.65)-(5.3.66), this results in

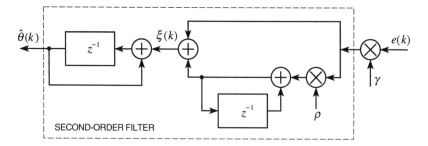

Figure 5.18. Second-order loop filter.

$$\phi(k+1) = \phi(k) + 2\pi f_d T - \xi(k) \tag{5.3.75}$$

Next, assuming that a steady-state condition has been achieved, say $\phi(k) = \phi_{ss}$, from (5.3.75) it follows that $\xi(k)$ must be constant:

$$\xi_{ss} = 2\pi f_d T \tag{5.3.76}$$

Then, from (5.3.74) it is seen that

$$e(k) = \frac{1}{1+\rho} e(k-1) \tag{5.3.77}$$

whose steady-state solution is zero (for $\rho>0$). Thus, letting $N(k)=0$ in (5.3.63) yields

$$S(\phi_{ss}) = 0 \tag{5.3.78}$$

and comparing with (5.3.69) it is concluded that static phase errors have been eliminated.

From the above discussion it appears that a second-order loop will eventually lock on the incoming carrier with no static error, whatever the value of $\rho$. In practice this is only true with very small values of $f_d$ (on the order of the loop noise bandwidth). Furthermore, even in these conditions, the system response does depend on $\rho$ and, therefore, it is of interest to look for $\rho$ values that correspond to relatively short settling times. This problem is approached here under the assumption that the loop noise bandwidth is small compared with the symbol rate. Under these circumstances the variables $\phi(k)$, $\xi(k)$ and $e(k)$ are slowly varying in time and the digital loop can be approximated by an analog system for which a well-established theory is available.

The analog system is arrived at postulating that its state variables $\phi(t)$,

$\xi(t)$ and $e(t)$ are related to $\phi(k)$, $\xi(k)$ and $e(k)$ by relations of the type

$$\phi(t) \approx \phi(k) \tag{5.3.79}$$

$$\frac{d\phi(t)}{dt} \approx \frac{\phi(k+1) - \phi(k)}{T} \tag{5.3.80}$$

for $kT + \tau \le t \le (k+1)T + \tau$. Making these approximations in (5.3.74)-(5.3.75) and letting

$$e(k) \approx A\phi(k) \tag{5.3.81}$$

after some manipulations it is found that

$$\frac{d\xi(t)}{dt} = \gamma \frac{de(t)}{dt} + \frac{\gamma\rho}{T} e(t+T)$$

$$\approx \gamma(1+\rho)\frac{de(t)}{dt} + \frac{\gamma\rho}{T} e(t) \tag{5.3.82}$$

$$\frac{d\phi(t)}{dt} = 2\pi f_d - \frac{1}{T}\xi(t) \tag{5.3.83}$$

$$e(t) = A\phi(t) \tag{5.3.84}$$

A solution for $\phi(t)$ is now derived by Laplace transform methods, setting to zero the initial conditions. This yields the Laplace transform for $\phi(t)$

$$\Phi(s) = \frac{2\pi f_d}{s^2 + 2\varsigma\omega_n s + \omega_n^2} \tag{5.3.85}$$

where the parameters $\varsigma$ and $\omega_n$ are defined as

$$\varsigma \triangleq \frac{(1+\rho)\sqrt{\gamma A}}{2\sqrt{\rho}} \tag{5.3.86}$$

$$\omega_n \triangleq \frac{\sqrt{\gamma A \rho}}{T} \tag{5.3.87}$$

In the parlance of servomechanism theory, $\varsigma$ is the *damping factor* and $\omega_n$ the *natural frequency* of the loop. Next, the inverse Laplace transform of (5.3.85) is obtained from transform pairs as available in many textbooks. Figure 5.19 illustrates the effect of the damping factor on the response to a step in the frequency error. It is apparent that, as time increases, $\phi(t)$ tends to zero anyway.

# Carrier Phase Recovery with Linear Modulations

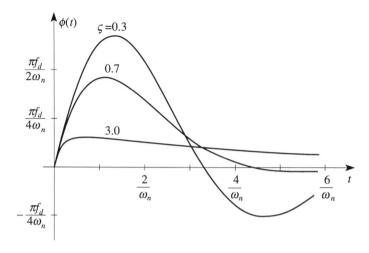

Figure 5.19. Transient phase error due to a frequency error step.

Short transients are observed with damping factors in the range $0.6 \leq \varsigma \leq 1.0$. These are the $\varsigma$ values that are commonly adopted in practice.

Returning to the loop filter, one wonders how to choose $\rho$ so as to properly control the integrator's action. This problem can be addressed considering the loop tracking performance. Briefly, let us start with the loop model as summarized by the equations

$$\phi(k+1) = \phi(k) - \xi(k) \tag{5.3.88}$$

$$\xi(k) = \xi(k-1) + \gamma(1+\rho)e(k) - \gamma e(k-1) \tag{5.3.89}$$

$$e(k) = A\phi(k) + N(k) \tag{5.3.90}$$

In writing (5.3.88) we have dropped $f_d$ since frequency errors are compensated in the steady-state. From these equations $\phi(k)$ is found as the response to $N(k)$ of

$$H(z) = -\frac{\gamma[(1+\rho)z - 1]}{(z-1)^2 + \gamma A[(1+\rho)z - 1]} \tag{5.3.91}$$

The noise bandwidth of this filter is found with the methods in [32] and is given by

$$B_L T = \frac{2\rho + \gamma A(2+\rho)}{2[4 - \gamma A(2+\rho)]} \tag{5.3.92}$$

Hence, the loop phase error variance is given by

$$\sigma^2 \approx \frac{S_N(0)}{A^2} 2B_L \qquad (5.3.93)$$

where the spectral density in the origin, $S_N(0)$, is computed as indicated in Section 5.3.6. Once $S_N(0)$ and $A$ are given, the choice of $\rho$ proceeds by first choosing the loop bandwidth value so as to guarantee small phase errors (in general, computer simulations are needed to establish how small they must be to keep bit error rate degradations within reasonable limits). Then, equations (5.3.86) and (5.3.92) are solved for $\rho$, taking $\varsigma$ in the range $0.6 \leq \varsigma \leq 1.0$.

### 5.3.9. Phase Noise

So far a constant carrier frequency has been assumed. In practice the carrier frequency undergoes slow fluctuations due to imperfections in the transmitter and receiver oscillators. Correspondingly the demodulated carrier becomes a phase modulated sinewave of the type $e^{j[2\pi v t + \theta(t)]}$ wherein the modulation $\theta(t)$ is referred to as *phase noise*. In this section we characterize the statistics of phase noise and establish its effects on the tracking performance of a Costas loop.

As a first step in this direction we concentrate on the loop equivalent model. With the same arguments used in Section 5.3.4 it can be shown that the scheme in Figure 5.9 is still valid, except that the constant input $\theta$ must be replaced by the samples $\theta(k)$ of the phase noise process. In particular, assuming that the phase errors $\phi(k) \triangleq \theta(k) - \hat{\theta}(k)$ are sufficiently small to allow the approximation $S[\phi(k)] \approx A\phi(k)$ leads to the block diagram depicted in Figure 5.20 wherein, for greater generality, we have adopted a generic loop filter $\mathcal{F}(z)$ instead of a first-order one. In particular, $\mathcal{F}(z)$ has the form

$$\mathcal{F}(z) = \frac{1}{z-1} \qquad (5.3.94)$$

for a first-order and

$$\mathcal{F}(z) = \frac{z(1+\rho)-1}{(z-1)^2} \qquad (5.3.95)$$

for a second-order loop.

As is seen, there are two inputs to the scheme in Figure 5.20, $N(k)$ and $\theta(k)$. The former accounts for thermal noise and, possibly, intersymbol interference; the latter for phase noise. Accordingly, $\phi(k)$ can be thought of as

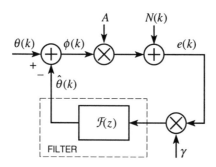

Figure 5.20. Loop equivalent model in the presence of phase noise.

the sum of two uncorrelated terms, $\phi_N(k)$ and $\phi_\theta(k)$, corresponding to $N(k)$ and $\theta(k)$, respectively. By inspection it is easily seen that $\phi_N(k)$ is the response to $N(k)$ of the filter

$$\mathcal{H}_N(z) = -\frac{\gamma \mathcal{F}(z)}{1 + \gamma A \, \mathcal{F}(z)} \qquad (5.3.96)$$

Similarly, $\phi_\theta(k)$ may be viewed as the response to $\theta(k)$ of the filter

$$\mathcal{H}_\theta(z) = \frac{1}{1 + \gamma A \, \mathcal{F}(z)} \qquad (5.3.97)$$

The phase error variance is given by the sum of the separate variances of $\phi_N(k)$ and $\phi_\theta(k)$, which are expressed by

$$\sigma_N^2 = \int_{-1/2T}^{1/2T} S_N(f) |H_N(f)|^2 \, df \qquad (5.3.98)$$

$$\sigma_\theta^2 = \int_{-1/2T}^{1/2T} S_\theta^{(d)}(f) |H_\theta(f)|^2 \, df \qquad (5.3.99)$$

where $H_N(f)$ and $H_\theta(f)$ denote the right-hand sides of (5.3.96)-(5.3.97) for $z = e^{j2\pi fT}$ and $S_\theta^{(d)}(f)$ represents the power spectral density of $\theta(k)$. Another way of expressing $\sigma_N^2$ is by means of the loop bandwidth

$$B_L \triangleq \frac{1}{2|H_N(0)|^2} \int_{-1/2T}^{1/2T} |H_N(f)|^2 \, df \qquad (5.3.100)$$

In fact, assuming that $S_N(f)$ is flat over the range where $|H_N(f)|$ takes significant values, equation (5.3.98) becomes

$$\sigma_N^2 = 2|H_N(0)|^2 S_N(0) B_L \qquad (5.3.101)$$

Now we turn our attention to the phase noise spectral density $S_\theta^{(d)}(f)$ involved in (5.3.99). It should be stressed that $S_\theta^{(d)}(f)$ corresponds to the time discrete process $\theta(k)$ whereas many references (see [28], [33]-[34], for example) provide experimental data for the spectral density $S_\theta^{(c)}(f)$ of the continuous-time process $\theta(t)$. Let us postpone for a moment the question of the relation between the two and concentrate on the characterization of $S_\theta^{(c)}(f)$.

One popular representation of this spectral density consists of dividing the frequency axis into separate segments and expressing $S_\theta^{(c)}(f)$ as a function of the type $|f|^{-(2+\delta)}$ on each segment. In particular, $\delta = 0$ corresponds to *Wiener* phase noise, whereas $\delta = 1$ and $\delta = 2$ are related to *flicker* noise and *random frequency walk*, respectively. In many practical cases flicker noise is the dominant disturbance.

The following alternative representation of $S_\theta^{(c)}(f)$ has been accepted as a reference in the European digital terrestrial television broadcasting (dTTb) project [34]:

$$S_\theta^{(c)}(f) = 10^{-c} + \begin{cases} 10^{-a} & |f| \le f_1 \\ 10^{-a} \cdot 10^{-b(|f|-f_1)/(f_2-f_1)} & |f| > f_1 \end{cases} \qquad (5.3.102)$$

In this equation the PSD is expressed in seconds and the parameters $a$, $b$, $c$, $f_1$ and $f_2$ are related to the quality of the transmit and receive oscillators. Typical values are $a = 6.5$, $b = 4$, $c = 10.5$, $f_1 = 1$ kHz and $f_2 = 10$ kHz. A plot of $S_\theta^{(c)}(f)$ with these parameter values is shown in Figure 5.21. Note that $S_\theta^{(c)}(f)$ must be set to zero for sufficiently high frequencies to account for the bandlimiting effects of the intermediate-frequency (IF) filter in the receiver.

The relation between $S_\theta^{(d)}(f)$ and $S_\theta^{(c)}(f)$ is established as follows. By definition we have

$$S_\theta^{(d)}(f) = T \sum_{m=-\infty}^{\infty} R_\theta(mT) e^{-j2\pi mfT} \qquad (5.3.103)$$

where $R_\theta(\tau) \triangleq E\{\theta(t+\tau)\theta(t)\}$ is the inverse Fourier transform of $S_\theta^{(c)}(f)$. On the other hand, application of the Poisson sum formula [37, p. 395] yields

$$T \sum_{m=-\infty}^{\infty} R_\theta(mT) e^{-j2\pi mfT} = \sum_{m=-\infty}^{\infty} S_\theta^{(c)}(f - m/T) \qquad (5.3.104)$$

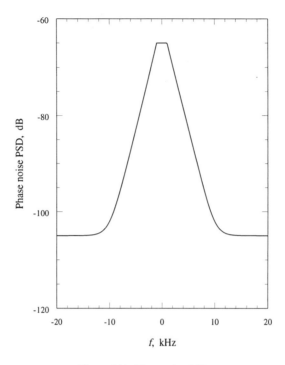

Figure 5.21. Phase noise PSD.

Hence, comparing with (5.3.103) produces the result sought

$$S_\theta^{(d)}(f) = \sum_{m=-\infty}^{\infty} S_\theta^{(c)}(f - m/T) \qquad (5.3.105)$$

Figure 5.22 illustrates the total error variance $\sigma^2 = \sigma_\theta^2 + \sigma_N^2$ as a function of the loop bandwidth for a second-order loop with a damping factor of $\varsigma = 0.7$. The phase noise PSD has the shape indicated in Figure 5.21, with parameter $a = 6.5$, $b = 4$, $c = 10.5$, $f_1 = 1$ kHz and $f_2 = 10$ kHz. The modulation pulses are root-raised-cosine-rolloff with a 50% rolloff. As $B_L$ decreases we see that $\sigma^2$ tends to a value independent of the SNR. This is expected because the thermal noise contribution becomes negligible with a small $B_L$. When $B_L$ increases, the loop tracking capability improves and the phase noise contribution decreases. At the same time, however, the thermal noise contribution increases. For high SNR values an optimum loop bandwidth exists that minimizes the error variance.

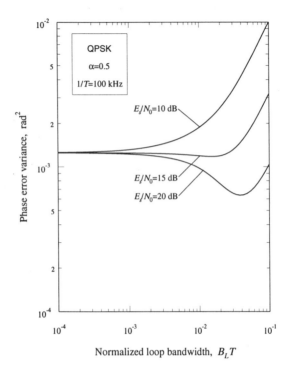

Figure 5.22. Total phase error variance as a function of the noise bandwidth.

## 5.4. Decision-Directed Phase Recovery with Offset Modulation

### 5.4.1. Phase Estimation Loop

Phase recovery with offset formats can be addressed with the same methods adopted in the previous section. In view of this similarity, the ensuing treatment is considerably abbreviated. As happens with non-offset signaling, two approaches may be followed, one based on the estimation equation (5.2.29), the other on the maximization of the log-likelihood function (5.2.28). The former leads to a structure similar to that shown in Figure 5.6, the latter produces a Costas loop. For the sake of brevity we concentrate on the latter.

We begin with the derivative of the log-likelihood. From (5.2.28) we have

$$\frac{d}{d\tilde{\theta}} \ln \Lambda(r|\tilde{\theta}) = \sum_{k=0}^{L_0-1} \left[ a_k Q(k,\tilde{\theta}) - b_k I(k+1/2,\tilde{\theta}) \right] \qquad (5.4.1)$$

## Carrier Phase Recovery with Linear Modulations

where the functions $I$ and $Q$ are defined as

$$I(k,\tilde{\theta}) + jQ(k,\tilde{\theta}) \triangleq x(k)e^{-j\tilde{\theta}} \qquad (5.4.2)$$

$$I(k+1/2,\tilde{\theta}) + jQ(k+1/2,\tilde{\theta}) \triangleq x(k+1/2)e^{-j\tilde{\theta}} \qquad (5.4.3)$$

Next, the true data in (5.4.1) are replaced by their estimates taken from the detector. Finally, the generic term in the sum is computed setting $\tilde{\theta}$ equal to the current phase estimate $\hat{\theta}(k)$ and the result is used as an error signal to update this estimate. Formally we have

$$\hat{\theta}(k+1) = \hat{\theta}(k) + \gamma e(k) \qquad (5.4.4)$$

with

$$e(k) \triangleq \hat{a}_k Q[k,\hat{\theta}(k)] - \hat{b}_k I[k+1/2,\hat{\theta}(k)] \qquad (5.4.5)$$

Figure 5.23 illustrates a block diagram for this algorithm. As is seen, the output $x(t)$ from the matched filter is sampled at twice the symbol rate to produce $x(k) \triangleq x(kT + \tau)$ and $x(k+1/2) \triangleq x(kT + T/2 + \tau)$. A series-to-parallel converter (S/P) separates $\{x(k)e^{-j\hat{\theta}(k)}\}$ from $\{x(k+1/2)e^{-j\hat{\theta}(k)}\}$. The block labelled I/Q computes the *in-phase* and *quadrature* components of $x(k)$ and $x(k+1/2)$ with respect to $e^{j\hat{\theta}(k)}$ which are denoted $I(k)$, $Q(k)$, $I(k+1/2)$ and $Q(k+1/2)$.

With uncoded OQPSK modulation the detector makes the following decisions:

$$\hat{a}_k = \text{sgn}[I(k)] \qquad (5.4.6)$$

$$\hat{b}_k = \text{sgn}[Q(k+1/2)] \qquad (5.4.7)$$

where sgn[$z$] equals $\pm 1$, depending on the sign of $z$.

S-curves for the phase detector (5.4.5) are difficult to derive analytically. The following qualitative arguments provide some insight into the form of these curves for uncoded OQPSK and an overall channel response $h(t)$ satisfying the first Nyquist criterion. To proceed, we open the feedback loop in Figure 5.23 and rotate the samples $x(k)$ and $x(k+1/2)$ by a fixed angle $\tilde{\theta}$. This produces

$$I(k) + jQ(k) = e^{j\phi}\left[a_k + j\sum_i b_i h(k-i-1/2)\right] + n'(k) \qquad (5.4.8)$$

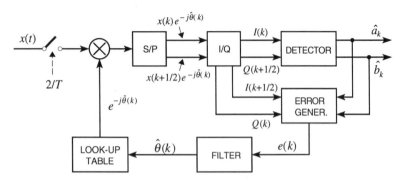

Figure 5.23. Block diagram for the phase tracking loop.

$$I(k+1/2) + jQ(k+1/2) = e^{j\phi}\left[jb_k + \sum_i a_i h(k-i+1/2)\right] + n''(k+1/2) \quad (5.4.9)$$

where $\phi \triangleq \theta - \hat{\theta}$ and $n'(k)$ and $n''(k+1/2)$ are phase rotated versions of $n(k)$ and $n(k+1/2)$ respectively. From (5.4.6)-(5.4.7) it follows that

$$\hat{a}_k = \text{sgn}\left[a_k \cos\phi - \sin\phi \sum_i b_i h(k-i+1/2) + \text{noise}\right] \quad (5.4.10)$$

$$\hat{b}_k = \text{sgn}\left[b_k \cos\phi + \sin\phi \sum_i a_i h(k-i+1/2) + \text{noise}\right] \quad (5.4.11)$$

In general, the values of $\hat{a}_k$ and $\hat{b}_k$ are difficult to predict from (5.4.10)-(5.4.11). Nevertheless, for $\phi$ close to either 0, or $\pm\pi$ or $\pm\pi/2$, a few simple conclusions can be drawn. For example, consider the terms in square brackets in (5.4.10). The first is the useful signal component, the second accounts for intersymbol interference from the quadrature channel and the last is contributed by thermal noise. For $\phi$ close to zero we have $\cos\phi \approx 1$, $\sin\phi \approx 0$ and (5.4.10) yields (at high SNR)

$$\hat{a}_k \approx a_k, \quad \phi \approx 0 \quad (5.4.12)$$

Similarly, for $\phi$ close to $\pm \pi$, it is found that

$$\hat{a}_k \approx -a_k, \quad \phi \approx \pm\pi \quad (5.4.13)$$

Finally, for $\phi$ close to $\pm\pi/2$, the signal component in (5.4.10) is negligible and $\hat{a}_k$ becomes a random variable (independent of $a_k$) taking values $\pm 1$ with the same probability

$$\hat{a}_k \approx \pm 1, \quad \phi \approx \pm \pi/2 \qquad (5.4.14)$$

With similar arguments it is concluded that

$$\hat{b}_k \approx b_k, \quad \phi \approx 0 \qquad (5.4.15)$$

$$\hat{b}_k \approx -b_k, \quad \phi \approx \pm \pi \qquad (5.4.16)$$

$$\hat{b}_k \approx \pm 1, \quad \phi \approx \pm \pi/2 \qquad (5.4.17)$$

With these results at hand it is a simple matter to get the S-curve for $\phi$ about 0, $\pm\pi$ and $\pm\pi/2$. In fact, substituting $\hat{\theta}(k) = \hat{\theta}$ into (5.4.5) and letting $\hat{a}_k \approx a_k$ and $\hat{b}_k \approx b_k$ (which correspond to $\phi \approx 0$) yields

$$e(k) \approx \mathrm{Im}\left\{e^{j\phi}\left[1 + ja_k \sum_i b_i h(k-i-1/2)\right] + \text{noise}\right\}$$

$$- \mathrm{Re}\left\{e^{j\phi}\left[j + b_k \sum_i a_i h(k-i+1/2)\right] + \text{noise}\right\} \qquad (5.4.18)$$

Thus, bearing in mind that $\{a_k\}$ and $\{b_k\}$ are zero mean and uncorrelated sequences, the expectation of (5.4.18) results in

$$S(\phi) = 2\sin\phi, \quad \phi \approx 0 \qquad (5.4.19)$$

Similarly, for $\phi$ about $\pm\pi$, it is found that

$$S(\phi) = -2\sin\phi, \quad \phi \approx \pm\pi \qquad (5.4.20)$$

Finally, for $\phi \approx \pi/2$, equation (5.4.5) becomes

$$e(k) \approx \mathrm{Im}\left\{e^{j\phi}\left[a_k\hat{a}_k + j\hat{a}_k \sum_i b_i h(k-i-1/2)\right] + \text{noise}\right\}$$

$$- \mathrm{Re}\left\{e^{j\phi}\left[jb_k\hat{b}_k + \hat{b}_k \sum_i a_i h(k-i+1/2)\right] + \text{noise}\right\} \qquad (5.4.21)$$

Taking the expectation and bearing in mind that $\hat{a}_k$ is independent of $a_k$ and $\hat{b}_k$ is independent of $b_k$, it is concluded that $S(\pi/2)=0$. With the same arguments it

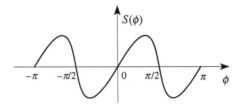

Figure 5.24. S-curve for OQPSK.

is shown that the S-curve vanishes at $\phi = -\pi/2$.

The above results are qualitatively summarized in Figure 5.24. Comparing with the S-curve in Figure 5.12(b) (which corresponds to QPSK) it is seen that there are now only two stable points in a period ($\phi = 0$ and $\phi = \pi$). Symbol decisions have the wrong sign for $\phi = \pi$. This drawback may be overcome by differential encoding/decoding.

### 5.4.2. Tracking Performance with Offset Formats

For the sake of simplicity we consider only uncoded OQPSK modulation. Also, we assume that the channel response $h(t)$ is Nyquist and the filtering is equally split between transmitter and receiver. In these conditions it is easily checked that:

(i) The signal energy is unity (see also Appendix 2.A) and, accordingly, the noise power spectral density may be written as

$$N_0 = \frac{1}{E_s/N_0} \qquad (5.4.22)$$

(ii) The samples $x(k)$ and $x(k+1/2)$ from the matched filter have the form

$$x(k) = e^{j\theta}\left[a_k + j\sum_i b_i h(k-i-1/2)\right] + n(k) \qquad (5.4.23)$$

$$x(k+1/2) = e^{j\theta}\left[jb_k + \sum_i a_i h(k-i+1/2)\right] + n(k+1/2) \qquad (5.4.24)$$

Here, $n(k)$ and $n(k+1/2)$ are zero-mean complex-valued random variables with the same variance $2N_0$. Also, the real component of $n(k)$ is independent of the imaginary component of $n(k+1/2)$ and, similarly, the imaginary component of $n(k)$ is independent of the real component of $n(k+1/2)$.

To proceed, let us first concentrate on the error signal in (5.4.5). In the steady state the phase errors are small so that $e^{j\phi(k)} \approx 1$. Furthermore, most decisions are correct and we have

$$\hat{a}_k \approx a_k \quad \text{and} \quad \hat{b}_k \approx b_k \tag{5.4.25}$$

Thus, substituting (5.4.23)-(5.4.25) into (5.4.5) and rearranging yields

$$e(k) = 2\sin\phi(k) + N_{SN}(k) + N_{TN}(k) \tag{5.4.26}$$

where $N_{SN}(k)$ is a *self-noise* term

$$N_{SN}(k) \approx \sum_i a_k b_i h(k-i-1/2) - \sum_i b_k a_i h(k-i+1/2) \tag{5.4.27}$$

while $N_{TN}(k)$ is contributed by thermal noise

$$N_{TN}(k) \triangleq n'_I(k) - n''_R(k+1/2) \tag{5.4.28}$$

with

$$n'_R(k) + jn'_I(k) \triangleq a_k n(k) e^{-j\hat{\theta}(k)} \tag{5.4.29}$$

$$n''_R(k+1/2) + jn''_I(k+1/2) \triangleq b_k n(k+1/2) e^{-j\hat{\theta}(k)} \tag{5.4.30}$$

Next we compute the correlation of the overall noise in (5.4.26). Taking the expectation of (5.4.26) yields the S-curve for $\phi$ about the origin:

$$S(\phi) = 2\sin\phi \tag{5.4.31}$$

Note that the slope of the S-curve at the origin is $A=2$.

From (5.4.27) it appears that $N_{SN}(k)$ has zero mean. Also, bearing in mind that $a_k$ and $b_k$ take the values $\pm 1$ independently, the autocorrelation of $N_{SN}(k)$ is found to be

$$R_{SN}(m) = \begin{cases} 2\sum_{k=-\infty}^{\infty} h^2(k-1/2) - 2h^2(1/2) & m=0 \\ -h^2(m+1/2) - h^2(m-1/2) & m \neq 0 \end{cases} \tag{5.4.32}$$

As for $N_{TN}(k)$, it can be shown that

$$R_{TN}(m) = \begin{cases} \dfrac{2}{E_s/N_0} & m = 0 \\ 0 & m \neq 0 \end{cases} \qquad (5.4.33)$$

Finally, since $N_{SN}(k)$ and $N_{TN}(k)$ are uncorrelated, the autocorrelation of the overall noise process $N(k) \triangleq N_{SN}(k) + N_{TN}(k)$ in (5.4.26) is given by

$$R_N(m) = R_{SN}(m) + R_{TN}(m) \qquad (5.4.34)$$

With these results at hand the error variance is computed with the methods illustrated in Section 5.3.6. In particular, equation (5.3.41) is still valid and reads (recall that $A=2$)

$$\sigma^2 = \frac{\gamma}{4(1-\gamma)} \sum_{m=-\infty}^{\infty} R_N(m)(1-2\gamma)^{|m|} \qquad (5.4.35)$$

where the step size $\gamma$ is related to the loop noise bandwidth by (see (5.3.46))

$$\gamma \approx 2B_L T \qquad (5.4.36)$$

**Exercise 5.4.1.** Compute the phase error variance of a Costas loop when $h(t)$ is a raised cosine Nyquist pulse with rolloff factor $\alpha=1$.
*Solution.* For $\alpha = 1$ the samples of $h(t)$ at $t = kT - T/2$ are given by

$$h(k - 1/2) = \begin{cases} 1/2 & k = 0, 1 \\ 0 & \text{elsewhere} \end{cases} \qquad (5.4.37)$$

Substituting into (5.4.32) yields

$$R_{SN}(m) = \begin{cases} 0.5 & m = 0 \\ -0.25 & m = \pm 1 \\ 0 & \text{elsewhere} \end{cases} \qquad (5.4.38)$$

and from (5.4.33)-(5.4.34) we obtain

$$R_N(m) = \begin{cases} 0.5 + \dfrac{2}{E_s/N_0} & m = 0 \\ -0.25 & m = \pm 1 \\ 0 & \text{elsewhere} \end{cases} \qquad (5.4.39)$$

# Carrier Phase Recovery with Linear Modulations

Finally, collecting (5.4.35)-(5.4.36) and assuming $\gamma \ll 1$ gives the desired result

$$\sigma^2 \approx B_L T \cdot (E_s/N_0)^{-1} + B_L^2 T^2 \qquad (5.4.40)$$

Figure 5.25 illustrates simulation results corresponding to $B_L T = 5 \cdot 10^{-3}$. The modified Cramer-Rao bound is also shown for comparison. This bound has been computed letting

$$L_0 = \frac{1}{2 B_L T} \qquad (5.4.41)$$

As is seen, the simulations are quite close to the bound at signal-to-noise ratios about 10 dB. It is also apparent that, as $E_s/N_0$ increases, the error variance has a floor corresponding to the self-noise term $B_L^2 T^2$ in (5.4.40).

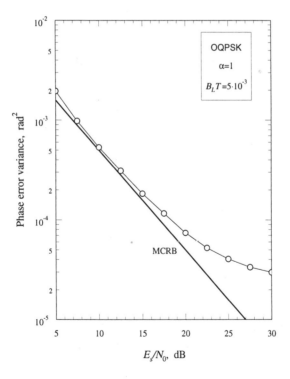

Figure 5.25. Estimation variance with Nyquist pulses with $\alpha = 1$.

### 5.4.3. Effects of Phase Noise and Frequency Errors

The effects of phase noise and frequency errors can be investigated with the same methods used with non-offset formats and essentially the same conclusions are arrived at. In particular, in the presence of a constant error $f_d$, the loop equivalent model in Figure 5.20 is still valid provided that we set

$$\theta(k) = \theta(k-1) + 2\pi f_d T \qquad (5.4.42)$$

## 5.5. Multiple Phase-Recovery with Trellis-Coded Modulations

### 5.5.1. Tentative Decisions

In discussing the tracking performance of Costas loops we did not distinguish between coded and uncoded modulations. We only assumed that the detector decisions are (almost always) correct in either case. With trellis-coded modulation, however, a question arises insofar as the detector reliability depends on the decision delay and it is not clear where the break-even point is between having good decisions or short delays. As we shall see, this issue will prompt some important qualifications on the results of the previous sections and, ultimately, will open the way to improvements on the decision-directed algorithms discussed earlier.

Let us start with some definitions. Denote by $\{...,c_{k-3},c_{k-2},c_{k-1}\}$ the transmitted sequence up to time $kT$, and let $S_k \in \{0,1,...,Q-1\}$ be the generic node in the encoder trellis at that time. Also, call $\{...,\hat{c}_{k-3}^{(m)},\hat{c}_{k-2}^{(m)},\hat{c}_{k-1}^{(m)}\}$ the sequence associated with the survivor path entering the $m$-th node (see Figure 5.26) and $\lambda(m,\tilde{c}_k)$ the metric for the branch stemming from $S_k=m$ and bearing the symbol label $\tilde{c}_{k+1}$. If the Nyquist condition is satisfied, the samples at the matched-filter output are expressed by

$$x(k) = c_k e^{j\theta} + n(k) \qquad (5.5.1)$$

and the metric $\lambda(m,\tilde{c}_k)$ takes the form [30]

$$\lambda(m,\tilde{c}_k) = \left| x(k) - \tilde{c}_k e^{j\hat{\theta}(k)} \right|^2 \qquad (5.5.2)$$

where $\hat{\theta}(k)$ is the carrier phase estimate at $kT$.

According to the formulation in the previous section the sequence $\{\hat{\theta}(k)\}$ is computed from equations (5.3.18)-(5.3.19) which are repeated here for convenience

# Carrier Phase Recovery with Linear Modulations

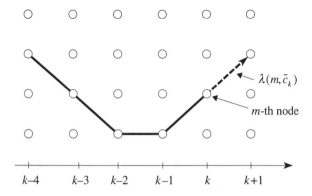

Figure 5.26. Encoder trellis for a four-state code.

$$\hat{\theta}(k+1) = \hat{\theta}(k) + \gamma e(k) \quad (5.5.3)$$

$$e(k) = \text{Im}\left\{\hat{c}_k^* x(k) e^{-j\hat{\theta}(k)}\right\} \quad (5.5.4)$$

Looking at (5.5.4), however, a first problem arises with $\hat{c}_k^*$ insofar as the Viterbi detector makes final decisions with some delay $\Delta$ and, in consequence, only the estimates $\{\ldots, \hat{c}_{k-\Delta-3}, \hat{c}_{k-\Delta-2}, \hat{c}_{k-\Delta-1}\}$ are available at time $kT$. A possible solution is to replace $e(k)$ in (5.5.3) by its delayed version

$$e(k-\Delta) = \text{Im}\left\{\hat{c}_{k-\Delta}^* x(k-\Delta) e^{-j\hat{\theta}(k-\Delta)}\right\} \quad (5.5.5)$$

In this way a delay is introduced in the tracking loop, however. As $\Delta$ may be of the order of 15-20 symbol intervals or greater, this would degrade the tracking performance and could even make the loop unstable, depending on the noise bandwidth.

As a compromise, a smaller delay $D$ may be employed, which means updating $\hat{\theta}(k)$ according to

$$\hat{\theta}(k+1) = \hat{\theta}(k) + \gamma e(k-D) \quad (5.5.6)$$

$$e(k-D) = \text{Im}\left\{\hat{c}_{k-D}^* x(k-D) e^{-j\hat{\theta}(k-D)}\right\} \quad (5.5.7)$$

Naturally, the problem arises as to where $\hat{c}_{k-D}^*$ is to be taken from and which is the best value for $D$. A reasonable answer to the first question comes from presuming that the sequence $\{\ldots, \hat{c}_{k-3}, \hat{c}_{k-2}, \hat{c}_{k-1}\}$ associated with the *best* *survivor* at $t=kT$ is sufficiently reliable for synchronization purposes. In other words, any one of its elements can be used to compute (5.5.7). In the parlance of Viterbi decoding, decisions $\hat{c}_{k-D}$ with a small $D$ are referred to as *tentative* decisions. They should be distinguished from the *final* decisions which are delivered with a larger delay and are taken from the subpath in common with all the survivors (see Figure 5.27).

The second question is concerned with the optimum $D$. Intuitively, a large $D$ corresponds to more reliable decisions but, as we shall see shortly, it produces larger tracking errors. This may be seen by deriving the loop equivalent model for the new equations (5.5.6)-(5.5.7) and computing the variance of the phase errors as a function of $D$. The first step is accomplished with the methods of Section 5.3.6 and leads to the block diagram in Figure 5.28. Comparing it with Figure 5.9, it is seen that the only change is in the loop delay; all the other parameters are the same. Also, assuming negligible phase noise, the phase error variance is given by

$$\sigma^2 \approx \frac{S_N(0)}{A^2} 2B_L \qquad (5.5.8)$$

As is now explained, the dependence of $\sigma^2$ on the parameter $D$ is hidden in the expression of the loop equivalent bandwidth. In fact, the loop transfer function is readily derived from Figure 5.28 and reads

$$H(z) = -\frac{\gamma}{z^{D+1} - z^D + \gamma A} \qquad (5.5.9)$$

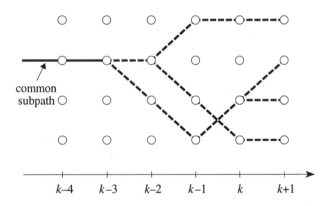

Figure 5.27. Survivors merge in a common path.

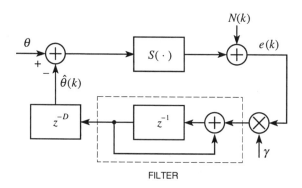

Figure 5.28. Mathematical model with delayed decisions.

Then, with the methods indicated in [32] the loop bandwidth is found to be

$$B_L T = \begin{cases} \dfrac{K}{2(2-K)} & \text{for } D=0 \\[6pt] \dfrac{K(1+K)}{2(1-K)(2+K)} & \text{for } D=1 \\[6pt] \dfrac{K(1+K-K^2)}{2(2-3K-K^2+K^3)} & \text{for } D=2 \\[6pt] \dfrac{K(1+2K-K^2-K^3)}{2(2-3K-4K^2+K^3+K^4)} & \text{for } D=3 \end{cases} \quad (5.5.10)$$

with $K \triangleq \gamma A$. For example, for $K=0.04$ we have

$$B_L T = \begin{cases} 1.00 \cdot 10^{-2} & \text{for } D=0 \\ 1.06 \cdot 10^{-2} & \text{for } D=1 \\ 1.11 \cdot 10^{-2} & \text{for } D=2 \\ 1.15 \cdot 10^{-2} & \text{for } D=3 \end{cases} \quad (5.5.11)$$

These results show that increasing $D$ makes the loop bandwidth grow larger and the tracking performance deteriorates (since the error variance is proportional to $B_L$). For instance, in passing from $D=0$ to $D=3$ the error variance increases by 0.6 dB.

Strictly speaking the foregoing argument might be flawed as, in deriving

(5.5.8), the symbol decisions have been assumed correct while they need not be so. In other words, one might argue that increasing $D$ would improve the decision reliability and, correspondingly, the tracking performance. Simulations show that this is not the case, however. Rather, they indicate that equation (5.5.8) is accurate at reasonably high SNR values, which implies that the optimum decision delay is $D=0$. It should be pointed out, however, that this conclusion is only true with negligible phase noise. In the presence of phase noise the optimum delay may not be zero.

### 5.5.2. Multiple Synchronizers

Returning to the notion of tentative decisions, we have seen that a reasonable approach is to take them from the best survivor. An intuitive motivation for this is that the best survivor provides the most reliable decisions *at the current time*. On occasions, however, the accumulated metric of the best survivor may be only slightly different from the others, as happens when the Viterbi algorithm is operating in "difficult" conditions (for example, when trying to recover from a loss-of-lock). Under these circumstances the motivation in favor of any specific survivor is weak and one wonders whether more clever strategies can be found. The answer is difficult as long as we think of a single phase estimator. But what if several estimators are allowed? A possible answer is as follows.

Suppose we have as many estimators as the number of survivors. The generic survivor, say the one arriving at the $m$-th node at time $kT$, generates a phase estimate based on its own tentative decisions. This is performed by updating the previous estimate according to (see also (5.5.3)-(5.5.4))

$$\hat{\theta}^{(m)}(k+1) = \hat{\theta}^{(m)}(k) + \gamma e^{(m)}(k) \qquad (5.5.12)$$

$$e^{(m)}(k) = \text{Im}\left\{\hat{c}_k^{(m)*} x(k) e^{-j\hat{\theta}^{(m)}(k)}\right\} \qquad (5.5.13)$$

where $\{...,\hat{c}_{k-3}^{(m)},\hat{c}_{k-2}^{(m)},\hat{c}_{k-1}^{(m)}\}$ is the sequence associated with that survivor. Then, the metrics

$$\lambda(m,\tilde{c}_k) = \left|x(k) - \tilde{c}_k e^{j\hat{\theta}^{(m)}(k)}\right|^2 \qquad (5.5.14)$$

are employed for the branches stemming from the $m$-th node in the detection algorithm.

Clearly, the above scheme implies an array of synchronizers. This concept has been proposed in [36]-[38] and may be viewed as an application of a general principle dubbed *per-survivor-processing* (PSP) [39]. The PSP concept aims at reducing the complexity of a Viterbi detector whenever the received

# Carrier Phase Recovery with Linear Modulations 241

signal has unknown parameters. The crucial idea is that these parameters (the carrier phase, in our case) must be estimated on the basis of the data sequence associated with each survivor and the estimates must be incorporated into the Viterbi algorithm.

A natural question arises as to the pros and cons of multiple tracking. One obvious drawback is that multiple tracking needs extra processing. The bright side is that there are significant advantages when operating under "difficult" conditions, such as those occurring in the presence of phase noise or during re-acquisitions. Marginal gains are expected, instead, during "normal" tracking operations.

Simulations shown below illustrate this superior performance of multiple phase trackers; further confirmations may be found in [37]. The comparisons to follow focus on three performance indicators: (*i*) bit error rate (BER) versus $E_s/N_0$; (*ii*) mean time to slip; (*iii*) acquisition time. 8PSK trellis-coded modulation is considered, with four- or eight-state coding schemes described in Ungerboeck's paper [30] and indicated as UNG-4 and UNG-8 respectively. An equivalent noise bandwidth of $B_L T = 10^{-2}$ is assumed and the tentative decision delay $D$ is set to zero, both with single and multiple tracking. Finally, the phase noise is modeled as a Wiener process

$$\theta(k+1) = \theta(k) + \Delta(k) \qquad (5.5.15)$$

where the $\Delta(k)$ are independent and zero-mean Gaussian random variables with the same standard deviation $\sigma_\Delta$.

Figure 5.29 illustrates BER versus $E_s/N_0$ in the absence of phase noise. The solid curve corresponds to feeding the Viterbi detector with an ideal phase reference. It appears that single tracking (ST) and multiple tracking (MT) have approximately the same performance, with differences of only 0.1 dB in terms of $E_s/N_0$. By contrast, Figure 5.30 indicates that multiple tracking allows power savings of 2-3 dB in the presence of phase noise with $\sigma_\Delta = 1.3°$.

Cycle slipping has been mentioned in Section 5.3.4 in the context of synchronization failures of Costas loops and is resumed here to outline the differences between single and multiple tracking. The detector fails when the phase errors deviate enough from a stable point in the S-curve. This degrades the estimation process and may induce even larger phase errors, to the point that the detector decisions becomes totally random. Under such circumstances the estimation mechanism stops working properly and we say that a loss-of-lock has occurred.

Out-of-lock operation of a Costas loop may be described as a random walk. Recovery from a loss-of-lock takes place when, by chance, the current phase estimate comes close to some stable point. At that stage the detector decisions improve and the phase estimate is steered toward a steady-state condition.

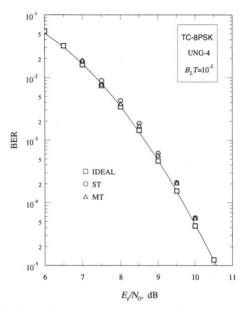

Figure 5.29. Single tracking versus multiple tracking in the absence of phase noise.

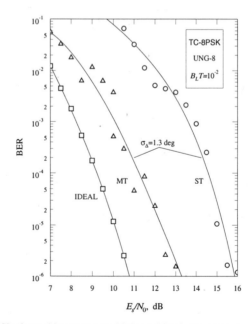

Figure 5.30. Single tracking versus multiple tracking in the presence of phase noise.

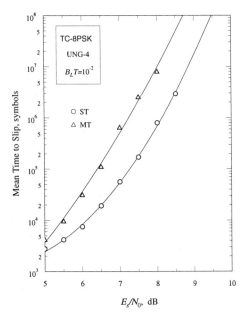

Figure 5.31. Mean time to slip with UNG-4 coding.

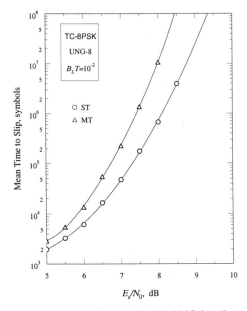

Figure 5.32. Mean time to slip with UNG-8 coding.

Now, let us see what happens with a multiple synchronizer. Several phase estimators are operating simultaneously and a cycle slip takes place when *all the estimators* fail at the same time. Conversely, the cycle slip finishes when *one of the estimators*, whichever it is, approaches a stable operating point.

Differences between multiple and single tracking should be clear at this stage: (*i*) the chances of losing lock seem slimmer when several trackers are pursuing the carrier phase; (*ii*) assuming that a loss-of-lock has occurred, the chances of approaching a stable point seem better when several trackers are searching around. Based on these intuitive considerations we expect that multiple tracking may be useful to reduce slip rates (or, equivalently, to increase the time separation between slips) and to shorten reacquisitions. These conclusions are confirmed by the computer simulations in Figures 5.31-5.32.

They show the mean time to slip (MTS) versus $E_s/N_0$. To give an example, consider phase recovery for the single-channel-per-carrier INTELSAT system, wherein slip rates less than one per minute are required at $E_s/N_0 = 10$ dB. With a symbol rate of 64 kb/s, this means an MTS of $2 \cdot 10^6$ symbols or greater. From Figures 5.31-5.32 we see that a multiple tracker gains about 0.7 dB at MTS=$2 \cdot 10^6$.

Figures 5.33-5.34 illustrate acquisition times with single and multiple trackers. Acquisition time (AT) is defined as follows. Suppose that the (single/multiple) phase recovery algorithm is running in its tracking mode. At time zero a constant offset $\Delta\phi$ is applied to the carrier phase and the resulting system evolution is monitored until the best phase estimate enters a strip of

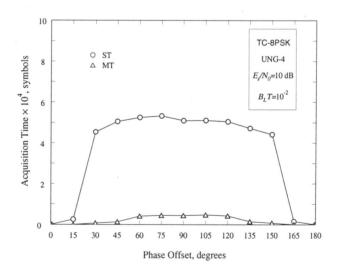

Figure 5.33. Acquisition time with UNG-4.

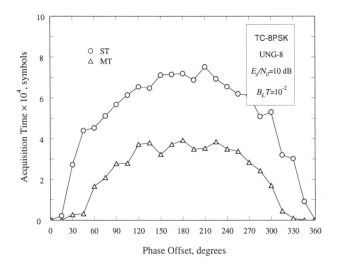

Figure 5.34. Acquisition time with UNG-8.

±10° around a stable point in the S-curve. The number of symbol periods elapsed from the start is taken as AT. Since AT is a random variable, its average over 100 acquisitions has been computed to draw each point. Especially with UNG-4 it appears that multiple tracking allows significant acquisition time reductions.

## 5.6. Phase Tracking with Frequency-Flat Fading

### 5.6.1. Channel Estimation Problem

So far only additive white Gaussian noise (AWGN) channels have been considered. In this section we investigate phase recovery for transmission over flat-fading channels [40]-[42]. An overview of such channels is first provided for completeness.

We assume that the transmission medium has multiple propagation paths, with associated delays located in a small range compared with the inverse of the signal bandwidth. In these conditions the channel transfer function is constant over the signal bandwidth and the channel is said to be *frequency-flat* for it affects the signal only through a (complex) multiplying factor. However, as the structure of the medium is time-varying, the factor varies in time and (assuming many propagation paths) becomes a complex-valued Gaussian random process

$$a(t) = \rho(t)e^{j\theta(t)} \tag{5.6.1}$$

When the paths are only due to reflections from *randomly moving* scatterers, the process has zero mean and, at any given time, its envelope has a Rayleigh probability density function

$$p(\rho) = 2\rho e^{-\rho^2} \quad \rho \geq 0 \tag{5.6.2}$$

where, without loss of generality, it has been assumed that $E\{\rho^2\}=1$. Under these circumstances we speak of a frequency-flat *Rayleigh channel*. In the presence of *fixed* scatterers, vice versa, the process has a non-zero mean, its envelope is Rice distributed, and the channel is said to be *Ricean*. The probability density function of $\rho(t)$ takes the form (assuming again $E\{\rho^2\}=1$)

$$p(\rho) = 2\rho(K+1)e^{-(K+1)\rho^2 - K} I_0\left(2\rho\sqrt{K(K+1)}\right), \quad \rho \geq 0 \tag{5.6.3}$$

In this equation $I_0(x)$ is the zero-order modified Bessel function of the first kind and $K$ is the ratio of the power from the fixed scatterers to the (average) power from the moving scatterers. Equation (5.6.3) reduces to a Rayleigh distribution for $K=0$, whereas it tends to a delta function $\delta(\rho-1)$ as $K$ tends to infinity. This is intuitively clear since $K=\infty$ means that there are no moving scatterers and the channel is just AWGN.

The fading autocorrelation function

$$R(\tau) = E\{a(t+\tau)a^*(t)\} \tag{5.6.4}$$

gives a measure of the rapidity of the channel variations. For example, when the channel varies slowly, $R(\tau)$ is spread over a rather long interval. Vice versa, when it varies quickly, $R(\tau)$ fades away very soon. The duration of $R(\tau)$, as indicated in Figure 5.35, is referred to as the *coherence time* of the channel.

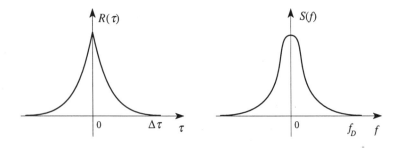

Figure 5.35. $R(\tau)$ and $S(f)$ are Fourier transform pairs.

## Carrier Phase Recovery with Linear Modulations

Another way to express the rapidity of the channel variations is by means of the Fourier transform of $R(\tau)$, $S(f)$, which is referred to as the *Doppler spectrum*. The bandwidth $f_D$ of this spectrum is the *Doppler bandwidth*. This is approximately the inverse of the coherence time $\Delta\tau$.

Simple Doppler spectra are adopted in theoretical studies. With satellite mobile channels and/or HF channels a useful spectrum shape is obtained by passing white noise through either Butterworth or Gaussian shaping filters. This results in either

$$S(f) = \frac{S(0)}{1+(f/f_D)^{2n}} \tag{5.6.5}$$

($n$ denotes the number of filter poles) or

$$S(f) = S(0)\exp\left\{-\frac{f^2}{2f_D^2}\right\} \tag{5.6.6}$$

With land cellular mobile communications it has been found [43] that the Doppler spectrum is well approximated by

$$S(f) = \frac{S(0)}{\sqrt{1-(f/f_D)^2}}, \quad |f| \leq f_D \tag{5.6.7}$$

In the above equations $f_D$ is given by

$$f_D = \frac{v}{\lambda} \tag{5.6.8}$$

where $v$ is the mobile vehicle speed and $\lambda$ the transmission wavelength.

When a PAM signal is transmitted over a frequency-flat channel, the received waveform has the form

$$r(t) = s(t) + w(t) \tag{5.6.9}$$

with

$$s(t) = a(t)e^{j2\pi vt}\sum_{i=0}^{N-1} c_i g(t-iT-\tau) \tag{5.6.10}$$

Note that the carrier phase does not appear in (5.6.10) as it has been incorporated into the complex fading process.

In many practical cases the symbol rate is large enough (compared with the

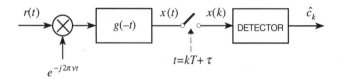

Figure 5.36. Receiver block diagram.

Doppler bandwidth) that $a(t)$ is nearly constant over the duration of the generic pulse $g(t-iT-\tau)$. In these conditions (*slow fading*) the following approximation can be made:

$$a(t)g(t - iT - \tau) \approx a(i)g(t - iT - \tau) \quad (5.6.11)$$

where $a(i)$ is short for $a(iT+\tau)$. Then, equation (5.6.10) becomes

$$s(t) = e^{j2\pi vt} \sum_{i=0}^{N-1} a(i)c_i g(t - iT - \tau) \quad (5.6.12)$$

Assuming that both frequency offset and timing epoch are known and the convolution $g(t) \otimes g(-t)$ satisfies the Nyquist condition, the received waveform may be processed as indicated in Figure 5.36 to obtain

$$x(k) = a(k)c_k + n(k) \quad (5.6.13)$$

If the channel gains $\{a(k)\}$ were known, the data sequence could be optimally detected in one of the following manners [44].

### 5.6.1.1. Uncoded modulations

The detector makes symbol-by-symbol decisions according to the rule

$$\hat{c}_k = \arg\left\{\min_{\tilde{c}_k}\left\{|x(k) - a(k)\tilde{c}_k|^2\right\}\right\} \quad (5.6.14)$$

where $\tilde{c}_k$ is the generic symbol from the signal alphabet.

### 5.6.1.2. Coded modulations

The optimum receiver is a maximum likelihood (ML) sequence detector. Denoting $\tilde{c} \triangleq \{\tilde{c}_0, \tilde{c}_1, \ldots, \tilde{c}_{N-1}\}$ the generic coded sequence, its task is to look for that sequence that minimizes the Euclidean distance of $\{a(k)\tilde{c}_k\}$ to $\{x(k)\}$:

# Carrier Phase Recovery with Linear Modulations

$$\hat{c} = \arg\left\{\min_{\tilde{c}}\left\{\sum_{k=0}^{N-1}|x(k) - a(k)\tilde{c}_k|^2\right\}\right\} \quad (5.6.15)$$

### 5.6.1.3. Coded modulations with interleaving

Interleaving/de-interleaving techniques are often used with coded modulation to scramble the symbols and protect the decoding process against deep fades. With interleaving, the received samples $x(k)$ are first descrambled and then are fed to a maximum likelihood detector. Calling $\{\bar{x}(k)\}$ the descrambled sequence and $\{\bar{a}(k)\}$ the corresponding descrambled channel gains, the detector looks for that sequence $\tilde{c} \triangleq \{\tilde{c}_0, \tilde{c}_1, \ldots, \tilde{c}_{N-1}\}$ that minimizes the Euclidean distance of $\{\bar{a}(k)\tilde{c}_k\}$ to $\{\bar{x}(k)\}$

$$\hat{c} = \arg\left\{\min_{\tilde{c}}\left\{\sum_{k=0}^{N-1}|\bar{x}(k) - \bar{a}(k)\tilde{c}_k|^2\right\}\right\} \quad (5.6.16)$$

From the above discussion it is seen that channel gains are needed for optimum detection in any case. As they are unknown, however, they must be estimated in some manner. Clearly, channel gain estimation is a generalization of the phase recovery problem encountered with AWGN channels where the channel phase shift is the only parameter of interest (attenuation is compensated for by some automatic gain control circuit).

Interestingly enough, channel amplitude is not always necessary with fading channels. For example, consider uncoded $M$-ary PSK symbols, i.e.,

$$c_k = e^{j\alpha_k}, \quad \alpha_k \in \{0, 2\pi/M, \ldots, 2\pi(M-1)/M\} \quad (5.6.17)$$

and call $\tilde{c}_k = e^{j\tilde{\alpha}_k}$ a trial value of $c_k$. Then, letting $a(k) = \rho(k)e^{j\theta(k)}$ and bearing in mind that

$$\left|x(k) - \rho(k)e^{j\theta(k)}e^{j\tilde{\alpha}_k}\right|^2 = |x(k)|^2 + |\rho(k)|^2 - 2\rho(k)\operatorname{Re}\left\{x(k)e^{-j[\theta(k)+\tilde{\alpha}_k]}\right\} \quad (5.6.18)$$

equation (5.6.14) becomes

$$\hat{\alpha}_k = \arg\left\{\max_{\tilde{\alpha}_k}\left\{\operatorname{Re}\left\{x(k)e^{-j[\theta(k)+\tilde{\alpha}_k]}\right\}\right\}\right\} \quad (5.6.19)$$

which does not involve $\rho(k)$. Thus, only channel phase information is needed with uncoded PSK.

Unfortunately, the disappearance of $\rho(k)$ from the detection rule does not necessarily imply that carrier phase can be derived with the same methods discussed in the context of AWGN channels. Actually, phase fluctuations have rather different dynamics in the presence of fading. To explain this point, consider the occurrence of a deep fade. Figure 5.37(a) depicts a possible trajectory of $a(t)$ during a fade, whereas Figure 5.37(b) shows the corresponding variations of $\rho(t)$ and $\theta(t)$. It is seen that, as $\rho(t)$ approaches a minimum, $\theta(t)$ varies quite rapidly and, in fact, it undergoes a jump of about 180° in a short time. Unfortunately, such a jump may be difficult to track with a Costas loop since the loop bandwidth must be kept small to limit tracking errors under unfaded conditions. A deep fade will likely make the Costas loop lose lock.

Intuitively, the occurrence of deep fades depends on the Rice parameter $K$. With $K$ values on the order of 20 dB or so, deep fades are so rare that the behavior of a Costas loop is much the same as with no fading. In many mobile radio applications, however, $K$ may be far smaller. For example, most channels in urban areas are Rayleigh and even some satellite links operating over rural areas have $K$ values as small as 5-8 dB, as a consequence of shadowing effects due to foliage. In these circumstances the performance of Costas loops is poor and the bit-error-rate curve exhibits a high floor due to synchronization losses [45].

In conclusion, standard phase recovery schemes are not useful with most fading channels. An overview of alternative methods is provided in the following. They are *pilot-tone assisted* and *pilot-symbol assisted* schemes which, in some way, provide an embedded reference for the receiver. More recently, improved methods have been proposed where channel gain is estimated through per-survivor-processing techniques.

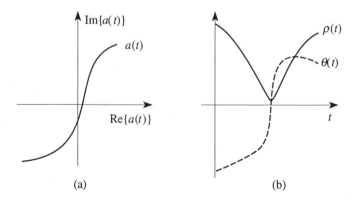

Figure 5.37. Deep fade phenomenology.

## 5.6.2. Pilot-Tone Assisted Modulation

We limit ourselves to the most successful of the pilot-tone assisted methods, the so-called *transparent tone-in-band* (TTIB) method [46]. In essence the TTIB notion is to split the original signal spectrum into two subbands so as to create a gap for transmitting a pilot tone (PT). Figure 5.38 illustrates the band splitting, as obtained by suitable filtering and frequency translation of the signal components.

The tone location at the center of the spectrum intends to make the PT undergo the same distortions as the signal components. Assuming that this goal is achieved and normalizing the tone level to unity produces the received waveform

$$r(t) = a(t)[s(t)+1] + w(t) \qquad (5.6.20)$$

which indicates that the complex envelope of the received PT is just $a(t)$. Thus, separating $a(t)$ from the data subbands (by low-pass filtering) and computing the ratio $r(t)/a(t)$ yields the sum of the undistorted signal plus noise

$$\frac{r(t)}{a(t)} = s(t) + \frac{w(t)}{a(t)} \qquad (5.6.21)$$

One important issue about TTIB is the width of the gap between subbands. This must be sufficiently large to accomodate the PT Doppler spreading and satisfy some practical requirements imposed by the subband recombination mechanism [46]. Figure 5.39 illustrates the separation of the PT from the signal components and the compensation for the channel fading. It also indicates that subband recombination is needed prior to detection. It turns

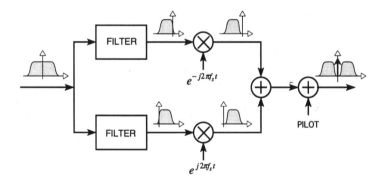

Figure 5.38. Transmitter block diagram.

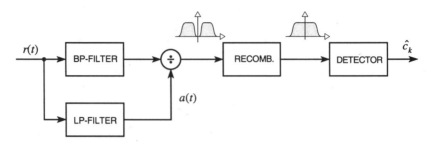

Figure 5.39. Sub-band recombination at the receiver.

out that TTIB needs a bandwidth increase of twice the Doppler bandwidth for the tone, plus something of comparable size for proper recombination. For example, with a Doppler bandwidth equal to 5% of the symbol rate, TTIB needs a fractional increase of about 20% in signal bandwidth.

Another issue about TTIB is the peak to average power ratio (P/A) in the transmitted waveform. The *peak power* cannot exceed the saturation power of the RF amplifier. On the other hand, it is the signal *average* power that establishes the receiver performance against thermal noise. So, a high P/A translates into a penalty in terms of SNR. It turns out that the P/A is 3-4 dB larger than in conventional systems as a consequence of the band splitting and the pilot-tone addition. For this and other reasons it has been concluded that TTIB is not a good solution in most applications [47].

### 5.6.3. Pilot-Symbol Assisted Modulation

Pilot symbol assisted modulation (PSAM) has been discussed in [48]-[51] as an alternative to TTIB. PSAM systems can be described as indicated in the block diagram of Figure 5.40. At the transmitter, known symbols are multiplexed with the data sequence in a ratio of 1 to $L-1$ data. This results in a framed structure, each frame formed by $L$ symbols. The sequence is filtered and transmitted over the channel. For simplicity we consider Rayleigh fading and, for the time being, uncoded PSK. Extensions to Rice fading and other uncoded modulation schemes are straightforward. Coded schemes are addressed afterwards.

At the receiver the incoming waveform is filtered and sampled at the symbol rate. Assuming perfect timing information and an overall Nyquist pulse shaping, the samples from the matched filter have the form indicated in (5.6.13). They are split into two streams: a data stream (upper branch) and a reference stream (lower branch). The latter is decimated and only the samples corresponding to known data are kept. The useful samples are denoted

# Carrier Phase Recovery with Linear Modulations

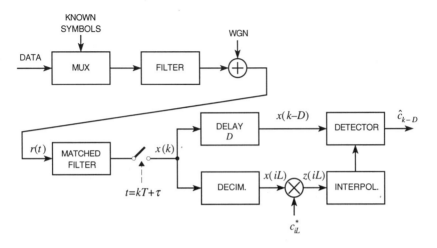

Figure 5.40. Block diagram of a PSAM system.

$$x(iL) = a(iL)c_{iL} + n(iL), \quad i = 0, 1, 2, \ldots \quad (5.6.22)$$

Next, modulation is removed from these samples by performing the products $x(iL)c_{iL}^* \triangleq z(iL)$ (recall that $c_{iL}c_{iL}^* = 1$). This results in

$$z(iL) = a(iL) + n'(iL) \quad (5.6.23)$$

where $n'(iL) \triangleq n(iL)c_{iL}^*$ is white Gaussian noise equivalent to $n(iL)$. Equation (5.6.23) indicates that $z(iL)$ are noisy measurements of $a(iL)$, one measurement every $L$ symbols. They are fed to an interpolator which provides fading estimates *at symbol rate*.

Calling $D$ the delay inherent in the estimation process, the estimates $\hat{a}(k-D)$ are fed to the detector. Decisions are made according to the rule

$$\hat{c}_{k-D} = \arg\left\{\min_{\tilde{c}_{k-D}}\left\{\left|x(k-D) - \hat{a}(k-D)\tilde{c}_{k-D}\right|^2\right\}\right\} \quad (5.6.24)$$

Extension to trellis-coded modulation is straightforward. When no interleaving is used, the communication scheme is still as in Figure 5.40, but the detector consists of a Viterbi decoder operating according to (5.6.15) wherein the true channel gains are replaced by their estimates. With interleaved systems, the encoded data are interleaved prior to multiplexing. Also, two separate de-interleavers are used at the receiver as indicated in Figure 5.41. Calling $\{\bar{x}(k)\}$ the de-interleaved sequence and $\{\bar{a}(k)\}$ the de-interleaved

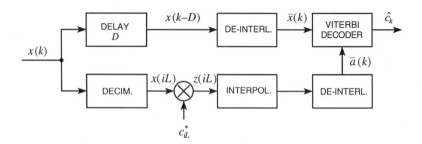

Figure 5.41. Block diagram of a PSAM receiver with de-interleaving.

channel estimates, the detector operates according to (5.6.16) where, again, estimated gains are used in place of the true gains.

There are two major issues with PSAM. One is the delay involved in the interpolation process. Since several channel measurements are needed to get good estimates, the receiver must wait and buffer them for several frames. Whether this is a serious drawback or not depends on the application. For speech transmission purposes, for example, the modem delay should be limited to 50 ms. The other problem is the interpolation technique and its implementation complexity. These two issues are now overviewed.

Before proceeding, however, we concentrate on the frame length $L$. An obvious question is about the $L$ values that allow satisfactory channel estimates. Clearly, decreasing $L$ tends to shorten the estimation delay but wastes energy in unnecessary pilot symbols. Sampling theory says that $a(t)$ can be recovered from the samples $\{a(iL)\}$ provided that $1/(LT)$ exceeds twice the Doppler bandwidth, i.e.,

$$L \le \frac{1}{2f_D T} \qquad (5.6.25)$$

For example, with a symbol rate of 2400 symbols/s and $f_D=0.05/T$, the frame length must not exceed 10. Choosing $L=10$ results in a fractional increase of signal bandwidth of 10% of the symbol rate. As noted earlier, the increase would be about twice as large with TTIB.

Several schemes have been proposed for interpolating noisy channel measurements. In [49] a low-pass filter is first used to smooth out the noise. Then, adjacent filtered samples are linearly interpolated. The filter is fixed and is designed for the worst fading conditions (the largest predictable Doppler bandwidth). The interpolation delay is not mentioned in [49] but it is expected to be a few frames.

A more powerful method is proposed in [50]. The sequence $z(iL)$ is fed to a Kalman smoother (KS) and a linear filter is used to interpolate between smoothed samples. If the fading model were perfectly known, the KS would provide optimum channel estimation (in the minimum mean-square-error sense). In practice, fading is time varying and the scheme becomes suboptimum since the KS structure is fixed. An adaptive (extended) Kalman smoother could be designed but the resulting complexity would be quite considerable. More realistically, the KS may be optimized for one operating point (perhaps, the most demanding one) and then used anyway, even if conditions change.

A third method makes use of a bank of Wiener filters [51]. Assume that the gains $a(iL+l)$, $l = 0,1,\ldots,L-1$, must be estimated on the basis of $2M+1$ fading measurements $z[(i+m)L]$, $m = 0,\pm 1,\ldots,\pm M$, as indicated in Figure 5.42. To this end a linear combination of the measurements is formed

$$\hat{a}(iL+l) = \sum_{m=-M}^{M} \gamma(m,l) z[(i+m)L] \qquad (5.6.26)$$

and the coefficients $\gamma(m,l)$ are chosen so as to minimize the mean square error

$$J(l) = \mathrm{E}\left\{\left|a(iL+l) - \sum_{m=-M}^{M} \gamma(m,l) z[(i+m)L]\right|^2\right\} \qquad (5.6.27)$$

Application of the orthogonality principle yields the optimum coefficients as the solution to the set of equations [41]

$$\mathrm{E}\left\{\left[a(iL+l) - \sum_{m=-M}^{M} \gamma(m,l) z[(i+m)L]\right] z^*[(i+p)L]\right\} = 0, \quad p = 0,\pm 1,\ldots,\pm M \qquad (5.6.28)$$

for $l = 0,1,\ldots,L-1$.

They may be put in a more suitable form by introducing the autocorrelation of $z(iL)$, $R_{zz}(pL) \triangleq \mathrm{E}\{z[(i+p)L]z^*(iL)\}$, and the cross-correlation of $a(iL)$ and $z(iL)$, $R_{az}(pL) \triangleq \mathrm{E}\{a[(i+p)L]z^*(iL)\}$. Then, (5.6.28) becomes

$$\sum_{m=-M}^{M} \gamma(m,l) R_{zz}(mL - pL) = R_{az}(l - pL), \quad p = 0,\pm 1,\ldots,\pm M \qquad (5.6.29)$$

On the other hand, letting $R_{aa}(pL) \triangleq \mathrm{E}\{a[(i+p)L]a^*(iL)\}$, it is seen from (5.6.23) that

$$R_{zz}(pL) = R_{aa}(pL) + 2N_0\delta(p) \qquad (5.6.30)$$

$$R_{az}(pL) = R_{aa}(pL) \qquad (5.6.31)$$

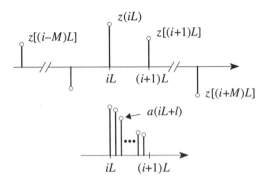

Figure 5.42. Estimation of $a(iL+l)$ based on $2M+1$ fading measurements.

where $\delta(p)$ is the Kronecker delta function

$$\delta(p) = \begin{cases} 1 & p=0 \\ 0 & p \neq 0 \end{cases} \quad (5.6.32)$$

Then, substituting into (5.6.29) yields

$$\sum_{m=-M}^{M} \gamma(m,l)\rho_{aa}(mL-pL) + \frac{2N_0}{R_{aa}(0)}\gamma(p,l) = \rho_{aa}(l-pL), \quad p=0,\pm 1,\ldots,\pm M$$
(5.6.33)

where $\rho_{aa}(pL) \triangleq R_{aa}(pL)/R_{aa}(0)$ is the fading autocorrelation coefficient. Solving (5.6.33) for $l = 0,1,\ldots,L-1$ gives the taps of the Wiener filters. Equation (5.6.33) indicates that the taps depend on $\rho_{aa}(pL)$ (a function of the fading rate) and the ratio $R_{aa}(0)/N_0$ which is proportional to SNR. As fading rate and SNR are time varying, Cavers [51] suggests computing the taps for the worst-case operating conditions.

The number of taps in each filter is an important issue as it affects the computing load, the channel estimation delay and the estimation performance. From Figure 5.42 it is seen that the estimation delay equals approximately the product $ML$. Accordingly, one would be tempted to keep this number as small as possible. On the other hand, a too small value would degrade performance as few channel measurements would be available for channel estimation. Thus, a trade-off is needed. Looking for the break-even point, observe that increasing $ML$ beyond a certain limit is useless as channel measurements far from the current estimation time $iL+l$ are uncorrelated with $a(iL+l)$ and therefore do not provide useful information. As the decorrelation distance is approximately $1/f_D T$, we expect that $ML$ values beyond this figure are unnecessary. This suggests the limitation

$$ML < \frac{1}{f_D T} \qquad (5.6.34)$$

Simulation results indicate that (5.6.34) is a reasonable rule-of-thumb formula. For example, with a Doppler bandwidth of $5 \cdot 10^{-2}/T$ the estimation delay is about 20 symbol intervals.

Returning to general PSAM schemes, one wonders how they perform in terms of BER. For simplicity we concentrate on uncoded PSK but the same conclusions are valid with trellis-coded modulation. To put the problem into perspective bear in mind that differential PSK (DPSK) is an obvious alternative to PSAM as it does not need channel information and is simpler to implement. In other words, DPSK is a natural reference in PSAM performance assessments. Thus, a few words about DPSK performance are in order.

With DPSK the information data $\{\eta_k\}$ are differentially encoded into PSK symbols as follows

$$c_k = c_{k-1} e^{j\eta_k} \qquad (5.6.35)$$

and the matched-filter output $x(k)$ is used to compute the statistics

$$z(k) = x(k) x^*(k-1) \qquad (5.6.36)$$

These are fed to the detector which makes decisions according to the rule

$$\hat{\eta}_k = \arg\left\{ \min_{\tilde{\eta}_k} \left\{ \left| z(k) - e^{j\tilde{\eta}_k} \right|^2 \right\} \right\} \qquad (5.6.37)$$

One drawback with DPSK is that, unless the Doppler bandwidth is very small compared with symbol rate, BER curves exhibit a floor which may be too high even for digital voice services [52]-[53]. The reason for this is readily seen from the decision statistic $z(k)$. In fact, collecting (5.6.13) and (5.6.35)-(5.6.36) yields

$$z(k) = a(k) a^*(k-1) e^{j\eta_k} + N(k) \qquad (5.6.38)$$

where $N(k)$ is a noise term. This equation says that the signal component is rotated by an angle $\psi(k) = \arg\{a(k) a^*(k-1)\}$ with respect to its ideal value $e^{j\eta_k}$. With slow fading $\psi(k)$ is small (as $a(k) \approx a(k-1)$) and the detector performance is dominated by thermal noise. Correspondingly, the BER curve decreases as $E_s/N_0$ increases. With fast fading, vice versa, $\psi(k)$ may be large and decision errors may occur even in the absence of noise. In these conditions BER curves exhibit a floor.

A comparison between PSAM and DPSK is illustrated in Figure 5.43 where simulation results are shown for QPSK with Rayleigh fading. The lower curve corresponds to an ideal coherent detector operating with perfect channel state information (CSI). The PSAM system employs 9 tap Wiener filters. Known symbols are multiplexed in the ratio 1/10. The fading process is generated by filtering two independent white Gaussian noise (WGN) sequences with two identical fourth-order Butterworth filters, as indicated in Figure 5.44. The 6-dB bandwidth of the filters is used as a measure of the fading bandwidth $f_D$. For convenience the expectation of $a(t)$ is chosen equal to unity so that the average received signal power coincides with the transmitted power. It appears that PSAM performs better at moderate values of the Doppler bandwidth (no floor is still visible at $f_D T=10^{-2}$). With fast fading PSAM and DPSK have the same floor, however.

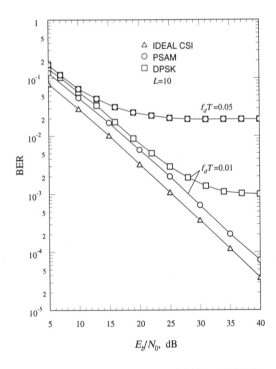

Figure 5.43. Comparison between PSAM and DPSK.

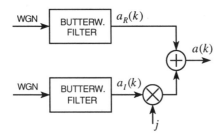

Figure 5.44. Generation of the fading process.

### 5.6.4. Per-Survivor Channel Estimation

Per-survivor processing (PSP) has been applied to maximum likelihood joint fading estimation and data detection [54]-[55]. Here we illustrate these techniques following the heuristic approach adopted in reference [54].

As an introduction to more efficient solutions, consider the following conceptually simple method. Modulation is PSK and the matched-filter output reads

$$x(k) = a(k)c_k + n(k), \quad k = 0,1,2,... \quad (5.6.39)$$

with

$$c_k \in \{e^{j2\pi l/M}, \quad l = 0,1,...,M-1\} \quad (5.6.40)$$

Suppose that the sequence $\{...,\hat{c}_{k-3},\hat{c}_{k-2},\hat{c}_{k-1}\}$ is available at time $k-1$ and we want to make a decision on the next symbol $c_k$. To this end we need an estimate $\hat{a}(k)$ of the channel gain for use in the decision rule

$$\hat{c}_k = \arg\left\{\min_{\tilde{c}_k}\left\{|x(k) - \hat{a}(k)\tilde{c}_k|^2\right\}\right\} \quad (5.6.41)$$

A possible procedure is as follows. First, past samples are multiplied by the corresponding conjugate decisions to form the products $x(k-i)\hat{c}^*_{k-i}$, $i = 1,2,...$ . Assuming that the detector decisions are correct so that $\hat{c}_{k-i}\hat{c}^*_{k-i} = 1$, this yields a sequence of channel gain estimates

$$x(k-i)\hat{c}^*_{k-i} = a(k-i) + n'(k-i), \quad i = 1,2,... \quad (5.6.42)$$

where $n'(k-i)$ is white Gaussian noise. These estimates are fed to a one-step Wiener predictor which provides the desired gain estimate

$$\hat{a}(k) = \sum_{i=1}^{N} \gamma(i) x(k-i) \hat{c}_{k-i}^{*} \qquad (5.6.43)$$

Optimum predictor coefficients are found by minimizing the mean square error between the true gain and its estimate. Paralleling the arguments in the foregoing section it is found that they satisfy the equations

$$\sum_{i=1}^{N} \gamma(i) \rho_{aa}(l-i) + \frac{2N_0}{R_{aa}(0)} \gamma(l) = \rho_{aa}(l), \qquad 1 \le l \le N \qquad (5.6.44)$$

Unfortunately, the above method has two drawbacks. One is that the estimates $\hat{a}(k)$ have a phase ambiguity by multiples of $\Delta\phi = 2\pi/M$. This is recognized by observing that, if $\{\hat{c}_k\}$ is a solution to (5.6.41) and (5.6.43), then $\{\hat{c}_k e^{jm\Delta\phi}\}$ is also a solution, for any integer $m$. In other words, sequences of the type $\{\hat{c}_k e^{jm\Delta\phi}\}$ are all legitimate solutions to (5.6.41) and (5.6.43) and the receiver cannot distinguish one from the other. This problem could be solved either by differential encoding/decoding or by multiplexing known symbols with the data (as is done with PSAM systems).

The second problem is more serious and is concerned with the algorithm stability. Computer simulations indicate that, even starting with correct decisions, sooner or later the detector makes errors. This deteriorates the channel predictions and, in consequence, induces further errors which, in their turn, worsen the predictions even further ... and so on until a breakdown in the algorithm takes place. These phenomena are referred to as "run-aways." They can be counteracted by multiplexing known symbols with the data so as to get "clean" channel measurements. Indeed, simulations indicate that high percentages of known symbols do mitigate the run-away problem but the overall detection process remains far poorer than with simple DPSK.

Things may be improved with multiple predictors, however, just as we have seen with multiple phase trackers in Section 5.5. To describe how this comes about consider $M$ predictors, as many as the points of the PSK constellation and concentrate on Figure 5.45 where a PSK constellation is repeated at multiples of the symbol period. Suppose that $M$ candidate sequences $\{...,\hat{c}_{k-3}^{(m)}, \hat{c}_{k-2}^{(m)}, \hat{c}_{k-1}^{(m)}\}$, $m = 1,2,...,M$, are already available at time $k-1$. They represent our best guess on the transmitted sequence $\{...,c_{k-3},c_{k-2},c_{k-1}\}$. Note that each sequence corresponds to a path in the trellis-like diagram in the figure and each path (*survivor*) terminates in a constellation point. For each path an estimate of the fading gain may be computed with a one-step predictor

$$\hat{a}^{(m)}(k) = \sum_{i=1}^{N} \gamma(i) x(k-i) \hat{c}_{k-i}^{(m)*}, \qquad m = 1,2,...,M \qquad (5.6.45)$$

# Carrier Phase Recovery with Linear Modulations

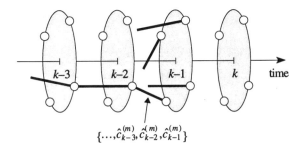

Figure 5.45. Survivor sequences at time $k-1$.

Also, the cumulative metric

$$\Lambda_{k-1}^{(m)} = \sum_{i=1}^{k-1} \left| x(i) - \hat{a}^{(m)}(i)\, \hat{c}_i^{(m)*} \right|^2 \qquad (5.6.46)$$

may be assigned to the $m$-th survivor to measure its agreement with the sample sequence $\{\ldots, x(k-3), x(k-2), x(k-1)\}$.

Assume that pilot symbols are multiplexed with the data in the ratio $1/L$. We want to extend the survivors one step further. To this end we distinguish according to whether the $k$-th symbol is a data or a pilot symbol. In the former case, the extension is made so as to minimize the accumulated metrics of the extended paths. For example (see Figure 5.46), the path arriving at the constellation point $e^{j2\pi(i-1)/M}$, $i = 1, 2, \ldots, M$, is selected as the extension of that survivor, say the $l$-th, such that

$$l = \arg\left\{ \min_{1 \le m \le M} \left\{ \Lambda_{k-1}^{(m)} + \left| x(k) - \hat{a}^{(m)}(k) e^{j2\pi(i-1)/M} \right|^2 \right\} \right\} \qquad (5.6.47)$$

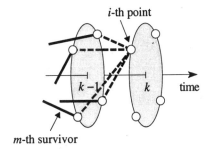

Figure 5.46. Extension rule when the $k$-th symbol is unknown.

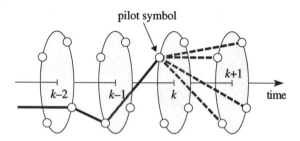

Figure 5.47. Extending the only survivor.

Vice versa, if $c_k$ is a pilot symbol, we know that the true path will pass through node $c_k$ at time $k$ and, accordingly, we take the path arriving at $c_k$ with the minimum accumulated metric as the *only survivor* at time $k$. This is the extension of the *l*-th survivor, where *l* is derived from

$$l = \arg\left\{\min_{1 \leq m \leq M}\left\{\Lambda^{(m)}_{k-1} + \left|x(k) - \hat{a}^{(m)}(k)c_k\right|^2\right\}\right\} \quad (5.6.48)$$

Note that there will again be $M$ survivors at $k+1$ (see Figure 5.47); they extend the only survivor at time $k$ to each point of the constellation at time $k+1$.

In summary, at each step either one or $M$ survivors are retained, depending on whether a pilot symbol or a data symbol is transmitted. From Figure 5.47 it is apparent that final decisions can be released every $L$ steps. This is so because one survivor only is selected every $L$ steps and the associated symbol sequence can be put out as a firm decision.

An obvious question arises about the value of $L$. At first sight a trade-off seems to be needed between the following contrasting requirements:

(*i*) high transmission efficiency ($L$ large);

(*ii*) run-away suppression ($L$ small);

(*iii*) short decision delays ($L$ small).

Actually, point (*iii*) is not a crucial issue as a reasonably short decision delay can be attained anyway by truncating the survivors to some reasonable length (as is done with traditional Viterbi detectors). Thus, the true obstacle to increasing $L$ (and achieving high transmission efficiency) is the risk of run-aways. As we shall see later, run-away suppression requires that $L$ be decreased as the Doppler bandwidth increases.

Next, we consider trellis coded PSK. As we mentioned earlier, interleaving/de-interleaving (I/D) is often adopted to make the decoding process more robust against deep fades. In the ensuing discussion we assume I/D but we

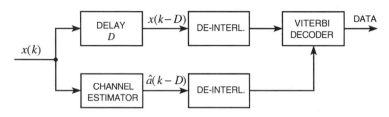

Figure 5.48. Block diagram of the receiver.

point out that PSP techniques may also be employed without I/D. It has been found, however, that I/D provides better performance in terms of power efficiency at the cost of some extra channel delay [55].

The block diagram of the receiver is shown in Figure 5.48. Here $x(k)$ represents the matched-filter output. Note that, as far as channel estimation is concerned, the received symbols may be viewed as independent, due to the scrambling action of the interleaver. Thus, PSP channel estimation may be performed as with uncoded modulation and this yields preliminary symbol decisions, say $\{\hat{c}_k\}$. Such decisions are *coarse* as no use is made of the encoding rule in their derivation. Their only purpose is to provide channel estimates for the Viterbi decoder. These estimates are obtained through a $(2N+1)$-tap interpolating filter, i.e.,

$$\hat{a}(k) = \sum_{i=-N}^{N} \gamma(i) x(k-i) \hat{c}^*_{k-i} \qquad (5.6.49)$$

where the coefficients $\{\gamma(i)\}$ are computed from the equations

$$\sum_{i=-N}^{N} \gamma(i) \rho_{aa}(l-i) + \frac{2N_0}{R_{aa}(0)} \gamma(l) = \rho_{aa}(l), \quad -N \le l \le N \qquad (5.6.50)$$

Finally, channel estimates $\hat{a}(k-D)$ and samples $x(k-D)$ are de-interleaved and fed to the Viterbi decoder. Note that a time alignment is needed to compensate for the overall channel estimation delay $D$. This is the sum of two terms, $D=L+N$, where $L$ accounts for the coarse decision computation and $N$ for the interpolating filter.

One important issue about the application of PSP methods (with or without coding) is the adaptivity of the Wiener filters to time-varying fading conditions. In the foregoing discussion we have assumed that filter taps can be computed solving either (5.6.44) (with uncoded modulation) or (5.6.50) (with trellis coding). It is clear from these equations, however, that tap values depend on the noise level (actually, the SNR) and the fading correlation coefficient

$\rho_{aa}(l)$. Thus, both noise level and correlation coefficient must be estimated as a function of time in a non-stationary environment. Assuming that this can be done in some way, adaptivity is accomplished, for example, by pre-computing various coefficient sets for the channel of interest and choosing the most suitable one according to the current estimates of SNR and $\rho_{aa}(l)$.

Alternative solutions have been indicated in [54]-[55]. Here, initial estimates of SNR and $\rho_{aa}(l)$ are derived from a known preamble at the transmission start-up. Then, their changes are tracked with suitable algorithms.

Simulation results are now shown to illustrate the performance of PSP methods. In these simulations a Rayleigh fading is generated by filtering two independent Gaussian random sequences in fourth-order Butterworth filters, as indicated in Figure 5.44. The 6-dB filter bandwidth is taken as the Doppler bandwidth. Eight-tap predictors are adopted with uncoded modulation while a seven-tap interpolator is used with coded modulation. Tap values are computed assuming perfect knowledge of SNR and $\rho_{aa}(l)$. Performance is rather tolerant of differences between assumed and actual values of these parameters, however. For example, computing $\{\gamma(i)\}$ for a Doppler bandwidth $f_D T = 5 \cdot 10^{-2}$, while the actual bandwidth is either $4 \cdot 10^{-2}$ or $6 \cdot 10^{-2}$, entails only penalties of a fraction of dB in the BER curves.

As mentioned earlier, the choice of the multiplexing ratio is an important issue. As the fading rate increases, run-aways take place unless $L$ is decreased adequately. For example, with $f_D T = 10^{-2}$, run-aways are eliminated taking $L \leq 10$.

In principle, either block or convolutional interleavers may be used. Simulations presented here are performed with convolutional interleavers. As is discussed in [56], they introduce a channel delay of $\lambda(\lambda-1)$ symbols, where $\lambda$ is the *interleaver depth*. Modulation is either uncoded 4-PSK or eight-state trellis-coded 8-PSK [30]. A decoding delay of 15 symbols is adopted.

Figure 5.49 shows BER curves for uncoded 4-PSK. The lowest curve indicates performance with perfect channel state information (CSI). The second lowest curve illustrates PSP operation with a multiplexing parameter $L=10$ and Doppler bandwidth of either $10^{-2}$ or $5 \cdot 10^{-2}$ (experimental points are overlapped). As is seen, PSP loses only 2 dB with respect to an ideal detector. Finally, the remainder two curves represent differential detection. They exhibit a BER floor which increases with the Doppler bandwidth.

Results for coded-modulation are illustrated in Figure 5.50. Again, the lowest curve represents ideal detection, while the second and third lowest curves correspond to PSP operation with different Doppler bandwidths and multiplexing ratios. Finally, the two upper curves indicate trellis-coded 8-DPSK [52]. A significant influence of the fading rate is now observed even with PSP. A high rate of pilot symbols (20%) is employed with $f_D T = 5 \cdot 10^{-2}$ to avoid run-aways.

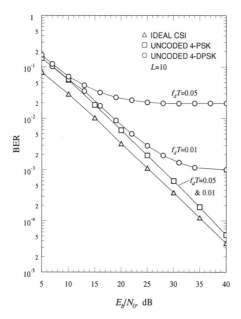

Figure 5.49. Performance of PSP algorithm with uncoded modulation.

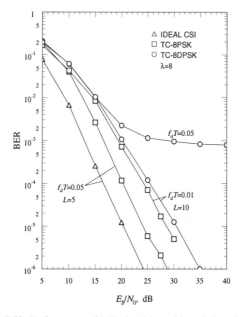

Figure 5.50. Performance of PSP algorithm with coded modulation.

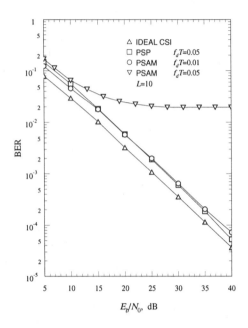

Figure 5.51. Comparison between PSP and PSAM.

Comparisons between PSP and PSAM (implemented with Wiener interpolators) is shown in Figure 5.51 with uncoded 4-PSK. Nine-tap interpolators are used for PSAM. The multiplexing parameter is $L=10$, both with PSP and PSAM. At the highest Doppler rate $(f_D T = 5 \cdot 10^{-2})$ PSAM exhibits a floor, while PSP is still very close to ideal performance. The difference between PSP and PSAM is significant also in terms of channel delay which is $D=40$ with PSAM and $D=10$ with PSP.

## 5.7. Clock-Aided but Non-Data-Aided Phase Recovery with Non-Offset Formats

### 5.7.1. Likelihood Function

Now we return to AWGN channels and investigate carrier phase recovery under the assumption that timing and frequency have already been accurately acquired. Data symbols are not available for phase synchronization, however, and this makes the difference with respect to the treatment in Sections 5.2-5.4 where timing was ideal and symbols were known or taken from the detector.

One obvious question is why, in the absence of known symbols, one should obviate data decisions. Indeed, decision-directed (DD) methods seem intuitively superior to non-data-aided (NDA) schemes at low error rates. One answer is that DD methods involve feedback loops which may exhibit too long acquisitions when information is transmitted in short bursts or in applications where fast re-acquisitions from deep fades are required. NDA open-loop (feedforward) algorithms are better suited in these circumstances.

The ensuing discussion has the following outline. We start with the computation of the likelihood function $\Lambda(r|\tilde{\theta})$ and then we concentrate on its maximization at high and low SNR. Suitable asymptotic analysis and judicious interpretations will allow us to arrive at physically implementable phase estimators. Other estimation schemes, based on heuristic reasoning, will be introduced when the opportunity presents itself. The treatment up to Section 5.7.4 focuses on feedback methods. Feedforward schemes are considered later.

Let us begin with the baseband signal component

$$s(t) = e^{j(2\pi vt + \theta)} \sum_i c_i g(t - iT - \tau) \qquad (5.7.1)$$

where $\theta$ and $\{c_i\}$ are unknown. For the time being we take the frequency offset $v$ as a known quantity but we shall return to this assumption later to see how frequency errors affect phase recovery. The joint likelihood function for $\theta$ and $\{c_i\}$ is obtained with the methods of Chapter 2 and reads

$$\Lambda(r|\tilde{\theta},\tilde{c}) = \exp\left\{\frac{1}{N_0}\int_0^{T_0} \mathrm{Re}\{r(t)\tilde{s}^*(t)\}dt - \frac{1}{2N_0}\int_0^{T_0}|\tilde{s}(t)|^2 dt\right\} \qquad (5.7.2)$$

with $\tilde{c} \triangleq \{\tilde{c}_i\}$ and

$$\tilde{s}(t) = e^{j(2\pi vt + \tilde{\theta})} \sum_i \tilde{c}_i g(t - iT - \tau) \qquad (5.7.3)$$

Paralleling the arguments of Section 3.2.1 and performing ordinary manipulations it is easily found that

$$\int_0^{T_0}|\tilde{s}(t)|^2 dt \approx \sum_{k=0}^{L_0-1}\sum_{m=0}^{L_0-1} \tilde{c}_k \tilde{c}_m^* h[(k-m)T] \qquad (5.7.4)$$

$$\int_0^{T_0} \mathrm{Re}\{r(t)\tilde{s}^*(t)\}dt \approx \sum_{k=0}^{L_0-1} \mathrm{Re}\{x(k)\tilde{c}_k^* e^{-j\tilde{\theta}}\} \qquad (5.7.5)$$

where, as usual, $L_0 = T_0/T$ denotes the length of the observation interval in symbol periods, $h(t)$ is the convolution $g(t) \otimes g(-t)$ and $x(k)$ is the sample at $t=kT+\tau$ of the matched-filter output, i.e.,

$$x(t) \triangleq \left[ r(t) e^{-j2\pi vt} \right] \otimes g(-t) \tag{5.7.6}$$

So as to make the problem analytically manageable, we choose not to deal with intersymbol interference and, in fact, we take

$$h(kT) = \begin{cases} 1 & k=0 \\ 0 & k \neq 0 \end{cases} \tag{5.7.7}$$

Then, (5.7.4) reduces to

$$\int_0^{T_0} |\tilde{s}(t)|^2 dt \approx \sum_{k=0}^{L_0-1} |\tilde{c}_k|^2 \tag{5.7.8}$$

Hence, substituting (5.7.8) and (5.7.5) into (5.7.2) produces

$$\Lambda(r|\tilde{\theta}, \tilde{c}) = \exp\left\{ \frac{1}{N_0} \sum_{k=0}^{L_0-1} \mathrm{Re}\left\{ x(k) \tilde{c}_k^* e^{-j\tilde{\theta}} \right\} - \frac{1}{2N_0} \sum_{k=0}^{L_0-1} |\tilde{c}_k|^2 \right\} \tag{5.7.9}$$

This result may be put in a more convenient form as follows. First, we note that the right-hand side can be multiplied by any positive constant $C$, independent of $\tilde{\theta}$ and $\tilde{c} \triangleq \{\tilde{c}_i\}$, without consequences for our purposes. In particular, taking

$$C \triangleq \exp\left\{ -\frac{1}{2N_0} \sum_{k=0}^{L_0-1} |x(k)|^2 \right\} \tag{5.7.10}$$

and rearranging yields

$$\Lambda(r|\tilde{\theta}, \tilde{c}) = \exp\left\{ -\frac{1}{2N_0} \sum_{k=0}^{L_0-1} \left| x(k) e^{-j\tilde{\theta}} - \tilde{c}_k \right|^2 \right\} \tag{5.7.11}$$

Second, assuming independent and equiprobable symbols, it can be shown (see Appendix 2.A) that the signal energy equals $C_2/2$, where $C_2 \triangleq \mathrm{E}\{|c_i|^2\}$ is the mean-square value of the data symbols. Correspondingly, the ratio $E_s/N_0$ equals $C_2/(2N_0)$, which means that

$$2N_0 = \frac{C_2}{E_s/N_0} \tag{5.7.12}$$

and (5.7.11) becomes

$$\Lambda(r|\tilde{\theta},\tilde{c}) = \exp\left\{-\frac{E_s}{C_2 N_0}\sum_{k=0}^{L_0-1}\left|x(k)e^{-j\tilde{\theta}} - \tilde{c}_k\right|^2\right\} \quad (5.7.13)$$

or, equivalently

$$\Lambda(r|\tilde{\theta},\tilde{c}) = \prod_{k=0}^{L_0-1}\exp\left\{-\frac{E_s}{C_2 N_0}\left|x(k)e^{-j\tilde{\theta}} - \tilde{c}_k\right|^2\right\} \quad (5.7.14)$$

At this point we average the likelihood function with respect to the data. In doing so we denote $\{P_m,\ m = 0,1,\ldots,M-1\}$ the constellation points. As the symbols are assumed independent, the expectation of (5.7.14) can be performed on a factor-by-factor basis and results in

$$\Lambda(r|\tilde{\theta}) = \prod_{k=0}^{L_0-1}\left\{\frac{1}{M}\sum_{m=0}^{M-1}\exp\left\{-\frac{E_s}{C_2 N_0}\left|x(k)e^{-j\tilde{\theta}} - P_m\right|^2\right\}\right\} \quad (5.7.15)$$

The ML estimate is that $\tilde{\theta}$ that maximizes $\Lambda(r|\tilde{\theta})$. Unfortunately there is no apparent way to maximize this likelihood function. Only approximate methods have been found that assume either high or low SNR. In the following we concentrate on these methods as they lead to readily implementable schemes.

## 5.7.2 High SNR

At high SNR the sum in (5.7.15) is dominated by that term corresponding to the minimum of $\left|x(k)e^{-j\tilde{\theta}} - P_m\right|^2$. Thus, letting

$$\hat{m}_k \triangleq \arg\left\{\min_{0 \leq m \leq M-1}\left\{\left|x(k)e^{-j\tilde{\theta}} - P_m\right|^2\right\}\right\} \quad (5.7.16)$$

and

$$\hat{c}_k \triangleq P_{\hat{m}_k} \quad (5.7.17)$$

equation (5.7.15) becomes

$$\Lambda(r|\tilde{\theta}) \approx \left(\frac{1}{M}\right)^{L_0}\exp\left\{-\frac{E_s}{C_2 N_0}\sum_{k=0}^{L_0-1}\left|x(k)e^{-j\tilde{\theta}} - \hat{c}_k\right|^2\right\} \quad (5.7.18)$$

and our problem reduces to minimizing the sum

$$\sum_{k=0}^{L_0-1} \left| x(k)e^{-j\tilde{\theta}} - \hat{c}_k \right|^2 \tag{5.7.19}$$

or, equivalently, to maximizing the function

$$F(\tilde{\theta}) \triangleq \sum_{k=0}^{L_0-1} \text{Re}\left\{ \hat{c}_k^* x(k)e^{-j\tilde{\theta}} \right\} - \frac{1}{2}\sum_{k=0}^{L_0-1} |\hat{c}_k|^2 \tag{5.7.20}$$

To this end the following remarks are useful. First, from (5.7.16)-(5.7.18) it is realized that $\hat{c}_k$ represents the detector decision corresponding to the input $x(k)e^{-j\tilde{\theta}}$ (see Figure 5.52). In particular, if $\tilde{\theta}$ is close to the true carrier phase, then $\{\hat{c}_k\}$ coincides with the transmitted symbol sequence (apart from noise induced errors). More generally, with a signal constellation having a $2\pi/N$ rotation symmetry, if $\tilde{\theta}$ is close to $\theta + 2\pi n/N, 0 \le n \le N-1$, then $\{\hat{c}_k\}$ equals the rotated sequence $\{c_k e^{-j2\pi n/N}\}$. Second, it is physically clear that (5.7.19) has $N$ minima around the points $\theta + 2\pi n/N, 0 \le n \le N-1$. Third, in the vicinity of these minima the second sum in (5.7.20) reads

$$\sum_{k=0}^{L_0-1} |\hat{c}_k|^2 \approx \sum_{k=0}^{L_0-1} \left| c_k e^{-j2\pi n/N} \right|^2 = \sum_{k=0}^{L_0-1} |c_k|^2 \tag{5.7.21}$$

and, therefore, it is independent of $\tilde{\theta}$. Fourth, the derivative of (5.7.20) around $\theta + 2\pi n/N, 0 \le n \le N-1$, is given by

$$\frac{d}{d\tilde{\theta}} F(\tilde{\theta}) = \sum_{k=0}^{L_0-1} \text{Im}\left\{ \hat{c}_k^* x(k)e^{-j\tilde{\theta}} \right\} \tag{5.7.22}$$

Putting all these facts together it is concluded that the maxima of $F(\tilde{\theta})$ may be computed looking for the nulls of (5.7.22). This is recognized as the problem addressed in Section 5.3.3 and can be approached with the same methods proposed there. In particular the generic term in the summation (5.7.22) can be turned into a signal error

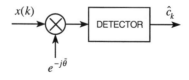

Figure 5.52. Interpretation of $\hat{c}_k$.

# Carrier Phase Recovery with Linear Modulations

$$e(k) = \text{Im}\left\{\hat{c}_k^* x(k) e^{-j\hat{\theta}(k)}\right\} \qquad (5.7.23)$$

and exploited to update the current phase estimate $\hat{\theta}(k)$. Interestingly enough, we have started this section assuming totally unknown symbols and, for high SNR, we have arrived at a Costas loop, the same loop we found in Section 5.3.3 when reliable detector decisions were available. Further discussions on these topics and performance evaluations of Costas loops are available in several papers and books [10-11].

### 5.7.3. Low SNR

For simplicity we concentrate on PSK modulation but, as is shown in Exercise 5.7.3, the resulting algorithms can be applied to more general signaling formats. Denoting $\{e^{j2\pi m/M}\}$, $m = 0, 1, \ldots, M-1$, the PSK constellation points and performing normal manipulations on (5.7.15) yields

$$\Lambda(r|\tilde{\theta}) = C\prod_{k=0}^{L_0-1}\left[\frac{1}{M}\sum_{m=0}^{M-1}\exp\left\{\frac{2E_s}{N_0}\text{Re}\left\{x(k)e^{-j\tilde{\theta}}e^{j2\pi m/M}\right\}\right\}\right] \qquad (5.7.24)$$

where $C$ is a positive constant independent of $\tilde{\theta}$. Instead of $\Lambda(r|\tilde{\theta})$ we choose to maximize its logarithm

$$\ln\Lambda(r|\tilde{\theta}) = \sum_{k=0}^{L_0-1}\ln\left[\frac{1}{M}\sum_{m=0}^{M-1}\exp\left\{\frac{2E_s}{N_0}\text{Re}\left\{x(k)e^{-j\tilde{\theta}}e^{j2\pi m/M}\right\}\right\}\right] \qquad (5.7.25)$$

where the additive term $\ln C$ has been dropped for simplicity.

In computing the internal sum in (5.7.25) it is expedient to assume an even $M$ and take the terms in pairs, associating the index $m$ to $m+M/2$. Bearing in mind that

$$e^{j2\pi(m+M/2)/M} = -e^{j2\pi m/M} \qquad (5.7.26)$$

this results in

$$\exp\left\{\frac{2E_s}{N_0}\text{Re}\left\{x(k)e^{-j\tilde{\theta}}e^{j2\pi m/M}\right\}\right\} + \exp\left\{\frac{2E_s}{N_0}\text{Re}\left\{x(k)e^{-j\tilde{\theta}}e^{j2\pi(m+M/2)/M}\right\}\right\}$$

$$= 2\cosh\left\{\frac{2E_s}{N_0}\text{Re}\left\{x(k)e^{-j\tilde{\theta}}e^{j2\pi m/M}\right\}\right\}$$

$$(5.7.27)$$

and, in consequence, (5.7.25) becomes

$$\ln \Lambda(r|\tilde{\theta}) = \sum_{k=0}^{L_0-1} \ln\left[\frac{2}{M}\sum_{m=0}^{M/2-1}\cosh\left\{\frac{2E_s}{N_0}\text{Re}\left\{x(k)e^{-j\tilde{\theta}}e^{j2\pi m/M}\right\}\right\}\right] \quad (5.7.28)$$

A necessary condition for a local maximum of $\ln\Lambda(r|\tilde{\theta})$ is that its derivative be zero. This derivative is found to be

$$\frac{d}{d\tilde{\theta}}\ln \Lambda(r|\tilde{\theta}) = \frac{2E_s}{N_0}\sum_{k=0}^{L_0-1}\frac{\sum_{m=0}^{M/2-1}\text{Im}\{X_m(k,\tilde{\theta})\}\sinh\left[\frac{2E_s}{N_0}\text{Re}\{X_m(k,\tilde{\theta})\}\right]}{\sum_{m=0}^{M/2-1}\cosh\left[\frac{2E_s}{N_0}\text{Re}\{X_m(k,\tilde{\theta})\}\right]} \quad (5.7.29)$$

with

$$X_m(k,\tilde{\theta}) \triangleq x(k)e^{-j\tilde{\theta}}e^{j2\pi m/M} \quad (5.7.30)$$

It is worth noting that (5.7.29) is a general result and, in particular, it is not subject to restrictions on the size of the signal constellation and the SNR. Henceforth, however, we find it useful to distinguish between binary modulation (BPSK) and polyphase modulation ($M>2$).

### 5.7.3.1. BPSK

Assume that $E_s/N_0$ is sufficiently small so that $\sinh(x)$ and $\cosh(x)$ in (5.7.29) can be approximated as

$$\sinh(x) \approx x \quad (5.7.31)$$

$$\cosh(x) \approx 1 \quad (5.7.32)$$

Accordingly, (5.7.29) becomes (neglecting an immaterial factor)

$$\frac{d}{d\tilde{\theta}}\ln \Lambda(r|\tilde{\theta}) = \sum_{k=0}^{L_0-1} I(k,\tilde{\theta})Q(k,\tilde{\theta}) \quad (5.7.33)$$

where $I(k,\tilde{\theta})$ and $Q(k,\tilde{\theta})$ are the so-called *in-phase* and *quadrature* components of $x(k)$ with respect to $e^{j\tilde{\theta}}$, i.e.,

$$x(k)e^{-j\tilde{\theta}} \triangleq I(k,\tilde{\theta}) + jQ(k,\tilde{\theta}) \quad (5.7.34)$$

# Carrier Phase Recovery with Linear Modulations

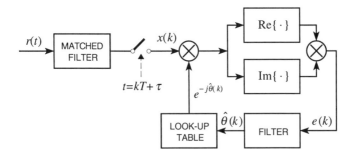

Figure 5.53. Costas loop for BPSK.

The null of $d[\ln \Lambda(r|\tilde{\theta})]/d\tilde{\theta}$ can be iteratively computed by turning the generic term in the sum (5.7.33) into an error signal

$$e(k) \triangleq I[k,\hat{\theta}(k)]Q[k,\hat{\theta}(k)] \qquad (5.7.35)$$

and updating the current phase estimate $\hat{\theta}(k)$ according to

$$\hat{\theta}(k+1) = \hat{\theta}(k) + \gamma e(k) \qquad (5.7.36)$$

Feedback will drive the average of $e(k)$ toward zero. The resulting scheme is a Costas loop as illustrated in Figure 5.53. Its performance has been thoroughly investigated by Lindsey and Simon [11].

### 5.7.3.2. Polyphase PSK

Assume again a low $E_s/N_0$ but, in computing $\sinh(x)$ and $\cosh(x)$ in (5.7.29), write

$$\sinh(x) \approx x - \frac{x^3}{3} \qquad (5.7.37)$$

$$\cosh(x) \approx 1 \qquad (5.7.38)$$

The reason for approximating $\sinh(x)$ as in (5.7.37) as opposed to (5.7.31) is that, as is now shown, (5.7.31) would make $d[\ln \Lambda(r|\tilde{\theta})]/d\tilde{\theta}$ identically zero and, therefore, of no use for our purposes. Inserting (5.7.37)-(5.7.38) into (5.7.29) yields the rather cumbersome formula

$$\frac{d}{d\tilde{\theta}}\ln\Lambda(r|\tilde{\theta}) = \frac{2}{M}\left(\frac{2E_s}{N_0}\right)^2 \sum_{k=0}^{L_0-1}\sum_{m=0}^{M/2-1} \text{Re}\{X_m(k,\tilde{\theta})\}\text{Im}\{X_m(k,\tilde{\theta})\}$$

$$-\frac{2}{3M}\left(\frac{2E_s}{N_0}\right)^4 \sum_{k=0}^{L_0-1}\sum_{m=0}^{M/2-1} \left[\text{Re}\{X_m(k,\tilde{\theta})\}\right]^3 \text{Im}\{X_m(k,\tilde{\theta})\} \quad (5.7.39)$$

We maintain, however, that the first term in the right is zero. This is easily checked by writing

$$\text{Re}\{X_m(k,\tilde{\theta})\} = \frac{1}{2}x(k)e^{-j\tilde{\theta}}e^{j2\pi m/M} + \frac{1}{2}x^*(k)e^{j\tilde{\theta}}e^{-j2\pi m/M} \quad (5.7.40)$$

$$\text{Im}\{X_m(k,\tilde{\theta})\} = \frac{1}{2j}x(k)e^{-j\tilde{\theta}}e^{j2\pi m/M} - \frac{1}{2j}x^*(k)e^{j\tilde{\theta}}e^{-j2\pi m/M} \quad (5.7.41)$$

and noting that, for $M$ even and greater than 2, the following formula applies:

$$\sum_{m=0}^{M/2-1} e^{j4\pi m/M} = 0 \quad (5.7.42)$$

Thus, stripping off an immaterial factor from the remaining term in (5.7.39) yields

$$\frac{d}{d\tilde{\theta}}\ln\Lambda(r|\tilde{\theta}) = -\sum_{k=0}^{L_0-1}\sum_{m=0}^{M/2-1}\left[\text{Re}\{X_m(k,\tilde{\theta})\}\right]^3 \text{Im}\{X_m(k,\tilde{\theta})\} \quad (5.7.43)$$

At this stage the zero of $d[\ln\Lambda(r|\tilde{\theta})]/d\tilde{\theta}$ is obtained in an iterative fashion taking the error signal

$$e(k) \triangleq -\sum_{m=0}^{M/2-1}\left[\text{Re}\{X_m[k,\hat{\theta}(k)]\}\right]^3 \text{Im}\{X_m[k,\hat{\theta}(k)]\} \quad (5.7.44)$$

and updating $\hat{\theta}(k)$ according to

$$\hat{\theta}(k+1) = \hat{\theta}(k) + \gamma e(k) \quad (5.7.45)$$

This leads to the Costas loop illustrated in Figure 5.54. Again, in depth analysis of this loop has been carried out by Lindsey and Simon [11].

**Exercise 5.7.1.** Express the error signal (5.7.44) as a function of the in-phase and quadrature components of $x(k)$ with respect to $e^{j\hat{\theta}(k)}$.

*Solution.* Collecting (5.7.30) and (5.7.34) yields

# Carrier Phase Recovery with Linear Modulations

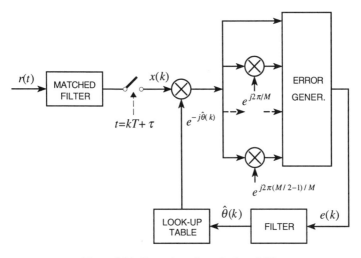

Figure 5.54. Costas loop for polyphase PSK.

$$\text{Re}\{X_m[k,\hat{\theta}(k)]\} = I\cos\alpha - Q\sin\alpha \qquad (5.7.46)$$

$$\text{Im}\{X_m[k,\hat{\theta}(k)]\} = I\sin\alpha + Q\cos\alpha \qquad (5.7.47)$$

where the following shorthand notations have been used:

$$I \triangleq I[k,\hat{\theta}(k)], \quad Q \triangleq Q[k,\hat{\theta}(k)], \quad \alpha \triangleq 2\pi m/M \qquad (5.7.48)$$

Next, substituting into (5.7.44) and performing standard (but lengthy) manipulations gives the desired result

$$e(k) = -(I^3 Q - Q^3 I) \qquad (5.7.49)$$

**Exercise 5.7.2.** Compute the S-curve for the phase detector (5.7.35) assuming BPSK modulation and Nyquist pulses.

*Solution.* From (5.7.6) it is easily seen that the samples $x(k)$ from the matched filter are

$$x(k) = c_k e^{j\theta} + n(k) \qquad (5.7.50)$$

with $c_k = \pm 1$. The S-curve is computed taking the expectation of the error signal for $\hat{\theta}(k) = \hat{\theta}$=constant, i.e.,

$$S(\theta - \hat{\theta}) \triangleq E\{I(k,\hat{\theta})Q(k,\hat{\theta})\} \tag{5.7.51}$$

To go further we need the expressions of $I(k,\hat{\theta})$ and $Q(k,\hat{\theta})$. Collecting (5.7.34) and (5.7.50) yields

$$I(k,\hat{\theta}) + jQ(k,\hat{\theta}) = c_k e^{j(\theta-\hat{\theta})} + n'(k) \tag{5.7.52}$$

where $n'(k)$ is a noise term statistically equivalent to $n(k)$. In particular, its real and imaginary components $n'_R(k)$ and $n'_I(k)$ are zero-mean and independent Gaussian random variables. It follows from (5.7.52) that

$$I(k,\hat{\theta}) = c_k \cos(\theta - \hat{\theta}) + n'_R(k) \tag{5.7.53}$$

$$Q(k,\hat{\theta}) = c_k \sin(\theta - \hat{\theta}) + n'_I(k) \tag{5.7.54}$$

and substituting into (5.7.51) produces the desired result

$$S(\phi) = \frac{1}{2}\sin 2\phi \tag{5.7.55}$$

with $\phi \triangleq \theta - \tilde{\theta}$. Note that $S(\phi)$ has nulls with positive slope at $\phi=0$ and $\phi=\pi$, which implies that the phase detector has a 180° ambiguity.

**Exercise 5.7.3.** Compute the S-curve for the phase detector (5.7.44) assuming QPSK modulation and Nyquist pulses.
*Solution.* Collecting (5.7.46)-(5.7.48) yields

$$\text{Re}\{X_m(k,\hat{\theta})\} = I(k,\hat{\theta})\cos(2\pi n/M) - Q(k,\hat{\theta})\sin(2\pi n/M) \tag{5.7.56}$$

$$\text{Im}\{X_m(k,\hat{\theta})\} = I(k,\hat{\theta})\sin(2\pi n/M) + Q(k,\hat{\theta})\cos(2\pi n/M) \tag{5.7.57}$$

Substituting into (5.7.44) produces

$$e(k) = I(k,\hat{\theta})Q^3(k,\hat{\theta}) - Q(k,\hat{\theta})I^3(k,\hat{\theta}) \tag{5.7.58}$$

from which the S-curve is derived taking the expectation

$$S(\phi) = E\{I(k,\hat{\theta})Q^3(k,\hat{\theta}) - Q(k,\hat{\theta})I^3(k,\hat{\theta})\} \tag{5.7.59}$$

with $\phi \triangleq \theta - \hat{\theta}$.

Next we observe that (5.7.52) is still valid except that the symbols have the form $c_i = e^{j\alpha_i}$, with $\alpha_i$ uniformly spaced over $(-\pi,\pi)$. Choosing $\alpha_i \in \{\pm\pi/4, \pm 3\pi/4\}$ yields

$$I(k,\hat{\theta}) = \cos(\alpha_i + \phi) + n'_R(k) \quad (5.7.60)$$

$$Q(k,\hat{\theta}) = \sin(\alpha_i + \phi) + n'_I(k) \quad (5.7.61)$$

and inserting into (5.7.59), after long manipulations leads to

$$S(\phi) = \frac{1}{4}\sin(4\phi) \quad (5.7.62)$$

The S-curve is independent of the noise level and has a 90° phase ambiguity.

**Exercise 5.7.4.** Compute the S-curve for the phase detector (5.7.58) assuming QAM modulation and Nyquist pulses.

*Solution.* Equations (5.7.52) and (5.7.59) are still valid except that, in the former, the symbols take the form $c_i = a_i + jb_i$, where $a_i$ and $b_i$ are zero-mean and independent random variables with second- and fourth-order moments $E\{a_i^2\} = E\{b_i^2\} \triangleq C_2$ and $E\{a_i^4\} = E\{b_i^4\} \triangleq C_4$. Then, $I(k,\hat{\theta})$ and $Q(k,\hat{\theta})$ take the form

$$I(k,\hat{\theta}) = a_i \cos\phi - b_i \sin\phi + n'_R(k) \quad (5.7.63)$$

$$Q(k,\hat{\theta}) = a_i \sin\phi + b_i \cos\phi + n'_I(k) \quad (5.7.64)$$

and substituting into (5.7.59) gives

$$S(\phi) = \frac{1}{2}\left(3C_2^2 - C_4\right)\sin(4\phi) \quad (5.7.65)$$

It can be checked that (5.7.65) reduces to (5.7.62) if QPSK is seen as a particular form of QAM with $a_i = \pm 1/\sqrt{2}$ and $b_i = \pm 1/\sqrt{2}$. In general, (5.7.65) is a well-shaped S-curve for arbitrary QAM formats, which means that phase detector (5.7.58) may be used even with such modulations. This idea has been pointed out by Foschini [57] following a different reasoning.

### 5.7.4. Feedforward Estimation with PSK

Feedforward schemes are preferable over feedback ones when short acquisition times are required. In the following we derive a few feedforward schemes for PSK modulation.

Start with the log-likelihood function (5.7.25), which is rewritten as

$$\ln \Lambda(r|\tilde{\theta}) = \sum_{k=0}^{L_0-1} \ln T(k,\tilde{\theta}) \qquad (5.7.66)$$

with

$$T(k,\tilde{\theta}) \triangleq \frac{1}{M} \sum_{m=0}^{M-1} \exp\left\{ \frac{2E_s}{N_0} \operatorname{Re}\left\{ x(k)e^{-j\tilde{\theta}} e^{j2\pi m/M} \right\} \right\} \qquad (5.7.67)$$

Observing that

$$2\operatorname{Re}\left\{ x(k)e^{-j\tilde{\theta}} e^{j2\pi m/M} \right\} = x(k)e^{-j\tilde{\theta}} e^{j2\pi m/M} + x^*(k)e^{j\tilde{\theta}} e^{-j2\pi m/M} \qquad (5.7.68)$$

$$\left[ \frac{2E_s}{N_0} \operatorname{Re}\left\{ x(k)e^{-j\tilde{\theta}} e^{j2\pi m/M} \right\} \right]^p$$

$$= \left( \frac{E_s}{N_0} \right)^p \sum_{q=0}^{p} \binom{p}{q} x^q(k) \left[ x^*(k) \right]^{p-q} e^{j(p-2q)\tilde{\theta}} e^{-j2\pi m(p-2q)/M} \qquad (5.7.69)$$

and expanding the exponential in (5.7.67) into a power series yields

$$\exp\left\{ \frac{2E_s}{N_0} \operatorname{Re}\left\{ x(k)e^{-j\tilde{\theta}} e^{j2\pi m/M} \right\} \right\}$$

$$= \sum_{p=0}^{\infty} \frac{1}{p!} \left( \frac{E_s}{N_0} \right)^p \sum_{q=0}^{p} \binom{p}{q} x^q(k) \left[ x^*(k) \right]^{p-q} e^{j(p-2q)\tilde{\theta}} e^{-j2\pi m(p-2q)/M} \qquad (5.7.70)$$

Finally, substituting into (5.7.67) produces

$$T(k,\tilde{\theta}) = \sum_{p=0}^{\infty} \frac{1}{p!} \left( \frac{E_s}{N_0} \right)^p \sum_{q=0}^{p} \binom{p}{q} x^q(k) \left[ x^*(k) \right]^{p-q} e^{j(p-2q)\tilde{\theta}} A(p-2q) \qquad (5.7.71)$$

where the following definition has been made:

$$A(p-2q) \triangleq \frac{1}{M} \sum_{m=0}^{M-1} e^{-j2\pi m(p-2q)/M} \qquad (5.7.72)$$

## Carrier Phase Recovery with Linear Modulations

It should be noted that $A(\cdot)$ is zero, unless $p-2q$ is a multiple of $M$, i.e.,

$$p - 2q = lM \quad l = 0, \pm 1, \pm 2, \ldots \quad (5.7.73)$$

Also, $p$ and $q$ are both non-negative and $q$ is less than or equal to $p$. In conclusion, the pairs $(q, p)$ corresponding to non-zero terms in the right-hand side of (5.7.72) are limited. For example, for $l=0$ we have the only pair $(0,0)$ whereas, for $l=1$ we have

$$(0, M), (1, M+2), (2, M+4), \ldots \quad (5.7.74)$$

and for $l=-1$

$$(M, M), (M+1, M+2), (M+2, M+4), \ldots \quad (5.7.75)$$

To go further we assume that $E_s/N_0$ is sufficiently small so that powers of $E_s/N_0$ greater than $M$ can be neglected. Then, equation (5.7.71) reduces to

$$T(k, \tilde{\theta}) \approx 1 + \frac{2}{M!} \left(\frac{E_s}{N_0}\right)^M \text{Re}\left\{x^M(k) e^{-jM\tilde{\theta}}\right\} \quad (5.7.76)$$

Inserting (5.7.76) into (5.7.66) and approximating $\ln(1+\varepsilon) \approx \varepsilon$ gives

$$\ln \Lambda(r|\tilde{\theta}) \approx \text{Re}\left\{e^{-jM\tilde{\theta}} \sum_{k=0}^{L_0-1} x^M(k)\right\} \quad (5.7.77)$$

where a positive factor independent of $\tilde{\theta}$ has been dropped for simplicity.

In summary, within the above approximations, the ML carrier phase estimate is that $\tilde{\theta}$ that maximizes (5.7.77). This estimate is expressed by

$$\hat{\theta} = \frac{1}{M} \arg\left\{\sum_{k=0}^{L_0-1} x^M(k)\right\} \quad (5.7.78)$$

This is the familiar $M$-power synchronizer and its block diagram is illustrated in Figure 5.55. Note that (5.7.78) is well suited for digital implementation. As the arg-function takes values in the range $\pm \pi$, the estimate is restricted within $\pm \pi/M$. This corresponds to an $M$-fold ambiguity in the phase estimates which can be dealt with by resorting to differential encoding/decoding.

For later reference the following interpretation of the $M$-power synchronizer is of interest. Computing $x^M(k)$ from (5.7.50) and recognizing that $c_k^M = 1$ for any PSK yields

$$x^M(k) = e^{jM\theta} + N(k) \quad (5.7.79)$$

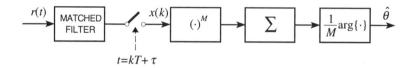

Figure 5.55. *M*-power phase recovery scheme.

where $N(k)$ is a zero-mean noise term resulting from the products Signal×Noise and Noise×Noise. As is seen, the modulation has been removed from $x^M(k)$ and, therefore, taking the average of $x^M(k)$ over the observation interval amounts to smoothing out $N(k)$ and producing a vector which, hopefully, deviates little from the direction of $e^{jM\theta}$. Hence, the argument of such a vector (scaled by a factor *M*) can be taken as an estimate of the carrier phase.

Figure 5.56 shows simulation results for the variance of the *M*-power with QPSK. Because of the Nyquist assumption these results are independent of the rolloff factor. As is seen, the modified Cramer-Rao bound is attained at high SNR.

An interesting question about the *M*-power estimator is its performance at intermediate/low SNR values and whether better methods can be found when operating in these conditions (which are typical with coded modulations). This issue has been addressed by A.J.Viterbi and A.M.Viterbi [58] who have investigated a variant to the *M*-power synchronizer. Their estimation scheme, henceforth denoted by V&V algorithm, is a generalization of (5.7.78) and may be described by writing $x(k)$ in the form $x(k) = \rho(k)e^{j\phi(k)}$ and replacing $x^M(k)$ in (5.7.78) by

$$y(k) = F[\rho(k)]e^{jM\phi(k)} \qquad (5.7.80)$$

where $F[\rho(k)]$ is an appropriately chosen function of $\rho(k)$. This produces

$$\hat{\theta} = \frac{1}{M}\arg\left\{\sum_{k=0}^{L_0-1} F[\rho(k)]e^{jM\phi(k)}\right\} \qquad (5.7.81)$$

Clearly, taking $F[\rho(k)]=\rho^M(k)$ leads back to the *M*-power algorithm. It turns out, however, that better performance is obtained raising $\rho(k)$ to a smaller power, say $F[\rho(k)]=\rho^2(k)$ or $F[\rho(k)]=1$, depending on the SNR and the number of constellation points. For example, with QPSK and $E_s/N_0 = 6$ dB, the loss with respect to the MCRB is about 4 dB with $F[\rho(k)]=\rho^4(k)$, 3 dB with $F[\rho(k)]=1$ and 2.6 dB with $F[\rho(k)]=\rho^2(k)$.

In the foregoing discussion a perfect frequency reference has been assumed and, therefore, one wonders how the result changes in the presence of

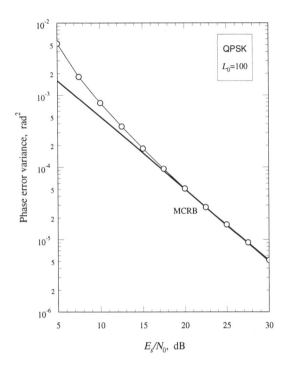

Figure 5.56. Performance of the $M$-power phase estimator with QPSK.

a carrier frequency error $f_d$. The sensitivity of the V&V algorithm to frequency errors is assessed in [58] and the results are similar to those established in Section 5.2.5 when dealing with data-aided phase recovery. In essence, degradations take place in two forms: (*i*) estimation error variance increases with $|f_d|$, much more so than with the data-aided schemes; (*ii*) estimates are biased for symbols away from the center of the observation interval. In particular, the bias equals $2\pi m M f_d T$, where $m$ is the distance in symbol intervals from the center. It is worth recalling that the bias is $M$ times smaller with data-aided phase estimation. This is due to the fact that the modulation removal in the V&V scheme is accomplished by raising the signal samples to the $M$-power, which amplifies phase errors by a factor $M$.

### 5.7.5. Feedforward Estimation with QAM

In pursuing feedforward phase estimation with QAM signals we adopt a heuristic approach. Return to (5.7.50) which gives the samples $x(k)$ from the matched filter. With QAM the symbols have the form $c_k = a_k + jb_k$, where $a_k$ and

$b_k$ are zero-mean and independent random variables with second- and fourth-order moments $E\{a_i^2\} = E\{b_i^2\} \triangleq C_2$ and $E\{a_i^4\} = E\{b_i^4\} \triangleq C_4$. The expectation of $x^4(k)$ is easily found to be

$$E\{x^4(k)\} = 2(C_4 - 3C_2^2)e^{j4\theta} \qquad (5.7.82)$$

As $C_4 - 3C_2^2$ turns out to be positive, it follows from (5.7.82) that $E\{x^4(k)\}$ provides information on the carrier phase. In fact one has

$$\theta = \frac{1}{4}\arg\{E\{x^4(k)\}\} \qquad (5.7.83)$$

In practice, the expectation operation can be approximated by the sample average of $x^4(k)$ and (5.7.83) is transformed into the estimator

$$\hat{\theta} = \frac{1}{4}\arg\left\{\sum_{k=0}^{L_0-1} x^4(k)\right\} \qquad (5.7.84)$$

This algorithm coincides with (5.7.78) for QPSK and its performance achieves the MCRB at high SNR. With general QAM modulation, instead, its estimation accuracy falls further and further from the bound as the number of constellation points increases [60]. Sensitivity to carrier frequency errors has not been explored in the literature but, by analogy with the V&V algorithm, it is expected that phase estimates will be biased by the quantity $8\pi m f_d T$ for symbols at a distance $m$ from the center of the observation interval.

### 5.7.6. Ambiguity Resolution

As we mentioned earlier, phase estimation schemes give phase estimates with a $2\pi/M$ ambiguity, $M$ being the symmetry angle of the signal constellation. One way of coping with this problem is to resort to differential encoding/decoding. Alternately, a unique word (UW) appended to the data can be exploited by an ambiguity resolution (AR) circuit, as is now explained.

For simplicity let us assume: (*i*) PSK signaling; (*ii*) perfect frequency recovery; (*iii*) a UW consisting of the first $L_{UW}$ symbols $\{c_k\}$, $0 \le k \le L_{UW} - 1$, in the transmitted sequence. Also, suppose that the phase estimation algorithm operates over the whole data sequence $\{x(k)\}$, $0 \le k \le L_0 - 1$, yielding

$$\hat{\theta} = \theta + \Delta\phi + 2\pi m/M \qquad (5.7.85)$$

where $\Delta\phi$ accounts for (hopefully small) estimation errors and $2\pi m/M$ represents the phase ambiguity. The task of the AR circuit is to employ the samples $\{x(k)\}$ and the UW symbols $\{c_k\}$, $0 \le k \le L_{UW} - 1$, to establish the value of the integer $m$, $0 \le m \le M - 1$.

As is now shown, this can be accomplished using the statistics

$$y(k) \triangleq x(k) c_k^* e^{-j\hat{\theta}}, \quad 0 \le k \le L_{UW} - 1 \tag{5.7.86}$$

which can be computed from $\hat{\theta}$ and the UW. To this end, combining (5.7.50) and (5.7.85)-(5.7.86) under the assumption $|c_k|^2 = 1$ yields

$$y(k) = e^{-j2\pi m/M} e^{-j\Delta\phi} + n'(k), \quad 0 \le k \le L_{UW} - 1 \tag{5.7.87}$$

where $\{n'(k)\}$ are independent complex-valued Gaussian random variables with real and imaginary parts of variance $N_0$. If the phase error $\Delta\phi$ is small, we have approximately $e^{-j\Delta\phi} \approx 1$ and the first term in the right-hand side may be viewed as a complex symbol from an $M$-ary PSK constellation. Then, ambiguity resolution reduces to a detection problem in which $L_{UW}$ independent measurements of the sum *symbol+noise* are given. As is known, the optimum detection strategy [41] is to take the sample average of $y(k)$, say

$$Y \triangleq \frac{1}{L_{UW}} \sum_{k=0}^{L_{UW}-1} y(k) \tag{5.7.88}$$

and choose $m$ as that integer $\tilde{m}$ which maximizes the real part of $Y e^{j2\pi \tilde{m}/M}$:

$$\hat{m} = \arg\left\{\max_{\tilde{m}}\left\{\text{Re}\left\{Y e^{j2\pi \tilde{m}/M}\right\}\right\}\right\} \tag{5.7.89}$$

The probability of false ambiguity resolution, $P_{FAR}$, equals the error probability in the above detection problem and is computed as follows. Collecting (5.7.87)-(5.7.89) and neglecting $\Delta\phi$ yields

$$Y = e^{-j2\pi m/M} + \frac{1}{L_{UW}} \sum_{k=0}^{L_{UW}-1} n'(k) \tag{5.7.90}$$

On the other hand, it is easily checked that real and imaginary parts of the noise term in (5.7.90) have the same variance

$$\sigma^2 = \frac{1}{2 L_{UW}} \frac{1}{(E_s/N_0)} \tag{5.7.91}$$

Hence, using a well-known formula [44, p. 265] for the symbol error probability with $M$-ary PSK ($M \ge 4$) gives

$$P_{FAR} \approx 2Q\left(\sin(\pi/M)\sqrt{2L_{UW}\frac{E_s}{N_0}}\right) \qquad (5.7.92)$$

For $M=2$, instead, we have

$$P_{FAR} = Q\left(\sqrt{2L_{UW}\frac{E_s}{N_0}}\right) \qquad (5.7.93)$$

One important problem in designing the AR circuit is the choice of $L_{UW}$. Let $P_{ID}$ be the symbol error probability that is achieved with ideal phase recovery and perfect ambiguity resolution. In the presence of AR failures, the *average* error probability is shown to be $P_{ID}(1-P_{FAR}) + P_{FAR} \approx P_{ID} + P_{FAR}$ (assuming pessimistically that all the decisions are incorrect when a failure occurs). Thus, $P_{FAR}$, should be negligible as compared with $P_{ID}$.

For example, with uncoded QPSK and $E_s/N_0 = 10$ dB, the error probability $P_{ID}$ equals $1.6 \cdot 10^{-3}$. Suppose we want $P_{FAR} = 10^{-5}$, two orders of magnitude less than $P_{ID}$. Then, from (5.7.92) we obtain $L_{UW}=2$.

### 5.7.7. The Unwrapping Problem

A single estimate per data block is sufficient in burst transmissions with relatively short blocks. With longer blocks or continuous data transmission, vice versa, multiple estimates are needed because carrier phase does not remain constant (due to oscillator instabilities and/or imperfect frequency offset compensation). With multiple estimates, however, the problem arises as to how resolve phase ambiguity making use of a single UW. In fact, while the UW is useful to resolve ambiguity in the first estimate, it is not clear how to manage with the subsequent estimates.

The question may be viewed as indicated in Figure 5.57 where, to ease the reading, the estimates $\hat{\theta}(t)$ are represented on a continuous-time scale (in reality they are issued at intervals of $L_0 T$ seconds from each other). As a consequence of the modulo $2\pi/M$ operation in the estimator, $\hat{\theta}(t)$ exhibits a jump every time the true carrier phase $\theta(t)$ crosses an odd multiple of $\pi/M$. Thus, the problem is to: (*i*) resolve the ambiguity in the first phase estimate; (*ii*) stitch together the various segments of $\hat{\theta}(t)$.

This is the so-called *unwrapping* problem [61]-[63] which may be approached as indicated by Oerder and Meyr in [61]. Let $\{\hat{\theta}(k)\}$, $k=1,2,\ldots$, be the (still wrapped) estimates and $2\pi\hat{m}/M$ the ambiguity phase, as derived from the observation of the first data sub-block. Also, denote by $\{\hat{\theta}_f(k)\}$, $k=1,2,\ldots$, the *final* unwrapped phase trajectory and $\hat{\theta}_p(k) \triangleq \hat{\theta}_f(k) - 2\pi\hat{m}/M$ a *provisional* trajectory, which differs from $\hat{\theta}_f(k)$ by a constant angle $2\pi\hat{m}/M$. Provisional and final trajectories are recursively computed as

# Carrier Phase Recovery with Linear Modulations

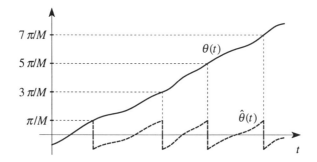

Figure 5.57. Explaining the unwrapping problem.

$$\hat{\theta}_p(k) = \hat{\theta}_p(k-1) + \alpha SAW\left[\hat{\theta}(k) - \hat{\theta}_p(k-1)\right] \quad (5.7.94)$$

$$\hat{\theta}_f(k) = \hat{\theta}_p(k) + 2\pi\hat{m}/M \quad (5.7.95)$$

where $SAW[\phi] \triangleq [\phi]_{-\pi/M}^{+\pi/M}$ is a sawtooth nonlinearity that reduces $\phi$ to the interval $[-\pi/M, \pi/M)$ and $\alpha$ is a parameter in the range $0 < \alpha \leq 1$. Intuitively, $\hat{\theta}_p(k)$ serves to stitch together the various segments of $\hat{\theta}(k)$ (see Figure 5.57) so as to form a continuous curve extending beyond the interval $\pm \pi/M$. Once this is done, the phase ambiguity that still affects $\hat{\theta}_p(k)$ is canceled by adding $2\pi\hat{m}/M$. Clearly, no phase ambiguity compensation is needed when differential encoding/decoding is used.

Figure 5.58 illustrates equations (5.7.94)-(5.7.95). The performance of this unwrapping scheme is significantly affected by the parameter $\alpha$. In the steady state a small $\alpha$ provides good final estimates $\hat{\theta}_f(k)$. On the other hand, a small $\alpha$ may result in a too slow updating of $\hat{\theta}_p(k)$, meaning that $\hat{\theta}_f(k)$ may be unable to adequately track the carrier phase dynamics.

Another problem associated with the unwrapping method in Figure 5.58 is

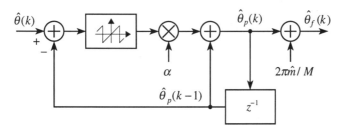

Figure 5.58. Unwrapping algorithm.

the insurgence of cycle slips. During normal operation $\hat{\theta}_f(k)$ undergoes small fluctuations around the true phase $\theta$. Due to the feedback mechanism, however, noise disturbances can make $\hat{\theta}_f(k)$ slip away from this operating condition and move toward another stable point $\theta \pm 2\pi/M$. Clearly, the detrimental effect of a cycle slip on the symbol error probability is limited to the slip duration when differential encoding/decoding is adopted. With coherent detection, on the contrary, it lasts until the arrival of a new UW. Intuitively, the slip rate increases with the parameter $\alpha$. De Jonghe and Moeneclaey [63] have worked out approximate formulas expressing this rate as a function of $\alpha$ and other system parameters.

## 5.8. Clock-Aided but Non-Data-Aided Phase Recovery with OQPSK

### 5.8.1. Likelihood Function

In this section we address non-data aided phase recovery with OQPSK under the assumption of ideal timing and a Nyquist channel response. Paralleling the treatment in Section 5.7, we first investigate algorithms based on ML estimation methods and, subsequently, we discuss *ad hoc* schemes.

Start with the mathematical model for OQPSK

$$s(t) = e^{j(2\pi v t + \theta)} \left[ \sum_i a_i g(t - iT - \tau) + j \sum_i b_i g(t - iT - T/2 - \tau) \right] \quad (5.8.1)$$

Here, $v$ and $\tau$ are known parameters whereas $a_i$ and $b_i$ are unknown (and independent) data symbols taking values $\pm 1$ with the same probability. Letting $\tilde{a} \triangleq \{\tilde{a}_i\}$, $\tilde{b} \triangleq \{\tilde{b}_i\}$ and

$$\tilde{s}(t) \triangleq e^{j(2\pi v t + \tilde{\theta})} \left[ \sum_i \tilde{a}_i g(t - iT - \tau) + j \sum_i \tilde{b}_i g(t - iT - T/2 - \tau) \right] \quad (5.8.2)$$

the joint likelihood function for phase and the data reads

$$\Lambda\left(r|\tilde{\theta},\tilde{a},\tilde{b}\right) = \exp\left[ \frac{1}{N_0} \int_0^{T_0} \text{Re}\{r(t)\tilde{s}^*(t)\} dt - \frac{1}{2N_0} \int_0^{T_0} |\tilde{s}(t)|^2 dt \right] \quad (5.8.3)$$

Performing the usual manipulations it is found that

## Carrier Phase Recovery with Linear Modulations

$$\int_0^{T_0} |\tilde{s}(t)|^2 dt \approx \sum_{k=0}^{L_0-1}\sum_{m=0}^{L_0-1} \tilde{a}_k \tilde{a}_m h[(k-m)T] + \sum_{k=0}^{L_0-1}\sum_{m=0}^{L_0-1} \tilde{b}_k \tilde{b}_m h[(k-m)T] \quad (5.8.4)$$

$$\int_0^{T_0} \text{Re}\{r(t)\tilde{s}^*(t)\}dt \approx \sum_{k=0}^{L_0-1} \tilde{a}_k \text{Re}\{x(k)e^{-j\tilde{\theta}}\} + \sum_{k=0}^{L_0-1} \tilde{b}_k \text{Im}\{x(k+1/2)e^{-j\tilde{\theta}}\} \quad (5.8.5)$$

where $L_0 = T_0/T$, $h(t) \triangleq g(t) \otimes g(-t)$ is the overall system response and $x(k)$ and $x(k+1/2)$ are samples from the matched-filter output at $t=kT+\tau$ and $t=kT+T/2+\tau$, respectively.

Assume that $h(t)$ satisfies the Nyquist criterion

$$h(kT) = \begin{cases} 1 & k=0 \\ 0 & k \neq 0 \end{cases} \quad (5.8.6)$$

Then, substituting into (5.8.4) and bearing in mind that $|\tilde{a}_i|^2 = |\tilde{b}_i|^2 = 1$ yields

$$\int_0^{T_0} |\tilde{s}(t)|^2 dt = 2L_0 \quad (5.8.7)$$

Inserting (5.8.5) and (5.8.7) into (5.8.3) and dropping a factor independent of $\tilde{\theta}, \tilde{a}, \tilde{b}$ results in

$$\Lambda(r|\tilde{\theta},\tilde{a},\tilde{b}) = \prod_{k=0}^{L_0-1} \exp\left[\frac{1}{N_0}\tilde{a}_k \text{Re}\{x(k)e^{-j\tilde{\theta}}\}\right] \cdot \prod_{k=0}^{L_0-1} \exp\left[\frac{1}{N_0}\tilde{b}_k \text{Im}\{x(k+1/2)e^{-j\tilde{\theta}}\}\right]$$
$$(5.8.8)$$

It is useful to let the ratio $E_s/N_0$ appear explicitly in (5.8.8). This is accomplished bearing in mind that, under the present assumptions, the signal energy per symbol is unity (see Appendix 2.A) and therefore $1/N_0$ equals $E_s/N_0$. Hence

$$\Lambda(r|\tilde{\theta},\tilde{a},\tilde{b}) = \prod_{k=0}^{L_0-1} \exp\left[\frac{E_s}{N_0}\tilde{a}_k \text{Re}\{x(k)e^{-j\tilde{\theta}}\}\right] \cdot \prod_{k=0}^{L_0-1} \exp\left[\frac{E_s}{N_0}\tilde{b}_k \text{Im}\{x(k+1/2)e^{-j\tilde{\theta}}\}\right]$$
$$(5.8.9)$$

Next, the *marginal* likelihood function $\Lambda(r|\tilde{\theta})$ is obtained by averaging (5.8.9) over the data. As the symbols are independent and equiprobable, the expectation is performed on each factor separately and produces

$$\Lambda(r|\tilde{\theta}) = \prod_{k=0}^{L_0-1} \cosh\left[\frac{E_s}{N_0}\text{Re}\{x(k)e^{-j\tilde{\theta}}\}\right] \cdot \prod_{k=0}^{L_0-1} \cosh\left[\frac{E_s}{N_0}\text{Im}\{x(k+1/2)e^{-j\tilde{\theta}}\}\right] \quad (5.8.10)$$

from which the following expression of the log-likelihood function is obtained:

$$\ln \Lambda(r|\tilde{\theta}) = \sum_{k=0}^{L_0-1} \ln\left\{\cosh\left[\frac{E_s}{N_0}\text{Re}\{x(k)e^{-j\tilde{\theta}}\}\right]\right\}$$
$$+ \sum_{k=0}^{L_0-1} \ln\left\{\cosh\left[\frac{E_s}{N_0}\text{Im}\{x(k+1/2)e^{-j\tilde{\theta}}\}\right]\right\} \quad (5.8.11)$$

The ML estimate is that $\tilde{\theta}$ that maximizes (5.8.11). As happens with non-offset modulation there is no way to arrive at an explicit solution unless either high or low SNR is assumed. Paralleling the arguments in Section 7.2 it can be shown that, in the first instance, maximization of (5.8.11) is approximately performed by means of a Costas loop. As this result is intuitively clear, it is not further discussed in the sequel. Instead, assuming a low SNR, we first consider a feedback estimation loop and, later, some feedforward schemes.

### 5.8.2. Feedback Estimation Method

As the derivation is similar to that performed in Section 5.7.3 with non-offset formats, we only highlight the major steps. The starting point is the derivative of the log-likelihood function, which is found to be

$$\frac{d}{d\tilde{\theta}}\ln \Lambda(r|\tilde{\theta}) = \frac{E_s}{N_0}\sum_{k=0}^{L_0-1} Q(k,\tilde{\theta})\tanh\left[\frac{E_s}{N_0}I(k,\tilde{\theta})\right]$$
$$- \frac{E_s}{N_0}\sum_{k=0}^{L_0-1} I(k+1/2,\tilde{\theta})\tanh\left[\frac{E_s}{N_0}Q(k+1/2,\tilde{\theta})\right] \quad (5.8.12)$$

where the following notations have been used:

$$x(k)e^{-j\tilde{\theta}} \triangleq I(k,\tilde{\theta}) + jQ(k,\tilde{\theta}) \quad (5.8.13)$$

$$x(k+1/2)e^{-j\tilde{\theta}} \triangleq I(k+1/2,\tilde{\theta}) + jQ(k+1/2,\tilde{\theta}) \quad (5.8.14)$$

Next, we make the approximation $\tanh(x) \approx x$ which is valid for $E_s/N_0 \ll 1$. As a result (5.8.12) becomes

# Carrier Phase Recovery with Linear Modulations

$$\frac{d}{d\tilde{\theta}} \ln \Lambda(r|\tilde{\theta}) \approx \left(\frac{E_s}{N_0}\right)^2 \sum_{k=0}^{L_0-1} \left\{ I(k,\tilde{\theta})Q(k,\tilde{\theta}) - I(k+1/2,\tilde{\theta})Q(k+1/2,\tilde{\theta}) \right\} \quad (5.8.15)$$

Finally, the generic term in the sum (5.8.15) is exploited as an error signal to update the current phase estimate $\hat{\theta}(k)$. This produces

$$\hat{\theta}(k+1) = \hat{\theta}(k) + \gamma e(k) \quad (5.8.16)$$

with

$$e(k) \triangleq I[k,\hat{\theta}(k)]Q[k,\hat{\theta}(k)] - I[k+1/2,\hat{\theta}(k)]Q[k+1/2,\hat{\theta}(k)] \quad (5.8.17)$$

Figure 5.59 illustrates a block diagram for this estimation loop where abbreviations of the type $I(k)$ in place of $I[k,\hat{\theta}(k)]$ have been made. Comparing it with the decision-directed scheme in Figure 5.23, it is recognized that the only difference is that $I[k,\hat{\theta}(k)]$ and $Q[k+1/2,\hat{\theta}(k)]$ are used in place of the hard decisions sgn$\{I[k,\hat{\theta}(k)]\}$ and sgn$\{Q[k+1/2,\hat{\theta}(k)]\}$ in computing the error signal.

The S-curve for (5.8.17) is derived as follows. The right-hand sides of (5.8.13)-(5.8.14) are found to be

$$I(k,\hat{\theta}) + jQ(k,\hat{\theta}) = e^{j\phi}\left[a_k + j\sum_i b_i h(k-i-1/2)\right] + n'(k) \quad (5.8.18)$$

$$I(k+1/2,\hat{\theta}) + jQ(k+1/2,\hat{\theta}) = e^{j\phi}\left[jb_k + \sum_i a_i h(k-i+1/2)\right] + n''(k+1/2) \quad (5.8.19)$$

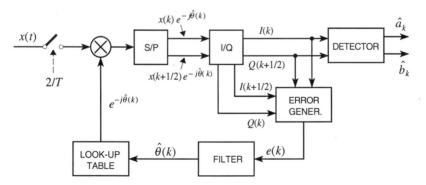

Figure 5.59. Non-data-aided phase recovery loop.

where $\phi \triangleq \theta - \hat{\theta}$ and $n'(k)$ and $n''(k+1/2)$ are phase rotated versions of $n(k)$ and $n(k+1/2)$ Hence, letting $n'(k) \triangleq n'_R(k) + jn'_I(k)$ and $n''(k) \triangleq n''_R(k) + jn''_I(k)$, it follows that

$$I(k,\hat{\theta}) = a_k \cos\phi - \sin\phi \sum_i b_i h(k-i-1/2) + n'_R(k) \qquad (5.8.20)$$

$$Q(k,\hat{\theta}) = a_k \sin\phi + \cos\phi \sum_i b_i h(k-i-1/2) + n'_I(k) \qquad (5.8.21)$$

$$I(k+1/2,\hat{\theta}) = -b_k \sin\phi + \cos\phi \sum_i a_i h(k-i+1/2) + n''_R(k+1/2) \qquad (5.8.22)$$

$$Q(k+1/2,\hat{\theta}) = b_k \cos\phi + \sin\phi \sum_i a_i h(k-i+1/2) + n''_I(k+1/2) \qquad (5.8.23)$$

Finally, substituting into (5.8.17) with $\hat{\theta}(k) = \hat{\theta}$ and taking the expectation with respect to data and noise produces the desired result

$$S(\phi) = \left[1 - \sum_{m=-\infty}^{\infty} h^2(m-1/2)\right] \sin 2\phi \qquad (5.8.24)$$

As is seen, the S-curve is proportional to $\sin 2\phi$, which means that the phase detector has a 180° ambiguity.

It is worth noting that the amplitude of $S(\phi)$ depends on the signal bandwidth. To illustrate this point we assume that the Fourier transform of $h(t)$ is a raised-cosine rolloff function with rolloff $\alpha$ and show that the amplitude of the S-curve is proportional to the rolloff. To proceed, let us write the sum in (5.8.24) in the form (Poisson sum formula)

$$\sum_{m=-\infty}^{\infty} h^2(m-1/2) = \frac{1}{T} \sum_{m=-\infty}^{\infty} H_2(m/T) e^{-jm\pi} \qquad (5.8.25)$$

where $H_2(f)$, the Fourier transform of $h^2(t)$, is related to the Fourier transform of $h(t)$ by

$$H_2(f) = \int_{-\infty}^{\infty} H(v) H(f-v) dv \qquad (5.8.26)$$

As $H(f)$ is bandlimited to $\pm(1+\alpha)/(2T)$, $H_2(f)$ is confined within $\pm(1+\alpha)/T$ and the only nonzero terms in (5.8.25) are those with $m=0$ and $m=\pm 1$. Carrying out the calculations it is found that

$$\sum_{m=-\infty}^{\infty} h^2(m-1/2) = 1 - \frac{\alpha}{2} \qquad (5.8.27)$$

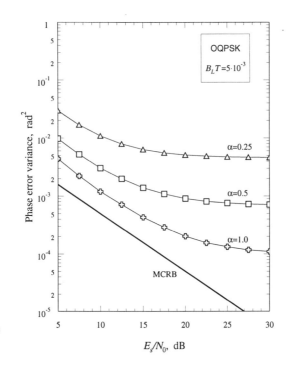

Figure 5.60. Phase error variance with rolloff $\alpha$ as a parameter.

Thus, substituting into (5.8.24) yields the result sought

$$S(\phi) = \frac{\alpha}{2}\sin 2\phi \qquad (5.8.28)$$

The tracking performance of the phase recovery loop in Figure 5.59 can be analytically assessed by paralleling the arguments in Section 5.4.2. This subject is not pursued here, however, as the calculations are too lengthy. Figure 5.60 illustrates simulation results for a loop bandwidth $B_L T = 5 \cdot 10^{-3}$. The modified Cramer-Rao bound is also shown as a baseline. As is seen, performance degrades as $\alpha$ decreases. All the curves exhibit a floor as the SNR increases. Of course this is a manifestation of the self noise originated in the phase detector.

### 5.8.3. ML-Oriented Feedforward Method

In this section we derive a feedforward estimation scheme based on ML methods. Later, in Section 5.8.4, a variant to this algorithm is proposed which arises from analogy with the Viterbi&Viterbi method in Section 5.7.4. To begin, assume $E_s/N_0 \ll 1$ and approximate $\ln[\cosh(x)]$ with the first term of its power series

$$\ln[\cosh(x)] \approx \frac{x^2}{2} \tag{5.8.29}$$

Substituting into (5.8.12) yields (within an immaterial factor)

$$\ln \Lambda(r|\tilde{\theta}) \approx \sum_{k=0}^{L_0-1} \left\{ \left[ \text{Re}\left\{ x(k)e^{-j\tilde{\theta}} \right\} \right]^2 + \left[ \text{Im}\left\{ x(k+1/2)e^{-j\tilde{\theta}} \right\} \right]^2 \right\} \tag{5.8.30}$$

Next, write

$$\text{Re}\left\{ x(k)e^{-j\tilde{\theta}} \right\} = \frac{1}{2}\left[ x(k)e^{-j\tilde{\theta}} + x^*(k)e^{j\tilde{\theta}} \right] \tag{5.8.31}$$

$$\text{Im}\left\{ x(k+1/2)e^{-j\tilde{\theta}} \right\} = \frac{1}{2j}\left[ x(k+1/2)e^{-j\tilde{\theta}} - x^*(k+1/2)e^{j\tilde{\theta}} \right] \tag{5.8.32}$$

and insert these expressions into (5.8.30). Ignoring constants independent of $\tilde{\theta}$, it turns out that maximizing (5.8.30) is equivalent to maximizing

$$F(\tilde{\theta}) \triangleq \text{Re}\left\{ e^{-j2\tilde{\theta}} \sum_{k=0}^{L_0-1} \left[ x^2(k) - x^2(k+1/2) \right] \right\} \tag{5.8.33}$$

The maximum is clearly achieved setting $2\tilde{\theta}$ equal to the argument of the sum in (5.8.33) and this leads to the estimation rule

$$\hat{\theta} = \frac{1}{2}\arg\left\{ \sum_{k=0}^{L_0-1} \left[ x^2(k) - x^2(k+1/2) \right] \right\} \tag{5.8.34}$$

Note that, as the arg-function takes values in the interval $\pm\pi$, the estimate is restricted within $\pm\pi/2$. This generates a phase ambiguity of 180° which can be resolved either with differential encoding/decoding or the methods indicated in Section 5.7.6.

The following interpretation of (5.8.34) is of interest. As the channel response is Nyquist, the samples $x(k)$ and $x(k+1/2)$ from the matched filter are given by

$$x(k) = e^{j\theta}\left\{ a_k + j\sum_i b_i h(k-i-1/2) \right\} + n(k) \tag{5.8.35}$$

## Carrier Phase Recovery with Linear Modulations

$$x(k+1/2) = e^{j\theta}\left\{\sum_i a_i h(k-i+1/2) + jb_k\right\} + n(k+1/2) \quad (5.8.36)$$

Squaring, taking the average with respect to symbols and noise and using (5.8.27) yields

$$E\{x^2(k)\} = \frac{\alpha}{2}e^{j2\theta} \quad (5.8.37)$$

$$E\{x^2(k+1/2)\} = -\frac{\alpha}{2}e^{j2\theta} \quad (5.8.38)$$

Thus, $x^2(k)$ and $x^2(k+1/2)$ may be written in the form

$$x^2(k) = \frac{\alpha}{2}e^{j2\theta} + N_1(k) \quad (5.8.39)$$

$$x^2(k+1/2) = -\frac{\alpha}{2}e^{j2\theta} + N_2(k+1/2) \quad (5.8.40)$$

where $N_1(k)$ and $N_2(k+1/2)$ are zero-mean random variables. It follows that, *on average*, the difference $x^2(k)-x^2(k+1/2)$ is a complex number parallel to the phasor $e^{j2\theta}$ and, therefore, the sample average of $x^2(k)-x^2(k+1/2)$ has an argument close to $2\theta$. This justifies the estimation procedure indicated in (5.8.34).

Performance analysis of the above estimator may be pursued with a method proposed by Meyrs and Franks [15]. Their key idea is to take a Taylor series expansion of $F(\tilde{\theta})$ around $\tilde{\theta} = \theta$

$$F(\tilde{\theta}) \approx F(\theta) + F^{(1)}(\theta)(\tilde{\theta} - \theta) + \frac{1}{2}F^{(2)}(\theta)(\tilde{\theta} - \theta)^2 \quad (5.8.41)$$

where $F^{(1)}(\theta)$ and $F^{(2)}(\theta)$ denote first and second derivatives. It can be checked that this approximation is valid at high SNR and with long observation intervals. Then, the maximum of $F(\tilde{\theta})$ occurs for

$$\hat{\theta} = \theta - \frac{F^{(1)}(\theta)}{F^{(2)}(\theta)} \quad (5.8.42)$$

Next, the quantity $F^{(2)}(\theta)$ in the denominator is replaced by its statistical expectation over symbols and noise. Again, this is a valid approximation for $L_0 \gg 1$, in which case the variance of $F^{(2)}(\theta)$ turns out to be much smaller than the squared expected value $[E\{F^{(2)}(\theta)\}]^2$. As a result we get

$$\hat{\theta} \approx \theta - \frac{F^{(1)}(\theta)}{E\{F^{(2)}(\theta)\}} \quad (5.8.43)$$

At this point the expected value and the variance of the estimator are easily computed. In fact, taking the derivative $F^{(1)}(\theta)$ from (5.8.33) and making use of (5.8.39)-(5.8.40) gives

$$E\{F^{(1)}(\theta)\} = 0 \quad (5.8.44)$$

In view of (5.8.43), this means that the estimator is unbiased, i.e., $E\{\hat{\theta}\} = \theta$. It follows from (5.8.43) that

$$\text{Var}\{\hat{\theta}\} = \frac{E\left\{\left[F^{(1)}(\theta)\right]^2\right\}}{\left[E\{F^{(2)}(\theta)\}\right]^2} \quad (5.8.45)$$

Performing the expectations in (5.8.45) is a lengthy process which leads to rather awkward formulas. Figure 5.61 illustrates simulation results for the

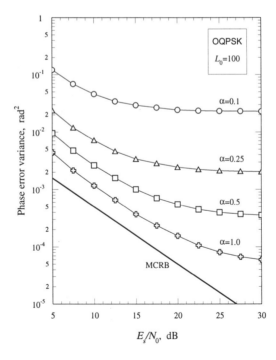

Figure 5.61. Phase error variance versus $E_s/N_0$ with rolloff $\alpha$ as a parameter.

### Carrier Phase Recovery with Linear Modulations

estimation variance versus $E_s/N_0$, taking $L_0=100$ and the rolloff $\alpha$ as a parameter. We see that performance degrades as the rolloff decreases due to self noise. This same situation is observed with NDA closed-loop estimation (see Figure 5.60).

**Exercise 5.8.1.** Derive a feedback algorithm to maximize $F(\tilde{\theta})$ in (5.8.33) and show that it coincides with the estimation loop in Figure 5.59.
*Solution.* The derivative of $F(\tilde{\theta})$ is given by

$$\frac{dF(\tilde{\theta})}{d\tilde{\theta}} = 2\sum_{k=0}^{L_0-1} \text{Im}\left\{e^{-j2\tilde{\theta}}\left[x^2(k) - x^2(k+1/2)\right]\right\} \quad (5.8.46)$$

The desired algorithm is obtained taking the generic term in the summation as an error signal to update the current phase estimate

$$\hat{\theta}(k+1) = \hat{\theta}(k) + \gamma e(k) \quad (5.8.47)$$

with

$$e(k) \triangleq \text{Im}\left\{e^{-j2\hat{\theta}(k)}\left[x^2(k) - x^2(k+1/2)\right]\right\} \quad (5.8.48)$$

Inserting (5.8.39)-(5.8.40) into (5.8.48) and performing the expectation operation yields the S-curve of the phase detector

$$S(\phi) = \alpha \sin 2\phi \quad (5.8.49)$$

Next we show that (5.8.48) coincides with (5.8.17) within an immaterial factor. This follows from the fact that, as a consequence of (5.8.13)-(5.8.14), the quantities $I$ and $Q$ in (5.8.17) may be written in the form

$$I = \text{Re}\left\{xe^{-j\hat{\theta}}\right\} = \frac{1}{2}\left[xe^{-j\hat{\theta}} + x^*e^{j\hat{\theta}}\right] \quad (5.8.50)$$

$$Q = \text{Im}\left\{xe^{-j\hat{\theta}}\right\} = \frac{1}{2j}\left[xe^{-j\hat{\theta}} - x^*e^{j\hat{\theta}}\right] \quad (5.8.51)$$

Hence, substituting into (5.8.17) makes the right-hand side equal to (5.8.48), within a factor 1/2.

### 5.8.4. Viterbi-Like Method

A variant to the estimator (5.8.34) has been proposed by Moeneclaey and Ascheid [64] from analogy with the V&V method described in Section 5.7.4. It consists of expressing $x(k)$ and $x(k+1/2)$ in the form $x \triangleq \rho e^{j\phi}$ and making the following replacements in (5.8.34):

$$x^2(k) \rightarrow F[\rho(k)]e^{j2\phi(k)} \qquad (5.8.52)$$

$$x^2(k+1/2) \rightarrow F[\rho(k+1/2)]e^{j2\phi(k+1/2)} \qquad (5.8.53)$$

where $F(\rho)$ is some suitable function of $\rho$. This produces

$$\hat{\theta} = \frac{1}{2}\arg\left\{\sum_{k=0}^{L_0-1}\left\{F[\rho(k)]e^{j2\phi(k)} - F[\rho(k+1/2)]e^{j2\phi(k+1/2)}\right\}\right\} \qquad (5.8.54)$$

Clearly, for $F(\rho)=\rho^2$ the estimator (5.8.54) reduces to (5.8.34).

The performance of (5.8.54) is difficult to assess analytically and simulations seem the only viable route. This has been done in [64] where the cases $F(\rho)=1$, $F(\rho)=\rho$ and $F(\rho)=\rho^2$ are compared. It turns out that, with rolloff factors greater than 0.5, the effect of the nonlinearity $F(\rho)$ is limited. For smaller

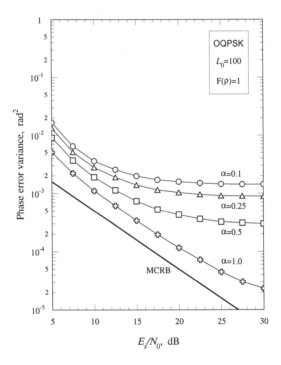

Figure 5.62. Phase error variance with $F(\rho)=1$.

rolloffs, vice versa, it is quite significant and, in fact, $F(\rho)=1$ and $F(\rho)=\rho$ are better than $F(\rho)=\rho^2$. This may be seen comparing the results with $F(\rho)=1$ given in Figure 5.62 with those with $F(\rho)=\rho^2$ in Figure 5.61.

## 5.9. Clockless Phase Recovery with PSK

### 5.9.1 ML-Based Feedforward Estimation

Carrier phase recovery has been investigated in Section 5.7 under the assumption of prior establishment of symbol timing. To speed up the overall synchronization process it is natural to wonder whether carrier phase can be estimated independently of timing, perhaps in parallel with it. This problem is investigated in this section in the case of PSK signaling. In doing so we assume unknown data symbols, perfect carrier frequency recovery, and a Nyquist channel response.

Our analysis draws heavily from the discussion in Section 5.7.4 wherein timing phase $\tau$ is perfectly known. In the present situation, on the contrary, $\tau$ is unknown and is modelled as a random variable uniformly distributed on the interval $\pm T/2$. Then, paralleling the arguments leading to (5.7.66) it is found that

$$\Lambda(r|\tilde{\theta},\tilde{\tau}) = \prod_{k=0}^{L_0-1} T(k,\tilde{\theta},\tilde{\tau}) \tag{5.9.1}$$

where

$$T(k,\tilde{\theta},\tilde{\tau}) \triangleq \frac{1}{M}\sum_{m=0}^{M-1}\exp\left\{\frac{2E_s}{N_0}\mathrm{Re}\left\{x(kT+\tilde{\tau})e^{-j\tilde{\theta}}e^{j2\pi m/M}\right\}\right\} \tag{5.9.2}$$

and $x(t)$ is the matched-filter output.

Next, suppose $E_s/N_0 \ll 1$ and expand the right-hand side of (5.9.2) into a power series. Again, following the arguments leading to (5.7.76), it turns out that the following approximation holds:

$$T(k,\tilde{\theta},\tilde{\tau}) \approx 1 + \frac{2}{M!}\left(\frac{E_s}{N_0}\right)^M \mathrm{Re}\left\{x^M(kT+\tilde{\tau})e^{-jM\tilde{\theta}}\right\} \tag{5.9.3}$$

Substituting into (5.9.1) and keeping only terms containing powers of $E_s/N_0$ up to $M$ yields

$$\Lambda(r|\tilde{\theta},\tilde{\tau}) \approx 1 + \frac{2}{M!}\left(\frac{E_s}{N_0}\right)^M \mathrm{Re}\left\{e^{-jM\tilde{\theta}}\sum_{k=0}^{L_0-1} x^M(kT+\tilde{\tau})\right\} \qquad (5.9.4)$$

At this point the marginal likelihood $\Lambda(r|\tilde{\theta})$ is computed taking the expectation of (5.9.4) with respect to $\tilde{\tau}=\tau$, which entails computing a time average over $-T/2 \leq \tilde{\tau} \leq T/2$. Ignoring immaterial constants independent of $\tilde{\theta}$ it is found that

$$\Lambda(r|\tilde{\theta}) \approx \mathrm{Re}\left\{e^{-jM\tilde{\theta}}\sum_{k=0}^{L_0-1}\int_{-T/2}^{T/2} x^M(kT+\tilde{\tau})d\tilde{\tau}\right\}$$

$$= \mathrm{Re}\left\{e^{-jM\tilde{\theta}}\int_0^{T_0} x^M(t)dt\right\} \qquad (5.9.5)$$

which indicates that the maximum of $\Lambda(r|\tilde{\theta})$ is achieved for $\tilde{\theta}$ equal to

$$\hat{\theta} = \frac{1}{M}\arg\left\{\int_0^{T_0} x^M(t)dt\right\} \qquad (5.9.6)$$

There is a clear similarity between (5.9.6) and the corresponding estimator (5.7.78) for clock-aided phase recovery. In both cases a phase estimate is obtained by averaging the $M$-th power of the matched-filter output. The purpose of raising $x(t)$ to the $M$-th power is to cancel the modulation. However, this goal can only be achieved from the samples $x^M(kT+\tau)$, which implies perfect knowledge of the symbol timing. With clockless phase recovery, on the contrary, intersymbol interference is inevitable and, in consequence, the estimation errors are expected to be larger.

In a digital implementation of (5.9.6) the integral is computed by accumulating the samples of $x(t)$ taken at some rate $N/T$. Correspondingly (5.9.6) becomes

$$\hat{\theta} = \frac{1}{M}\arg\left\{\sum_{k=0}^{NL_0-1} x^M(kT/N)\right\} \qquad (5.9.7)$$

The choice of the oversampling factor $N$ is related to the size of the signal constellation, $M$, and the rolloff factor $\alpha$. As indicated in Appendix 2.A, the sampling rate must exceed $M(1+\alpha)/(2T)$, the bandwidth of $x^M(t)$. Hence

$$N \geq \frac{M(1+\alpha)}{2} \qquad (5.9.8)$$

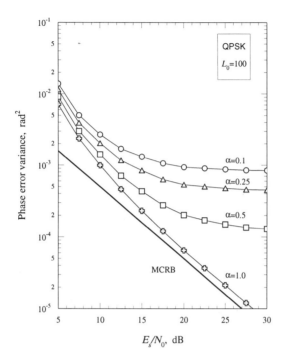

Figure 5.63. Phase error variance for QPSK with oversampling factor $N=4$.

For example, an oversampling $N \geq 3$ is needed with QPSK signals and a rolloff of 0.5.

Figure 5.63 illustrates the performance of estimator (5.9.7) as obtained by simulation with QPSK modulation and an oversampling factor $N=4$. Comparing it with Figure 5.56, which corresponds to perfect timing knowledge, it appears that (5.9.7) is slightly inferior. This is an expected result as the samples $x^M(kT/N)$ are affected by intersymbol interference.

## 5.10. Clockless Phase Recovery with OQPSK

### 5.10.1. Ad Hoc Method

The motivation to pursue clockless phase recovery with QPSK is to shorten the overall acquisition time. While this same motivation holds with OQPSK as well, there is one further reason here because clock recovery is difficult to obtain in the absence of carrier phase knowledge. In the following we provide a clockless phase estimation scheme for OQPSK based on heuristic reasonings.

The starting point is the integral

$$\mathcal{J} \triangleq \int_0^{L_0 T} x^4(t)\,dt \qquad (5.10.1)$$

where $x(t)$ is the matched-filter output

$$x(t) = e^{j\theta}\left\{\sum_i a_i h(t - iT - \tau) + j\sum_i b_i h(t - iT - T/2 - \tau)\right\} + n(t) \qquad (5.10.2)$$

With long manipulations it can be shown that the expected value of $\mathcal{J}$ (over data and noise) has the form

$$\mathrm{E}\{\mathcal{J}\} = -K e^{j4\theta} \qquad (5.10.3)$$

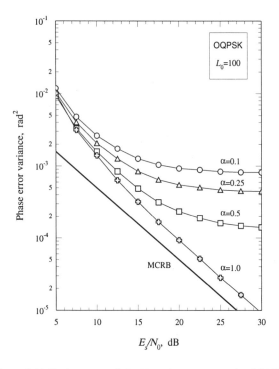

Figure 5.64. Performance of clockless phase recovery for OQPSK.

where $K$ is a positive constant independent of $\theta$. Equation (5.10.3) says that, *on average*, the argument of $\mathcal{J}$ equals $4\theta-\pi$ and this suggests using the estimation formula

$$\hat{\theta} = \frac{\pi}{4} + \frac{1}{4}\arg\left\{\int_0^{L_0 T} x^4(t)dt\right\} \tag{5.10.4}$$

In a digital implementation the integral can be computed through the samples of $x^4(t)$ taken at some rate $N/T$. As $x(t)$ is bandlimited within $\pm(1+\alpha)/(2T)$ and $x^4(t)$ is confined within $\pm 2(1+\alpha)/T$, an oversampling factor $N \geq 2(1+\alpha)$ is sufficient. Correspondingly (5.10.4) becomes

$$\hat{\theta} = \frac{\pi}{4} + \frac{1}{4}\arg\left\{\sum_{k=0}^{4NL_0-1} x^4(kT/N)\right\} \tag{5.10.5}$$

Figure 5.64 shows simulation results for the variance of this estimator. Comparing it with Figure 5.62, it is clear that (5.10.5) is slightly superior to the V&V estimator (5.8.54) for any rolloff.

## 5.11. Key Points of the Chapter

- Optimum clock-aided and data-aided carrier phase estimators are obtained by straightforward application of ML methods. They employ the samples from the matched filter taken either at symbol rate or at twice that rate, depending on the modulation format (non-offset or offset).
- In the absence of frequency errors their performance achieves the Cramer-Rao bound.
- Frequency errors produce degradations. In particular, the estimates are biased whenever the center of the observation interval does not coincide with the phase estimation time.
- Decision-directed phase recovery is best implemented by means of Costas loops. The heart of a Costas loop is the phase detector which generates an error signal making use of signal samples and symbol decisions from the detector.
- The error signal provides a measure of the difference between the carrier phase and its current estimate. The error signal serves to steer the loop and keep that difference as small as possible.
- The average of the error signal, conditioned on a fixed phase error, is

- referred to as the S-curve and provides information about the loop acquisition capability. In particular, its nulls with positive slope represent stable equilibrium points while those with negative slope are unstable points.

- S-curves may have several stable equilibrium points, separated by some fixed quantity from each other. Thus, the loop may settle on phase estimates that differ from the carrier phase by multiples of that quantity.

- Phase ambiguity affects Costas loops as well as other non-data-aided phase estimators. A common method to cope with phase ambiguity is to employ coherent differential detection.

- Tracking performance of a Costas loop is close to MCRB with PSK and sufficiently high SNR. With OQPSK the bound is nearly achieved at intermediate SNR. At higher values, instead, the estimation variance exhibits a floor.

- Phase noise involves some trade-off in the design of the Costas loop bandwidth. Phase noise cannot be tracked adequately if the bandwidth is too small. On the other hand, a large bandwidth makes too much noise enter the loop with degrading effects on the tracking accuracy.

- In the presence of uncompensated carrier frequency offsets a first-order Costas loop exhibits steady-state phase errors which degrade the BER performance. The drawback is avoided with a second-order loop.

- Multiple tracking comes about as an application of per-survivor-processing methods and consists of endowing each survivor in the decoder with a separate Costas loop. Multiple tracking is useful to reduce slip rates and acquisition times and is quite effective against phase noise.

- In general, Costas loops are inadequate to track the fast phase variations encountered on fading channels. In these circumstances differential detection is attractive due to its implementation simplicity. With fast fading, however, BER curves exhibit a floor which may be too high even for digital voice applications.

- As an alternative, pilot-symbol assisted modulation (PSAM) may be adopted. In PSAM systems known symbols are multiplexed with the data to allow periodic measurements of the channel gain. Interpolation of these measurements yields channel estimates at the symbol rate. PSAM methods are superior to differential detection for Doppler bandwidths up to about 1% of the symbol rate. With faster fading, however, BER curves show a floor even with PSAM.

- Per-survivor methods are effective with Doppler bandwidths ranging from

1% to 5% of the symbol rate. The fundamental idea with these methods is that the channel gain must be estimated on a per-survivor basis and the estimates must be incorporated into the detection algorithm.

- Clock-aided but non-data-aided (NDA) phase estimation for AWGN channels leads to feedforward structures that are interesting in burst mode transmissions. A variety of such structures is available for either non-offset or offset formats. Their performance is largely dependent on the modulation format and the rolloff factor (the latter only for offset modulations).

- NDA phase recovery may be pursued even without symbol timing. Feedforward schemes are useful whenever a short estimation time is needed. Performance with OQPSK is even better than with clock-aided methods. The price to pay is a higher sampling rate, however.

# References

[1] A.J.Viterbi and J.K.Omura, *Principles of Digital Communication and Coding*. New York: McGraw-Hill, 1979.
[2] G.Lorden, R.J.McElice and L.Swanson, Node Synchronization for the Viterbi Decoder, *IEEE Trans. Commun.*, **COM-32**, 524-531, May 1984.
[3] M.Moeneclaey and P.Sanders, Syndrome-Based Viterbi Decoder Node Synchronization and Out-of-Lock Detection, Conf. Rec. GLOBECOM'90, San Diego, CA, Dec. 1990, paper 407.4.
[4] U.Mengali, R.Pellizzoni and A.Spalvieri, Soft-Decision-Based Node Synchronization for Viterbi Decoders, *IEEE Trans. Commun.*, **COM-43**, 2532-2539, Sept. 1995.
[5] F.M.Gardner, *Demodulator Reference Recovery Techniques Suited for Digital Implementation*, European Space Agency, Final Report, ESTEC Contract No 6847/86/NL/DG, August, 1988.
[6] P.Y.Kam, Maximum Likelihood Carrier Phase Recovery for Linearly Suppressed-Carrier Digital Data Modulations, *IEEE Trans. Commun.*, **COM-34**, 522-527, June 1986.
[7] G.Asheid, M.Oerder, J.Stahl and H.Meyr, An All Digital Receiver for Bandwidth Efficient Transmission at High Data Rates, *IEEE Trans. Commun.*, **COM-37**, 804-813, Aug. 1989.
[8] F.Daffara and J.Lamour, Comparison Between Digital Phase Recovery Techniques in the Presence of a Frequency Shift, Proc. ICC'94, May 1-5, New Orleans, Louisiana, 1994.
[9] R.De Gaudenzi, V.Vanghi and T.Garde, Feed-Forward Decision-Feedback Carrier Phase Synchronizer for M-PSK Signals, Proc. ICC'95, June 18-22, Seattle, Washington, 1995.
[10] F.M.Gardner, *Phaselock Techniques*, 2nd Edition, New York: Wiley, 1979.
[11] W.C.Lindsey and M.K.Simon, *Telecommunication Systems Engineering*, Englewood Cliffs, New Jersey: Prentice-Hall, 1973.
[12] H.Kobayashi, Simultaneous Adaptive Estimation and Decision Algorithms for Carrier Modulated Data Transmission Systems, *IEEE Trans. Commun.*, **COM-19**, 268-280, June 1971.
[13] D.D.Falconer and J.Salz, Optimal Reception of Digital Data over the Gaussian Channel with Unknown Delay and Phase Jitter, *IEEE Trans. Inf. Theory*, **IT-23**, 117-126, Jan. 1977.
[14] W.C.Lindsey, *Synchronization Systems in Communication and Control*, Englewood Cliffs: Prentice-Hall, 1972.

[15] M.H.Meyrs and L.E.Franks, Joint Carrier Phase and Symbol Timing Recovery for PAM Systems, *IEEE Trans. Commun.*, **COM-28**, 1121-1129, Aug. 1980.
[16] L.E.Franks, "Synchronization Subsystems: Analysis and Design," in K.Feher (ed.), *Digital Communications: Satellite/Earth Station Engineering*, Englewood Cliffs: Prentice Hall, 1983.
[17] M.K.Simon and J.G.Smith, Carrier Synchronization and Detection of QASK Signal Sets, *IEEE Trans. Commun.*, **COM-22**, 98-106, Feb. 1974.
[18] M.K.Simon, Optimum Receiver Structures for Phase-Multiplexed Modulations, *IEEE Trans. Commun.*, **COM-26**, 865-872, June 1978.
[19] D.D.Falconer, Jointly Adaptive Equalization and Carrier Recovery in Two-Dimensional Digital Communication Systems, *Bell Syst. Tech. J.*, **55**, 317-334, March 1976.
[20] U.Mengali, Joint Phase and Timing Acquisition in Data Transmission, *IEEE Trans. Commun.*, **COM-25**, 1174-1185, Oct. 1977.
[21] A.Leclerc and P.Vandamme, Universal Carrier Recovery Loop for QASK and PSK Signal Sets, *IEEE Trans. Commun.*, **COM-31**, 130-136, Jan. 1983.
[22] S.Moridi and H.Sari, Analysis of Four Decision Feedback Carrier Recovery Loops in the Presence of Intersymbol Interference, *IEEE Trans. Commun.*, **COM-33**, 543-550, June 1985.
[23] F.M.Gardner, Hangup in Phase-Lock Loops, *IEEE Trans. Commun.*, **COM-25**, 1210-1214, Oct. 1977.
[24] H.Meyr and L.Popken, Phase Acquisition Statistics for Phase-Locked Loops, *IEEE Trans. Commun.*, **COM-28**, 1365-1372, Aug. 1980.
[25] F.M.Gardner, Equivocation as a Cause of PLL Hangup, *IEEE Trans. Commun.*, **COM-30**, 2242-2243, Oct. 1982.
[26] G.Asheid and H.Meyr, Cycle Slips in Phase-Locked Loops: A Tutorial Survey, *IEEE Trans. Commun.*, **COM-30**, 2228-2241, Oct. 1982.
[27] M.Moeneclaey, The Influence of Phase-Dependent Loop Noise on the Cycle Slipping of Symbol Synchronizers, *IEEE Trans. Commun.*, **COM-33**, 1234-1239, Dec. 1985.
[28] H.Meyr and G.Asheid, *Synchronization in Digital Communications*, New York: John Wiley & Sons, 1990.
[29] T.Jesupret, M.Moeneclaey and G.Asheid, *Digital Demodulator Synchronization*, ESA Final Report, ESTEC Contract No 8437/89/NL/RE, June 1991.
[30] G.Ungerboeck, Channel Coding with Multilevel Phase Signals, *IEEE Trans. Inf. Theory*, **IT-28**, 55-67, Jan. 1982.
[31] S.S.Pietrobon *et al.*, Trellis Coded Multidimensional Phase Modulation, *IEEE Trans. Inf. Theory*, **IT-36**, 63-89, Jan. 1990.
[32] E.I.Jury, *Theory and Application of the z-Transform Method*, Huntingdon, New York: R.E.Krieger Publishing Co, 1964.
[33] V.F.Kroupa, Noise Properties in PLL Systems, *IEEE Trans. Commun.*, **COM-30**, 2244-2252, Oct. 1982.
[34] Members of Module 3, Reducing the Effects of Phase Noise, Race-dTTb/WP3.33, Oct. 1994.
[35] A.Papoulis, *Probability, Random Variables, and Stochastic Processes*, New York: McGraw-Hill, Third Edition, 1991.
[36] A.J.Macdonald and J.B.Anderson, PLL Synchronization for Coded Modulation, Conf. Rec. ICC'91, pp. 52.6.1-52.6.5, Denver, Colorado, June 23-25, 1991.
[37] A.N.D'Andrea, U.Mengali and G.Vitetta, Approximate ML Decoding of Coded PSK with no Explicit Carrier Phase Reference, *IEEE Trans. Commun.*, **COM-42**, 1033-1039, Feb./March/April 1994.
[38] A.N.D'Andrea, U.Mengali and G.Vitetta, Multiple Phase Synchronization in Continuous Phase Modulation, Digital Signal Processing vol. 3, Academic Press, pp. 188-198, 1993.
[39] A.Polydoros, R.Raheli and C-K.Tzou, Per-Survivor-Processing: A General Approach to MLSE in Uncertain Environments, *IEEE Trans. Commun.*, **COM-43**, 354-364,

Feb./March/April 1995.
[40] R.S.Kennedy, *Fading Dispersive Communication Channels*, New York: Wiley-Interscience, 1960.
[41] J.G.Proakis, *Digital Communications*, New York: McGraw-Hill Inc., Second Edition, 1989.
[42] P.A.Bello, Characterization of Randomly Time-Variant Linear Channels, *IEEE Trans. Commun. Systems*, 360-393, Dec. 1965.
[43] W.Jakes (ed.), *Microwave Mobile Communications*, New York: John Wiley and Sons, 1974.
[44] E.Biglieri, D.Divsalar, P.J.McLane and M.K.Simon, *Introduction to Trellis-Coded Modulation with Applications*, New York: Macmillan Publishing Company, 1991.
[45] R.De Gaudenzi and V.Vieri, Performance of Coherent and Differential Detection for Trellis-Coded 8-PSK over Satellite Flat Rician Fading Channels, European Trans. Telecom., vol. 6, March-April 1995, pp. 141-152.
[46] A.Bateman, Feedforward Transparent Tone-In-Band: Its Implementations and Applications, *IEEE Trans. Vehic. Tech.*, **VT-39**, 235-243, Aug. 1990.
[47] J.K.Cavers and M.Liao, A Comparison of Pilot Tone and Pilot Symbol Techniques for Digital Mobile Communication, Proc. IEEE GLOBECOM'92, Orlando, FL, Dec. 1992, pp. 915-921.
[48] M.L.Moher and J.H.Lodge, TCMP–a Modulation and Coding Strategy for Fading Channels, *IEEE J. Selected Areas Commun.*, **SAC-7**, 1347-1355, Dec. 1989.
[49] G.T.Irvine and P.J.McLane, Symbol-Aided plus Decision-Directed Reception for PSK/TCM Modulation on Shadowed Mobile Satellite Fading Channels, *IEEE J. Select. Areas Commun.*, **SAC-10**, 1289-1299, Oct. 1992.
[50] A.Aghamohammadi, H.Meyr and G.Asheid, A New Method for Phase Synchronization and Automatic Gain Control of Linearly Modulated Signals on Frequency-Flat Fading Channels, *IEEE Trans. Commun.*, **COM-39**, 25-29, Jan. 1991.
[51] J.K.Cavers, An Analysis of Pilot Symbol Assisted Modulation for Rayleigh Fading Channels, *IEEE Trans. Vehic. Tech.*, **VT-40**, 686-693, Nov. 1991.
[52] F.Edbauer, Performance of Interleaved Trellis-Coded Differential 8-PSK Modulation Over Fading Channels, *IEEE J. Select. Areas Commun.*, **SAC-7**, 1340-1346, Dec. 1989.
[53] D.Divsalar and M.K.Simon, Performance of Trellis Coded MPSK on Fast Fading Channels, Proc. IEEE ICC'89, Boston, MA, June 1989, pp. 261-267.
[54] A.N.D'Andrea, A.Diglio and U.Mengali, Symbol-Aided Channel Estimation with Nonselective Rayleigh Fading Channels, *IEEE Trans. Vehic. Technol.*, **VT-44**, 41-49, Feb. 1995.
[55] G.M.Vitetta and D.Taylor, Maximum Likelihood Decoding of Uncoded and Coded PSK Signals Sequences Transmitted over Rayleigh Flat-Fading Channels, *IEEE Trans. Commun.*, **COM-43**, 2750-2758, Nov. 1995.
[56] A.C.M.Lee and P.J.McLane, Convolutionally Interleaved PSK and DPSK Trellis Codes for Shadowed, Fast Fading Mobile Satellite Communication Channels, *IEEE Trans. Vehic. Tech.*, **VT-39**, Feb. 1990.
[57] G.J.Foschini, Equalizing Without Altering or Detecting Data, *Bell Syst. Tech. J.*, **64**, 1885-1911, October 1985.
[58] A.J.Viterbi and A.M.Viterbi, Nonlinear Estimation of PSK-Modulated Carrier Phase with Application to Burst Digital Transmission, *IEEE Trans. Inf. Theory*, **IT-29**, 543-551, July 1983.
[59] A.N.D'Andrea, U.Mengali and R.Reggiannini, Carrier Phase Recovery for Narrow-Band Polyphase Shift Keyed Signals, *Alta Frequenza*, **LVII**, 575-581, Dec. 1988.
[60] M.Moeneclaey and G. de Jonghe, ML-Oriented NDA Carrier Synchronization for General Rotationally Symmetric Signal Constellations, *IEEE Trans. Commun.*, **COM-42**, 2531-2533, Aug. 1994.
[61] M.Oerder and H.Meyr, Digital Filter and Square Timing Recovery, *IEEE Trans.*

*Commun.*, **COM-36**, 605-612, May 1988.
[62] M.P.Fitz, Equivocation in Nonlinear Digital Carrier Synchronizers, *IEEE Trans. Commun.*, **COM-39**, 1672-1682, Nov. 1991.
[63] G.De Jonghe and M.Moeneclaey, Cycle Slip Analysis of the NDA-FF Carrier Synchronizer Based on the Viterbi & Viterbi Algorithm, Proc. IEEE ICC'94, New Orleans, Louisiana, May 1994, pp. 880-884.
[64] M.Moeneclaey and G.Ascheid, Extension of the Viterbi and Viterbi carrier Synchronization Algorithm to OQPSK Transmission, Proc. 2nd International Workshop on DSP Applied to Space Communications, Politecnico di Torino, 24-25 Sept. 1990, paper 1.7.

# 6

# Carrier Phase Recovery with CPM Modulations

## 6.1. Introduction

In this chapter we consider carrier phase recovery for CPM signaling. By and large the presentation follows the same route taken with linear modulations. In particular, data-aided (DA) estimation is considered first, then we concentrate on decision-directed (DD) methods and, finally, on non-data-aided (NDA) techniques (either clock-aided or non-clock-aided).

As we pointed out in Chapter 2, CPM formats with good bandwidth and energy efficiencies often require an ML sequence detector implemented by means of a Viterbi algorithm. As the complexity of this algorithm may be considerable, simplified detection schemes are of practical interest. One popular arrangement is the so-called MSK-type receiver [1]-[2] which requires only two real filters and a limited amount of processing.

In the next section we provide an overview of MSK-type receivers and discuss related DA phase estimation methods. DA phase recovery with general CPM modulation (multilevel signaling and arbitrary modulation index) seems unlikely to be of practical interest in the near future and therefore will not be addressed in this book. In Section 6.3 we concentrate on DD phase recovery for MSK-type modulation and, in Section 6.4, we extend the discussion to general CPM formats. Section 6.5 deals with joint detection and channel estimation for CPM signaling over flat-fading channels. In Section 6.6 we return to AWGN channels and discuss NDA but clock-aided phase estimation. Finally, in Section 6.7, we address non-clock-aided phase recovery.

## 6.2. Data-Aided Phase Estimation with MSK-Type Modulation

### 6.2.1. MSK-Type Receivers

We first overview optimum ML sequence detection for MSK and then we concentrate on general binary modulations (with arbitrary phase responses) with modulation index 1/2 (MSK-type signaling). In either case the signal model is

$$s(t) = \sqrt{\frac{2E_s}{T}} e^{j\psi(t,\alpha)} \qquad (6.2.1)$$

with

$$\psi(t,\alpha) \triangleq \pi \sum_{i=0}^{N-1} \alpha_i q(t-iT) \qquad (6.2.2)$$

The starting point is the representation of an MSK signal as an OQPSK waveform. This subject has been discussed in Exercise 4.2.1, where an expression of the following type has been derived for $s(t)$:

$$s(t) = \sqrt{\frac{2E_s}{T}} e^{j\phi} \left[ \sum_{i=1}^{\infty} a_{2i} h'_0(t-2iT) + j \sum_{i=0}^{\infty} a_{2i+1} h'_0[t-(2i+1)T] \right] \qquad (6.2.3)$$

with

$$h'_0(t) \triangleq \begin{cases} \cos\dfrac{\pi t}{2T} & -T \leq t \leq T \\ 0 & \text{elsewhere} \end{cases} \qquad (6.2.4)$$

For convenience we have chosen a pulse $h'_0(t)$ centered around the origin. This has been obtained by shifting the pulse $h_0(t)$ in Chapter 4 by $T$ seconds rightward. All other notation has been left unchanged. In particular, the parameter $\phi$ takes values $\{0, \pi/2, \pi, 3\pi/2\}$, depending on the data pattern prior to the time origin, and the *pseudo-symbols* $\{a_i\}$ are related to the modulation symbols $\{\alpha_i\}$ by the relations

$$a_{2i} \triangleq \cos\left(\frac{\pi}{2} \sum_{l=1}^{2i} \alpha_l\right) \qquad (6.2.5)$$

$$a_{2i+1} \triangleq \sin\left(\frac{\pi}{2} \sum_{l=1}^{2i+1} \alpha_l\right) \qquad (6.2.6)$$

## Carrier Phase Recovery with CPM Modulations

It is worth noting that the pseudo-symbols and the modulation symbols all belong to the same alphabet $\{-1,1\}$.

Next we concentrate on the received waveform

$$r(t) = \sqrt{\frac{2E_s}{T}} e^{j(2\pi vt + \theta)} \left[ \sum_{i=1}^{\infty} a_{2i} h'_0(t - 2iT - \tau) \right.$$

$$\left. + j \sum_{i=0}^{\infty} a_{2i+1} h'_0[t - (2i+1)T - \tau] \right] + w(t) \quad (6.2.7)$$

where, for simplicity, the phase $\phi$ appearing in (6.2.3) has been absorbed into $\theta$. If the synchronization parameters $(v, \theta, \tau)$ were known, ML sequence detection would be performed by maximizing the likelihood function $\Lambda(r|\tilde{a})$ over the allowed sequences $\tilde{a} \triangleq \{\tilde{a}_1, \tilde{a}_2, \ldots, \tilde{a}_N\}$, as is now explained.

Start from the likelihood function which is given by

$$\Lambda(r|\tilde{a}) = \exp\left\{ \frac{1}{N_0} \sqrt{\frac{2E_s}{T}} \left[ \sum_{k=1}^{N/2} \tilde{a}_{2k} \operatorname{Re}\{x(2k)e^{-j\theta}\} + \sum_{k=0}^{N/2-1} \tilde{a}_{2k+1} \operatorname{Im}\{x(2k+1)e^{-j\theta}\} \right] \right\}$$
(6.2.8)

where the shorthand notation $x(k) \triangleq x(kT + \tau)$ has been used and $x(t)$ represents the response to $r(t)e^{-j2\pi vt}$ of the matched filter $h'_0(-t)$, i.e.,

$$x(t) = \left[ r(t)e^{-j2\pi vt} \right] \otimes h'_0(-t) \quad (6.2.9)$$

Clearly, maximizing (6.2.8) amounts to maximizing the quantity

$$\sum_{k=1}^{N/2} \tilde{a}_{2k} \operatorname{Re}\{x(2k)e^{-j\theta}\} + \sum_{k=0}^{N/2-1} \tilde{a}_{2k+1} \operatorname{Im}\{x(2k+1)e^{-j\theta}\} \quad (6.2.10)$$

This task can be pursued through a symbol-by-symbol decision procedure wherein $\hat{a}_{2k}$ is set equal to $\pm 1$, according to the sign of $\operatorname{Re}\{x(2k)e^{-j\theta}\}$ and, similarly, $\hat{a}_{2k+1}$ is chosen according to the sign of $\operatorname{Im}\{x(2k+1)e^{-j\theta}\}$. In other words, letting

$$x(k)e^{-j\theta} \triangleq I(k) + jQ(k) \quad (6.2.11)$$

and denoting by $\{\hat{a}_k\}$ the detected sequence, the decision rule takes the form

$$\hat{a}_{2k} = \operatorname{sgn}[I(2k)] \quad (6.2.12)$$

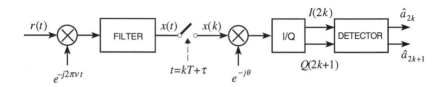

Figure 6.1. MSK-type receiver.

$$\hat{a}_{2k+1} = \text{sgn}[Q(2k+1)] \quad (6.2.13)$$

Figure 6.1 illustrates the block diagram of the receiver. The impulse response of the filter is $h'_0(-t)$. As expected from the signal representation (6.2.3), the receiver structure coincides with that of a conventional OQPSK system. In the latter, however, optimum detection requires that pulses at the filter output be Nyquist. This is no longer true with MSK, however, as may be checked by computing $h'_0(t) \otimes h'_0(-t)$ from (6.2.4).

One question about this scheme is concerned with the computation of the information symbols from the pseudo-symbols released by the detector. This problem is usually approached by differential encoding/decoding as follows. Call $\{\eta_k\}$ the information symbols (taking values ±1) and suppose they are differentially encoded into modulation symbols as follows:

$$\alpha_k = \alpha_{k-1} \eta_k \quad (6.2.14)$$

The problem is to derive $\{\eta_k\}$ from the detected sequence $\{\hat{a}_k\}$. Simple manipulations on (6.2.5)-(6.2.6) show that

$$\alpha_{2k} = -a_{2k} a_{2k-1} \quad (6.2.15)$$

$$\alpha_{2k+1} = a_{2k+1} a_{2k} \quad (6.2.16)$$

or, in compact form,

$$\alpha_k = (-1)^{k+1} a_k a_{k-1} \quad (6.2.17)$$

Then, writing (6.2.14) as

$$\eta_k = \alpha_k \alpha_{k-1} \quad (6.2.18)$$

and making use of (6.2.17) yields

$$\eta_k = -a_k a_{k-2} \quad (6.2.19)$$

## Carrier Phase Recovery with CPM Modulations

which suggests computing the estimate of $\eta_k$ as $\hat{\eta}_k = -\hat{a}_k\hat{a}_{k-2}$.

Next we turn our attention to MSK-type signals. As we pointed out in Chapter 4, they can be approximated as indicated in (6.2.3), with the qualification that $h'_0(t)$ is no longer an arc of a sinusoid. It follows that a *suboptimum* receiver may be derived making use of ML-oriented arguments of the type adopted earlier. Clearly, this leads again to the structure indicated in Figure 6.1, wherein the filter must be properly chosen on the basis of the actual phase response of the modulator. In summary, the block diagram in Figure 6.1 is an optimum receiver for MSK while it is suboptimum for MSK-type modulation.

An interesting issue is whether the filter in Figure 6.1 should be really matched to $h'_0(t)$. One argument against this choice is that the matching condition arises from taking (6.2.3) as the true signal model. Since (6.2.3) is only an approximation, however, it is quite possible that better results might be obtained with other filters. This problem has been addressed in the literature and references [3]-[5] provide a good sample of the proposed solutions. In particular, El-Tanany and Mahmoud [5] try to minimize the mean-square intersymbol interference at the filter output while Galko and Pasupathy [3] and Svensson and Sundberg [4] look for the minimum receiver error probability. More practical design criteria are discussed in [6]-[7].

### 6.2.2. Data-Aided Phase Estimation with MSK-Type Modulation

We first concentrate on true MSK because, in this case, ML data-aided estimation can be solved in a straightforward manner. Assume that the signal parameters in (6.2.7) are all known, with the exception of the carrier phase $\theta$. Then, ML phase estimation is performed by maximizing the likelihood function

$$\Lambda(r|\tilde{\theta}) = \exp\left\{\frac{1}{N_0}\sqrt{\frac{2E_s}{T}}\operatorname{Re}\left\{Xe^{-j\tilde{\theta}}\right\}\right\} \quad (6.2.20)$$

with

$$X \triangleq \int_0^{T_0} r(t)e^{-j2\pi vt}\left[\sum_{i=1}^{\infty} a_{2i}h'_0(t-2iT-\tau) - j\sum_{i=0}^{\infty} a_{2i+1}h'_0[t-(2i+1)T-\tau]\right]dt \quad (6.2.21)$$

As we now explain, $X$ can be approximated as

$$X \approx \sum_{k=1}^{L_0/2} x(2k)a_{2k} - j\sum_{k=0}^{L_0/2-1} x(2k+1)a_{2k+1} \quad (6.2.22)$$

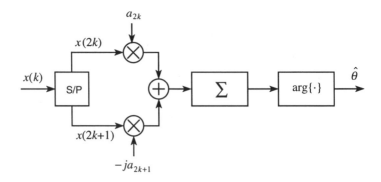

Figure 6.2. Block diagram for ML data-aided phase estimation.

where $L_0 \triangleq T_0/T$, and $x(k)$ are the samples from the matched filter. Clearly, equation (6.2.22) would hold exactly if the pulses $\{h'_0(t-iT-\tau),\ 1 \le i \le L_0\}$ were all confined within the observation interval $0 \le t \le L_0 T$ while the other pulses were zero on that interval. Unfortunately, this is only approximately true. For example, $h'_0(t - L_0 T - \tau)$ extends beyond $t = L_0 T$. It is easily realized, however, that these edge effects become negligible as $L_0$ increases. Thus, assuming $L_0$ sufficiently large, it follows that $\Lambda(r|\tilde{\theta})$ achieves a maximum when $\tilde{\theta}$ equals the argument of $X$ and the ML estimator becomes

$$\hat{\theta} = \arg\left\{ \sum_{k=1}^{L_0/2} x(2k)a_{2k} - j \sum_{k=0}^{L_0/2-1} x(2k+1)a_{2k+1} \right\} \quad (6.2.23)$$

Figure 6.2 illustrates a block diagram for this estimator. The block S/P represents a serial-to-parallel converter. As expected, the estimator structure is the same as for OQPSK.

Extension to MSK-type modulation is pursued by approximating the signal as in (6.2.3) and reasoning as with MSK. This leads again to the estimator (6.2.23) where the $x(k)$ are samples taken from some shaping filter. As before, the question arises as to whether this filter should be really $h'_0(-t)$. This is left as an open question, even though it would seem sensible to employ the same filter for both phase recovery and data detection [3]-[7].

### 6.2.3. Estimator Performance with MSK

The performance of estimator (6.2.23) can be assessed as follows. Taking $r(t)$ from (6.2.7) and letting

$$m(t) \triangleq \sum_{i=1}^{\infty} a_{2i} h_0'(t - 2iT - \tau) + j \sum_{i=0}^{\infty} a_{2i+1} h_0'[t - (2i+1)T - \tau] \quad (6.2.24)$$

from (6.2.21) we have

$$X = \sqrt{\frac{2E_s}{T}} e^{j\theta} \int_0^{T_0} |m(t)|^2 dt + \int_0^{T_0} w(t) e^{-j2\pi vt} m^*(t) dt \quad (6.2.25)$$

On the other hand, from (6.2.1) and (6.2.3) it is recognized that $|m(t)| = 1$. Hence

$$X = T_0 \sqrt{\frac{2E_s}{T}} e^{j\theta} (1 + N_R + jN_I) \quad (6.2.26)$$

where $N_R$ and $N_I$ are random variables defined as

$$N_R + jN_I \triangleq \frac{1}{T_0} \sqrt{\frac{T}{2E_s}} \int_0^{T_0} w(t) e^{-j(2\pi vt + \theta)} m^*(t) dt \quad (6.2.27)$$

Putting these facts together and recalling that $\theta = \arg\{X\}$ yields

$$\hat{\theta} = \theta + \arg\{1 + N_R + jN_I\} \quad (6.2.28)$$

The estimator performance is now computed from the statistics of $N_R$ and $N_I$. Straightforward arguments indicate that $N_R$ and $N_I$ have zero mean and the same variance

$$\sigma^2 = \frac{1}{2L_0} \frac{1}{E_s/N_0} \quad (6.2.29)$$

Also, assuming $L_0 \gg 1$ so that $N_R$ and $N_I$ are statistically much smaller than unity, equation (6.2.28) reduces to

$$\hat{\theta} \approx \theta + N_I \quad (6.2.30)$$

from which it is concluded that the estimator is unbiased and has variance

$$\mathrm{Var}\{\hat{\theta}\} = \frac{1}{2L_0} \frac{1}{E_s/N_0} \quad (6.2.31)$$

This coincides with the MCRB and the true CRB. Indeed, the two bounds coincide as there are no unknown parameters involved in the present estimation problem.

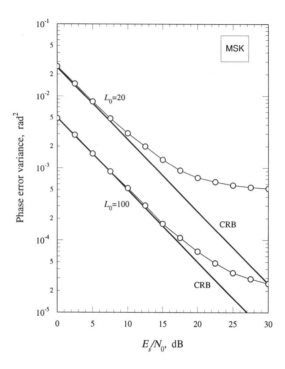

Figure 6.3. Estimation variance versus $E_s/N_0$.

Figure 6.3 illustrates simulation results for the estimator variance with two values of $L_0$. The corresponding CRBs are also shown for comparison. The simulations depart from the bounds as the signal-to-noise ratio increases. Intuitively this is due to edge effects (self noise) in the approximation (6.2.22). As expected, self noise becomes less important as $L_0$ increases.

## 6.3. Decision-Directed Estimation with MSK-Type Modulation

### 6.3.1. Decision-Directed Estimation with MSK

We shall limit our discussion to decision-directed phase estimation with MSK but, as mentioned earlier, the same methods may be utilized with MSK-type signaling. In particular, they involve approximating the signal model as in (6.2.3), with a proper choice of $h'_0(t)$. The question of the optimum $h'_0(t)$ is not addressed in this book but it is believed that most of the solutions indicated in

[3]-[7] should be adequate.

Collecting (6.2.20)-(6.2.22) and rearranging yields

$$\Lambda(r|\tilde{\theta}) = \exp\left\{\frac{1}{N_0}\sqrt{\frac{2E_s}{T}}\left[\sum_{k=1}^{L_0/2} a_{2k} \operatorname{Re}\left\{x(2k)e^{-j\tilde{\theta}}\right\}\right.\right.$$
$$\left.\left. + \sum_{k=0}^{L_0/2-1} a_{2k+1} \operatorname{Im}\left\{x(2k+1)e^{-j\tilde{\theta}}\right\}\right]\right\} \quad (6.3.1)$$

The decision-directed approach assumes that the symbols $\{a_i\}$ are replaced by decisions $\{\hat{a}_i\}$ taken from the detector. Thus, maximizing $\Lambda(r|\tilde{\theta})$ amounts to maximizing

$$F(\tilde{\theta}) \triangleq \sum_{k=1}^{L_0/2} \hat{a}_{2k} \operatorname{Re}\left\{x(2k)e^{-j\tilde{\theta}}\right\} + \sum_{k=0}^{L_0/2-1} \hat{a}_{2k+1} \operatorname{Im}\left\{x(2k+1)e^{-j\tilde{\theta}}\right\} \quad (6.3.2)$$

which can be done by looking for the zeroes of the derivative

$$\frac{dF(\tilde{\theta})}{d\tilde{\theta}} = \sum_{k=1}^{L_0/2} \hat{a}_{2k} \operatorname{Im}\left\{x(2k)e^{-j\tilde{\theta}}\right\} - \sum_{k=0}^{L_0/2-1} \hat{a}_{2k+1} \operatorname{Re}\left\{x(2k+1)e^{-j\tilde{\theta}}\right\} \quad (6.3.3)$$

Toward this end we adopt the usual recursive method which consists of computing the $k$-th terms in (6.3.3) for $\tilde{\theta}$ equal to the current phase estimate and using their sum as an error signal to update the estimate. Formally, we adopt the following error signal:

$$e(2k) \triangleq \frac{1}{K}\left[\hat{a}_{2k} \operatorname{Im}\left\{x(2k)e^{-j\hat{\theta}(k)}\right\} - \hat{a}_{2k+1} \operatorname{Re}\left\{x(2k+1)e^{-j\hat{\theta}(k)}\right\}\right] \quad (6.3.4)$$

where

$$K \triangleq \sqrt{2E_s T} \quad (6.3.5)$$

is a normalizing factor. It should be noted that the phase estimates are computed at intervals of $2T$ according to

$$\hat{\theta}(2k+2) = \hat{\theta}(2k) + \gamma e(2k) \quad (6.3.6)$$

Introducing the shorthand notations

$$I(2k) + jQ(2k) \triangleq x(2k)e^{-j\hat{\theta}(2k)} \quad (6.3.7)$$

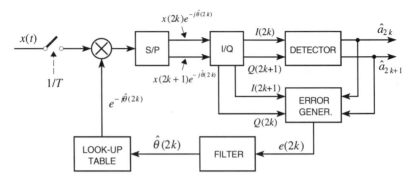

Figure 6.4. Block diagram for the DD estimation algorithm.

$$I(2k+1) + jQ(2k+1) \triangleq x(2k+1)e^{-j\hat{\theta}(2k)} \qquad (6.3.8)$$

equation (6.3.4) becomes

$$e(2k) = \frac{1}{K}\left[\hat{a}_{2k}Q(2k) - \hat{a}_{2k+1}I(2k+1)\right] \qquad (6.3.9)$$

and the estimation algorithm takes the form illustrated in Figure 6.4. This scheme is identical to the decision-directed phase recovery loop we brought out in Chapter 5 for OQPSK.

Analysis of this loop may be carried out by paralleling the arguments used with OQPSK and the results are qualitatively the same. In particular the S-curve has stable points at $\phi = 0$ and $\phi = \pm\pi$, which implies a 180° phase ambiguity. Such an ambiguity has no consequences for data detection, however, since a phase error of 180° affects only the sign of the individual decisions $\hat{a}_k$ but not the products $-\hat{a}_k\hat{a}_{k-2}$ from which the information data are retrieved (see (6.2.19)).

**Exercise 6.3.1.** Compute the S-curve of detector (6.3.9) over a small interval around the origin.

*Solution.* Substituting (6.2.7) into (6.2.9) and performing straightforward manipulations yields

$$x(t) = \sqrt{\frac{2E_s}{T}}e^{j\theta}\left[\sum_{i=1}^{\infty} a_{2i}h'_{00}(t - 2iT - \tau) \right.$$
$$\left. + j\sum_{i=0}^{\infty} a_{2i+1}h'_{00}[t - (2i+1)T - \tau]\right] + n(t) \qquad (6.3.10)$$

# Carrier Phase Recovery with CPM Modulations

where

$$h'_{00}(t) \triangleq h'_0(t) \otimes h'_0(-t) \tag{6.3.11}$$

and $n(t)$ is a thermal noise component

$$n(t) \triangleq \left[ w(t) e^{-j2\pi v t} \right] \otimes h'_0(-t) \tag{6.3.12}$$

Next, $x(t)$ is sampled at $2kT+\tau$ and $(2k+1)T+\tau$ to produce $x(2k)$ and $x(2k+1)$. Inserting the results into (6.3.7)-(6.3.8) gives the outputs from the I/Q block in Figure 6.4. For $\hat{\theta}(k) = \hat{\theta}$ it is found that

$$I(2k) = \frac{K}{T} \left[ \cos\phi \sum_{i=1}^{\infty} a_{2i} h'_{00}[2(k-i)] \right.$$
$$\left. -\sin\phi \sum_{i=0}^{\infty} a_{2i+1} h'_{00}[2(k-i)-1] \right] + n_I(2k) \tag{6.3.13}$$

$$Q(2k) = \frac{K}{T} \left[ \sin\phi \sum_{i=1}^{\infty} a_{2i} h'_{00}[2(k-i)] \right.$$
$$\left. +\cos\phi \sum_{i=0}^{\infty} a_{2i+1} h'_{00}[2(k-i)-1] \right] + n_Q(2k) \tag{6.3.14}$$

$$I(2k+1) = \frac{K}{T} \left[ \cos\phi \sum_{i=1}^{\infty} a_{2i} h'_{00}[2(k-i)+1] \right.$$
$$\left. -\sin\phi \sum_{i=0}^{\infty} a_{2i+1} h'_{00}[2(k-i)] \right] + n_I(2k+1) \tag{6.3.15}$$

$$Q(2k+1) = \frac{K}{T} \left[ \sin\phi \sum_{i=1}^{\infty} a_{2i} h'_{00}[2(k-i)+1] \right.$$
$$\left. +\cos\phi \sum_{i=0}^{\infty} a_{2i+1} h'_{00}[2(k-i)] \right] + n_Q(2k+1) \tag{6.3.16}$$

In these equations $\phi \triangleq \theta - \hat{\theta}$ is the phase error, $n_I$ and $n_Q$ are I/Q noise components and $h'_{00}(k)$ is short for $h'_{00}(kT)$. From (6.2.4) and (6.3.11) it can be seen that

$$h'_{00}(k) = \begin{cases} T & k=0 \\ T/\pi & k=\pm 1 \\ 0 & \text{elsewhere} \end{cases} \tag{6.3.17}$$

Then, for $E_s/N_0 \gg 1$ and $\phi \approx 0$ we have

$$\hat{a}_{2k} \triangleq \text{sgn}[I(2k)] \approx a_{2k} \tag{6.3.18}$$

$$\hat{a}_{2k+1} \triangleq \text{sgn}[Q(2k+1)] \approx a_{2k+1} \tag{6.3.19}$$

meaning that the detector decisions are (almost always) correct. Thus, substituting into (6.3.9) produces

$$e(2k) \approx 2\sin\phi + N_{SN}(2k) + N_{TN}(2k) \tag{6.3.20}$$

where $N_{SN}(2k)$ is a self-noise term

$$N_{SN}(2k) \approx \frac{1}{\pi}\left(a_{2k}a_{2k-1} - a_{2k+2}a_{2k+1}\right) \tag{6.3.21}$$

while $N_{TN}(2k)$ is thermal noise

$$N_{TN}(2k) = \frac{1}{K}\left[a_{2k}n_Q(2k) - a_{2k+1}n_I(2k+1)\right] \tag{6.3.22}$$

Finally, taking the expectation of (6.3.20) for a fixed $\phi$ yields the S-curve around the origin

$$S(\phi) \approx 2\sin\phi \tag{6.3.23}$$

**Exercise 6.3.2.** Compute the error variance $\sigma^2 \triangleq \text{E}\{[\theta - \hat{\theta}(2k)]^2\}$ as a function of the ratio $E_s/N_0$ and the loop noise bandwidth.

*Solution.* Bearing in mind that the slope of the S-curve at the origin equals 2 (see (6.3.23)) and the phase estimates are updated at multiples of $2T$ (instead of $T$), we get (assuming $\gamma A \ll 1$)

$$B_L T \approx \frac{\gamma}{4} \tag{6.3.24}$$

Similarly, the phase error variance is computed from (3.5.60) of Chapter 3 as

$$\sigma^2 \approx \frac{\gamma}{4} \sum_{m=-\infty}^{\infty} R_N(2m)(1-2\gamma)^{2|m|} \qquad (6.3.25)$$

where $R_N(2m)$ represents the autocorrelation of the total noise $N_{SN}(2k)+N_{TN}(2k)$ in (6.3.20). As $N_{SN}(2k)$ and $N_{TN}(2k)$ are uncorrelated, $R_N(2m)$ is the sum of the separate autocorrelations of $N_{SN}(2k)$ and $N_{TN}(2k)$, i.e.,

$$R_N(2m) = R_{SN}(2m) + R_{TN}(2m) \qquad (6.3.26)$$

Performing further calculations it is found that

$$R_{SN}(2m) = \begin{cases} 2/\pi^2 & m = 0 \\ -1/\pi^2 & m = \pm 1 \\ 0 & m \neq 0 \end{cases} \qquad (6.3.27)$$

$$R_{TN}(2m) = \begin{cases} \dfrac{1}{E_s/N_0} & m = 0 \\ 0 & m \neq 0 \end{cases} \qquad (6.3.28)$$

Hence, substituting into (6.3.24)-(6.3.25) yields the desired result

$$\sigma^2 \approx \frac{B_L T}{E_s/N_0} + \frac{32}{\pi^2}(B_L T)^2(1-B_L T) \qquad (6.3.29)$$

The first term in (6.3.29) is recognized as the MCRB($\theta$) while the second represents self noise.

## 6.4. Decision-Directed Estimation with General CPM

### 6.4.1. ML Receivers for CPM

Before addressing decision-directed phase estimation it is useful to overview ML sequence detection with general CPM modulation. To this end we start from the signal written in the form

$$s(t) = e^{j(2\pi\nu t + \theta)}\sqrt{\frac{2E_s}{T}}e^{j\psi(t-\tau,\alpha)} \qquad (6.4.1)$$

with

$$\psi(t,\boldsymbol{\alpha}) = 2\pi h \sum_{i=0}^{N-1} \alpha_i q(t-iT) \qquad (6.4.2)$$

and assume that the data symbols are $M$-ary, equally likely and independent. Also, we take a rational modulation index $h=K/P$, with $P$ and $K$ relatively prime integers. Finally, we normalize the phase response of the modulator so that

$$q(t) = \begin{cases} 0 & t \leq 0 \\ 1/2 & t \geq LT \end{cases} \qquad (6.4.3)$$

where $L$ represents the correlation length. Letting $\boldsymbol{\alpha}_k \triangleq \{\alpha_0, \alpha_1, \ldots, \alpha_k\}$, it is easily seen that (6.4.2) may be written in the form

$$\psi(t,\boldsymbol{\alpha}) = \eta(t, C_k, \alpha_k) + \Phi_k, \qquad kT \leq t < (k+1)T \qquad (6.4.4)$$

with

$$\eta(t, C_k, \alpha_k) \triangleq 2\pi h \sum_{i=k-L+1}^{k} \alpha_i q(t-iT) \qquad (6.4.5)$$

$$C_k \triangleq (\alpha_{k-L+1}, \ldots, \alpha_{k-2}, \alpha_{k-1}) \qquad (6.4.6)$$

$$\Phi_k \triangleq \pi h \sum_{i=0}^{k-L} \alpha_i \quad \text{mod } 2\pi \qquad (6.4.7)$$

Here, $C_k$ and $\Phi_k$ are referred to as the *correlative state* and the *phase state* of the modulator. It can be shown that $\Phi_k$ takes $P$ values.

From (6.4.4)-(6.4.7) it appears that $\psi(t,\boldsymbol{\alpha})$ is uniquely defined by $C_k$, $\Phi_k$ and the current symbol $\alpha_k$. As there are $M^{L-1}$ correlative states and $P$ phase states, $\psi(t,\boldsymbol{\alpha})$ has a total of $PM^{L-1}$ states which correspond to the $L$-tuples $S_k \triangleq (\alpha_{k-L+1}, \ldots, \alpha_{k-2}, \alpha_{k-1}; \Phi_k)$. Therefore, the CPM modulator may be viewed as the cascade of an encoder (with states $S_k$) and a mapper which transforms the encoder output into an exponential waveform with argument (6.4.4).

Assuming that the synchronization parameters $\{v, \theta, \tau\}$ are all known, ML sequence detection is performed by maximizing the likelihood function $\Lambda(r|\tilde{\boldsymbol{\alpha}})$ with respect to the trial sequence $\tilde{\boldsymbol{\alpha}} \triangleq \{\tilde{\alpha}_0, \tilde{\alpha}_1, \ldots, \tilde{\alpha}_{N-1}\}$. With the usual methods the likelihood function is found to be

$$\Lambda(r|\tilde{\boldsymbol{\alpha}}) = \exp\left\{ \frac{1}{N_0} \sqrt{\frac{2E_s}{T}} \sum_{k=0}^{L_0-1} \text{Re}\left\{ Z_k(\tilde{C}_k, \tilde{\alpha}_k) e^{-j(\theta + \tilde{\Phi}_k)} \right\} \right\} \qquad (6.4.8)$$

# Carrier Phase Recovery with CPM Modulations

with

$$Z_k(\tilde{C}_k, \tilde{\alpha}_k) \triangleq \int_{\tau+kT}^{\tau+(k+1)T} r(t) e^{-j2\pi\nu t} e^{-j\eta(t-\tau,\tilde{C}_k,\tilde{\alpha}_k)} dt \tag{6.4.9}$$

$$\tilde{C}_k \triangleq (\tilde{\alpha}_{k-L+1},\ldots,\tilde{\alpha}_{k-2},\tilde{\alpha}_{k-1}) \tag{6.4.10}$$

$$\tilde{\Phi}_k \triangleq \pi h \sum_{i=0}^{k-L} \tilde{\alpha}_i \mod 2\pi \tag{6.4.11}$$

As explained in [1]-[2], there are $M^L$ different values of $Z_k(\tilde{C}_k, \tilde{\alpha}_k)$ which can be computed by feeding $r(t)e^{-j2\pi\nu t}$ into a bank of filters and sampling the filter outputs at $(k+1)T+\tau$. The filters' impulse responses are given by

$$h^{(l)}(t) \triangleq \begin{cases} e^{-j\eta_l(T-t, C_0^{(l)}, \alpha_0^{(l)})} & 0 \leq t \leq T \\ 0 & \text{elsewhere} \end{cases} \tag{6.4.12}$$

with $(l=1,2,\ldots,M^L)$. In this equation $(C_0^{(l)}, \alpha_0^{(l)}) = (\alpha_{-L+1}^{(l)},\ldots,\alpha_{-1}^{(l)},\alpha_0^{(l)})$ is a generic realization of $(\alpha_{-L+1},\ldots,\alpha_{-1},\alpha_0)$ and $\eta_l(t, C_0^{(l)}, \alpha_0^{(l)})$ is given by

$$\eta_l(t, C_0^{(l)}, \alpha_0^{(l)}) = 2\pi h \sum_{i=-L+1}^{0} \alpha_i^{(l)} q(t-iT) \tag{6.4.13}$$

Returning to (6.4.8), it is clear that ML sequence detection amounts to locating the maximum of

$$\Gamma(\tilde{\alpha}) \triangleq \sum_{k=0}^{L_0-1} \text{Re}\left\{ Z_k(\tilde{C}_k, \tilde{\alpha}_k) e^{-j(\theta+\tilde{\Phi}_k)} \right\} \tag{6.4.14}$$

which can be efficiently performed by means of the Viterbi algorithm as follows. Let $\overline{\alpha}_{k-1} \triangleq (\overline{\alpha}_0, \overline{\alpha}_1, \ldots, \overline{\alpha}_{k-1})$ be the survivor path in the encoder trellis terminating in the node $\overline{S}_k = (\overline{\alpha}_{k-L+1},\ldots,\overline{\alpha}_{k-2},\overline{\alpha}_{k-1}; \overline{\Phi}_k)$. Then, the metrics

$$\lambda_k(\overline{S}_k, \tilde{\alpha}_k) \triangleq \text{Re}\left\{ Z_k(\overline{C}_k, \tilde{\alpha}_k) e^{-j(\theta+\overline{\Phi}_k)} \right\}, \quad \tilde{\alpha}_k \in \{\pm 1, \pm 3, \ldots, \pm(M-1)\} \tag{6.4.15}$$

are assigned to the branches stemming from $\overline{S}_k$ and the Viterbi algorithm looks for the path with the maximum accumulated metric. Figure 6.5 illustrates a block diagram for the ML detector. Note that the final decisions are delivered with some delay $DD$ with respect to the current processing time.

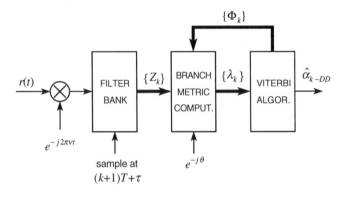

Figure 6.5. Block diagram of the ML receiver.

### 6.4.2. Decision-Directed Phase Estimation

At this stage we are ready to address decision-directed phase estimation. Initially we assume that the symbols are known, with the understanding that they will later be replaced by the decisions taken from the Viterbi decoder. The likelihood function for the unknown phase is given by

$$\Lambda(r|\tilde{\theta}) = \exp\left\{\frac{1}{N_0}\sqrt{\frac{2E_s}{T}} \sum_{k=0}^{L_0-1} \text{Re}\left\{Z_k(C_k,\alpha_k)e^{-j(\tilde{\theta}+\Phi_k)}\right\}\right\} \quad (6.4.16)$$

Hence, the ML estimate of $\theta$ is sought by setting to zero the derivative of the sum in (6.4.16) with respect to $\tilde{\theta}$:

$$\sum_{k=0}^{L_0-1} \text{Im}\left\{Z_k(C_k,\alpha_k)e^{-j(\tilde{\theta}+\phi_k)}\right\} = 0 \quad (6.4.17)$$

To solve (6.4.17) we resort to the usual argument in which the $k$-th term in the sum is computed for $\tilde{\theta}$ equal to the current phase estimate $\hat{\theta}(k)$ and then is used as an error signal to update this estimate. One question here is how to choose the sequence $\alpha_k \triangleq (\ldots,\alpha_{k-2},\alpha_{k-1},\alpha_k)$ involved in the computation of $Z_k(C_k,\alpha_k)$ and $\Phi_k$. A reasonable answer is to take the best survivor sequence $\overline{\alpha}_k$ at time $kT$. As is explained soon, an even better solution is to keep only that portion of $\overline{\alpha}_k$ up to time $(k-D)T$ (which means discarding the symbols from $\overline{\alpha}_{k-D+1}$ to $\overline{\alpha}_k$). Denoting by $\overline{C}_{k-D}$ and $\overline{\Phi}_{k-D}$ the corresponding values of $C_{k-D}$ and $\Phi_{k-D}$, this amount to updating the phase estimates as follows:

## Carrier Phase Recovery with CPM Modulations

$$\hat{\theta}(k+1) = \hat{\theta}(k) + \gamma e(k) \qquad (6.4.18)$$

with

$$e(k) \triangleq \mathrm{Im}\left\{ Z_{k-D}(\overline{C}_{k-D}, \overline{\alpha}_{k-D}) e^{-j\left[\hat{\theta}(k-D) + \overline{\Phi}_{k-D}\right]} \right\} \qquad (6.4.19)$$

An intuitive reason for choosing $D>0$ is that more reliable decisions are expected to be found going backward along the best survivor. In doing so, however, a delay is introduced in the loop which tends to degrade the response to rapid carrier phase changes. This issue has been investigated in [8] where it has been concluded that $D=1$ is the best choice in the presence of phase noise while $D=0$ is preferable with negligible noise.

The feedback algorithm (6.4.18)-(6.4.19) represents the conventional approach to decision-directed phase recovery for CPM [9]. At any given time a single phase estimate is employed to compute the branch metrics in the Viterbi algorithm. An alternative method is to resort to multiple phase tracking, as we did in Section 4.5 of Chapter 4. The idea is to associate a phase estimator with each survivor in the decoder [8], [10]. For each estimator the error signal has the form (6.4.19) and is computed using the decisions taken from its own survivor. Clearly, there are as many phase estimates as the number of survivors. This implies an increased computational load which is justified only in particular cases. One such case occurs when dealing with phase-noise channels, as is now illustrated.

From the discussion in [8] it appears that multiple trackers achieve optimum performance when the delay $D$ equals zero, independently of the phase noise intensity. Accordingly, in the simulations shown in Figures 6.6-6.8 the delay $D$ has been set to zero with multiple tracking (MT) while it has been chosen as either $D=0$ or $D=1$ with single tracking (ST), depending on whether phase noise is present or not. Phase noise is modeled as a Wiener process

$$\theta(k+1) = \theta(k) + \Delta(k) \qquad (6.4.20)$$

where the $\Delta(k)$ are independent zero-mean Gaussian random variables of variance $\sigma_\Delta^2$.

Figure 6.6 compares symbol error rate (SER) as obtained with MT and ST in the absence of phase noise. Binary modulation is assumed with $h=1/2$ and 3RC frequency pulses. The lower curve, denoted *coherent*, corresponds to ideal phase recovery, i.e., $\hat{\theta}(k) = \theta$. It appears that single and multiple trackers have the same performance and, therefore, the latter are not justified in view of their greater complexity. Figures 6.7 and 6.8 show, vice versa, that multiple trackers are superior in the presence of significant phase noise. Again, the lower curve is obtained providing the branch metric computation unit with an ideal phase reference, $\hat{\theta}(k) = \theta(k)$.

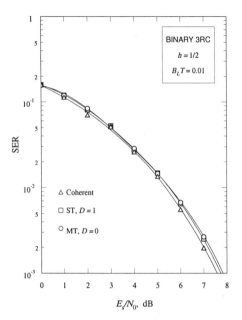

Figure 6.6. Comparison between ST and MT in the absence of phase noise.

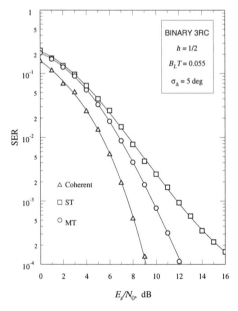

Figure 6.7. Comparison between ST and MT in the presence of phase noise.

# Carrier Phase Recovery with CPM Modulations

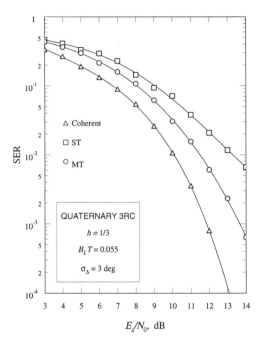

Figure 6.8. Comparison between ST and MT in the presence of phase noise.

**Exercise 6.4.1.** Compute the variance of the phase estimates as obtained with the error signal (6.4.19) under the following conditions: (*i*) carrier phase is constant (no phase noise or frequency errors); (*ii*) phase errors are small; (*iii*) decisions from the best survivor are correct.

*Solution.* If the decisions from the best survivor are correct, we have from (6.4.9)-(6.4.11)

$$Z_{k-D}(S_{k-D}, \alpha_{k-D})e^{-j\Phi_{k-D}} = e^{j\theta}\left[\sqrt{2E_sT} + n(k)\right] \quad (6.4.21)$$

where $n(k) \triangleq n_R(k) + jn_I(k)$ is zero-mean Gaussian noise with independent components, each with variance $N_0T$. Substituting into (6.4.19) and assuming that the phase errors $\phi(k-D) \triangleq \theta - \hat{\theta}(k-D)$ are sufficiently small yields

$$e(k) = \sqrt{2E_sT}\phi(k-D) + n'_I(k) \quad (6.4.22)$$

where $n'_I(k)$ is the imaginary part of $n'(k) \triangleq n(k)e^{-j\phi(k-D)}$. Then, inserting into (6.4.18) produces

$$\phi(k+1) = \phi(k) - \gamma\sqrt{2E_sT}\phi(k-D) - \gamma n'_l(k) \qquad (6.4.23)$$

from which the variance of the phase errors can be computed as follows.

The phase error $\phi(k)$ is viewed as the response to $n'_l(k)$ of a filter with transfer function

$$\mathcal{H}(z) = -\frac{\gamma}{z + \gamma\sqrt{2E_sT}z^{-D} - 1} \qquad (6.4.24)$$

Next, calling $H(f)$ the right-hand side of (6.4.24) for $z = e^{j2\pi fT}$, i.e.,

$$H(f) = -\frac{\gamma}{e^{j2\pi fT} + \gamma\sqrt{2E_sT}e^{-j2D\pi fT} - 1} \qquad (6.4.25)$$

and bearing in mind that the spectral density of $n'_l(k)$ is constant, i.e.,

$$S_{n'_l}(f) = N_0 T^2 \qquad (6.4.26)$$

it is concluded that the phase error variance is given by

$$\sigma^2 = B_L T \frac{1}{E_s/N_0} \qquad (6.4.27)$$

where $B_L$, the filter noise bandwidth, is defined as

$$B_L \triangleq \frac{1}{2|H(0)|^2} \int_{-1/(2T)}^{1/(2T)} |H(f)|^2 df \qquad (6.4.28)$$

Note that (6.4.27) coincides with the MCRB.

Clearly, $B_L$ depends on the delay parameter $D$. In particular, letting $K \triangleq \gamma\sqrt{2E_sT}$ and performing the integral in (6.4.28) it is found that

$$B_L T = \begin{cases} \dfrac{K}{2(2-K)} & \text{for } D=0 \\[2mm] \dfrac{K(1+K)}{2(1-K)(2+K)} & \text{for } D=1 \\[2mm] \dfrac{K(1+K-K^2)}{2(2-3K-K^2+K^3)} & \text{for } D=2 \end{cases} \qquad (6.4.29)$$

## 6.5. CPM Signaling Over Frequency-Flat Fading Channels

### 6.5.1. ML Receiver

In this section we investigate optimum and suboptimum receivers for CPM transmissions over frequency-flat fading channels. The approach closely parallels the discussion in Section 6 of Chapter 5 with PAM modulations. Here, we are concerned with general CPM signals of the type

$$s(t) = a(t)\sqrt{\frac{2E_s}{T}}e^{j2\pi vt}e^{j\psi(t-\tau,\alpha)} \qquad (6.5.1)$$

with

$$\psi(t,\alpha) = 2\pi h \sum_{i=0}^{N-1} \alpha_i q(t-iT) \qquad (6.5.2)$$

It should be noted that (6.5.1) differs from (6.4.1) in two respects: ($i$) the presence of the multiplicative distortion $a(t)$; ($ii$) the absence of the carrier phase $\theta$, which has been absorbed into $a(t)$. The multiplicative distortion is modeled as a complex-valued Gaussian random process with some power spectral density in the range $\pm f_D$. The parameter $f_D$ is the Doppler bandwidth. Without loss of generality the expectation of $|a(t)|^2$ is set to unity so that $E_s$ may be viewed either as the *transmitted energy* or the *average received energy* per symbol.

As a first step in our discussion we wonder what the optimum receiver would be like if the distortion $a(t)$ were known. As we shall see, the answer opens the way to a simple sub-optimal structure wherein $a(t)$ is estimated in a decision-directed manner and the estimates are exploited for data detection. Assuming that the parameters $\{v,\tau\}$ and the actual realization of $a(t)$ are known, optimum detection is performed looking for the maximum of the likelihood function $\Lambda(r|\tilde{\alpha})$ with respect to the trial sequence $\tilde{\alpha} \triangleq \{\tilde{\alpha}_0, \tilde{\alpha}_1, \ldots, \tilde{\alpha}_{N-1}\}$. With normal manipulations this function is found to be

$$\Lambda(r|\tilde{\alpha}) = \exp\left\{\frac{1}{N_0}\sqrt{\frac{2E_s}{T}}\sum_{k=0}^{N-1}\text{Re}\left\{Z'_k(\tilde{C}_k,\tilde{\alpha}_k)e^{-j\tilde{\Phi}_k}\right\}\right\} \qquad (6.5.3)$$

where

$$Z'_k(\tilde{C}_k,\tilde{\alpha}_k) \triangleq \int_{\tau+kT}^{\tau+(k+1)T} a^*(t)r(t)e^{-j2\pi vt}e^{-j\eta(t-\tau,\tilde{C}_k,\tilde{\alpha}_k)}dt \qquad (6.5.4)$$

and $\tilde{C}_k$, $\tilde{\Phi}_k$ are still as in (6.4.10)-(6.4.11). Comparing with (6.4.8)-(6.4.9), it is concluded that the optimum receiver has basically the same structure as with the AWGN channel. In fact it can be implemented by means of the Viterbi algorithm wherein the metrics

$$\lambda_k(\overline{S}_k, \tilde{\alpha}_k) \triangleq \text{Re}\left\{ Z'_k(\overline{C}_k, \tilde{\alpha}_k) e^{-j\overline{\Phi}_k} \right\} \tag{6.5.5}$$

are associated with the branches stemming from the generic node $\overline{S}_k = (\overline{C}_k, \overline{\Phi}_k)$. Clearly, if $a(t)$ equals $e^{j\theta}$, we are back to the optimum receiver for AWGN channels.

### 6.5.2. Approximate ML Receiver Based on Per-Survivor-Processing Methods

In many practical cases the fading bandwidth is sufficiently small (compared with the symbol rate) that $a(t)$ is approximately constant over one symbol interval. Then, letting

$$a(k) \triangleq a(kT + T/2 + \tau) \tag{6.5.6}$$

equation (6.5.4) reduces to

$$Z'_k(\tilde{C}_k, \tilde{\alpha}_k) \approx a^*(k) Z_k(\tilde{C}_k, \tilde{\alpha}_k) \tag{6.5.7}$$

with

$$Z_k(\tilde{C}_k, \tilde{\alpha}_k) \triangleq \int_{\tau+kT}^{\tau+(k+1)T} r(t) e^{-j2\pi\nu t} e^{-j\eta(t-\tau, \tilde{C}_k, \tilde{\alpha}_k)} dt \tag{6.5.8}$$

Correspondingly, the branch metrics become

$$\lambda_k(\overline{S}_k, \tilde{\alpha}_k) \triangleq \text{Re}\left\{ a^*(k) Z_k(\overline{C}_k, \tilde{\alpha}_k) e^{-j\overline{\Phi}_k} \right\} \tag{6.5.9}$$

As is seen, they differ from those for the Gaussian channel in that the phasor $e^{-j\theta}$ is replaced by the channel gain $a^*(k)$.

From (6.5.9) it is clear that the search for the most likely path in the trellis requires knowledge of the channel gains $\{a(k)\}$. A method to estimate these gains in a decision-directed manner is now described. We initially assume that the true path $\{..., \alpha_{k-3}, \alpha_{k-2}, \alpha_{k-1}\}$ is known up to epoch $k$. Letting $\{C_i, i \leq k\}$ and $\{\Phi_i, i \leq k\}$ be the sequences of correlative states and phase states associated with $\{\ldots,\alpha_{k-3},\alpha_{k-2},\alpha_{k-1}\}$ and $S_k$ the final node of the path, consider the statistics

$$z(i,S_k) \triangleq \frac{e^{-j\Phi_i}}{\sqrt{2E_s T}} Z_i(C_i,\alpha_i), \quad i \le k-1 \tag{6.5.10}$$

Recalling that

$$r(t) = a(i)e^{j2\pi vt}\sqrt{\frac{2E_s}{T}}e^{j\eta(t-\tau,C_i,\alpha_i)}e^{j\Phi_i} + w(t) \tag{6.5.11}$$

for $iT + \tau \le t < (i+1)T + \tau$, it is easily checked that

$$z(i,S_k) = a(i) + n(i) \tag{6.5.12}$$

where $\{n(i)\}$ is zero-mean white Gaussian noise. Equation (6.5.12) indicates that $z(i,S_k)$ may be viewed as a noisy measurement of $a(i)$ and, in consequence, can be employed to predict future channel gains. This can be done by means of a Wiener predictor

$$\hat{a}(k) = \sum_{i=1}^{L_p} \gamma(i) z(k-i, S_k) \tag{6.5.13}$$

whose coefficients minimize the mean square error between $a(k)$ and its estimate. Performing standard calculations it can be shown that the predictor coefficients satisfy the equations

$$\sum_{i=1}^{L_p} \gamma(i) R_{aa}(l-i) + \frac{\gamma(l)}{E_s/N_0} = R_{aa}(l), \quad 1 \le l \le L_p \tag{6.5.14}$$

where $R_{aa}(l) \triangleq E\{a(i+l)a^*(i)\}$ represents the fading autocorrelation.

Equation (6.5.14) says that the parameters $\{\gamma(i)\}$ depend on the ratio $E_s/N_0$ and the fading autocorrelation. Clearly, in a practical non-stationary environment they must be tracked in some way. Assuming that this is properly done, receiver adaptivity may be accomplished by pre-computing various coefficient sets for the channel of interest and applying the most suitable one at a given time.

Equation (6.5.13) indicates how to estimate $a(k)$ if the data $\{\ldots,\alpha_{k-3},\alpha_{k-2},\alpha_{k-1}\}$ were known. In practice they are not known, however, and must be approximated in some manner. One possibility is to resort to per-survivor-processing (PSP) estimation [11] methods. In their simplest form

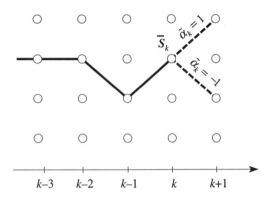

Figure 6.9. Explaining PSP estimation. A binary alphabet is assumed ($\tilde{\alpha}_k = \pm 1$).

these methods operate directly on the encoder trellis as follows. Call $\{\ldots, \overline{\alpha}_{k-3}, \overline{\alpha}_{k-2}, \overline{\alpha}_{k-1}\}$ the generic survivor at epoch $k$ and let $\overline{S}_k$ be its terminal node (see Figure 6.9). A prediction of $a(k)$ is associated with this survivor through the formula

$$\hat{a}(k) = \sum_{i=1}^{L_p} \gamma(i) z(k-i, \overline{S}_k) \qquad (6.5.15)$$

and the branch metrics are computed as

$$\lambda_k(\overline{S}_k, \tilde{\alpha}_k) \triangleq \mathrm{Re}\left\{\hat{a}^*(k) Z_k(\overline{C}_k, \tilde{\alpha}_k) e^{-j\overline{\Phi}_k}\right\} \qquad (6.5.16)$$

for $\tilde{\alpha}_k \in \{\pm 1, \pm 3, \ldots, \pm(M-1)\}$.

Better results are obtained by operating on a *super-trellis*, whose states $S_k^{(sup)}$ are collections of $Q+1$ successive decoder states. Formally a superstate is defined as

$$S_k^{(sup)} \triangleq (S_{k-Q}, S_{k-Q+1}, \ldots, S_k) \qquad (6.5.17)$$

with the understanding that the transitions $\{S_{k-i} \Rightarrow S_{k-i+1}, 1 \le i \le Q\}$ are all permissible. Bearing in mind that $S_k$ represents the $L$-tuple $(\alpha_{k-L+1}, \ldots, \alpha_{k-2}, \alpha_{k-1}; \Phi_k)$, it is realized that a super-state may also be viewed as a $(Q+L)$-tuple

$$S_k^{(sup)} = (\alpha_{k-L-Q+1}, \ldots, \alpha_{k-2}, \alpha_{k-1}; \Phi_k) \qquad (6.5.18)$$

An intuitive motivation for using a super-trellis instead of the encoder

# Carrier Phase Recovery with CPM Modulations

trellis is as follows. If the channel gains $\{a(k)\}$ were known, the best trellis would be the encoder trellis and, at each time and at each state, the Viterbi algorithm would retain just one path for extension. The other paths would be deleted. When the channel is unknown, however, no path may be dropped in favor of another. Then, it makes sense to retain a fixed number of paths for each state. For example, retaining $M$ paths (recall that $M$ is the alphabet size) amounts to letting $S_k^{(sup)} = (S_{k-1}, S_k)$. More generally, keeping $M^Q$ paths corresponds to choosing the super-states (6.5.17). This concept has been proposed by Seshadri [12] in the more general context of approximate ML decoding with frequency selective fading channels.

Returning to (6.5.18), it is seen that there are $PM^{L+Q-1}$ distinct super-states, i.e., $M^Q$ times as many as the decoder states. This implies an increase in decoding computational load which, of course, is only justified by a corresponding improvement in detection performance. Simulations indicate that most of the potential benefits offered by a super-trellis are attained with $Q=1$; marginal extra gains are achieved with a larger $Q$. Whatever the case, for a given super-trellis the branch metrics stemming from $\overline{S}_k^{(sup)}$ are computed as

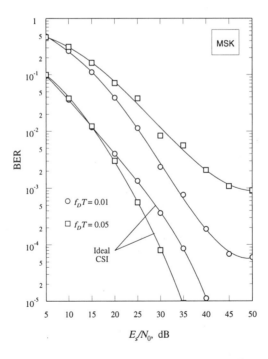

Figure 6.10. BER performance with no super-trellis.

indicated in (6.5.16), wherein $\overline{S}_k$ represents the rightmost component of $\overline{S}_k^{(sup)}$ and the channel prediction is computed along the survivor terminating in $\overline{S}_k^{(sup)}$.

Figure 6.10 illustrates BER curves with MSK and $Q=0$. In this case the trellis has four states. Simulations have been run under the following conditions: (*i*) 5-tap predictors ($L_p=5$); (*ii*) $E_s/N_0$ and fading autocorrelation are known so that the predictor coefficients can be computed exactly; (*iii*) the Doppler spectrum is shaped by a fourth-order Butterworth filter with a −3 dB bandwidth equal to $f_D$; (*iv*) the Viterbi algorithm has a decision delay of 10 symbols. Curves labelled "ideal CSI" (CSI stands for channel state information) are also shown for comparison. They correspond to optimum ML performance and are obtained by inserting true channel gains in (6.5.16) (instead of their estimates). As is seen, the BER curves develop a floor which rises with the fading rate.

BER curves with an 8-state super-trellis ($Q=1$) are shown in Figure 6.11. We see that the improvement with respect to Figure 6.10 is dramatic for $f_D T=0.01$. The loss from ideal is now reduced to 2-3 dB and no floor is visible. Improvement is significant also with $f_D T=0.05$, even though a floor is now apparent.

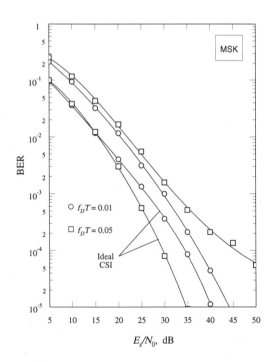

Figure 6.11. BER performance with super-trellis and $Q=1$.

### 6.5.3. Improved Methods for Fast-Fading Channels

We have seen that the BER curves exhibit a floor for fading rates exceeding about 1% of the symbol rate. In many practical cases $f_D$ is much smaller than 1% of $1/T$. For example, with a carrier frequency of 1 GHz and a vehicle speed of 100 km/h, the Doppler shift is about 100 Hz and $f_D$ is less than 1% of $1/T$ for symbol rates greater than 10 kHz. In some applications involving digital voice and data communications [13] however the aim is to operate at bit rates as low as 4800 b/s. This means that $f_D T$ is about 2% with a binary alphabet and twice as large with a quaternary alphabet. Thus, methods to cope with fading rates higher than 1% of $1/T$ are of practical interest.

To see how these situations can be tackled it is useful to concentrate on the floor-generating mechanism. As fading gets faster, the approximation (6.5.7) fails because $a(t)$ no longer remains constant over one symbol interval. In consequence, predictions become less and less accurate and the detector makes errors even though the signal-to-noise ratio is large. There seem to be two ways to get around this obstacle: either making closer predictions or taking the non-constant nature of $a(t)$ into account. The former route has been pursued by Lodge and Moher [14] and is discussed here; the latter is deferred to the next section.

Our description of the Lodge and Moher (L&M) approach is slightly different from the original treatment. As indicated in Figure 6.12, the received waveform is fed to an anti-alias filter (AAF) and sampled at $t=mT_s+\tau$. As usual, $1/T_s$ is large enough to satisfy the sampling theorem and the filter transfer function is rectangular and sufficiently large so as to pass the signal components undistorted. For convenience, the sampling period is taken to be a submultiple of $T$, say $T_s=T/N$. In these conditions (and assuming that possible frequency offsets have been perfectly compensated) the $m$-th sample takes the form

$$x(m) = a(m)\sqrt{\frac{2E_s}{T}}e^{j\eta(mT_s,C_k,\alpha_k)}e^{j\Phi_k} + n(m) \qquad (6.5.19)$$

where $a(m)$ is short for $a(mT_s+\tau)$, $n(m)$ is white Gaussian noise and $k$ is the

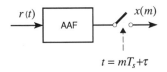

Figure 6.12. Anti-alias filtering and sampling.

*symbol index*, which is related to the *sample index m* by

$$k \triangleq \text{int}\left(\frac{m}{N}\right) \quad (6.5.20)$$

If the channel were known, an ML detector would maximize the log-likelihood function

$$\ln \Lambda(x|\tilde{\alpha}) = \sum_{m=0}^{NL_0-1} \text{Re}\left\{x(m)a^*(m)e^{-j\eta(mT_s, \tilde{C}_k, \tilde{\alpha}_k)}e^{-j\tilde{\Phi}_k}\right\} \quad (6.5.21)$$

where $L_0$ is the number of symbols in the sequence. This could be performed by means of a Viterbi detector with the following branch metrics

$$\lambda_k(\overline{S}_k, \tilde{\alpha}_k) \triangleq \sum_{m=kN}^{(k+1)N-1} \text{Re}\left\{x(m)a^*(m)e^{-j\eta(mT_s, \overline{C}_k, \tilde{\alpha}_k)}e^{-j\overline{\Phi}_k}\right\} \quad (6.5.22)$$

The channel gains are not known, however, and must be estimated. To see how this can be accomplished, we transform the encoder trellis into a super-trellis with nodes $\{S_k^{(sup)}\}$. Next, we assume that the true path $\{\ldots, \alpha_{k-3}, \alpha_{k-2}, \alpha_{k-1}\}$ is known and we call $\{S_i^{(sup)}, i \leq k\}$ its nodes. Also we define

$$z(i, S_k^{(sup)}) \triangleq \frac{1}{\sqrt{2E_s/T}} x(i) e^{-j\eta(iT_s, C_{\text{int}(i/N)}, \alpha_{\text{int}(i/N)})} e^{-j\Phi_{\text{int}(i/N)}}, \quad i \leq (k+1)N-1 \quad (6.5.23)$$

Combining (6.5.19) and (6.5.23) results in

$$z(i, S_k^{(sup)}) = a(i) + n'(i), \quad i \leq (k+1)N-1 \quad (6.5.24)$$

where $n'(i)$ is white Gaussian noise. Equation (6.5.24) indicates that $z(i, S_k^{(sup)})$ represents a noisy measurement of the channel gain and, as such, can be used in a Wiener predictor to estimate future gains:

$$\hat{a}(m) = \sum_{i=1}^{L_p} \gamma(i) z(m-i, S_k^{(sup)}) \quad (6.5.25)$$

Putting the above remarks together leads us to the following PSP-based estimation method. Denote by $\{\ldots, \overline{\alpha}_{k-3}, \overline{\alpha}_{k-2}, \overline{\alpha}_{k-1}\}$ the generic survivor arriving at node $\overline{S}_k^{(sup)}$ and let $\{\overline{C}_i, i \leq k\}$ and $\{\overline{\Phi}_i, i \leq k\}$ be the associated correlative states and phase states. Also, define

# Carrier Phase Recovery with CPM Modulations

$$z(i,\overline{S}_k^{(sup)}) \triangleq \begin{cases} \dfrac{1}{\sqrt{2E_s/T}} x(i) e^{-j\eta(iT_s,\overline{C}_k,\tilde{\alpha}_k)} e^{-j\overline{\Phi}_k} & kN \leq i \leq (k+1)N-1 \\ \dfrac{1}{\sqrt{2E_s/T}} x(i) e^{-j\eta(iT_s,\overline{C}_{\text{int}(i/N)},\tilde{\alpha}_{\text{int}(i/N)})} e^{-j\overline{\Phi}_{\text{int}(i/N)}} & i < kN \end{cases}$$

(6.5.26)

Then, the channel predictions are computed as

$$\hat{a}(m) = \sum_{i=1}^{L_p} \gamma(i) z(m-i, \overline{S}_k^{(sup)})$$ (6.5.27)

and are used in the branch metric as

$$\lambda_k(\overline{S}_k^{(sup)}, \tilde{\alpha}_k) \triangleq \sum_{m=kN}^{(k+1)N-1} \text{Re}\left\{ x(m) \hat{a}^*(m) e^{-j\eta(mT_s,\overline{C}_k,\tilde{\alpha}_k)} e^{-j\overline{\Phi}_k} \right\}$$ (6.5.28)

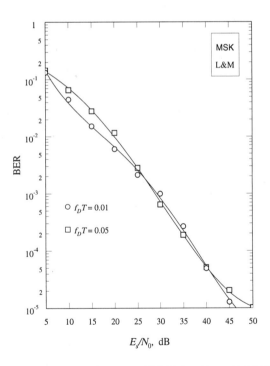

Figure 6.13. BER performance of L&M algorithm with MSK.

with $\tilde{\alpha}_k \in \{\pm 1, \pm 3, \ldots, \pm(M-1)\}$. It should be stressed that the prediction interval in (6.5.27) is $T_s$ seconds whereas it is $N$ times longer in (6.5.15). Thus, prediction accuracy in the former case should be superior when the fading becomes fast.

Figure 6.13 illustrates BER curves for the L&M detector under the following conditions. Modulation is MSK and the Doppler spectrum is shaped as in the simulations in Figures 6.10 and 6.11. Predictors with five taps are employed and the decision delay is still 10 symbols. Sampling is performed at twice the symbol rate and the AAF filter is implemented as a fourth-order Butterworth FIR structure of bandwidth $1.25/T$. We see that for $f_D T=0.01$ the curve is almost the same as in Figure 6.11. A floor shows up, however, when the fading rate grows to $f_D T=0.05$.

### 6.5.4. Linearly Time-Selective Channels

An alternative approach to data detection with fast fading channels is now described [15]. We start from equation (6.5.4), which is rewritten here for convenience:

$$Z'_k(\tilde{\mathbf{C}}_k, \tilde{\alpha}_k) \triangleq \int_{\tau+kT}^{\tau+(k+1)T} a^*(t) r(t) e^{-j2\pi vt} e^{-j\eta(t-\tau, \tilde{\mathbf{C}}_k, \tilde{\alpha}_k)} dt \qquad (6.5.29)$$

In Section 6.5.2 the channel distortion has been modeled as a constant over the integration interval. We now improve this approximation by taking $a(t)$ as a linear function of time around $kT+T/2+\tau$:

$$a(t) \approx a(kT+T/2+\tau) + (t-kT-T/2-\tau)\left[\frac{da(t)}{dt}\right]_{t=kT+T/2+\tau} \qquad (6.5.30)$$

Then, letting

$$a^{(0)}(k) \triangleq a(kT+T/2+\tau) \qquad (6.5.31)$$

$$a^{(1)}(k) \triangleq T\left[\frac{da(t)}{dt}\right]_{t=kT+T/2+\tau} \qquad (6.5.32)$$

$$Z_k^{(0)}(\tilde{\mathbf{C}}_k, \tilde{\alpha}_k) \triangleq \int_{\tau+kT}^{\tau+(k+1)T} r(t) e^{-j2\pi vt} e^{-j\eta(t-\tau, \tilde{\mathbf{C}}_k, \tilde{\alpha}_k)} dt \qquad (6.5.33)$$

$$Z_k^{(1)}(\tilde{\mathbf{C}}_k, \tilde{\alpha}_k) \triangleq \frac{1}{T}\int_{\tau+kT}^{\tau+(k+1)T} (t-kT-T/2-\tau) r(t) e^{-j2\pi vt} e^{-j\eta(t-\tau, \tilde{\mathbf{C}}_k, \tilde{\alpha}_k)} dt \qquad (6.5.34)$$

and inserting into (6.5.29) yields

$$Z'_k(\tilde{C}_k,\tilde{\alpha}_k) \approx a^{(0)*}(k)Z_k^{(0)}(\tilde{C}_k,\tilde{\alpha}_k) + a^{(1)*}(k)Z_k^{(1)}(\tilde{C}_k,\tilde{\alpha}_k) \qquad (6.5.35)$$

Next, reasoning as in Section 6.5.2 it is realized that, if the sequences $\{a^{(0)}(k)\}$ and $\{a^{(1)}(k)\}$ were known, optimum ML detection could be implemented by means of a Viterbi algorithm employing the metrics

$$\lambda_k(\overline{S}_k,\tilde{\alpha}_k) \triangleq \mathrm{Re}\left\{a^{(0)*}(k)Z_k^{(0)}(\overline{C}_k,\tilde{\alpha}_k)e^{-j\overline{\Phi}_k}\right\} \\ + \mathrm{Re}\left\{a^{(1)*}(k)Z_k^{(1)}(\overline{C}_k,\tilde{\alpha}_k)e^{-j\overline{\Phi}_k}\right\} \qquad (6.5.36)$$

The following comments can be made on this method:

(i) The *linearly time-selective* channel described in (6.5.30) has been proposed by P.A. Bello in [16].

(ii) The sequences $\{a^{(0)}(k)\}$ and $\{a^{(1)}(k)\}$ are Gaussian and their correlations depend on the fading spectrum $S_D(f)$. In particular it can be shown that

$$\mathrm{E}\left\{|a^{(0)}(k)|^2\right\} = \int_{-\infty}^{\infty} S_D(f)df \qquad (6.5.37)$$

$$\mathrm{E}\left\{|a^{(1)}(k)|^2\right\} = (2\pi T)^2 \int_{-\infty}^{\infty} f^2 S_D(f)df \qquad (6.5.38)$$

(iii) The quantities $Z_k^{(0)}(\tilde{C}_k,\tilde{\alpha}_k)$ coincide with the $Z_k(\tilde{C}_k,\tilde{\alpha}_k)$ defined in Section 6.5.2. They can be computed in the usual bank of filters with which any ML detector is endowed.

(iv) On the contrary, the computation of $Z_k^{(1)}(\tilde{C}_k,\tilde{\alpha}_k)$ requires extra circuitry.

Returning to the branch metrics (6.5.36), we now apply PSP methods to estimate $a^{(0)}(k)$ and $a^{(1)}(k)$. To this end we choose a super-trellis with nodes $S_k^{(sup)} \triangleq (S_{k-Q},S_{k-Q+1},\ldots,S_k)$. Next, supposing that the true path leading to $S_k^{(sup)}$ is known, call $\{C_i, i \le k\}$ and $\{\Phi_i, i \le k\}$ the associated correlative states and phase states and concentrate on the statistics

$$z^{(0)}(i,S_k^{(sup)}) \triangleq \frac{1}{\sqrt{2E_sT}} Z_i^{(0)}(C_i,\alpha_i)e^{-j\Phi_i}, \qquad i \le k-1 \qquad (6.5.39)$$

$$z^{(1)}(i,S_k^{(sup)}) \triangleq \frac{12}{\sqrt{2E_sT}} Z_i^{(1)}(C_i,\alpha_i)e^{-j\Phi_i}, \qquad i \le k-1 \qquad (6.5.40)$$

where $Z_i^{(0)}(C_i,\alpha_i)$ and $Z_i^{(1)}(C_i,\alpha_i)$ are defined as in (6.5.33)-(6.5.34). As the received waveform is given by

$$r(t) \approx \sqrt{\frac{2E_s}{T}} e^{j\Phi_i} \left[ a^{(0)}(i) + a^{(1)}(i) \left( \frac{t - iT - T/2 - \tau}{T} \right) \right]$$
$$\times e^{j2\pi\nu t} e^{j\eta(t-\tau,C_i,\alpha_i)} + w(t), \qquad iT + \tau \leq t \leq (i+1)T + \tau \quad (6.5.41)$$

it is easily checked that

$$z^{(0)}(i, S_k^{(sup)}) \triangleq a^{(0)}(i) + n^{(0)}(i), \quad i \leq k-1 \quad (6.5.42)$$

$$z^{(1)}(i, S_k^{(sup)}) \triangleq a^{(1)}(i) + n^{(1)}(i), \quad i \leq k-1 \quad (6.5.43)$$

where $n^{(0)}(i)$ and $n^{(1)}(i)$ are thermal noise contributions.

Equations (6.5.42)-(6.5.43) indicate that $z^{(0)}(i, S_k^{(sup)})$ and $z^{(1)}(i, S_k^{(sup)})$ represent noisy measurements of the channel gain and its time derivative. Also, since $a^{(0)}(i)$ and $a^{(1)}(i)$ are correlated, it follows that both $z^{(0)}(i, S_k^{(sup)})$ and $z^{(1)}(i, S_k^{(sup)})$ give information on the channel gain and its derivative. This suggests estimating future values of $a^{(0)}(k)$ and $a^{(1)}(k)$ through Wiener predictors of the type

$$\hat{a}^{(0)}(k) = \sum_{i=1}^{L_p^{(00)}} \gamma^{(00)}(i) z^{(0)}(k-i, S_k^{(sup)}) + \sum_{i=1}^{L_p^{(01)}} \gamma^{(01)}(i) z^{(1)}(k-i, S_k^{(sup)}) \quad (6.5.44)$$

$$\hat{a}^{(1)}(k) = \sum_{i=1}^{L_p^{(10)}} \gamma^{(10)}(i) z^{(0)}(k-i, S_k^{(sup)}) + \sum_{i=1}^{L_p^{(11)}} \gamma^{(11)}(i) z^{(1)}(k-i, S_k^{(sup)}) \quad (6.5.45)$$

At this point the structure of the PSP-based detector should be clear. It consists of a Viterbi algorithm wherein the branches stemming from $\overline{S}_k^{(sup)}$ are given the metrics

$$\lambda_k(\overline{S}_k^{(sup)}, \tilde{\alpha}_k) \triangleq \mathrm{Re}\left\{ \hat{a}^{(0)*}(k) Z_k^{(0)}(\overline{C}_k, \tilde{\alpha}_k) e^{-j\overline{\Phi}_k} \right\}$$
$$+ \mathrm{Re}\left\{ \hat{a}^{(1)*}(k) Z_k^{(1)}(\overline{C}_k, \tilde{\alpha}_k) e^{-j\overline{\Phi}_k} \right\} \quad (6.5.46)$$

and the channel estimates $\hat{a}^{(0)}(k)$ and $\hat{a}^{(1)}(k)$ are computed as

$$\hat{a}^{(0)}(k) = \sum_{i=1}^{L_p^{(00)}} \gamma^{(00)}(i) z^{(0)}(k-i, \overline{S}_k^{(sup)}) + \sum_{i=1}^{L_p^{(01)}} \gamma^{(01)}(i) z^{(1)}(k-i, \overline{S}_k^{(sup)}) \quad (6.5.47)$$

# Carrier Phase Recovery with CPM Modulations

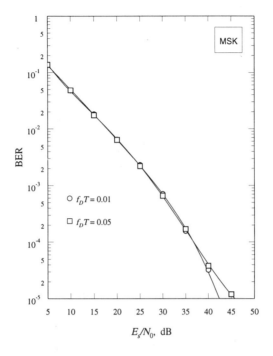

Figure 6.14. BER performance with MSK.

$$\hat{a}^{(1)}(k) = \sum_{i=1}^{L_p^{(10)}} \gamma^{(10)}(i) z^{(0)}(k-i, \overline{S}_k^{(sup)}) + \sum_{i=1}^{L_p^{(11)}} \gamma^{(11)}(i) z^{(1)}(k-i, \overline{S}_k^{(sup)}) \quad (6.5.48)$$

Note that the statistics $z^{(0)}(k-i, \overline{S}_k^{(sup)})$ and $z^{(1)}(k-i, \overline{S}_k^{(sup)})$ are derived from the survivor path terminating at $\overline{S}_k^{(sup)}$.

Figure 6.14 illustrates BER curves for MSK and the same channel model as in Sections 6.2 and 6.3. The super-trellis has 8 states ($Q=1$) and the predictor lengths are $L_p^{(00)} = 4$, $L_p^{(01)} = 2$ and $L_p^{(10)} = L_p^{(11)} = 3$. Comparing with Figure 6.13 it is seen that performance is slightly better than with the L&M receiver.

## 6.6. Clock-Aided but Non-Data-Aided Phase Estimation

### 6.6.1. 2P-Power Method for Full-Response Systems

Now we return to AWGN channels and concentrate on non-data-aided phase estimators. Under the assumption of ideal timing and carrier frequency recovery, we investigate two methods. The first (discussed here) is suitable for full response formats while the other (deferred to the next sub-section) applies to MSK-type signaling.

We begin with the signal model

$$s(t) = e^{j\theta} \sqrt{\frac{2E_s}{T}} e^{j\psi(t-\tau,\alpha)} \qquad (6.6.1)$$

where

$$\psi(t,\alpha) = 2\pi h \sum_i \alpha_i q(t-iT) \qquad (6.6.2)$$

With *full response* modulation the phase response satisfies the relations

$$q(t) = \begin{cases} 0 & t \le 0 \\ 1/2 & t \ge T \end{cases} \qquad (6.6.3)$$

Thus, taking $t=-i_0 T$ as the starting time and letting $h=K/P$, from (6.6.2) we get

$$2P\psi(kT) = 2\pi K \sum_{i=-i_0}^{k-1} \alpha_i \qquad (6.6.4)$$

which indicates that $2P\psi(kT)$ is a multiple of $2\pi$. Hence, sampling $s(t)$ at $t=kT+\tau$ and raising to the $2P$-th power yields

$$s^{2P}(kT+\tau) = \left(\frac{2E_s}{T}\right)^P e^{j2P\theta} \qquad (6.6.5)$$

Clearly, the sequence $\{s^{2P}(kT+\tau)\}$ contains information on the carrier phase. Unfortunately we have no direct access to $s(t)$ but only to the received waveform $r(t)$. Nevertheless, the latter can be filtered (to limit excess noise) and used in place of $s(t)$. Denoting $x(t)=s(t)+n(t)$ the filtered version of $r(t)$ and letting $x(k) \triangleq x(kT+\tau)$ produces

$$x^{2P}(k) = \left(\frac{2E_s}{T}\right)^P e^{j2P\theta} + N(k) \qquad (6.6.6)$$

where $N(k)$ is a zero-mean random term resulting from the products Signal×Noise and Noise×Noise in the binomial expansion of $(s+n)^{2P}$.

At this point a phase estimation algorithm is easily conceived. In fact, the average of $x^{2P}(k)$ over a window of $L_0$ samples is given by

# Carrier Phase Recovery with CPM Modulations

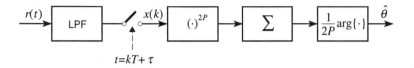

Figure 6.15. Block diagram of the 2P-power phase estimator.

$$\frac{1}{L_0}\sum_{k=0}^{L_0-1} x^{2P}(k) = \left(\frac{2E_s}{T}\right)^P e^{j2P\theta} + \frac{1}{L_0}\sum_{k=0}^{L_0-1} N(k) \tag{6.6.7}$$

If the last term were negligible, the argument of the left-hand side would equal $2P\theta$. This suggests the estimator

$$\hat{\theta} = \frac{1}{2P}\arg\left\{\sum_{k=0}^{L_0-1} x^{2P}(k)\right\} \tag{6.6.8}$$

whose block diagram is indicated in Figure 6.15. Note that the estimates have a 2P-fold phase ambiguity as $\hat{\theta}$ takes values in the range $\pm\pi/(2P)$.

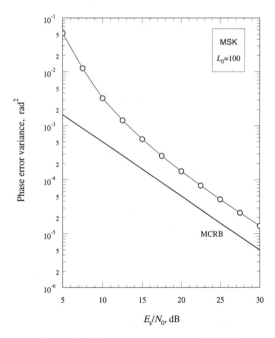

Figure 6.16. Phase error variance with MSK.

Figure 6.16 illustrates simulation results for the estimation error variance with MSK modulation ($P=2$). The low-pass filter (LPF) is implemented as a fourth-order Butterworth FIR with a bandwidth of $1.2/T$ and the observation length is of $L_0=100$ symbol intervals. As is seen, there is a loss of about 4.5 dB from the modified Cramer-Rao bound (MCRB) at high SNR.

### 6.6.2. ML-Oriented Phase Estimation

In deriving the $2P$-power estimator we have used heuristic arguments. In the following we adopt ML-based methods but, for simplicity, we restrict ourselves to MSK-type signaling. Again, we assume that timing is ideal and the frequency offset has been estimated and compensated for. In these conditions the received signal may be written as

$$s(t) = \sqrt{\frac{2E_s}{T}} e^{j\psi(t-\tau,\alpha)} \qquad (6.6.9)$$

with

$$\psi(t,\alpha) = \pi \sum_{i=-i_0}^{\infty} \alpha_i q(t-iT) \qquad (6.6.10)$$

and $\alpha \triangleq \{\alpha_i\}$. Note that $\tau$ is a known parameter and the index $-i_0$ indicates the first symbol in the sequence. Without loss of generality we take $i_0 > L+1$ so that the system encoder is in a steady-state condition as soon as $t \geq 0$.

Our estimation problem is best formulated in the discrete-time domain. Accordingly, the incoming waveform $r(t)$ is first fed to some anti-alias filter (AAF) and then sampled at the instants $t=kT_s+\tau$. The filter transfer function is such as to pass the signal undistorted and make the noise samples independent. For convenience, $T_s$ is chosen a sub-multiple of the symbol period, say $T_s=T/N$. The actual value of $N$ is discussed later.

Letting $x(t)$ be the AAF output and $x(k) \triangleq x(kT_s+\tau)$ its samples, the likelihood function for the unknown parameters has the form

$$\Lambda(r|\tilde{\theta},\tilde{\alpha}) = \exp\left\{\frac{1}{\sigma_n^2}\sqrt{\frac{2E_s}{T}} \operatorname{Re}\left\{e^{-j\tilde{\theta}} \sum_{k=0}^{NL_0-1} x(k) e^{-j\psi(kT_s,\tilde{\alpha})}\right\}\right\} \qquad (6.6.11)$$

where $L_0$ is the observation length in symbol intervals and $2\sigma_n^2$ is the variance of the noise samples. As we look for a non-data-aided estimation method, we must eliminate the symbols $\tilde{\alpha}$ from $\Lambda(r|\tilde{\theta},\tilde{\alpha})$. Unfortunately this turns out to be an insurmountable obstacle and, therefore, we resort to the usual low-SNR approximation which consists of expanding the exponential (6.6.11) into a power series and retaining only the leading terms. This produces

$$\Lambda(r|\tilde{\theta},\tilde{\alpha}) = 1 + \frac{1}{\sigma_n^2}\sqrt{\frac{2E_s}{T}} \mathrm{Re}\left\{e^{-j\tilde{\theta}}\sum_{k=0}^{NL_0-1} x(k)e^{-j\psi(kT_s,\tilde{\alpha})}\right\}$$

$$+ \frac{1}{\sigma_n^4}\frac{E_s}{T}\left[\mathrm{Re}\left\{e^{-j\tilde{\theta}}\sum_{k=0}^{NL_0-1} x(k)e^{-j\psi(kT_s,\tilde{\alpha})}\right\}\right]^2 \quad (6.6.12)$$

The averaging operation is lengthy and tedious and is not reported here. In [17] it is shown that the final result may be written as

$$\Lambda(r|\tilde{\theta}) \approx \mathrm{Re}\left\{(X_{+1} + X_{-1})e^{-j2\tilde{\theta}}\right\} \quad (6.6.13)$$

with

$$X_{+1} \triangleq \sum_{k_1=0}^{NL_0-1}\sum_{k_2=0}^{NL_0-1} x(k_1)x(k_2)h(k_1-k_2)e^{j\pi k_1/N} \quad (6.6.14)$$

$$X_{-1} \triangleq \sum_{k_1=0}^{NL_0-1}\sum_{k_2=0}^{NL_0-1} x(k_1)x(k_2)h^*(k_1-k_2)e^{-j\pi k_1/N} \quad (6.6.15)$$

In these equations $h(k)$ is the sample of

$$h(\Delta t) \triangleq \frac{1}{T}\int_0^T H(t,\Delta t)e^{-j\pi t/T}dt \quad (6.6.16)$$

at $\Delta t = kT_s$ and we have adopted the notations

$$H(t,\Delta t) \triangleq \prod_{i=-2L-1}^{L+1} \cos[\pi p(t-iT,\Delta t)] \quad (6.6.17)$$

$$p(t,\Delta t) \triangleq q(t) + q(t-\Delta t) \quad (6.6.18)$$

To give an example, consider 1REC pulses (MSK modulation). In this case $L=1$ and

$$q(t) = \begin{cases} 0 & t < 0 \\ t/(2T) & 0 \le t \le T \\ 1/2 & t > T \end{cases} \quad (6.6.19)$$

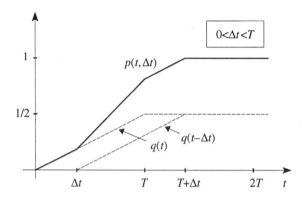

Figure 6.17. Function $p(t,\Delta t)$ with 1REC pulses.

Thus, $p(t,\Delta t)$ takes the form indicated in Figure 6.17 (for $0 \le \Delta t \le T$). Correspondingly, $\cos[\pi p(t,\Delta t)]$ is unity for $t \le 0$, it equals $-1$ for $t \ge \Delta t + T$, and takes intermediate values in between. Computing $H(t,\Delta t)$ from (6.6.17) and substituting into (6.6.16) yields $h(\Delta t)$ whose shape is shown in Figure 6.18. As is seen, $h(\Delta t)$ has a duration of about 4 symbol intervals.

Returning to (6.6.13), it is clear that the maximum of $\Lambda(r|\tilde{\theta})$ is achieved for

$$\hat{\theta} = \frac{1}{2}\arg\{X_{+1} + X_{-1}\} \qquad (6.6.20)$$

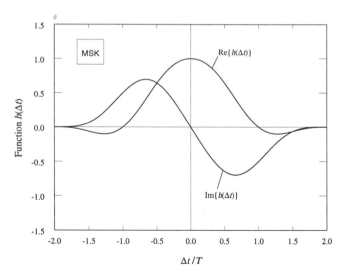

Figure 6.18. Function $h(\Delta t)$ with 1REC pulses.

# Carrier Phase Recovery with CPM Modulations

which gives the desired phase estimation formula. In practice the computation of the quantities $X_{+1}$ and $X_{-1}$ can be simplified as follows. Consider $X_{+1}$, for example, and rewrite (6.6.14) as

$$X_{+1} = \sum_{k=0}^{NL_0-1}\left[x(k)e^{j\pi k/N}\right]y(k) \tag{6.6.21}$$

with

$$y(k) \triangleq \sum_{k_2=0}^{NL_0-1} x(k_2)h(k-k_2) \tag{6.6.22}$$

As we mentioned earlier, $h(\Delta t)$ has a duration of a few symbol intervals. Since $L_0$ is normally much longer, we expect that the right-hand side in (6.6.22) will be marginally affected if the summation is extended to $+\infty$. In these conditions the signal $y(k)$ can be computed by feeding $x(k)$ into a filter with impulse response $h(k)$. Although $h(k)$ is noncausal, it can be made causal by an appropriate time shift, say of $ND$ steps rightward. The value $D=2$ is sufficient in most cases. On the other hand, this shift transforms the filter output into $y(k-ND)$. Thus, to get things straight in (6.6.21), the signal $x(k)$ must also be delayed by $ND$ and, in conclusion, $X_{+1}$ must be rearranged as follows

$$X_{+1} = \sum_{k=ND}^{N(L_0+D)-1}\left[x(k-ND)e^{j\pi(k-ND)/N}\right]y(k-ND) \tag{6.6.23}$$

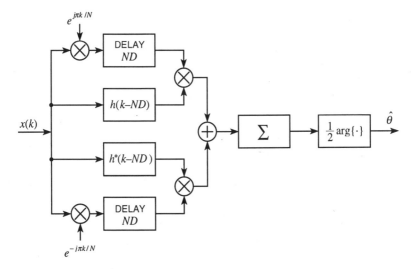

Figure 6.19. Block diagram of the phase estimator.

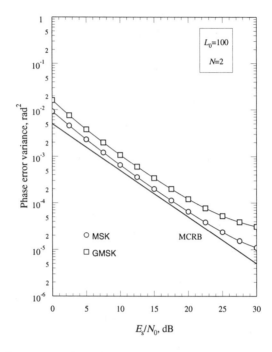

Figure 6.20. Phase error variance for MSK and GMSK with a sampling rate 2/T.

Expressing $X_{-1}$ in analogous form leads us to the estimation scheme indicated in Figure 6.19.

Computer simulations for MSK and GMSK modulation are now shown. A premodulation Gaussian filter of bandwidth $0.3/T$ is used with GMSK. The anti-aliasing filter consists of a fourth-order FIR structure of bandwidth $B_{AAF}=1.2/T$ and the parameter $L_0$ is set at 100. Two oversampling factors $N=2$ and $N=4$ have been tried without significant differences in estimation performance. In both cases the estimates are virtually unbiased and the estimation errors are close to the MCRB. This is apparent from Figure 6.20 which illustrates the case with $N=2$. For MSK, in particular, it is seen that the distance from the MCRB is a fraction of a dB.

## 6.7. Clockless Phase Estimation

We close this chapter with non-clock-aided phase estimation for MSK-type modulation. The treatment could be extended to more general formats but, for simplicity, we stick to binary signaling with a modulation index of 1/2. As

# Carrier Phase Recovery with CPM Modulations

in the previous section we assume that carrier frequency offset has been perfectly compensated so that the parameter $v$ can be set to zero.

To begin, we temporarily concentrate on the signal in isolation

$$s(t) = e^{j\theta} \sqrt{\frac{2E_s}{T}} e^{j\psi(t-\tau,\alpha)} \qquad (6.7.1)$$

where

$$\psi(t,\alpha) = \pi \sum_{i=-i_0}^{\infty} \alpha_i q(t - iT) \qquad (6.7.2)$$

and $-i_0$ correponds to the initial symbol in the data sequence. With no loss of generality in the sequel we assume that $i_0$ is greater than the correlation length $L$, so that $\psi(t,\alpha)$ has already achieved steady-state conditions at $t=0$. Squaring and rearranging (6.7.1) yields

$$s^2(t) = e^{j2\theta} \frac{2E_s}{T} \prod_{i=-i_0}^{\infty} \exp\{j2\pi\alpha_i q(t - iT - \tau)\} \qquad (6.7.3)$$

Next, averaging with respect to the symbols produces [18]

$$E\{s^2(t)\} = e^{j2\theta} \frac{2E_s}{T} \prod_{i=-i_0}^{\infty} \cos[2\pi q(t - iT - \tau)] \qquad (6.7.4)$$

In Exercise 6.7.1 it is shown that the function

$$F(t) \triangleq \prod_{i=-i_0}^{\infty} \cos[2\pi q(t - iT)] \qquad (6.7.5)$$

is alternating, i.e., it satisfies the property

$$F(t - T) = -F(t) \qquad (6.7.6)$$

This means that $F(t)$ is periodic of period $2T$ and, as such, can be expanded into a Fourier series of the type

$$F(t) = \sum_k F_k e^{j\pi kt/T} \qquad (6.7.7)$$

wherein the coefficients satisfy the relations

$$F_0 = 0 \tag{6.7.8}$$

$$F_{-k} = F_k^* \tag{6.7.9}$$

Thus, inserting (6.7.7) into (6.7.4) yields

$$E\{s^2(t)\} = \sum_k C_k e^{j\pi k(t-\tau)/T} \tag{6.7.10}$$

with

$$C_k \triangleq \frac{2E_s}{T} F_k e^{j2\theta} \tag{6.7.11}$$

Alternatively, $C_k$ can be expressed as a function of $E\{s^2(t)\}$ in the form

$$C_k \triangleq \frac{1}{2T} \int_{-T}^{T} E\{s^2(t)\} e^{-j\pi kt/T} dt \tag{6.7.12}$$

From (6.7.9) and (6.7.11) it appears that, if a pair of coefficients $(C_{-k}, C_k)$ were given, the carrier phase could be computed through the formula

$$\theta = \frac{1}{4} \arg\{C_{-k} C_k\} \tag{6.7.13}$$

It is clear from (6.7.12) that the computation of $(C_{-k}, C_k)$ requires knowledge of $s(t)$. However, what is available is $r(t)$, not $s(t)$. Thus, an approximation to $C_k$ is obtained by replacing $s(t)$ with some filtered version of $r(t)$, say $x(t)$. In practice this can be done as follows.

Let us concentrate on $C_1$, for example. As a first step we transform (6.7.12) into

$$C_1 \approx \frac{1}{L_0 T} \int_0^{L_0 T} x^2(t) e^{-j\pi t/T} dt \tag{6.7.14}$$

where $L_0$ is an even integer, much greater than unity. Clearly, the expectation operation has been replaced by a time average over $(0, L_0 T)$. The second step is called for by digital implementation considerations. In fact we want to express the integral in (6.7.14) in terms of samples of $x(t)$ taken at some rate $1/T_s$. Choosing $1/T_s = N/T$ produces

# Carrier Phase Recovery with CPM Modulations

$$C_1 \approx \frac{1}{L_0 N} \sum_{k=0}^{L_0 N-1} x^2(k) e^{-j\pi k/N} \qquad (6.7.15)$$

where $x(k) \triangleq x(kT_s)$. As shown in Appendix 2.C, the condition for (6.7.15) to hold is that $1/T_s$ exceed $f_{\max}$, the highest frequency component in $x^2(t)e^{-j\pi t/T}$. Since $x(t)$ has a bandwidth of about $1/T$, it is recognized that $f_{\max}$ is close to $2.5/T$ and a suitable oversampling factor is either 3 or 4. In the simulations discussed below we have taken $N=4$.

Reasoning in a similar manner with $C_{-1}$ yields

$$C_{-1} \approx \frac{1}{L_0 N} \sum_{k=0}^{L_0 N-1} x^2(k) e^{j\pi k/N} \qquad (6.7.16)$$

Hence, inserting into (6.7.13) yields the phase estimator

$$\hat{\theta} = \frac{1}{4} \arg \left\{ \left[ \sum_{k=0}^{L_0 N-1} x^2(k) e^{j\pi k/N} \right] \left[ \sum_{k=0}^{L_0 N-1} x^2(k) e^{-j\pi k/N} \right] \right\} \qquad (6.7.17)$$

whose block diagram is depicted in Figure 6.21.

Figure 6.22 shows simulation curves for the estimation error variance with MSK and GMSK modulation. The low-pass filter yielding $x(t)$ has been implemented as a fourth-order Butterworth having a bandwidth of $1.2/T$. Also, an observation length of 100 symbols has been chosen. Comparing with the results in Figure 6.20 it appears that the lack of timing information produces some loss in performance (in addition to requiring a higher sampling rate). For example, with MSK the distance from the MCRB is now about 3 dB, as opposed to less than 1 dB in Figure 6.20.

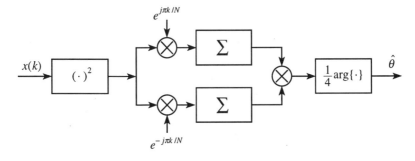

Figure 6.21. Block diagram of the phase estimator.

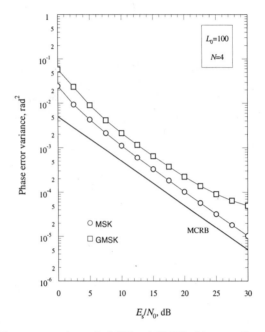

Figure 6.22. Phase error variance for MSK and GMSK with a sampling rate 4/T.

**Exercise 6.7.1.** Letting $i_0$ be some positive integer greater than the correlation length $L$, show that

$$F(t) \triangleq \prod_{i=-i_0}^{\infty} \cos[2\pi q(t - iT)] \qquad (6.7.18)$$

has an alternating pattern on the positive axis, i.e.,

$$F(t - T) = -F(t), \quad t \geq 0 \qquad (6.7.19)$$

*Solution.* From (6.7.18) we have in succession

$$\begin{aligned} F(t - T) &= \prod_{i=-i_0}^{\infty} \cos[2\pi q(t - (i+1)T)] \\ &= \prod_{i'=-i_0+1}^{\infty} \cos[2\pi q(t - i'T)] \\ &= \frac{1}{\cos[2\pi q(t + i_0 T)]} \prod_{i'=-i_0}^{\infty} \cos[2\pi q(t - i'T)] \\ &= \frac{F(t)}{\cos[2\pi q(t + i_0 T)]} \end{aligned} \qquad (6.7.20)$$

Bearing in mind that $i_0 > L$ and $q(t)=1/2$ for $t \geq LT$, it follows that the cosine in the last line equals $-1$ for $t \geq 0$ and this proves (6.7.19).

## 6.8. Key Points of the Chapter

- Data-aided phase estimation with MSK-type modulation is accomplished by means of feedforward schemes. Their performance with MSK modulation is close to the CRB at low/intermediate SNR. As SNR increases, however, some pattern noise shows up. The critical point at which this happens depends on the observation length. The longer the observation, the higher the SNR at which departure from the CRB begins.
- Decision-directed phase-lock loops are often employed with MSK-type modulations. At high/intermediate SNR their performance is close to the MCRB. Phase tracking loops fit well into simplified MSK-type receiver structures.
- Decision-directed phase recovery with general CPM modulations is accomplished by exploiting the decisions taken from the best survivor in the Viterbi detector. In the presence of phase noise, however, a better strategy is to resort to multiple phase tracking. This means associating a decision-directed phase estimator with each survivor in the trellis.
- With frequency-flat fading channels several approximate ML decoding schemes are available. They are based on per-survivor-processing methods by which channel distortion and data sequence are jointly estimated. The most efficient schemes need knowledge of the channel statistics, the SNR and the fading autocorrelation function. Extra receiver complexity is required to estimate these parameters, however.
- Two feedforward schemes are available for clock-aided (but non-data-aided) phase recovery with AWGN channels. The $2P$-power scheme is *ad hoc* and is suited for full response formats. The other scheme is based on ML arguments and applies to MSK-like modulations. Its performance is excellent. In particular, with MSK it is a fraction of a dB short of the MCRB.
- Non-clock-aided phase recovery is also possible. A feedforward structure has been indicated for MSK-type formats. Its performance is good, only a bit worse than that of the previous clock-aided and ML-oriented scheme.

## References

[1] J.B.Anderson and C-E.Sundberg, Advances in Constant Envelope Coded Modulation, *IEEE Communications Magazine*, **29**, No 12, 36-45, Dec. 1991.
[2] J.B.Anderson, T.Aulin and C.-E.Sundberg, *Digital Phase Modulation*, New York: Plenum, 1986.

[3] P.Galko and S.Pasupathy, Optimization of Linear Receivers for Data Communication Signals, *IEEE Trans. Inf. Theory*, **IT-34**, 79-92, Jan. 1988.
[4] A.Svensson and C.-E.Sundberg, Optimum MSK-type Receivers for CPM on Gaussian and Rayleigh Fading Channels, *IEE Proc. Pt. F*, **131**, 480-490, 1984.
[5] M.S.El-Tanany and S.A.Mahmoud, Mean-Square Error Optimization of Quadrature Receivers for CPM with Modulation Index 1/2, *IEEE J. Selec. Areas Commun.*, **SAC-5**, 896-905, June 1987.
[6] F.de Jager and C.B.Dekker, Tamed Frequency Modulation, A Novel Method to Achieve Spectrum Economy in Digital Transmission, *IEEE Trans. Commun.*, **COM-26**, 534-542, May 1978.
[7] K.Murota, K.Hirade, GMSK Modulation for Digital Mobile Radio Telephony, *IEEE Trans. Commun.*, **COM-29**, 1044-1050, July 1981.
[8] A.N.D'Andrea, U.Mengali and G.Vitetta, Multiple Phase Synchronization in Continuous Phase Modulation, Digital Signal Processing, Academic Press, vol. 3, 188-198, July 1993.
[9] A.Premji and D.P.Taylor, Receiver Structures for Multi-h Signaling Formats, *IEEE Trans. Commun.*, **COM-35**, 439-451, April 1987.
[10] A.J.Macdonald and J.B.Anderson, PLL Synchronization for Coded Modulation, Conf. Rec. ICC'91, Denver, Colorado, pp. 52.6.1-52.6.6, June 23-26.
[11] A.Polydoros, R.Raheli and C-K.Tzou, Per-Survivor-Processing: A General Approach to MLSE in Uncertain Environments, *IEEE Trans. Commun.*, **COM-43**, 354-364, Feb./March/April 1995.
[12] N.Seshadri, Joint Data and Channel Estimation Using Blind Trellis Search Techniques, *IEEE Trans. Commun.*, **COM-42**, 1000-1011, Feb./March/April 1994.
[13] G.T. Irvine and P.J.McLane, Symbol-Aided Plus Decision-Directed Reception for PSK/TCM Modulation on Shadowed Mobile Satellite Fading Channels, *IEEE J. Selec. Areas Commun.*, **SAC-10**, 1289-1299, Oct. 1992.
[14] J.H.Lodge and M.J.Moher, Maximum Likelihood Sequence Estimation of CPM Signals Transmitted over Rayleigh Flat-Fading Channel, *IEEE Trans. Commun.*, **COM-38**, 787-794, June 1990.
[15] G.M.Vitetta, U.Mengali and D.P.Taylor, Blind Detection of CPM Signals Transmitted over Frequency-Flat Fading Channels, accepted for publication in *IEEE Trans. Vehic. Tech.*
[16] P.A.Bello, Characterization of Randomly Time-Variant Linear Channels, *IEEE Trans. Commun. Systems*, **COM-9**, 360-393, Dec. 1963.
[17] A.D'Amico, A.N.D'Andrea and U.Mengali, Joint Phase and Timing Estimation for MSK-type Signaling, In preparation.
[18] A.N.D'Andrea, U.Mengali and R.Reggiannini, Carrier Phase and Clock Recovery for Continuous Phase Modulated Signals, *IEEE Trans. Commun.*, **COM-33**, 1285-1290, Oct. 1987.

# 7

# Timing Recovery in Baseband Transmission

## 7.1. Introduction

In this chapter we investigate timing recovery in PAM baseband transmission. Additional features that arise with modulated signals will be explored in Chapters 8 and 9 for PAM and CPM modulations, respectively. The reason for first concentrating on baseband signals is motivated by the need of avoiding any distractions caused by modulation matters. As we shall see in the later chapters, such matters are easily taken into account when the basic concepts relevant to baseband systems are understood.

The organization of the present chapter reflects the fact that timing recovery consists of two distinct operations: ($i$) estimation of the timing phase $\tau$; ($ii$) application of the estimate to the sampling process. The former is referred to as *timing measurement*, the latter as *timing correction* (or *adjustment*). Timing correction serves to provide the decision device with signal samples (*strobes*) with minimum intersymbol interference. In the sequel we first concentrate on timing correction and then move to timing measurement.

A second issue which influences the chapter profile is the distinction between feedback and feedforward schemes. Figures 7.1 and 7.2 illustrate the salient features of the two topologies in the case of a *fully digital* implementation. In both cases an anti-alias filter (AAF) limits the bandwidth of the received waveform. Sampling is controlled by a fixed clock whose ticks are not locked to the incoming data. In practice the click rate will be close but not equal to some rational multiple of the symbol rate.

The bulk of the pulse shaping is performed in the matched filter (MF) whose location is not necessarily that shown in the figures. For example, the

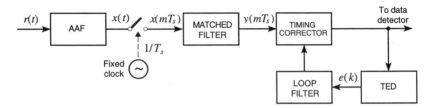

Figure 7.1. Feedback configuration.

MF may be moved inside the loop in Figure 7.1 or it may be shifted so as to have a common input with the timing estimator in Figure 7.2. Timing correction is akin to the operation of a voltage-controllable delay line and produces *synchronized* samples to be used for decision and synchronization purposes. In the feedback configuration of Figure 7.1, in particular, the timing corrector feeds a timing error detector (TED) whose purpose is to generate an error signal $e(k)$ proportional to the difference between $\tau$ and its current estimate. The error signal is then exploited to recursively update the timing estimates.

A third prominent issue we address in this chapter is the distinction between *synchronized* and *non-synchronized* sampling. In a fully digital implementation, sampling is not locked to the incoming pulses. This is referred to as *non-synchronous* sampling. On the other hand, sampling can be made *synchronous* by exploiting some error signal to adjust the timing phase of a number controlled oscillator (NCO), as is shown in Figure 7.3. Here the sampler is commanded by the NCO pulses at times $\{t_n\}$ (typically one or two pulses per symbol interval). Note that the *analog* MF in the figure may be replaced by a *digital* MF inside the loop.

The chapter is organized as follows. In the first part we discuss timing correction. In particular, synchronous sampling is investigated in Section 7.2 and non-synchronous sampling in Section 7.3. Next, in Section 7.4, we concentrate on decision-directed feedback timing recovery. Essentially, this involves looking for TED algorithms with good characteristics (low noise and

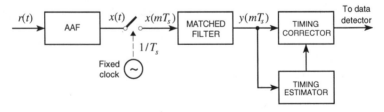

Figure 7.2. Feedforward configuration.

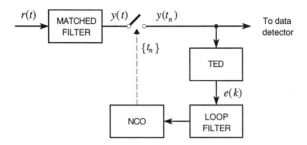

Figure 7.3. Synchronous sampling.

simple implementation). As we shall see, a number of such schemes can be derived either from application of ML criteria or from heuristic arguments. Section 7.5 is still concerned with feedback schemes but the TED algorithms are non-data-aided. Finally, in Section 7.6, a few feedforward timing estimation methods are discussed and compared.

## 7.2. Synchronous Sampling

In addressing synchronous sampling we assume that some TED circuit is available with an error signal $e(k)$ providing information on the difference between the timing parameter $\tau$ and its estimate $\hat{\tau}_k$ at time $kT$. More specifically, we suppose that the average of $e(k)$ for a fixed value of $\hat{\tau}_k$, say $\hat{\tau}$, is a regular function $S(\tau - \hat{\tau})$ passing through the origin with a positive slope. Accordingly, $e(k)$ may be written in the form

$$e(k) = S(\tau - \hat{\tau}_k) + N(k) \qquad (7.2.1)$$

where $\{N(k)\}$ is some zero-mean noise process resulting from interactions between signal and thermal noise inside the TED. In the sequel we first give an overview of the operation of a hybrid NCO and then we indicate how an NCO can be used for timing adjustment in a feedback timing loop.

### 7.2.1. Hybrid NCO

The block diagram of a hybrid NCO is depicted in Figure 7.4. It consists of two parts: a digital loop comprising a delay and a mod 1 adder (the so-called *digital* NCO), plus a look-up table and a digital-to-analog converter (DAC) that, together, transform the NCO output into a continuous-time waveform. To

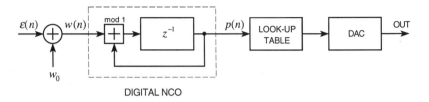

Figure 7.4. Block diagram of a hybrid NCO.

explain the operation of the overall circuit we first concentrate on the digital NCO. Let $w(n)$ be its input and $p(n)$ its output. In the sequel $w(n)$ and $p(n)$ are viewed as numbers between 0 and 1. At each tick of a clock (not shown in the figure) $w(n)$ is added to $p(n)$. This results in the following difference equation:

$$p(n+1) = p(n) + w(n) \quad \mod 1 \tag{7.2.2}$$

As indicated in the figure, $w(n)$ is the sum of some constant $w_0$ plus a zero-mean signal $\varepsilon(n)$. The constant establishes the "free running" period of the NCO whereas $\varepsilon(n)$ allows us to change this period. To see how this comes about suppose first that $\varepsilon(n)=0$. In these conditions the NCO output will recycle every $1/w_0$ ticks. Thus, calling $T_c$ the clock period, the free running period will be $T_c/w_0$ (in reality the NCO recycling time is not quite constant [1] but undergoes fluctuations around $T_c/w_0$ which fade away for $w_0$ sufficiently small). If $\varepsilon(n)$ is different from zero (but is slowly varying in time), then the recycling period will be

$$\begin{aligned} T_s(n) &= \frac{T_c}{w_0 + \varepsilon(n)} \\ &\approx \frac{T_c}{w_0}\left[1 - \frac{\varepsilon(n)}{w_0}\right] \end{aligned} \tag{7.2.3}$$

assuming $|\varepsilon(n)| \ll w_0$.

Returning to the scheme in Figure 7.4, a look-up table maps $p(n)$ into some function $f[p(n)]$ which, in turn, is transformed into a continuous-time function by the action of the DAC and some analog low-pass filter (not shown in the figure). Functions $f[p(n)]$ of different types are encountered in existing hardware. A sine table is often used, which means

$$f[p(n)] = \sin[2\pi p(n)] \tag{7.2.4}$$

In this case the hybrid NCO will generate a phase modulated sinewave.

## 7.2.2. Timing Adjustment for Synchronous Sampling

The question of timing correction for synchronous sampling is now addressed. To be specific we choose a hybrid NCO with a sine-table ROM and we use the *up-crossings* of the sinewave (zero crossings with positive slope) to generate command pulses for the sampler in Figure 7.3. Ideally, we would like the pulses be issued at the instants

$$t_n^{(id)} = n\frac{T}{N} + \tau, \quad n = 0, 1, 2, \ldots \quad (7.2.5)$$

where $N$ is an integer (*oversampling factor*) and $\tau$ is the timing phase (which is assumed to lie in the interval $0 \leq \tau < T$). The oversampling factor is typically 1 or 2 when the matched filter is external to the loop, as happens in Figure 7.3, while it ranges from 2 to 4 when the matched filter is internal. The goal is to steer the NCO so that its actual up-crossings occur at the times indicated in (7.2.5).

To address this problem we return to equation (7.2.1), whose right-hand side contains a deterministic component depending on $\tau - \hat{\tau}_k$. It is this component that will be exploited to drive the NCO. To see how, recollect that $e(k)$ is computed at symbol rate. Actually, $e(k)$ is recomputed at any $N$-th sinewave up-crossing. In other words, the time index $k$ for $e(k)$ (henceforth denoted *symbol index*) is related to the *sampling index* $n$ by

$$k = \mathrm{int}\left(\frac{n}{N}\right) \quad (7.2.6)$$

where $\mathrm{int}(x)$ means "the largest integer not exceeding $x$."

To proceed further we now introduce an auxiliary signal $e'(n)$ which is derived by upsampling $e(k)$ from 1 to $N$ (see Figure 7.5). Formally

$$e'(n) \triangleq e(k)\big|_{k=\mathrm{int}(n/N)} \quad (7.2.7)$$

Next, $e'(n)$ is reversed in sign, is scaled by some factor $K$ and is fed to the NCO. With the notations in Figure 7.4 we have

$$\varepsilon(n) \triangleq -Ke'(n) \quad (7.2.8)$$

We maintain that (under certain conditions to be presently established) the NCO up-crossing times

$$t_n = n\frac{T}{N} + \hat{\tau}'_n, \quad n = 0, 1, 2, \ldots \quad (7.2.9)$$

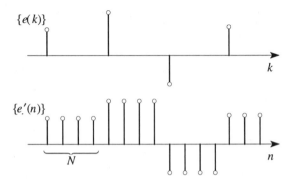

Figure 7.5. Relation between $e(k)$ and $e'(n)$ for $N=4$.

are close to the desired sampling instants $\{t_n^{(id)}\}$ in (7.2.5) or, which is the same, the timing phases $\{\hat{\tau}_n'\}$ are approximately equal to $\tau$, i.e.,

$$\hat{\tau}_n' \approx \tau, \quad n = 0,1,2,... \quad (7.2.10)$$

To achieve this goal we set the free running period of the NCO equal to $T/N$, which amounts to choosing $w_0$ in Figure 7.4 such that

$$\frac{T_c}{w_0} = \frac{T}{N} \quad (7.2.11)$$

Also, we define $\gamma \triangleq KT/w_0$. Then, inserting into (7.2.3) yields

$$T_s(n) \approx \frac{1}{N}[T + \gamma e'(n)] \quad (7.2.12)$$

from which the following relation between adjacent up-crossings is obtained:

$$t_{n+1} = t_n + \frac{T}{N} + \frac{\gamma}{N}e'(n) \quad (7.2.13)$$

If $\gamma$ is sufficiently small it is easily seen from (7.2.9) and (7.2.13) that the sequence $\{\hat{\tau}_n'\}$ varies slowly in time. Thus, we can limit ourselves to proving (7.2.10) for $n=Nk$. In other words, letting

$$\hat{\tau}_k \triangleq \hat{\tau}_{Nk}' \quad (7.2.14)$$

we want to show that

# Timing Recovery in Baseband Transmission

$$\hat{\tau}_k \approx \tau, \quad k = 0,1,2,\ldots \tag{7.2.15}$$

The proof proceeds as follows. Consider the up-crossing times $t_{Nk}$ and $t_{N(k+1)}$. From (7.2.9) and (7.2.14) we have

$$t_{N(k+1)} - t_{Nk} = T + \hat{\tau}_{k+1} - \hat{\tau}_k \tag{7.2.16}$$

On the other hand

$$t_{N(k+1)} - t_{Nk} = \sum_{n=Nk}^{N(k+1)-1} (t_{n+1} - t_n) \tag{7.2.17}$$

Thus, combining (7.2.16)-(7.2.17) and taking $t_{n+1} - t_n$ from (7.2.13) produces

$$\hat{\tau}_{k+1} = \hat{\tau}_k + \frac{\gamma}{N} \sum_{n=Nk}^{N(k+1)-1} e'(n) \tag{7.2.18}$$

As a consequence of (7.2.7), the $e'(n)$ appearing in (7.2.18) are all equal to $e(k)$. Hence

$$\hat{\tau}_{k+1} = \hat{\tau}_k + \gamma e(k) \tag{7.2.19}$$

Finally, taking $e(k)$ from (7.2.1) produces

$$\hat{\tau}_{k+1} = \hat{\tau}_k + \gamma S(\tau - \hat{\tau}_k) + \gamma N(k) \tag{7.2.20}$$

which is formally identical to the equation encountered in Section 3 of Chapter 5 when dealing with phase recovery loops. Based on the results in that section we conclude that $\hat{\tau}_k$ will fluctuate around $\tau$, in agreement with our previous claim.

To summarize, synchronous sampling can be implemented with the simple arrangement indicated in Figure 7.3. Because of the presence of the hybrid NCO, the overall circuit cannot be implemented in fully digital form. Whether this is a drawback depends on the circumstances. Current trends in modem design tend to prefer fully digital implementations.

## 7.3. Non-Synchronous Sampling

### 7.3.1. Feedback Recovery Scheme

As indicated in Figures 7.1-7.2, non-synchronous sampling can be used both with feedback and feedforward schemes. Here we concentrate on the former application; extensions to the latter are explored later. The major issue we shall be concerned with is the implementation of the timing correction block in Figure 7.1. This subject is first addressed in rather conceptual terms.

The starting point is the requirement that the samples $\{x(mT_s)\}$ contain the same information as the continuous-time waveform $x(t)$ from the anti-alias filter. For this to be true the sampling rate must satisfy the relation

$$\frac{1}{T_s} \geq 2B_X \quad (7.3.1)$$

where $B_X$ is the bandwidth of the waveform from the anti-alias filter. In these conditions the sequence $\{y(mT_s)\}$ from the matched filter is sufficient to reconstruct the underlying continuous-time waveform $y(t)$ through the interpolation formula

$$y(t) = \sum_{m=-\infty}^{\infty} h_I(t - mT_s) y(mT_s) \quad (7.3.2)$$

where

$$h_I(t) \triangleq \frac{\sin(\pi t/T_s)}{\pi t/T_s} \quad (7.3.3)$$

It is worth noticing that $y(t)$ is the same waveform indicated in Figure 7.3 (provided that the analog MF is equivalent to the digital MF in Figure 7.1). It follows that the clock recovery schemes in Figures 7.1 and 7.3 would be equivalent if the former were endowed with the *ideal interpolator* (7.3.2). In these conditions timing correction with non-synchronous sampling could be implemented using the circuit in Figure 7.1.

Unfortunately, this approach is difficult to follow because equation (7.3.2) involves an infinite summation whereas only a limited number of terms can be handled in practice. Truncating the sum in (7.3.2) will inevitably cause a certain amount of signal distortion. In the next section we investigate how to replace (7.3.3) by other interpolating functions so as to limit the sum (7.3.2) to a few terms.

**Exercise 7.3.1.** Consider the block diagram in Figure 7.6 (a) in which the input $x(t)$ from the anti-alias filter is fed to a filter $h(t)$ and then is sampled at some rate $1/T_I$. The sampling times $\{t_n\}$ are established by some controller not shown in the figure. Discuss the computation of $\{y(t_n)\}$ from the samples of $x(t)$ taken at some rate $1/T_s$. Assume that an ideal interpolator is available.

*Solution.* Figure 7.6(b)-(c) indicates two alternative solutions. In the former the filter precedes the interpolator; the reverse is true in the latter. In both

# Timing Recovery in Baseband Transmission

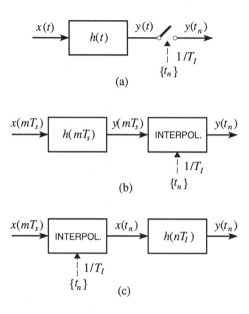

Figure 7.6. Different digital implementations of the same analog scheme.

cases the filter impulse response is derived by sampling $h(t)$ at proper rates (either $1/T_s$ or $1/T_I$). The question arises of the constraints to impose on $1/T_s$ and $1/T_I$ to guarantee that $\{y(t_n)\}$ not be distorted.

Start from the scheme in Figure 7.6(b) and denote $X(f)$ and $H(f)$ the Fourier transforms of $x(t)$ and $h(t)$. Also, call $B_H$ the bandwidth of the latter (see Figure 7.7(a)). Sampling theory says that the spectrum of $\{x(mT_s)\}$ has periodic images spaced at a frequency interval $1/T_s$, as shown in Figure 7.7 (b). Image tails lying on the interval $\pm B_H$ will cause random interference on the filter output and, in consequence, on the interpolator's. To avoid interference, the left edge of $X(f - 1/T_s)$ must be to the right of $f=B_H$. Hence

$$\frac{1}{T_s} \geq B_X + B_H \tag{7.3.4}$$

A similar argument applies to the scheme in Figure 7.6(c). Here, the spectrum of the sequence of interpolants $\{x(t_n)\}$ has images spaced at a frequency $1/T_I$, as indicated in Figure 7.7(c) and the condition to avoid distortion is

$$\frac{1}{T_I} \geq B_X + B_H \tag{7.3.5}$$

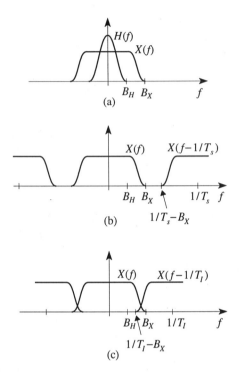

Figure 7.7. Signal spectra.

### 7.3.2. Piecewise Polynomial Interpolators

Now we turn our attention to the interpolator implementation and, to be specific, we concentrate on the clock recovery scheme in Figure 7.8. As in previous figures, $\{x(mT_s)\}$ are samples from the anti-alias filter. The interval $T_s$ is on the order of one-half to one-fourth of the symbol period $T$ but it is incommensurate with $T$. This is so because the sampling clock is derived from a local source that is independent of the transmitter's timing source. The controller's task is to tell the interpolator the desired interpolation times $\{t_n\}$. The interpolants are denoted $y_I(t_n)$. With an ideal interpolator $y_I(t_n)$ would coincide with $y(t_n)$, the sample of the waveform that underlies $\{y(mT_s)\}$. Note that the intervals between adjacent $t_n$ are not rigorously constant. They exhibit small fluctuations around some average value $T_I$, the interpolation interval. This is a submultiple of the symbol interval, say $T_I=T/N$, where $N$ is typically 1 or 2. Also notice that only one interpolant per symbol is needed for data detection whereas either one or two interpolants per symbol are required for synchronization purposes.

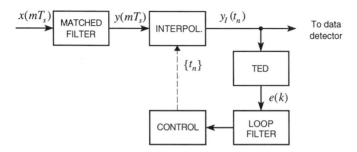

Figure 7.8. Block diagram of a non-synchronous clock recovery loop.

The error signal $e(k)$ provides information on the difference between the timing parameter $\tau$ and its current estimate $\hat{\tau}_k$ at time $kT$. Actually, $e(k)$ has the form

$$e(k) = S(\tau - \hat{\tau}_k) + N(k) \qquad (7.3.6)$$

where $S(\cdot)$ is the detector S-curve and $\{N(k)\}$ is some zero-mean noise process. The different role of the time indexes $m$, $n$ and $k$ should be stressed. The first index counts the signal samples from the anti-alias filter, the second counts the interpolants, the third counts the incoming data symbols. Since $T/T_s$ is an irrational number, there is no simple relation between $m$ and $n$. Vice versa, $n$ and $k$ are simply related by $k=\mathrm{int}(n/N)$.

At this stage the following question arises: how can a simple interpolator be implemented without running into severe signal distortions? This question is extensively covered in the digital signal processing literature [2]-[3]. The specific role of interpolation in timing adjustment is addressed in [4]-[7]. In this section we report on interpolation methods investigated in [6]-[7]. The interested reader is referred to these papers for further details and discussions.

One main conclusion achieved in [6]-[7] is that excellent interpolators can be implemented by replacing $h_I(t)$ in (7.3.3) by some piecewise polynomial functions of short duration (say, either two or four sampling intervals). These functions may be viewed as impulse responses of polynomial-based filters. Formulas for the most appealing impulse responses are provided soon.

As a preparation, let us return to (7.3.2) and suppose that $h_I(t)$ is some impulse response to be specified. Our task is to compute $y(t)$ at time $t_n$. For this purpose we first look for that sampling instant, say $l_n T_s$, occurring immediately before $t_n$. Its index $l_n$ is clearly given by

$$l_n \triangleq \mathrm{int}\left(\frac{t_n}{T_s}\right) \qquad (7.3.7)$$

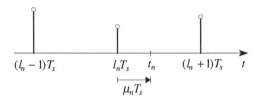

Figure 7.9. Definition of $l_n$ and $\mu_n$.

and is referred to as the *basepoint index*. It follows that $t_n$ may be written as the sum of $l_n T_s$ plus a positive fractional part

$$t_n = (l_n + \mu_n)T_s \qquad (7.3.8)$$

with $0 \leq \mu_n < 1$. The parameter $\mu_n$ is called the *fractional interval* and is expressed by

$$\mu_n = \mathrm{frc}\left(\frac{t_n}{T_s}\right) \qquad (7.3.9)$$

with $\mathrm{frc}(x) \triangleq x - \mathrm{int}(x)$. Figure 7.9 illustrates the meaning of $l_n T_s$ and $\mu_n T_s$.
Substituting (7.3.8) into (7.3.2) yields

$$y(t_n) = \sum_{m=l_n-I_2}^{l_n+I_1} h_I\big[(l_n - m)T_s + \mu_n T_s\big] y(mT_s) \qquad (7.3.10)$$

In writing this equation we have assumed that $h_I(t)$ takes significant values on the interval $-I_1 T_s \leq t \leq (I_2 + 1)T_s$ and, in consequence, the only samples $\{y(mT_s)\}$ contributing to the sum are those with index from $l_n - I_2$ to $l_n + I_1$. A more useful formula is obtained by changing the summation index into $i = l_n - m$. This leads to

$$y(t_n) = \sum_{i=-I_1}^{I_2} h_I(iT_s + \mu_n T_s) y\big[(l_n - i)T_s\big] \qquad (7.3.11)$$

or, more compactly,

$$y(t_n) = \sum_{i=-I_1}^{I_2} c_i(\mu_n) y\big[(l_n - i)T_s\big] \qquad (7.3.12)$$

with

$$c_i(\mu_n) \triangleq h_I(iT_s + \mu_n T_s) \qquad (7.3.13)$$

Equation (7.3.12) is the fundamental formula of the interpolator. It shows that two ingredients come into play when computing interpolants: the basepoint set $\{y(mT_s)\}$ ($l_n - I_2 \le m \le l_n + I_1$) and the coefficients $\{c_i(\mu_n)\}$ ($-I_1 \le i \le I_2$). The basepoints are located around $l_n T_s$. The coefficients $\{c_i(\mu_n)\}$ are just samples of the impulse response $h_I(t)$. Given $h_I(t)$, they are a function of the fractional index $\mu_n$ in (7.3.9).

Three sets of coefficients $\{c_i(\mu_n)\}$ are investigated in [7] and are reported here. They are characterized by: (i) the order of the polynomial in the piecewise representation of $h_I(t)$; (ii) the number of basepoints $I \triangleq I_1 + I_2 + 1$ in the interpolation formula. Two basepoints are sufficient with a piecewise linear interpolator while four basepoints are used with parabolic and cubic interpolators. The coefficients for a *linear interpolator* are

$$c_{-1}(\mu) = \mu \tag{7.3.14}$$

$$c_0(\mu) = 1 - \mu \tag{7.3.15}$$

Those for a *parabolic interpolator* are

$$c_{-2}(\mu) = \alpha\mu^2 - \alpha\mu \tag{7.3.16}$$

$$c_{-1}(\mu) = -\alpha\mu^2 + (1+\alpha)\mu \tag{7.3.17}$$

$$c_0(\mu) = -\alpha\mu^2 - (1-\alpha)\mu + 1 \tag{7.3.18}$$

$$c_1(\mu) = \alpha\mu^2 - \alpha\mu \tag{7.3.19}$$

where $\alpha$ is a design parameter that can be exploited to improve the interpolation accuracy. An $\alpha$ close to 0.5 seems nearly optimum [7]-[8]. It is worth noting that a parabolic interpolator reduces to a linear interpolator for $\alpha=0$. Finally, the coefficients for a *cubic interpolator* are

$$c_{-2}(\mu) = \mu^3/6 - \mu/6 \tag{7.3.20}$$

$$c_{-1}(\mu) = -\mu^3/2 + \mu^2/2 + \mu \tag{7.3.21}$$

$$c_0(\mu) = \mu^3/2 - \mu^2 - \mu/2 + 1 \tag{7.3.22}$$

$$c_1(\mu) = -\mu^3/6 + \mu^2/2 - \mu/3 \tag{7.3.23}$$

The computational load involved with the above interpolators depends on the order of the polynomial; it is minimal with a linear polynomial and maximum with a cubic polynomial. From extensive simulations discussed in

[7] it appears that linear interpolators provide adequate performance in many situations. For more critical applications, parabolic interpolation with $\alpha=0.5$ provides excellent performance with only a moderate complexity increase.

### 7.3.3. Timing Adjustment with Non-Synchronous Sampling

We now discuss interpolator control. Our aim is to keep the actual interpolation times $\{t_n\}$ as close as possible to the ideal times

$$t_n^{(id)} = nT_I + \tau \tag{7.3.24}$$

where $T_I = T/N$. As a first step in this direction we introduce an auxiliary error signal $e'(n)$, which is derived by reiterating ($N$ times) each sample $e(k)$ from the TED, i.e.,

$$e'(n) \triangleq e(k)\big|_{k=\text{int}(n/N)} \tag{7.3.25}$$

The auxiliary signal serves to update the current interpolation time $t_n$ according to

$$t_{n+1} = t_n + T_I + \frac{\gamma}{N}e'(n) \tag{7.3.26}$$

where $\gamma$ is a step-size parameter. Writing the interpolation times in the form

$$t_n = nT_I + \hat{\tau}'_n \tag{7.3.27}$$

we maintain that timing phases $\{\hat{\tau}'_n\}$ will fluctuate around $\tau$, i.e,

$$\hat{\tau}'_n \approx \tau, \quad n = 0,1,2,\ldots \tag{7.3.28}$$

To prove our claim we assume that $\gamma$ is sufficiently small so that the differences $t_{n+1} - t_n$ are close to $T_I$ (see (7.3.26)) or, which is the same, $\hat{\tau}'_n$ varies very slowly in time. Then, as we did in Section 7.2.2, we may limit ourselves to proving that (7.3.28) is true for $n=Nk$. In other words, letting

$$\hat{\tau}_k \triangleq \hat{\tau}'_{Nk} \tag{7.3.29}$$

we want to show that

$$\hat{\tau}_k \approx \tau, \quad k = 0,1,2,\ldots \tag{7.3.30}$$

# Timing Recovery in Baseband Transmission

To this end consider the times $t_{N(k+1)}$ and $t_{Nk}$. From (7.3.27) and (7.3.29) we have

$$t_{N(k+1)} - t_{Nk} = NT_I + \hat{\tau}_{k+1} - \hat{\tau}_k \quad (7.3.31)$$

Also, recognizing that $t_{N(k+1)} - t_{Nk}$ equals the sum of the $N$ consecutive differences $t_{n+1} - t_n$, $Nk \leq n \leq N(k+1) - 1$, from (7.3.26) we get

$$t_{N(k+1)} - t_{Nk} = NT_I + \frac{\gamma}{N} \sum_{n=Nk}^{N(k+1)-1} e'(n) \quad (7.3.32)$$

On the other hand from (7.3.25) it follows that $e'(n)$ equals $e(k)$ in the above sum. Hence, collecting (7.3.31)-(7.3.32) yields

$$\hat{\tau}_{k+1} = \hat{\tau}_k + \gamma e(k) \quad (7.3.33)$$

Finally, taking $e(k)$ from (7.3.6) produces

$$\hat{\tau}_{k+1} = \hat{\tau}_k + \gamma S(\tau - \hat{\tau}_k) + \gamma N(k) \quad (7.3.34)$$

which is formally identical to equation (7.2.20) for synchronous sampling. This means that $\hat{\tau}_k$ will fluctuate around the stable point $\hat{\tau}_k = \tau$ and, in consequence, the interpolation times will keep close to the ideal times $\{t_n^{(id)}\}$.

Having established a convenient mechanism to adjust interpolation times, we wonder whether it can be directly applied to interpolation control. To see where the problem is, consider equation (7.3.26) from which the sequence $\{t_n\}$ is to be recursively computed. Even though the correction term $\gamma e'(n)/N$ takes values on the order of a fraction of the symbol interval, the interpolation time $t_n$ grows unboundedly as $n$ increases. Thus, computation accuracy will soon deteriorate and the interpolation control will fail. This indicates that some equivalent but more suitable computational method is needed. Reference [6] proposes two alternatives. Here we report on the simplest one, but we point out that the other has certain hardware advantages that might be pre-eminent in some applications.

Take heed that two timing parameters are involved in the fundamental formula (7.3.12): the basepoint index $l_n$ and the fractional interval $\mu_n$. What we look for are equations to compute such parameters. A route toward this goal is now indicated. Substituting (7.3.8) into (7.3.26) and rearranging yields

$$l_{n+1} + \mu_{n+1} = l_n + \mu_n + r_I + \gamma \varepsilon(n) \quad (7.3.35)$$

where we have set

$$r_I \triangleq \frac{T_I}{T_s} \qquad (7.3.36)$$

$$\varepsilon(n) \triangleq \frac{e'(n)}{NT_s} \qquad (7.3.37)$$

Taking the integer parts of both sides in (7.3.35) produces

$$l_{n+1} = l_n + \text{int}[\mu_n + r_I + \gamma\varepsilon(n)] \qquad (7.3.38)$$

whereas, taking the fractional parts results in

$$\mu_{n+1} = \text{frc}[\mu_n + r_I + \gamma\varepsilon(n)] \qquad (7.3.39)$$

These are the desired equations. They derive the next interpolator control parameters from the current parameters and the error signal. In particular, writing (7.3.38) in the form

$$l_{n+1} - l_n = \text{int}[\mu_n + r_I + \gamma\varepsilon(n)] \qquad (7.3.40)$$

we may regard the right-hand side as the number of input samples to be shifted into the interpolator until the next interpolant is computed.

### 7.3.4. Timing Adjustment with Feedforward Schemes

The last topic is the interpolator control in feedforward schemes. To be specific, we assume continuous data transmission and denote by $r \triangleq T/T_s$ the ratio of symbol period to sampling period. Samples $\{y(mT_s)\}$ from the matched filter (see Figure 7.2) are grouped in blocks. The timing estimator produces estimates $\{\hat{\tau}_0, \hat{\tau}_1, \hat{\tau}_2, \ldots\}$, one per block. For convenience we assume that these estimates are limited within the interval $\pm T/2$. Note that there is no contradiction with the limitation to $(0,T)$ in the past sections as we can pass from one hypothesis to the other by adding or subtracting $T/2$ to $\hat{\tau}_n$.

The estimate $\hat{\tau}_n$ is used to compute the strobes in the $n$-th block, say those with indexes $K_n \leq k \leq K_{n+1} - 1$. Bearing in mind that strobes are separated by one symbol interval we write the desired interpolation times for the $n$-th block in the form

$$t_k \triangleq kT + T/2 + \hat{\tau}_n, \qquad K_n \leq k \leq K_{n+1} - 1 \qquad (7.3.41)$$

For each $t_k$ the interpolator computes the corresponding basepoint index $l_k$ and fractional interval $\mu_k$ as follows. Writing $t_k = (l_k + \mu_k)T_s$, comparing with

(7.3.41) and rearranging yields

$$l_k + \mu_k = kr + \frac{r}{2} + \frac{\hat{\tau}_n}{T_s} \qquad (7.3.42)$$

Also, substituting $k \to k+1$ gives

$$l_{k+1} + \mu_{k+1} = (k+1)r + \frac{r}{2} + \frac{\hat{\tau}_n}{T_s} \qquad (7.3.43)$$

Then subtracting (7.3.42) from (7.3.43) results in

$$l_{k+1} + \mu_{k+1} = l_k + \mu_k + r \qquad (7.3.44)$$

from which, taking fractional and integer parts produces

$$\mu_{k+1} = \text{frc}(\mu_k + r) \qquad (7.3.45)$$

and

$$l_{k+1} = l_k + \text{int}(\mu_k + r) \qquad (7.3.46)$$

In the foregoing discussion the timing phase has been taken as a constant. In practice this is not true as, in reality, $\tau$ varies in time due to the clock source instability at the transmitter. In consequence, as $\hat{\tau}_n$ is restricted to $\pm T/2$, it will occasionally exhibit jumps of $T$ seconds in passing from one block to the other. If the phenomenon is not properly recognized, some strobes will be missed or duplicated. To cope with this problem, the estimates $\{\hat{\tau}_n\}$ must be unwrapped. Following Oerder and Meyr [9], the unwrapped estimates $\{\hat{\tau}_n^{(u)}\}$ are derived from $\{\hat{\tau}_n\}$ through the following non-linear equation (see Figure 7.10):

$$\hat{\tau}_n^{(u)} = \hat{\tau}_{n-1}^{(u)} + \alpha \, \text{SAW}\!\left(\hat{\tau}_n - \hat{\tau}_{n-1}^{(u)}\right) \qquad (7.3.47)$$

Here, $\alpha$ is a design parameter and $\text{SAW}(x)$ is a sawtooth function that reduces $x$ to the interval $\pm T/2$. Formally

$$\text{SAW}(x) \triangleq (x + T/2)_{\text{mod } T} - T/2 \qquad (7.3.48)$$

Once the unwrapped estimates are available, timing control is performed as follows. Make the substitution $\hat{\tau}_n \to \hat{\tau}_n^{(u)}$ in (7.3.42)

$$l_k + \mu_k = kr + \frac{r}{2} + \frac{\hat{\tau}_n^{(u)}}{T_s} \qquad (7.3.49)$$

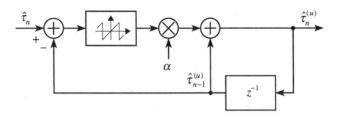

Figure 7.10. Block diagram of the unwrapping algorithm.

From the arguments leading to (7.3.45)-(7.3.46) it is clear that the computation of $l_k$ and $\mu_k$ over a given block proceeds as indicated earlier. On the contrary, when passing from one block to the next a readjustment is needed. In fact, letting $k = K_{n+1} - 1$ be the index for the last strobe in the $n$-th block and $k = K_{n+1}$ the index for the first strobe in the $(n+1)$-th block, we have

$$l_{K_{n+1}-1} + \mu_{K_{n+1}-1} = (K_{n+1} - 1)r + \frac{r}{2} + \frac{\hat{\tau}_n^{(u)}}{T_s} \qquad (7.3.50)$$

$$l_{K_{n+1}} + \mu_{K_{n+1}} = K_{n+1} r + \frac{r}{2} + \frac{\hat{\tau}_{n+1}^{(u)}}{T_s} \qquad (7.3.51)$$

Thus, subtracting (7.3.50) from (7.3.51) yields

$$l_{K_{n+1}} + \mu_{K_{n+1}} = l_{K_{n+1}-1} + \mu_{K_{n+1}-1} + r + \frac{\hat{\tau}_{n+1}^{(u)} - \hat{\tau}_n^{(u)}}{T_s} \qquad (7.3.52)$$

Finally, taking fractional and integer parts of both sides results in the desired continuity equations

$$\mu_{K_{n+1}} = \text{frc}\left( \mu_{K_{n+1}-1} + r + \frac{\hat{\tau}_{n+1}^{(u)} - \hat{\tau}_n^{(u)}}{T_s} \right) \qquad (7.3.53)$$

$$l_{K_{n+1}} = l_{K_{n+1}-1} + \text{int}\left( \mu_{K_{n+1}-1} + r + \frac{\hat{\tau}_{n+1}^{(u)} - \hat{\tau}_n^{(u)}}{T_s} \right) \qquad (7.3.54)$$

**Exercise 7.3.2.** The timing source at the transmitter has a frequency offset $\Delta f = 10^{-5}/T$ from its nominal value. In other words, the timing parameter $\tau$ varies by $10^{-5} T$ seconds at each symbol interval. Suppose that: (*i*) the signal samples are grouped into blocks of $L_0$ symbols; (*ii*) the timing estimate in a block is about the arithmetic mean of $\tau$ at the extremes of the block; (*iii*) the timing error cannot exceed 5% of $T$ for proper detection operation. What is the maximum allowed value of the block length?

*Solution.* It is easily seen that the timing errors at the extremes of a block are $\pm 5 \cdot 10^{-6} T L_0$. As they cannot exceed $5 \cdot 10^{-2} T$, the block length must be less than $10^4$ symbols.

## 7.4. Decision-Directed Timing Error Detectors

### 7.4.1. ML-Based Detectors

The foregoing discussion has concentrated on timing adjustment in clock recovery. As we have seen, either synchronous or non-synchronous sampling is possible with feedback schemes, on condition that some suitable timing error detector (TED) is available. The problem that comes next is to find appropriate TED algorithms. In this sub-section we address this question using ML estimation methods. Other approaches based on *ad hoc* reasoning are discussed later.

The conceptual path we choose may be summarized as follows. We assume that reliable data decisions are available. Correspondingly, we write the log-likelihood function $L(r|\tilde{\tau})$ for the timing parameter and look for its maximum as a function of $\tilde{\tau}$. A necessary condition for a maximum to occur is the vanishing of the derivative of $L(r|\tilde{\tau})$ with respect to $\tilde{\tau}$. It turns out that this derivative, $L'(r|\tilde{\tau})$, is given by the sum of several terms contributed by various segments of the observed waveform, say

$$L'(r|\tilde{\tau}) = \sum_k l'_k(r|\tilde{\tau}) \qquad (7.4.1)$$

Then, a recursive solution to the ML estimation problem can be attempted by exploiting the generic term in the summation (7.4.1) to update the current timing estimate $\hat{\tau}_k$. Formally

$$\hat{\tau}_{k+1} = \hat{\tau}_k + \gamma l'_k(r|\hat{\tau}_k) \qquad (7.4.2)$$

where $\gamma$ is a design parameter. But this equation is identical to (7.2.19) and (7.3.33) which govern the operation of feedback recovery systems, except that the notation $e(k)$ is used in those equations in place of $l'_k(r|\hat{\tau}_k)$. We conclude that the error signal we are looking for is just $l'_k(r|\hat{\tau}_k)$.

To derive the specific form of $e(k)$ let us start from the signal model

$$s(t) = \sum_i c_i g(t - iT - \tau) \qquad (7.4.3)$$

Here, the symbols belong to the alphabet $\{\pm 1, \pm 3, \ldots, \pm(M-1)\}$, $g(t)$ is a real-valued pulse, and $\tau$ is the channel delay in the range $-T/2 \leq \tau < T/2$. Noise is white and Gaussian, with two-sided spectral density $N_0/2$. Letting $\{\hat{c}_i\}$ be the detector decisions and $0 \leq t \leq T_0$ the observation interval, the log-likelihood function turns out to be

$$L(r|\tilde{\tau}) = \int_0^{T_0} r(t)\tilde{s}(t)dt - \frac{1}{2}\int_0^{T_0} \tilde{s}^2(t)dt \tag{7.4.4}$$

where $\tilde{s}(t)$ is the trial signal:

$$\tilde{s}(t) \triangleq \sum_i \hat{c}_i g(t - iT - \tilde{\tau}) \tag{7.4.5}$$

Substituting (7.4.5) into (7.4.4) yields

$$L(r|\tilde{\tau}) = \sum_i \hat{c}_i \int_0^{T_0} r(t)g(t - iT - \tilde{\tau})dt$$

$$-\frac{1}{2}\sum_i \sum_m \hat{c}_i \hat{c}_m \int_0^{T_0} g(t - iT - \tilde{\tau})g(t - mT - \tilde{\tau})dt \tag{7.4.6}$$

and taking the derivative with respect to $\tilde{\tau}$ results in

$$L'(r|\tilde{\tau}) = -\sum_i \hat{c}_i \int_0^{T_0} r(t)g'(t - iT - \tilde{\tau})dt$$

$$+\sum_i \sum_m \hat{c}_i \hat{c}_m \int_0^{T_0} g'(t - iT - \tilde{\tau})g(t - mT - \tilde{\tau})dt \tag{7.4.7}$$

where $g'(t)$ is the derivative of $g(t)$. Next, we make an approximation which is valid when the observation interval is much longer than the duration of $g(t)$. It consists of expanding the limits of the integrals in (7.4.7) to $\pm\infty$ while restricting the summations over $i$ from 0 to $L_0 - 1$, where $L_0$ is the observation length in symbols. Note that the limits on the summation over $m$ remain infinite. With simple manipulations we obtain

$$L'(r|\tilde{\tau}) = \sum_{i=0}^{L_0-1} \hat{c}_i \left[ y'(iT + \tilde{\tau}) - \sum_{m=-\infty}^{\infty} \hat{c}_m h'[(i-m)T] \right] \tag{7.4.8}$$

where $y'(t)$ is the response to $r(t)$ of the derivative matched filter

## Timing Recovery in Baseband Transmission

$dg(-t)/dt = -g'(-t)$, i.e.,

$$y'(t) \triangleq -\int_{-\infty}^{\infty} r(\xi)g'(\xi-t)d\xi \qquad (7.4.9)$$

and $h'(t)$ is the response of this filter to $g(t)$:

$$h'(t) \triangleq g(t) \otimes [-g'(-t)] \qquad (7.4.10)$$

The following remarks are useful:

(i) It is readily checked that $h'(t)$ equals the derivative of $h(t) \triangleq g(t) \otimes g(-t)$. As the latter is even and has a maximum at the origin, we have

$$h'(0) = 0 \qquad (7.4.11)$$

$$h'(t) = -h'(-t) \qquad (7.4.12)$$

(ii) $y'(t)$ may also be seen as the derivative of the matched-filter output

$$y(t) \triangleq \int_{-\infty}^{\infty} r(\xi)g(\xi-t)d\xi \qquad (7.4.13)$$

Returning to (7.4.8) it appears that the derivative of the log-likelihood function is the sum of several terms. In light of the previous remarks, the error signal $e(k)$ is derived by computing the $k$-th term for $\tilde{\tau} = \hat{\tau}_k$, i.e.,

$$e(k) = \hat{c}_k \left[ y'(kT + \hat{\tau}_k) - \sum_{m=-\infty}^{\infty} \hat{c}_m h'[(k-m)T] \right] \qquad (7.4.14)$$

Unfortunately this formula has some drawbacks. First, it requires an infinite sum, which is clearly impossible to realize. Luckily, the sum can be truncated since the coefficients $h'[(k-m)T]$ tend to fade out as the difference $|k-m|$ increases. This suggests the approximation

$$e(k) = \hat{c}_k \left[ y'(kT + \hat{\tau}_k) - \sum_{m=k-D}^{k+D} \hat{c}_m h'[(k-m)T] \right] \qquad (7.4.15)$$

A second problem is that, even in this form, $e(k)$ involves decisions $\{\hat{c}_{k+1}, \hat{c}_{k+2}, \dots \hat{c}_{k+D}\}$ which extend into the future. A simple remedy is to delay the right-hand side of (7.4.15) by $D$ steps. The adverse effects of delays on the

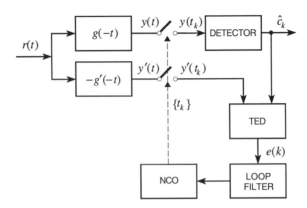

Figure 7.11. Timing recovery scheme with synchronous sampling.

loop performance are negligible if $D$ is limited to a small fraction of $1/(B_L T)$ ($B_L$ is the loop bandwidth). In practice, $D$ values of a few units will not have significant consequences since $1/(B_L T)$ is on the order of 100 or larger.

The error signal (7.4.15) has been brought out by Gardner in [10] and Bergmans and Wong-Lam in [11] and will be referred to as the *true* ML error detector. This will serve to distinguish it from the popular TED (see [10], [12]-[13]) which results from dropping the sum in (7.4.15), i.e.,

$$e(k) = \hat{c}_k y'(kT + \hat{\tau}_k) \qquad (7.4.16)$$

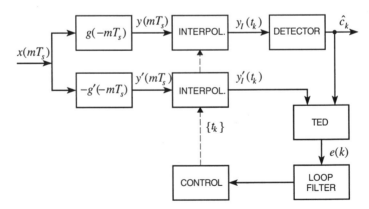

Figure 7.12. Timing recovery scheme with non-synchronous sampling.

Bearing in mind that $h'(0) = 0$, this detector may be seen as a particular case of (7.4.15) for $D=0$. The reason for setting $D=0$ is that the sum in (7.4.15) accounts for the second integral in (7.4.4). If this integral were independent of $\tilde{\tau}$, its derivative would vanish and so would the sum. Actually the integral does depend on $\tilde{\tau}$ but some people argue that the dependence is weak and can be ignored. The consequences of setting $D=0$ in (7.4.15) will be explored later. For now we simply observe that, whichever the value of $D$, the synchronizer takes the form indicated in Figure 7.11 for synchronous sampling, and in Figure 7.12 for non-synchronous sampling. The two schemes are equivalent when good interpolators are used.

### 7.4.2. S-Curve

In the next two sections we analyze the performance of timing recovery schemes endowed with the ML-based TED in (7.4.15). For simplicity we concentrate on the scheme in Figure 7.11, which corresponds to synchronous sampling. The discussion with non-synchronous sampling is more complex and will not be addressed in this book. The interested reader is referred to Bucket and Moeneclaey [14] for a full account of the subject. In that paper it is shown that degradations due to interpolation imperfections are negligible even with very simple interpolators, provided that the sampling rate is not close to multiples of the symbol rate.

As a first step in our analysis we concentrate on the S-curve. Recall that this is the expectation of $e(k)$ for $\hat{\tau}_k = \hat{\tau} =$ constant. Calculations are performed assuming $\hat{c}_k = c_k$, which implies that $\hat{\tau}$ is close to $\tau$ (otherwise intersymbol interference would cause decision errors).

Keeping in mind that

$$r(t) = \sum_i c_i g(t - iT - \tau) + w(t) \qquad (7.4.17)$$

from (7.4.9) we obtain

$$y'(t) = \sum_i c_i h'(t - iT - \tau) + n'(t) \qquad (7.4.18)$$

where

$$n'(t) \triangleq w(t) \otimes [-g'(-t)] \qquad (7.4.19)$$

is the response of the derivative matched filter to $w(t)$. Hence, substituting into (7.4.15) for $\hat{\tau}_k = \hat{\tau}$ and $\hat{c}_k = c_k$ and rearranging yields

$$e(k) = \sum_i c_k c_i h'[(k-i)T + \hat{\tau} - \tau]$$
$$- \sum_{m=k-D}^{k+D} c_k c_m h'[(k-m)T] + c_k n'(kT + \hat{\tau}) \qquad (7.4.20)$$

To proceed further, assume zero-mean and independent symbols so that

$$E\{c_k c_{k+m}\} = \begin{cases} C_2 & m = 0 \\ 0 & m \neq 0 \end{cases} \qquad (7.4.21)$$

Then, taking the expectation of (7.4.20) and bearing in mind (7.4.11)-(7.4.12) produces the S-curve

$$S(\delta) = -C_2 h'(\delta) \qquad (7.4.22)$$

where $\delta \triangleq \tau - \hat{\tau}$. As $h(t)$ has a maximum at the origin, it follows that $-h'(\delta)$ has a null with positive slope at $\delta = 0$. Thus, $\delta = 0$ is a stable operating point.

Timing errors are small in the steady state and the approximation $h'(\delta) \approx h''(0)\delta$ holds true. Correspondingly, (7.4.22) becomes

$$S(\delta) \approx A\delta \qquad (7.4.23)$$

with

$$A \triangleq -C_2 h''(0) \qquad (7.4.24)$$

For example, in some applications $h(t)$ is Nyquist and has the form

$$h(t) = \frac{\sin(\pi t/T)}{\pi t/T} \frac{\cos(\alpha \pi t/T)}{1 - (2\alpha t/T)^2} \qquad (7.4.25)$$

where $\alpha$ is the rolloff factor. Correspondingly, it is found that

$$h''(0) = -\frac{\pi^2}{T^2}\left[\frac{1}{3} + \alpha^2(1 - \frac{8}{\pi^2})\right] \qquad (7.4.26)$$

Figure 7.13 shows S-curves obtained by simulation. The alphabet is binary and $h(t)$ is Nyquist with 50% rolloff. The solid line labelled "theory" corresponds to equation (7.4.22). It appears that simulations agree well with theory for small timing errors. As the latter increase, however, the assumption that $\hat{c}_k = c_k$ is no longer valid and the theory fails.

# Timing Recovery in Baseband Transmission

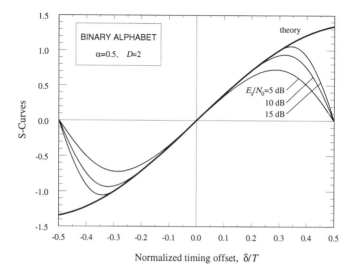

Figure 7.13. S-curves for the ML-based TED.

**Exercise 7.4.1.** In the foregoing discussion $h(t)$ has been assumed as the convolution $g(t) \otimes g(-t)$, which implies $h(t)=h(-t)$. Correspondingly, it has been shown that the S-curve has a null at the origin. Because of channel distortions, however, the pulses from the matched filter may not be even. How does this affect the S-curve?

*Solution.* Following the arguments leading to (7.4.22) we get

$$S(\delta) = C_2[h'(-\delta) - h'(0)] \qquad (7.4.27)$$

from which it is seen that the null in the S-curve still occurs at the origin, despite the pulse distortion.

## 7.4.3. Tracking Performance

The tracking performance of feedback synchronizers can be assessed with the methods of Section 3.5.5. The major results from that section are now summarized and are applied to the case at hand. Start with the loop equation

$$\hat{\tau}_{k+1} = \hat{\tau}_k + \gamma e(k) \qquad (7.4.28)$$

where

$$e(k) = A(\tau - \hat{\tau}_k) + N(k) \qquad (7.4.29)$$

In writing (7.4.29) we have assumed that $\hat{\tau}_k \approx \tau$ so that the approximation (7.4.23) holds true. The variance of the errors $\tau - \hat{\tau}_k$ can be expressed either in terms of the autocorrelation $R_N(m) = E\{N(k+m)N(k)\}$ or as a function of the power spectral density of $N(k)$

$$S_N(f) = T \sum_{m=-\infty}^{\infty} R_N(m) e^{-j2\pi mfT} \qquad (7.4.30)$$

Denoting by $\sigma^2$ the variance of the normalized errors $(\tau - \hat{\tau}_k)/T$, the former procedure leads to

$$\sigma^2 = \frac{2B_L T}{A^2} \cdot \frac{1}{T^2} \sum_{m=-\infty}^{\infty} R_N(m)(1-\gamma A)^{|m|} \qquad (7.4.31)$$

while the latter produces

$$\sigma^2 = \frac{1}{T^2} \int_{-1/2T}^{1/2T} S_N(f) |H(f)|^2 df \qquad (7.4.32)$$

where $B_L$ is the noise equivalent bandwidth of the loop

$$B_L T = \frac{\gamma A}{2(2 - \gamma A)} \qquad (7.4.33)$$

and $H(f)$ is the loop transfer function

$$H(f) = -\frac{\gamma}{e^{j2\pi fT} - (1 - \gamma A)} \qquad (7.4.34)$$

In some cases the spectral density $S_N(f)$ is flat over the frequency interval where $H(f)$ takes significant values and (7.4.32) reduces to

$$\sigma^2 = 2B_L \frac{S_N(0)}{A^2 T^2} \qquad (7.4.35)$$

The above results are now applied to synchronizers endowed with the error detector (7.4.15). For this purpose we observe that $N(k)$ in (7.4.29) coincides with $e(k)$ for $\hat{\tau}_k = \tau$. Thus, setting $\hat{\tau} = \tau$ in (7.4.20) and rearranging yields

$$N(k) = N_{SN}(k) + N_{TN}(k) \qquad (7.4.36)$$

# Timing Recovery in Baseband Transmission

with

$$N_{SN}(k) \triangleq \sum_{|i| \geq D+1} c_k c_{k+i} h'(iT) \qquad (7.4.37)$$

and

$$N_{TN}(k) \triangleq c_k n'(kT + \tau) \qquad (7.4.38)$$

The physical interpretation of these equations is clear: $N_{SN}(k)$ represents self noise while $N_{TN}(k)$ is contributed by thermal noise. Since the coefficient $h'[(k-m)T]$ decreases as $|k-m|$ increases, self noise becomes negligible when $D$ grows large. As the decrease of $h'[(k-m)T]$ is fast with a relatively large signal bandwidth, even a small $D$ is sufficient to make self noise negligible when $\alpha$ is close to unity.

The computation of the error variance $\sigma^2$ in (7.4.31) requires knowledge of the autocorrelation $R_N(m)$ which may be computed as follows. Since $N_{SN}(k)$ and $N_{TN}(k)$ are uncorrelated, we have

$$R_N(m) = R_{SN}(m) + R_{TN}(m) \qquad (7.4.39)$$

Let us first concentrate on $R_{TN}(m)$. From (7.4.38) we get (recall that the symbols are independent)

$$R_{TN}(m) = \begin{cases} C_2 \, \mathrm{E}\{n'^2(t)\} & m = 0 \\ 0 & m \neq 0 \end{cases} \qquad (7.4.40)$$

To evaluate $\mathrm{E}\{n'^2(t)\}$ we need the Fourier transform of $g'(-t)$. This is found to be $-j2\pi f G^*(f)$, where $G(f)$ is the Fourier transform of $g(t)$. Then, the variance of $n'(t)$ is given by [15, Ch. 10]

$$\mathrm{E}\{n'^2(t)\} = \frac{N_0}{2} \cdot 4\pi^2 \int_{-\infty}^{\infty} f^2 |G(f)|^2 \, df \qquad (7.4.41)$$

Next we turn our attention to $R_{SN}(m)$. After some manipulations from (7.4.37) we get

$$R_{SN}(m) = \sum_{|l_1| \geq D+1} \sum_{|l_2| \geq D+1} \mathrm{E}\{c_0 c_{l_1} c_m c_{m+l_2}\} h'(l_1 T) h'(l_2 T) \qquad (7.4.42)$$

For $|l_1| \geq D+1$ and $|l_2| \geq D+1$ it can be shown that $\mathrm{E}\{c_0 c_{l_1} c_m c_{m+l_2}\}$ is zero except in the following cases where it equals $C_2^2$:

$$m = 0 \text{ and } l_1 = l_2 \qquad (7.4.43)$$

$$|m| \geq D+1 \text{ and } l_1 = -l_2 = m \qquad (7.4.44)$$

Thus, equation (7.4.42) becomes

$$R_{SN}(m) = \begin{cases} C_2^2 \sum_{|l| \geq D+1} h'^2(lT) & m = 0 \\ -C_2^2 h'^2(mT) & |m| \geq D+1 \end{cases} \qquad (7.4.45)$$

At this point, collecting (7.4.39)-(7.4.40) and (7.4.45) yields the autocorrelation $R_N(m)$ and inserting the result into (7.4.31) produces the error variance. Exercise 7.4.3 shows that the synchronizer performance achieves the modified Cramer-Rao bound when $D$ is sufficiently large to make the self noise negligible.

Figures 7.14-7.15 illustrate simulation results obtained with a binary alphabet ($c_i = \pm 1$) and a loop noise bandwidth of $5 \cdot 10^{-3}/T$. In both figures $D$

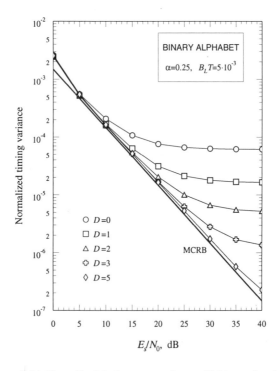

Figure 7.14. Normalized timing error variance with binary signaling.

# Timing Recovery in Baseband Transmission

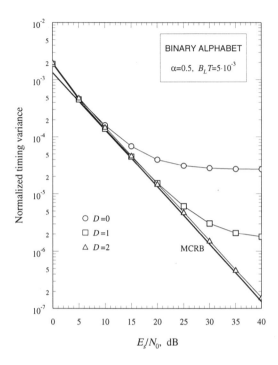

Figure 7.15. Normalized timing error variance with binary signaling.

is taken as a parameter and the bottom curve represents the MCRB. The floor in the curves is a manifestation of self noise; it says that the synchronizer performance cannot be further improved by increasing $E_s/N_0$. As is seen, the floor level diminishes as $D$ increases. The MCRB is practically attained for $D=5$ in Figure 7.14 and for $D=2$ in Figure 7.15. The larger $D$ which is needed in Figure 7.14 is due to the smaller rolloff being used. In fact a small $\alpha$ produces a longer channel response and, correspondingly, a relatively higher self noise level for a given $D$ (see (7.4.45)). Both figures point out that degradations incurred with $D=0$ are limited for $E_s/N_0$ less than 10 dB.

**Exercise 7.4.2.** In the previous section we have assumed that the S-curve has a null at the origin. How should the discussion be rearranged when the null is shifted from the origin?

*Solution.* Suppose that the null is shifted to the right by $\Delta\tau$ so that the timing estimates fluctuate around $\tau_0 \triangleq \tau - \Delta\tau$. Then, equation (7.4.29) must replaced by

$$e(k) = A(\tau_0 - \hat{\tau}_k) + N(k) \tag{7.4.46}$$

where $A$ is the slope of the S-curve at the null (not at the origin). It follows from (7.4.46) that $N(k)$ coincides with $e(k)$ for $\hat{\tau}_k = \tau_0$. Hence, setting $\hat{\tau} = \tau_0$ in (7.4.20) and rearranging yields

$$N(k) = N_{SN}(k) + N_{TN}(k) \qquad (7.4.47)$$

with

$$N_{SN}(k) \triangleq \sum_i c_k c_i h'[(k-i)T + \tau_0 - \tau] - \sum_{m=k-D}^{k+D} c_k c_m h'[(k-m)T] \qquad (7.4.48)$$

and

$$N_{TN}(k) \triangleq c_k n'(kT + \tau_0) \qquad (7.4.49)$$

The remainder of the discussion remains unchanged. The autocorrelations $R_{SN}(m)$ and $R_{TN}(m)$ have to be computed and their sum has to be inserted into (7.4.31).

**Exercise 7.4.3.** The MCRB for the timing error variance is given by

$$\frac{1}{T^2} \times MCRB(\tau) = \frac{1}{8\pi^2 \xi L_0} \frac{1}{E_s/N_0} \qquad (7.4.50)$$

where $L_0$ is the observation interval in symbol periods, $\xi$ is a parameter defined by

$$\xi \triangleq T^2 \frac{\int_{-\infty}^{\infty} f^2 |G(f)|^2 df}{\int_{-\infty}^{\infty} |G(f)|^2 df} \qquad (7.4.51)$$

and $G(f)$ is the Fourier transform of $g(t)$. Show that an ML-based synchronizer attains this bound when $h(t) = g(t) \otimes g(-t)$ and the parameter $D$ is sufficiently large to make self noise negligible.

*Solution.* When self noise is negligible, the autocorrelation $R_N(m)$ coincides with $R_{TN}(m)$ in (7.4.40). Then, substituting into (7.4.31) and using (7.4.24), (7.4.40)-(7.4.41) yields

$$\sigma^2 = \frac{N_0 B_L}{C_2 h''^2(0)T} 4\pi^2 \int_{-\infty}^{\infty} f^2 |G(f)|^2 df \qquad (7.4.52)$$

### Timing Recovery in Baseband Transmission

Now, bearing in mind that $h(t) = g(t) \otimes g(-t)$, it is easily seen that $h''(t)$ has Fourier transform $-4\pi^2 f^2 |G(f)|^2$ and, in consequence, $h''(0)$ may be written as

$$h''(0) = -4\pi^2 \int_{-\infty}^{\infty} f^2 |G(f)|^2 df \qquad (7.4.53)$$

Also, in Appendix 2.A it is shown that the signal energy is expressed by

$$E_s = C_2 \int_{-\infty}^{\infty} |G(f)|^2 df \qquad (7.4.54)$$

Then, substituting into (7.4.52) produces

$$\sigma^2 = \frac{B_L T}{4\pi^2 \xi} \frac{1}{E_s/N_0} \qquad (7.4.55)$$

This coincides with the MCRB, as is recognized using the relation

$$B_L T = \frac{1}{2L_0} \qquad (7.4.56)$$

between loop bandwidth and *equivalent* observation length.

### 7.4.4. Approximate-Derivative Method

The synchronizer in Figure 7.11 is rather complex to implement because of the presence of the derivative matched filter $-g'(-t)$ whose computational load is comparable with that of the matched filter $g(-t)$. While the latter cannot be avoided as it is indispensable for detection, we wonder whether the derivative matched filter can be obviated at the cost, perhaps, of some performance loss. This problem is discussed in [10] where some alternative solutions are offered. The simplest one consists of approximating the derivative of $y(t)$ in (7.4.16) with a finite difference as follows

$$y'(kT + \hat{\tau}_k) \approx \frac{1}{T}\left[y(kT + T/2 + \hat{\tau}_{k+1/2}) - y(kT - T/2 + \hat{\tau}_{k-1/2})\right] \qquad (7.4.57)$$

where $\hat{\tau}_{k\pm 1/2}$ are the timing estimates at $kT \pm T/2$. In practice, equation (7.4.57) must be retouched since the timing estimates are updated at multiples of $T$ and the only available quantities are $\{\hat{\tau}_k\}$. In consequence, $\hat{\tau}_{k+1/2}$ is replaced by $\hat{\tau}_k$ (the latest estimate), $\hat{\tau}_{k-1/2}$ by $\hat{\tau}_{k-1}$, and equation (7.4.16) becomes

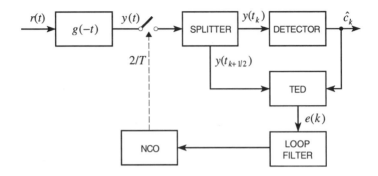

Figure 7.16. Block diagram of the early-late synchronizer.

$$e(k) = \hat{c}_k\left[y(kT + T/2 + \hat{\tau}_k) - y(kT - T/2 + \hat{\tau}_{k-1})\right] \qquad (7.4.58)$$

where the factor $1/T$ has been dropped for simplicity as it can be absorbed into the step size. Algorithm (7.4.58) is reminiscent of that employed in the *early-late* gate synchronizer discussed by Lindsey and Simon in [16, Ch. 9] and will be called *early-late detector* (ELD) in the sequel.

The block diagram of a synchronizer endowed with an ELD is illustrated in Figure 7.16. The splitter divides the sequence from the matched filter into two streams $\{y(t_k)\}$ and $\{y(t_{k+1/2})\}$, corresponding to the sampling times $t_{k+1/2} \triangleq kT + T/2 + \hat{\tau}_k$ and $t_k \triangleq kT + \hat{\tau}_k$. The tracking performance of this synchronizer is discussed later.

**Exercise 7.4.4.** Compute the S-curve for the ELD.
*Solution.* The matched-filter output is

$$y(t) = \sum_i c_i h(t - iT - \tau) + n(t) \qquad (7.4.59)$$

where $n(t) \triangleq w(t) \otimes g(-t)$. The S-curve is the expectation of (7.4.58) for $\hat{\tau}_k = \hat{\tau}_{k-1} = \hat{\tau}$. Substituting (7.4.59) into (7.4.58) and performing straightforward calculations (with the approximation $\hat{c}_k \approx c_k$) yields the desired result

$$S(\delta) = C_2\left[h(T/2 - \delta) - h(-T/2 - \delta)\right] \qquad (7.4.60)$$

with $\delta \triangleq \tau - \hat{\tau}$. If $h(t)$ is an even function, it is seen from (7.4.60) that $S(\delta)$ crosses the horizontal axis at the origin with a positive slope (provided that $h'(T/2) < 0$). In these conditions $\delta = 0$ is a stable tracking point.

### 7.4.5. Other Timing Error Detectors

Other simple timing detectors are available in the literature. Here we report on two algorithms that have been brought out with *ad hoc* reasoning. The first is the *zero-crossing detector* (ZCD) proposed by Gardner in [10, Ch. 10]

$$e(k) = (\hat{c}_{k-1} - \hat{c}_k) y(kT - T/2 + \hat{\tau}_{k-1}) \quad (7.4.61)$$

The other is due to Mueller and Mueller [17] and is expressed by

$$e(k) = \hat{c}_{k-1} y(kT + \hat{\tau}_k) - \hat{c}_k y[(k-1)T + \hat{\tau}_{k-1}] \quad (7.4.62)$$

Note that the Mueller and Mueller detector (MMD) operates on $T$-spaced samples, as opposed to ZCD which needs $T/2$ spacing.

ZCD and MMD have interesting features that are now discussed. Consider first the MMD and assume: (*i*) small tracking errors ($\hat{\tau}_k \approx \hat{\tau}_{k-1} \approx \tau$); (*ii*) correct decisions ($\hat{c}_k \approx c_k$); (*iii*) negligible thermal noise. Then, substituting the signal component of (7.4.59) into (7.4.62) yields the detector's self noise

$$e_{SN}(k) = \sum_i c_{k-1} c_i h[(k-i)T] - \sum_i c_k c_i h[(k-1-i)T] \quad (7.4.63)$$

We wonder whether $e_{SN}(k)$ can be made zero by a proper choice of $h(t)$. By inspection it is seen that $e_{SN}(k)$ vanishes when $h(t)$ is Nyquist, i.e., $h(0)=1$ and $h(kT)=0$ for $k \neq 0$. Thus, MMD has no self noise with Nyquist pulses. Of course, this does not necessarily mean that the detector has particularly good tracking performance. Thermal noise might be large. As we shall see, this is the case with large rolloff factors.

Next, consider ZCD. With the same arguments it is found that self noise is expressed by

$$e_{SN}(k) = (c_{k-1} - c_k) \sum_i c_i h[(k-i)T - T/2] \quad (7.4.64)$$

Assume again that $h(t)$ is Nyquist and, in particular, it has rolloff equal to unity. From (7.4.25) we have

$$h(kT/2) = \begin{cases} 1 & k = 0 \\ 1/2 & k = \pm 1 \\ 0 & \text{elsewhere} \end{cases} \quad (7.4.65)$$

Substituting into (7.4.64) yields

$$e_{SN}(k) = \frac{1}{2}(c_{k-1}^2 - c_k^2) \quad (7.4.66)$$

which says that $e_{SN}(k)$ is identically zero with a binary alphabet ($c_k=\pm1$). In summary, ZCD has no self noise with binary symbols and Nyquist pulses with 100% rolloff.

S-curves obtained by simulation are shown in Figures 7.17-7.18 for ZCD and MMD. The channel is Nyquist with 50% rolloff and the symbols belong to the $M$-ary alphabet $\{\pm1,\pm3,\ldots,\pm(M-1)\}$. The S-curve for the MMD with $M=4$ is of some concern as it exhibits a null for timing errors of about $\pm T/3$. This null may cause hangup effects if the system starts with errors on that order of magnitude. The situation may be even worse at higher SNR as is illustrated in Figure 7.19. For $E_s/N_0=23$ dB (which is realistic on coaxial cables) it is seen the S-curve with $M=4$ exhibits spurious lock points at about $\pm 0.37T$. A false lock with such large timing errors would have disabling effects on performance.

Tracking performance comparisons between ELD, ZCD and MMD are made in Figures 7.20-7.21 for Nyquist pulses and rolloff values of 0.75 and 0.25. It is seen that, as the signal bandwidth decreases, the performance of ELD and ZCD worsens while that of MMD improves. In particular it is confirmed that MMD has no self noise (there is no floor in the variance curve as $E_s/N_0$ increases) and ZCD has little self noise for $\alpha$ close to unity. Finally, comparing Figures 7.14 and 7.21 it appears that the approximate ML-detector with $D=0$ has almost the same performance as the ELD, which means that the approximation (7.4.57) to the derivative matched-filter output is effective.

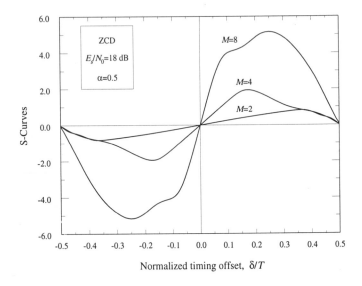

Figure 7.17. S-curves for ZCD.

# Timing Recovery in Baseband Transmission

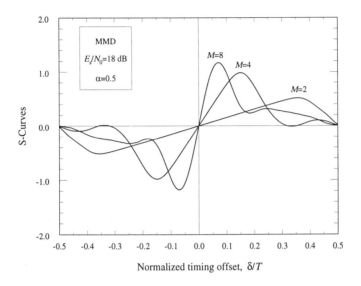

Figure 7.18. S-curves for MMD and SNR=18 dB.

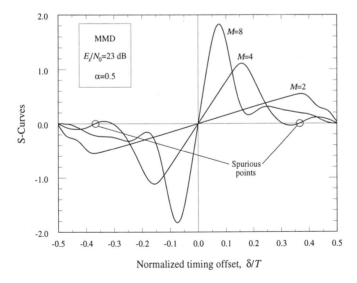

Figure 7.19. S-curves for MMD and SNR=23 dB.

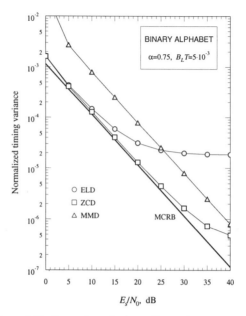

Figure 7.20. Comparisons between ELD, ZCD and MMD.

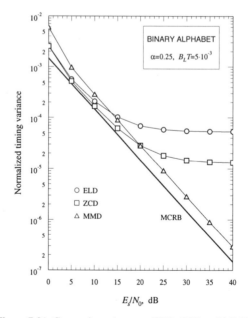

Figure 7.21. Comparisons between ELD, ZCD and MMD.

# Timing Recovery in Baseband Transmission

**Exercise 7.4.5.** Show that ZCD and ELD have the same S-curve about the origin.

*Solution.* From (7.4.59) we have

$$y(kT - T/2 + \hat{\tau}) = \sum_i c_i h[(k-i)T - T/2 + \hat{\tau} - \tau)] + n(kT - T/2 + \hat{\tau}) \quad (7.4.67)$$

Substituting into (7.4.61), letting $\hat{\tau}_{k-1} = \hat{\tau}, \hat{c}_k = c_k, \hat{c}_{k-1} = c_{k-1}$ and taking the expectation yields (7.4.60), which proves the statement.

**Exercise 7.4.6.** Compute the S-curve for MMD.

*Solution.* Letting $\hat{\tau}_k = \hat{\tau}_{k-1} = \hat{\tau}, \hat{c}_k = c_k, \hat{c}_{k-1} = c_{k-1}$ in (7.4.62) and taking the expectation produces

$$S(\delta) = C_2[h(T - \delta) - h(-T - \delta)] \quad (7.4.68)$$

with $\delta \triangleq \tau - \hat{\tau}$. As $h(t) = g(t) \otimes g(-t)$ is an even function, it is recognized that the S-curve vanishes for $\delta = 0$. Also, it can be shown that $S(\delta)$ has a positive slope at the origin when $h(t)$ is Nyquist. In these conditions $\delta = 0$ is a stable tracking point.

**Exercise 7.4.7.** In deriving (7.4.68) we have assumed that the received waveform $r(t)$ is filtered into a matched filter $g(-t)$. Suppose instead that $r(t)$ is passed through some generic filter $g_R(t)$. How does this affect the S-curve of the MMD?

*Solution.* Filtering $r(t)$ into $g_R(t)$ produces

$$y(t) = \sum_i c_i h_R(t - iT - \tau) + n_R(t) \quad (7.4.69)$$

with $h_R(t) \triangleq g(t) \otimes g_R(t)$ and $n_R(t) \triangleq w(t) \otimes g_R(t)$. Next, following the arguments leading to (7.4.68) yields

$$S(\delta) = C_2[h_R(T - \delta) - h_R(-T - \delta)] \quad (7.4.70)$$

which is formally identical to (7.4.68), except that $h(t)$ is replaced by $h_R(t)$. As the latter need not be an even function, the S-curve need not vanish at the origin. This means that the timing estimates may be biased. The same conclusion holds true with the other error detectors, ELD and ZCD.

**Exercise 7.4.8.** Compute the timing error variance for a synchronizer endowed with an MMD, assuming $h(t)$ pulses with a raised-cosine-rolloff Fourier transform $H(f)$.

*Solution.* Since $h(t)$ is Nyquist, the detector has no self noise. The thermal noise component is derived from (7.4.62) by replacing $y(t)$ with $n(t)$. Thus, letting $\hat{\tau}_k = \hat{\tau}_{k-1} = \tau$, $\hat{c}_k = c_k$, $\hat{c}_{k-1} = c_{k-1}$ (which is a valid assumption in the steady state) we have

$$e_{TN}(k) = c_{k-1} n(kT + \tau) - c_k n[(k-1)T + \tau] \quad (7.4.71)$$

Next, bearing in mind that $H(f) = |G(f)|^2$, it is readily shown that $\{n(kT + \tau)\}$ is a white sequence with

$$E\{n^2(kT + \tau)\} = \frac{N_0}{2} \quad (7.4.72)$$

It follows from (7.4.71) that the spectral density of $e_{TN}(k)$ is given by

$$S_{TN}(f) = TC_2 N_0 \quad (7.4.73)$$

Correspondingly, from (7.4.35) we obtain

$$\sigma^2 = 2B_L \frac{C_2 N_0}{A^2 T} \quad (7.4.74)$$

Let us compute the parameters $C_2$ and $A$. As $|G(f)|^2 = H(f)$, from (7.4.54) we get $C_2 = E_s$. The slope of the S-curve at the origin is derived from (7.4.68). Bearing in mind the expression of $h(t)$ in (7.4.25) we get

$$A = \frac{2C_2}{T} \frac{\cos(\alpha\pi)}{1 - 4\alpha^2} \quad (7.4.75)$$

Substituting into (7.4.74) and rearranging yields the result sought

$$\sigma^2 = \frac{B_L T}{4\pi^2 \eta} \frac{1}{E_s / N_0} \quad (7.4.76)$$

with

$$\eta \triangleq \frac{1}{2\pi^2} \left[ \frac{\cos(\alpha\pi)}{1 - 4\alpha^2} \right]^2 \quad (7.4.77)$$

Expression (7.4.76) has the same form as the MCRB in (7.4.55) except that the parameter $\xi$ in the bound is replaced by $\eta$. Through lengthy calculations it can be shown that

$$\xi = \frac{1}{4}\left[\frac{1}{3} + \alpha^2\left(1 - \frac{8}{\pi^2}\right)\right] \quad (7.4.78)$$

It is interesting to compare $\sigma^2$ with the MCRB. For a zero rolloff it is found that $\eta = 1/(2\pi^2)$ and $\xi = 1/12$. This corresponds to a loss from the bound of only

2.1 dB. For $\alpha=1$ we get $\eta = 1/(18\pi^2)$, $\xi=1/3-2/\pi^2$ and the loss grows to 13.6 dB. This confirms that MMD is better suited for small rolloff factors.

## 7.5. Non-Data-Aided Detectors

### 7.5.1. ML-Based Detector

In the previous section several timing detectors have been illustrated whose operation requires reliable symbol decisions. In this section we investigate non-data-aided detectors which, in principle, should do better when decisions are not available or not reliable. We first concentrate on an ML-based algorithm and then discuss an *ad hoc* approach.

Start with the likelihood function

$$\Lambda(r|\tilde{\tau},\tilde{c}) = \exp\left\{\frac{2}{N_0}\int_0^{T_0} r(t)\tilde{s}(t)dt - \frac{1}{N_0}\int_0^{T_0} \tilde{s}^2(t)dt\right\} \quad (7.5.1)$$

where $\tilde{c} \triangleq \{\tilde{c}_0, \tilde{c}_1, \tilde{c}_2, \ldots\}$ represents the generic data sequence and $\tilde{s}(t)$ is the trial signal

$$\tilde{s}(t) \triangleq \sum_i \tilde{c}_i g(t - iT - \tilde{\tau}) \quad (7.5.2)$$

Our goal is to estimate the timing epoch without bothering about data. To this purpose we should first average $\Lambda(r|\tilde{\tau},\tilde{c})$ with respect to $\tilde{c}$ and then look for the maximum of the marginal likelihood function $\Lambda(r|\tilde{\tau})$. Unfortunately, as we have seen other times in this book, this route leads to insurmountable difficulties. Thus, instead of pursuing a fruitless path we resort to the following approximations.

First, we drop the second integral in (7.5.1). From the discussion in the previous section we expect that this will make the self noise level grow but we have no choice. We shall see later how to fight self noise with other methods. Second, we assume the SNR sufficiently low so that the Taylor series expansion of the exponential in (7.5.1) can be truncated at the quadratic term, i.e.,

$$\Lambda(r|\tilde{\tau},\tilde{c}) \approx 1 + \frac{2}{N_0}\int_0^{T_0} r(t)\tilde{s}(t)dt + \frac{2}{N_0^2}\left[\int_0^{T_0} r(t)\tilde{s}(t)dt\right]^2 \quad (7.5.3)$$

Third, making use of the definition (7.5.2) we write

$$\int_0^{T_0} r(t)\tilde{s}(t)dt = \sum_i \tilde{c}_i \int_0^{T_0} r(t)g(t - iT - \tilde{\tau})dt \quad (7.5.4)$$

Then we expand the limits of the integrals in (7.5.4) to $\pm\infty$ and restrict the summation over $i$ from 0 to $L_0 - 1$, with $L_0 \triangleq T_0/T$. In other words we set

$$\int_0^{T_0} r(t)\tilde{s}(t)dt \approx \sum_{i=0}^{L_0-1} \tilde{c}_i y(iT + \tilde{\tau}) \qquad (7.5.5)$$

where

$$y(t) \triangleq \int_{-\infty}^{\infty} r(\xi)g(\xi - t)d\xi \qquad (7.5.6)$$

is the response of the matched filter to $r(t)$. Fourth, we substitute (7.5.5) into (7.5.3) and perform the expectation with respect to the symbols. Dropping some irrelevant terms we obtain [10], [13]

$$\Lambda(r|\tilde{\tau}) \approx \sum_{i=0}^{L_0-1} y^2(iT + \tilde{\tau}) \qquad (7.5.7)$$

This is the desired approximation to the marginal likelihood function $\Lambda(r|\tilde{\tau})$. It is a quadratic expression involving the samples from the matched filter. This formula says that the optimum timing estimate corresponds to the sampling epoch $\tilde{\tau}$ that maximizes the energy of the sequence $\{y(iT + \tilde{\tau})\}$. One way to pursue energy maximization is to employ a search algorithm, as indicated in Section 3.5.2. Alternately, the zero of the derivative of the likelihood function can be looked for. From (7.5.7) we see that the derivative has the form

$$\Lambda'(r|\tilde{\tau}) \approx 2 \sum_{i=0}^{L_0-1} y(iT + \tilde{\tau})y'(iT + \tilde{\tau}) \qquad (7.5.8)$$

Thus, the generic term in the sum may be computed for $\tilde{\tau} = \hat{\tau}_k$ and used as an error signal to update the timing estimate. This leads to

$$\hat{\tau}_{k+1} = \hat{\tau}_k + \gamma e(k) \qquad (7.5.9)$$

with

$$e(k) \triangleq y(kT + \hat{\tau}_k)y'(kT + \hat{\tau}_k) \qquad (7.5.10)$$

Comparing (7.5.10) with the decision-directed algorithm (7.4.16) it is seen that the $k$-th decision $\hat{c}_k$ has been replaced by the corresponding matched filter sample $y(kT + \hat{\tau}_k)$. The complexity of the two detectors is about the same, however. In particular they need two separate filters, the matched filter and the derivative matched filter. As in Section 7.4.4, a possible simplification is to approximate the derivative sample $y'(kT + \hat{\tau}_k)$ by a finite difference. Following that route the detector becomes

$$e(k) = y(kT + \hat{\tau}_k)[y(kT + T/2 + \hat{\tau}_k) - y(kT - T/2 + \hat{\tau}_{k-1})] \quad (7.5.11)$$

Note that two samples per symbol are needed in this detector. The similarity of (7.5.11) with the early-late detector (7.4.58) is apparent. The former is just the non-data-aided version of the latter and, accordingly, will be denoted as NDA-ELD.

### 7.5.2. The Gardner Detector

An alternative to the NDA-ELD has been devised by Gardner [18] following a heuristic reasoning. Although the arguments adopted in [18] are tailored for modulated carriers, they can be adapted to baseband transmission. Essentially, the Gardner detector (GAD) is obtained from the data-aided ZCD in (7.4.61) by replacing the decisions $\hat{c}_{k-1}$ and $\hat{c}_k$ by the corresponding samples of $y(t)$, i.e.,

$$e(k) = \{y[(k-1)T + \hat{\tau}_{k-1}] - y(kT + \hat{\tau}_k)\} y(kT - T/2 + \hat{\tau}_{k-1}) \quad (7.5.12)$$

Despite the strong similarity with (7.5.11), the reader is cautioned against concluding that the Gardner detector and NDA-ELD are equivalent. As a matter of fact they have the same S-curve but the former has a lower self noise level. In the remainder of this section we compute the S-curve for (7.5.12). The proof that this curve coincides with the S-curve for NDA-ELD is left as an exercise for the reader. Performance issues are discussed later.

Letting $\hat{\tau}_k = \hat{\tau}_{k-1} = \hat{\tau}$ in (7.5.12) and performing simple calculations yields

$$S(\delta) = C_2 \sum_i h(iT - T/2 - \delta)h(iT - T - \delta)$$

$$-C_2 \sum_i h(iT - \delta)h(iT - T/2 - \delta) \quad (7.5.13)$$

Next, recall the following relation (Poisson sum formula [15])

$$\sum_i w(iT) = \frac{1}{T} \sum_m W\left(\frac{m}{T}\right) \quad (7.5.14)$$

which applies to any finite-energy signal $w(t)$ with Fourier transform $W(f)$. Identifying $w(t)$ with $h(t-\delta)h(t-T/2-\delta)$ and bearing in mind that the Fourier transform of $h(t)h(t-T/2)$ is (Parseval theorem)

$$H_2(f) \triangleq \int_{-\infty}^{\infty} H(v)H(f-v)e^{-j\pi vT} dv \quad (7.5.15)$$

we get

$$\sum_i h(iT-\delta)h(iT-T/2-\delta) = \frac{1}{T}\sum_m H_2\left(\frac{m}{T}\right)e^{-j2\pi m\delta/T} \quad (7.5.16)$$

Now, assuming a signal bandwidth less than $1/T$ (so that $H(f)=0$ for $|f|\geq 1/T$), it is recognized that $H_2(m/T)$ is zero except for $m=0$ and $m=\pm 1$. Hence, (7.5.16) reduces to

$$\sum_i h(iT-\delta)h(iT-T/2-\delta) = \frac{1}{T}H_2(0) + \frac{2}{T}\text{Re}\left[H_2\left(\frac{1}{T}\right)e^{-j2\pi\delta/T}\right] \quad (7.5.17)$$

An analogous expression is found for the first summation in (7.5.13). Skipping the details, we have

$$\sum_i h(iT-T/2-\delta)h(iT-T-\delta) = \frac{1}{T}H_2(0) - \frac{2}{T}\text{Re}\left[H_2\left(\frac{1}{T}\right)e^{-j2\pi\delta/T}\right] \quad (7.5.18)$$

Finally, substituting into (7.5.13) yields

$$S(\delta) = -\frac{4C_2}{T}\text{Re}\left[H_2\left(\frac{1}{T}\right)e^{-j2\pi\delta/T}\right] \quad (7.5.19)$$

When $h(t)$ is an even function, $H(f)$ is real-valued and, as is now demonstrated, $H_2(1/T)$ is an imaginary number. In fact from (7.5.15) it can be shown that

$$H_2\left(\frac{1}{T}\right) = -jK \quad (7.5.20)$$

with

$$K \triangleq \int_{-\infty}^{\infty} H(\frac{1}{2T}+f)H(\frac{1}{2T}-f)\cos(\pi fT)df \quad (7.5.21)$$

Then, inserting into (7.5.19) yields

$$S(\delta) = \frac{4C_2 K}{T}\sin\left(\frac{2\pi\delta}{T}\right) \quad (7.5.22)$$

In particular, with a raised-cosine-rolloff function with rolloff $\alpha$ the constant K is given by

$$K = \frac{1}{4\pi(1-\alpha^2/4)}\sin\left(\frac{\pi\alpha}{2}\right) \quad (7.5.23)$$

Equation (7.5.22) indicates that the S-curve is sinusoidal of period $T$ and passes through the origin at $\delta = 0$. Its amplitude is proportional to $K$, which depends on the rolloff factor. As $\alpha$ decreases, $K$ gets smaller and smaller and the amplitude of the S-curve becomes inadequate for the tracking operation. Thus, the Gardner detector is ill-suited for narrow-band signaling.

### 7.5.3. Tracking Performance

The tracking performance of synchronizers endowed with GAD and NDA-ELD can be analytically assessed following [19]-[20] or the methods in Section 7.4.3. Unfortunately the passages are lengthy and the final results look messy. For this reason only simulation results are reported. In particular, Figures 7.22-7.23 illustrate the variance of the normalized error $(\tau - \hat{\tau}_k)/T$ as a function of $E_s/N_0$ for $h(t)$ satisfying the Nyquist criterion. As is seen, GAD is superior to NDA-ELD in that it has less self noise. As anticipated, the performance is strongly affected by the rolloff factor and degrades significantly as $\alpha$ decreases. This is in contrast with the behavior of decision-directed detectors which are much less sensitive to rolloff (see Figures 7.20-7.21). It is concluded that GAD and NDA-ELD are not useful with narrow-band signaling.

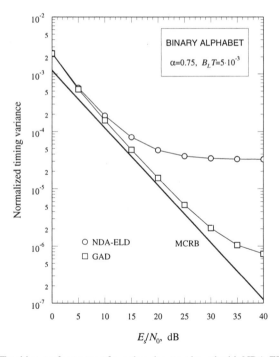

Figure 7.22. Tracking performance of synchronizers endowed with NDA-ELD and GAD.

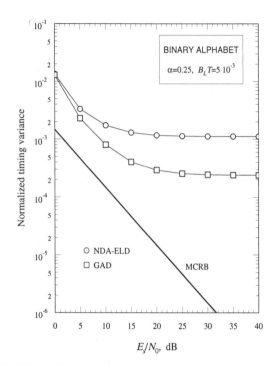

Figure 7.23. Tracking performance of synchronizers endowed with NDA-ELD and GAD.

### 7.5.4. Self Noise Elimination with the Gardner Detector

We have seen that the Gardner detector has significant self noise with strongly bandlimited signals. Thus, it would be useful to reduce this noise with simple methods. This problem has been addressed by D'Andrea and Luise in [19]. Their basic idea is borrowed from prefiltering techniques employed to minimize timing jitter in analog synchronizers [21]-[24]. Essentially, they adopt an error signal $e(k)$ that has still the form indicated in (7.5.12), except that the samples from the matched filter are first passed through an FIR *prefilter*. The synchronizer has the configuration indicated in Figure 7.24. As is intuitively clear, the tracking error variance $\sigma^2$ in the steady state depends on the prefilter taps and the question arises of looking for the tap values that minimize $\sigma^2$. The solution to this problem indicates that substantial improvements in tracking performance can be achieved even with very short prefilters (3-5 taps).

For example, Figure 7.25 compares the tracking performance of two

# Timing Recovery in Baseband Transmission

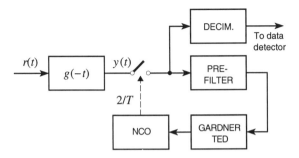

Figure 7.24. Block diagram of a synchronizer with prefilter.

synchronizers, one using a plain GAD and the other a GAD with a 5-tap prefilter. The alphabet is quaternary, the rolloff is 25% and the loop bandwidth equals $10^{-3}/T$. As is seen, prefiltering almost eliminates self noise. Indeed, the floor in the variance curve shows up at $E_s/N_0$ values greater than 40 dB.

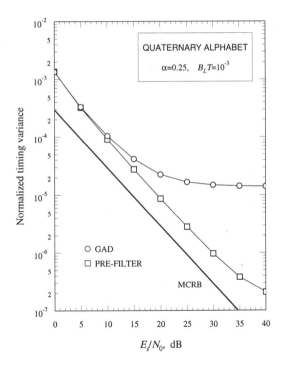

Figure 7.25. Performance comparisons between synchronizers with and without prefilter.

## 7.6. Feedforward Estimation Schemes

### 7.6.1. Non-Data-Aided ML-Based Algorithm

So far we have concentrated on timing detectors for use in feedback synchronizers. We have seen that several options are available, depending on the specific operating conditions. For example, decision-directed schemes have better tracking performance than quadratic methods when strongly bandlimited signals are used. It should be stressed, however, that tracking accuracy is not the only useful criterion. Implementation complexity is another important issue. A third one is acquisition time, especially in burst transmission systems where a fast timing recovery is needed. Unfortunately, feedback schemes are prone to hangup phenomena and, in consequence, may exhibit prolonged acquisitions. Thus, in certain applications feedforward recovery schemes are more appealing than feedback schemes.

In this section we investigate non-data-aided feedforward timing estimation using ML-oriented arguments. In doing so we adopt a discrete-time approach. Accordingly, the received waveform is first fed to an anti-alias filter (AAF) and then is sampled at some rate $1/T_s$, a multiple $N$ of the nominal symbol rate. Although a factor $N=2$ is sufficient in practice, in the sequel we keep $N$ generic. The following simplifying assumptions are made: (*i*) the AAF has a brick-wall-shaped transfer function of bandwidth $B_{AAF}$; (*ii*) this bandwidth is sufficiently large so as not to distort the signal components; (*iii*) the sampling rate equals $2B_{AAF}$. Also, we call $L_0$ the length of the observation interval in symbol periods and $x \triangleq \{x(kT_s)\}$ the sample sequence from the AAF. Note that $x$ has $NL_0$ components. Then, the likelihood function $\Lambda(x|\tilde{\tau},\tilde{c})$ for the unknown signal parameters takes the form

$$\Lambda(x|\tilde{\tau},\tilde{c}) = \exp\left\{\frac{2T_s}{N_0}\sum_{k=0}^{NL_0-1} x(kT_s)\tilde{s}(kT_s) - \frac{T_s}{N_0}\sum_{k=0}^{NL_0-1} \tilde{s}^2(kT_s)\right\} \qquad (7.6.1)$$

where $\tilde{s}(kT_s)$ are samples of the trial signal

$$\tilde{s}(t) \triangleq \sum_i \tilde{c}_i g(t - iT - \tilde{\tau}) \qquad (7.6.2)$$

Equation (7.6.1) is similar to the likelihood function (7.5.1) we have derived following a continuous-time approach. This means that the difficulties encountered earlier are still in our way. The only manner in which to get around them is to adopt the same approximations as in Section 7.5.1. This involves dropping the second sum in (7.6.1) and assuming a low $E_s/N_0$.

Skipping the details we obtain

$$\Lambda(r|\tilde{\tau}) \approx \sum_{k_1=0}^{NL_0-1} \sum_{k_2=0}^{NL_0-1} x(k_1 T_s) x(k_2 T_s) F(k_1, k_2, \tilde{\tau}) \tag{7.6.3}$$

where some immaterial constants have been suppressed and we have defined

$$F(k_1, k_2, \tilde{\tau}) \triangleq \sum_i g(k_1 T_s - iT - \tilde{\tau}) g(k_2 T_s - iT - \tilde{\tau}) \tag{7.6.4}$$

Next we note that $F(k_1, k_2, \tilde{\tau})$ is a periodic function of $\tilde{\tau}$ of period $T$. In fact

$$F(k_1, k_2, \tilde{\tau} - T) = \sum_i g[k_1 T_s - (i-1)T - \tilde{\tau}] g[k_2 T_s - (i-1)T - \tilde{\tau}]$$
$$= F(k_1, k_2, \tilde{\tau}) \tag{7.6.5}$$

Hence, $F(k_1, k_2, \tilde{\tau})$ can be expanded into a Fourier series

$$F(k_1, k_2, \tilde{\tau}) = \sum_m F_m(k_1, k_2) e^{j 2\pi m \tilde{\tau}/T} \tag{7.6.6}$$

with coefficients $F_m(k_1, k_2)$ given by

$$F_m(k_1, k_2) = \frac{1}{T} \int_0^T F(k_1, k_2, \tilde{\tau}) e^{-j 2\pi m \tilde{\tau}/T} d\tilde{\tau} \tag{7.6.7}$$

which satisfy the relation

$$F_{-m}(k_1, k_2) = F_m^*(k_1, k_2) \tag{7.6.8}$$

Returning to (7.6.3), we look for the location of the maximum of $\Lambda(r|\tilde{\tau})$. This will provide us with an (approximate) ML estimate of $\tau$. A simple method to attain this goal relies on the fact that, if the signal is bandlimited to $\pm 1/T$, then the coefficients $F_m(k_1, k_2)$ are all zero except those with index $m=0, \pm 1$ (see Appendix 7.A):

$$F_m(k_1, k_2) = 0, \quad |m| \geq 2 \tag{7.6.9}$$

In consequence, equation (7.6.6) becomes

$$F(k_1, k_2, \tilde{\tau}) = F_0(k_1, k_2) + 2 \text{Re}\left\{ F_1(k_1, k_2) e^{j 2\pi \tilde{\tau}/T} \right\} \tag{7.6.10}$$

Then, substituting into (7.6.3) and dropping again immaterial constants yields

$$\Lambda(r|\tilde{\tau}) \approx \text{Re}\left\{ e^{j2\pi\tilde{\tau}/T} \sum_{k_1=0}^{NL_0-1} \sum_{k_2=0}^{NL_0-1} x(k_1 T_s) x(k_2 T_s) F_1(k_1, k_2) \right\} \quad (7.6.11)$$

from which the desired estimator is derived noting that the right-hand side has a maximum at

$$\hat{\tau} = -\frac{T}{2\pi} \arg\left\{ \sum_{k_1=0}^{NL_0-1} \sum_{k_2=0}^{NL_0-1} x(k_1 T_s) x(k_2 T_s) F_1(k_1, k_2) \right\} \quad (7.6.12)$$

A more suitable form for (7.6.12) is obtained through the following steps. First, we exploit the fact that $F_1(k_1, k_2)$ is given by (see Appendix 7.A)

$$F_1(k_1, k_2) = \frac{1}{T} q[(k_1 - k_2) T_s] e^{-j\pi(k_1 + k_2)/N} \quad (7.6.13)$$

where $q(t)$ has the Fourier transform

$$Q(f) = G\left(f - \frac{1}{2T}\right) G^*\left(f + \frac{1}{2T}\right) \quad (7.6.14)$$

Inserting (7.6.13) into (7.6.12) and rearranging yields

$$\hat{\tau} = -\frac{T}{2\pi} \arg\left\{ \sum_{k=0}^{NL_0-1} y(kT_s) z(kT_s) \right\} \quad (7.6.15)$$

with

$$y(kT_s) \triangleq x(kT_s) e^{-j\pi k/N} \quad (7.6.16)$$

$$z(kT_s) \triangleq \sum_{k_2=0}^{NL_0-1} y(k_2 T_s) q[(k - k_2) T_s] \quad (7.6.17)$$

Second, from (7.6.14) it is seen that $Q(f)$ takes values within $\pm\alpha/(2T)$, where $\alpha$ is the rolloff factor associated with $G(f)$. Then, if $\alpha$ is not too small, $q(t)$ has a duration of a few symbol intervals, as is shown in Figure 7.26. On the other hand, since the estimation interval is usually much longer than $T$ (which means $L_0 \gg 1$), we expect that (7.6.17) is only marginally affected if the summation is extended from $-\infty$ to $+\infty$. In doing so the right-hand side becomes the output of a (non-causal) filter $q(kT_s)$ driven by $y(kT_s)$.

# Timing Recovery in Baseband Transmission

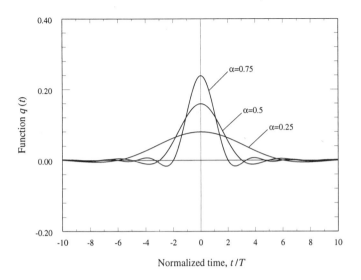

Figure 7.26. Function $q(t)$ for root-raised cosine pulses.

Third, the filter is made causal by shifting its impulse response rightward by $ND$ steps. This corresponds to delaying $q(t)$ by $DT$ seconds. For example, a $5T$ delay is adequate for $\alpha=0.5$ (see Figure 7.26). As the shift produces a delay in the filter output we have

$$z[(k-ND)T_s] = y(kT_s) \otimes q[(k-ND)T_s] \quad (7.6.18)$$

Fourth, $z[(k-ND)T_s]$ and $y[(k-ND)T_s]$ are eventually inserted in (7.6.15) to produce the desired estimator

$$\hat{\tau} = -\frac{T}{2\pi} \arg\left\{ \sum_{k=ND}^{N(L_0+D)-1} y[(k-ND)T_s] \, z[(k-ND)T_s] \right\} \quad (7.6.19)$$

Figure 7.27 gives a pictorial representation of the estimator. We see that the samples from the anti-alias filter are multiplied by $e^{-j\pi k/N}$ and fed to two parallel branches. The upper branch is made of the filter $q[(k-ND)T_s]$ while the lower branch is made of a simple delay. The branch outputs are multiplied together and the products are accumulated. The argument of the accumulator output gives the timing estimate within a factor $-T/(2\pi)$.

Performance analysis of this estimator may be carried out with the methods indicated in [9] and [25]. Unfortunately, passages and final formulas are very involved and are not reported here. Only simulations are shown in the next section.

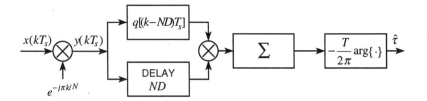

Figure 7.27. Block diagram of the ML-based estimator.

**Exercise 7.6.1.** Compute the inverse Fourier transform of $Q(f)$ when $G(f)$ has a root-raised-cosine rolloff.

*Solution.* Function $G(f)$ has the form

$$G(f) = \begin{cases} \sqrt{T} & |f| \leq \dfrac{1-\alpha}{2T} \\ \sqrt{T}\cos\left[\dfrac{\pi}{4\alpha}(|2fT|-1+\alpha)\right] & \dfrac{1-\alpha}{2T} \leq |f| \leq \dfrac{1+\alpha}{2T} \\ 0 & \text{otherwise} \end{cases} \quad (7.6.20)$$

Inserting into (7.6.14) and taking the inverse Fourier transform yields, after some manipulations,

$$q(t) = \frac{\alpha}{\pi} \frac{\cos(\pi\alpha t/T)}{1-(2\alpha t/T)^2} \quad (7.6.21)$$

### 7.6.2. Oerder and Meyr Algorithm

An alternative feedforward method has been proposed by Oerder and Meyr (O&M) in [9] based on heuristic arguments. The block diagram of the O&M estimator is depicted in Figure 7.28. The samples $x(kT_s)$ are taken from some low-pass filter (LPF) which, in particular, may be matched to the received pulses. As with the ML-based estimator, the sampling rate is a multiple $N$ of $1/T$. However, $N=4$ is needed in O&M, as opposed to $N=2$ in the ML-based estimator. The input $x(kT_s)$ is first squared, then is multiplied by $e^{-j2\pi k/N}$ and finally is integrated in an accumulator. The argument of the accumulator output gives the timing estimate within a factor $-T/(2\pi)$. Formally

$$\hat{\tau} = -\frac{T}{2\pi}\arg\left\{\sum_{k=0}^{NL_0-1} x^2(kT_s)e^{-j2\pi k/N}\right\} \quad (7.6.22)$$

# Timing Recovery in Baseband Transmission

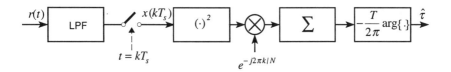

Figure 7.28. Block diagram of the O&M estimator.

where $L_0$ is the integration window in symbol intervals.

Comparing O&M with the ML-based estimator it appears that the former is a little simpler to implement as it does not need any filtering of $x(kT_s)$. On the other hand, it requires a sampling rate which is twice as large. Analysis indicates that both schemes are unbiased, which means that the average of $\hat{\tau}$ coincides with $\tau$. Their estimation accuracy depends on several system parameters such as the observation interval $L_0$, the signal-to-noise ratio $E_s/N_0$ and the rolloff factor $\alpha$. With both estimators the variance of the normalized errors $(\tau - \hat{\tau})/T$ can be put in the form [9], [25]

$$\sigma^2 = \frac{1}{L_0}\left[K_{SS} + K_{SN}\left(\frac{E_s}{N_0}\right)^{-1} + K_{NN}\left(\frac{E_s}{N_0}\right)^{-2}\right] \quad (7.6.23)$$

where the coefficients $K_{SS}$, $K_{SN}$ and $K_{NN}$ depend, in particular, on the symbol alphabet and the rolloff factor. Note that $K_{SS}/L_0$ represents the self noise term, as is seen from (7.6.23) letting $N_0$ decrease to zero.

Figures 7.29-7.31 show simulation results comparing the accuracy of the two methods. The following assumptions are made with the ML-based algorithm:

(*i*) the anti-alias filter is an eight-pole Butterworth with a −3 dB bandwidth of $1/T$;

(*ii*) the oversampling factor is $N=2$;

(*iii*) the pulse $g(t)$ corresponds to a root-raised-cosine-rolloff function;

(*iv*) the impulse response $q[(k-ND)T_s]$ is derived by truncating $q(t)$ to $\pm 5T$;

(*v*) the delay $ND$ is set equal to 10;

(*vi*) the observation length is $L_0=100$ symbols.

With the O&M algorithm it is assumed that:

(*i*) the low-pass filter is matched to $g(t)$ and is still of root-raised-cosine-rolloff type;

(*ii*) the oversampling factor is $N=4$;

(*iii*) the observation length is $L_0=100$.

Figures 7.29-7.30 show the tracking performance with a binary alphabet and two different rolloff factors. For $\alpha=0.75$ it is seen that the algorithms are essentially equivalent and their accuracy is close to the MCRB. For $\alpha=0.25$, however, the O&M algorithm is slightly inferior due to a larger self noise. The differences tend to increase with the alphabet sizes, as is shown in Figure 7.31 which illustrates the case with $M=8$.

A comparison between feedforward and NDA feedback algorithms is interesting. Indications in this regard may be gathered contrasting Figures 7.29-7.30 with Figures 7.22-7.23. Keeping in mind the relation

$$B_L T = \frac{1}{2L_0} \qquad (7.6.24)$$

it is recognized that the comparison is fair since $L_0=100$ in Figures 7.29-7.30 corresponds to $B_L T = 5 \cdot 10^{-3}$ in Figures 7.22-7.23. With large rolloff values it is seen that a feedback loop with Gardner detector has the same accuracy as the feedforward algorithms. As $\alpha$ decreases, however, feedforward schemes are definitely superior.

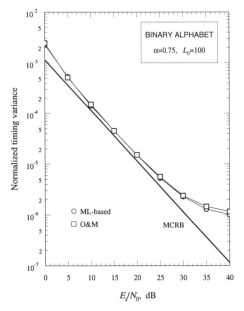

Figure 7.29. Performance comparisons with binary alphabet and $\alpha=0.75$.

# Timing Recovery in Baseband Transmission

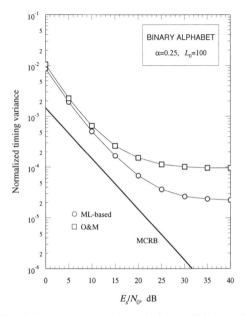

Figure 7.30. Performance comparisons with binary alphabet and $\alpha=0.25$.

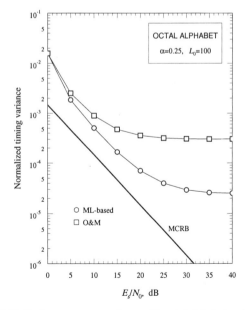

Figure 7.31. Performance comparisons with octal alphabet and $\alpha=0.25$.

## 7.7. Key Points of the Chapter

- Sampling is a crucial operation in digital modems as all the receiver processing is performed on samples. In some cases sampling is synchronized to the symbol rate, in others it is not.
- Synchronous sampling is adjusted by exploiting the information from the timing error detector (TED). The synchronizer necessarily has a feedback configuration insofar as the information from the TED must be fed back to correct the sampling phase.
- A hybrid NCO is needed to control synchronous sampling. As the NCO has analog components, the synchronizer cannot be implemented in fully digital form.
- Timing correction with non-synchronous sampling is performed by interpolators. They operate on non-synchronized samples to produce synchronized interpolants for use in the decision device and TED.
- Simple polynomial-based interpolators are available with good characteristics. A two-point linear interpolation is sufficient in many applications. In more critical cases, parabolic interpolating filters provide excellent results.
- The configuration of timing recovery schemes with non-synchronous sampling may be either feedback or feedforward. In the former case, the error signal is fed back to adjust the interpolation phase. In the latter, the timing parameter is estimated in some nonlinear device and is used to interpolate the strobes at the correct times.
- A great profusion of timing error detectors exists, either decision-directed (DD) or non-data-aided (NDA). A synchronizer endowed with a DD ML-based timing detector practically achieves the MCRB. Unfortunately an ML-based detector is complex to implement as it requires a derivative matched filter in addition to the matched filter. Thus, simpler solutions are of interest. The most interesting ones seem to be the zero-crossing detector and the Mueller&Mueller detector. The latter is especially recommended with small excess bandwidth factors.
- NDA timing detectors may be as good as DD detectors, provided that the excess bandwidth factor is not too small. The Gardner detector is a prominent case.
- NDA feedforward timing estimators are an alternative to feedback schemes, either DD or NDA. They have good performance with intermediate or large rolloff factors and are appealing in applications where fast acquisitions are needed.

## Appendix 7.A

In this Appendix we compute the coefficients

$$F_m(k_1,k_2) = \frac{1}{T}\int_0^T F(k_1,k_2,\tilde{\tau})e^{-j2\pi m\tilde{\tau}/T}d\tilde{\tau} \qquad (7.A.1)$$

where $F(k_1,k_2,\tilde{\tau})$ is defined as

$$F(k_1,k_2,\tilde{\tau}) \triangleq \sum_i g(k_1T_s - iT - \tilde{\tau})g(k_2T_s - iT - \tilde{\tau}) \qquad (7.A.2)$$

To begin, we substitute (7.A.2) into (7.A.1) and make the change of variable $t = k_1T_s - iT - \tilde{\tau}$. As a result we get

$$\begin{aligned}F_m(k_1,k_2) &= \frac{1}{T}e^{-j2\pi mk_1/N}\sum_i \int_{k_1T_s-(i+1)T}^{k_1T_s-iT} g(t)g[t-(k_1-k_2)T_s]e^{j2\pi mt/T}dt \\ &= \frac{1}{T}e^{-j2\pi mk_1/N}\int_{-\infty}^{\infty} g(t)g[t-(k_1-k_2)T_s]e^{j2\pi mt/T}dt \end{aligned} \qquad (7.A.3)$$

where we have used the definition $N \triangleq T/T_s$.

Next we observe that the integral in (7.A.3) may be viewed as the Fourier transform of $g(t)g[t-(k_1-k_2)T_s]$ at $f=-m/T$. As this transform is given by

$$\int_{-\infty}^{\infty} G(v)G(f-v)e^{-j2\pi(k_1-k_2)(f-v)T_s}dv \qquad (7.A.4)$$

it follows from (7.A.3) that

$$F_m(k_1,k_2) = \frac{1}{T}e^{-j2\pi mk_2/N}\int_{-\infty}^{+\infty} G(v)G^*(v+m/T)e^{j2\pi(k_1-k_2)vT_s}dv \qquad (7.A.5)$$

Inspection of (7.A.5) reveals that, if $G(f)$ is bandlimited to $\pm 1/T$, the integral is zero except for $m=0,\pm 1$. In particular, for $m=1$ we have

$$F_1(k_1,k_2) = \frac{1}{T}e^{-j2\pi k_2/N}\int_{-\infty}^{+\infty} G(v)G^*(v+1/T)e^{j2\pi(k_1-k_2)vT_s}dv \qquad (7.A.6)$$

or, making the change of variable $v=f-1/(2T)$,

$$F_1(k_1,k_2) = \frac{1}{T}e^{-j\pi(k_1+k_2)/N}\int_{-\infty}^{\infty}G[f-1/(2T)]G^*[f+1/(2T)]e^{j2\pi(k_1-k_2)fT_s}df \quad (7.A.7)$$

Alternatively, letting

$$Q(f) \triangleq G[f-1/(2T)]G^*[f+1/(2T)] \quad (7.A.8)$$

and denoting by $q(t)$ the inverse Fourier transform of $Q(f)$, $F_1(k_1,k_2)$ we obtain

$$F_1(k_1,k_2) = \frac{1}{T}q[(k_1-k_2)T_s]e^{-j\pi(k_1+k_2)/N} \quad (7.A.9)$$

## References

[1] F.M.Gardner, Frequency Granularity in Digital Phaselock Loops, *IEEE Trans. Commun.*, **COM-44**, 749-758, June 1996.
[2] R.E.Crochiere and L.R.Rabiner, *Multirate Digital Signal Processing*, Englewood Cliffs, NJ: Prentice Hall, 1983.
[3] R.W.Shafer and L.R.Rabiner, A Digital Signal Processing Approach to Interpolation, *Proc. IEEE,* Vol. 61, pp. 692-702, June 1973.
[4] F.Takahata, M.Yasunaga, Y.Hirata, T.Ohsawa and J.Namiki, A PSK Group Modem for Satellite Communication, *IEEE J. Select. Areas Commun.*, **SAC-5**, 648-661, May 1987.
[5] G.Asheid, M.Oerder, J.Stahl and H.Meyr, An All Digital Receiver Architecture for Bandwidth Efficient Transmission at High Data Rates, *IEEE Trans. Commun.*, **COM-37**, 804-813, Aug. 1989.
[6] F.M.Gardner, Interpolation in Digital Modems-Part I: Fundamentals, *IEEE Trans. Commun.*, **COM-41**, 501-507, March 1993.
[7] L.Erup, F.M.Gardner and R.A.Harris, Interpolation in Digital Modems-Part II: Implementation and Performance, *IEEE Trans. Commun.*, COM-41, 998-1008, June 1993.
[8] K.Bucket and M.Moeneclaey, Optimization of a Second-Order Interpolator for Bandlimited Direct-Sequence Spread-Spectrum Communication, *Electronics Letters,* Vol. 28, pp. 1029-1031, May 1992.
[9] M.Oerder and H.Meyr, Digital Filter and Square Timing Recovery, *IEEE Trans. Commun.*, **COM-36**, 605-612, May 1988.
[10] F.M.Gardner, *Demodulator Reference Recovery Techniques Suited for Digital Implementation*, European Space Agency, Final Report, ESTEC Contract No. 6847/86/NL/DG, Aug. 1988.
[11] J.M.Bergmans and H.W.Wong-Lam, A Class of Data-Aided Timing Recovery Schemes, *IEEE Trans. Commun.*, **COM-43**, 1819-1827, Feb./March/April 1995.
[12] H.Kobayashi, Simultaneous Adaptive Estimation and Decision Algorithm for Carrier Modulated Data Transmission Systems, *IEEE Trans. Commun.*, **COM-19**, 268-280, June 1971.
[13] M.H.Meyers and L.E.Franks, Joint Carrier and Symbol Timing Recovery for PAM Systems, *IEEE Trans. Commun.*, **COM-28**, 1121-1129, Aug. 1980.
[14] K.Bucket and M.Moeneclaey, The Effect of Non-Ideal Interpolation on Symbol Synchronizer Performance, *European Trans. Telecommun.*, Vol. 6, pp. 627-632,

Nov./Dec. 1995.
[15] A.Papoulis, *Probability, Random Variables, and Stochastic Processes*, New York: McGraw-Hill, 1991.
[16] W.C.Lindsey and M.K.Simon, *Telecommunication Systems Engineering*, Englewood Cliffs, NJ: Prentice Hall, 1972.
[17] K.H.Mueller and M.Mueller, Timing Recovery in Digital Synchronous Data Receivers, *IEEE Trans. Commun.*, **COM-24,** 516-531, May 1976.
[18] F.M.Gardner, A BPSK/QPSK Timing-Error Detector for Sampled Receivers, *IEEE Trans. Commun.*, **COM-34,** 423-429, May 1986.
[19] A.N.D'Andrea and M.Luise, Optimization of Symbol Timing Recovery for QAM Data Demodulators, *IEEE Trans. Commun.*, **COM-44,** 399-406, March 1996.
[20] W.G.Cowley, The Performance of Two Symbol Timing Recovery Algorithms for PSK Demodulators, *IEEE Trans. Commun.*, **COM-42,** 2345-2355, June 1994.
[21] L.E.Franks and J.P.Bubrouski, Statistical Properties of Timing Jitter in PAM Recovery Schemes, *IEEE Trans. Commun.*, **COM-22,** 913-920, July 1974.
[22] M.Moeneclaey, Comparisons of Two Types of Symbol Synchronizers for Which Self Noise Is Absent, *IEEE Trans. Commun.*, **COM-31,** 329-334, March 1983.
[23] A.N.D'Andrea, U.Mengali and M.Moro, Nearly Optimum Prefiltering in Clock Recovery, *IEEE Trans. Commun.*, **COM-34,** 1081-1088, Nov. 1986.
[24] A.N.D'Andrea and M.Luise, Design and Analysis of Jitter-Free Clock Recovery Scheme for QAM Systems, *IEEE Trans. Commun.*, **COM-41,** 1296-1299, Sept. 1993.
[25] M.Signorini, Feedforward Timing Estimation in PAM Modulated Signals, Master Thesis (in Italian), Dept. Information Engineering, June 1995.

# 8

# Timing Recovery with Linear Modulations

## 8.1. Introduction

In this chapter we address timing recovery with modulated PAM signals. For reasons that will soon be explained, the discussion is divided into several parts, depending on the specific modulation format and the operating conditions. For example, a distinction between non-offset and offset modulations is useful as different signal representations call for separate analyses and lead to different solutions. In the sequel we first discuss non-offset modulations and then consider offset QPSK modulation (OQPSK).

Other categories arise from the fact that the carrier phase plays a crucial role with modulated signals. Assuming that possible carrier frequency offsets have been compensated, the following scenarios may be envisaged:

(*i*) The carrier phase is known (as it has been preliminary recovered in a clockless manner).

(*ii*) The carrier phase is not known and is jointly recovered with timing in a decision-directed (DD) fashion.

(*iii*) The carrier phase is not known but is recovered later, in a clock-aided manner.

In Chapter 5 it has been shown that clockless phase recovery is a concrete possibility for PSK and OQPSK. Thus, the notion of phase-aided timing recovery, as implied in (*i*), deserves consideration. Its application consists of derotating the signal samples by the estimated phase and splitting the result into real and imaginary components. This produces two baseband sample sequences from which clock information can be derived with the methods indicated in the preceding chapter. As the technique is straightforward, it will not be further discussed in the sequel.

Case (*ii*) requires that timing be estimated along with phase. To do so some feedback schemes with excellent tracking performance are available. Their limitations are spurious locks and prolonged acquisitions caused by complex interactions between phase and timing correction algorithms. Luckily, spurious locks occur only with multi-amplitude and phase modulations and high SNRs. A prudent designer should make sure (typically, by extensive simulations) that the synchronizer under examination has no spurious locks under the expected operating conditions.

As an alternative, feedforward joint phase and timing estimation methods may be pursued. Feedforward schemes have no spurious locks (there is nothing to lock on) and have comparatively shorter acquisition times. So, they are particularly attractive for burst data transmissions.

Another way of avoiding interaction between phase and timing algorithms is to recover timing independently of the carrier phase, and then exploit this knowledge for carrier estimation. This corresponds to the third scenario indicated above.

The chapter is organized as follows. In the next section we discuss DD joint phase and timing recovery for non-offset formats. The synchronization schemes in this section all have a feedback structure and, in general, good tracking performance. This is also true for the algorithms in Section 8.3 which, however, are of NDA type. Taken together, the detectors in Sections 8.2 and 8.3 offer adequate synchronization methods for continuous transmissions. When it comes to burst transmissions, however, feedforward schemes are preferable. This subject is discussed in Section 8.4 for non-offset modulations. Section 8.5 addresses timing estimation for transmissions over frequency-flat channels. The last two sections of the chapter deal with OQPSK modulation for the AWGN channel. In particular, Section 8.6 discusses DD joint phase and timing recovery while Section 8.7 investigates feedforward methods.

## 8.2. Decision-Directed Joint Phase and Timing Recovery with Non-Offset Formats

Our task here is to jointly estimate carrier phase and timing from the incoming waveform. This approach is very popular and has been extensively explored in [1]-[3]. As we shall see, it leads to an overall synchronizer in which two interacting parts can be identified, a phase-recovery loop and a timing-recovery loop. The phase detector in the former is the same as that proposed for Costas loops in Chapter 5. In the following we first address joint phase and timing estimation with ML methods. An *ad hoc* algorithm is discussed later.

### 8.2.1. ML-Based Joint Phase and Timing Estimation

Suppose that possible carrier frequency offsets have been compensated and reliable decisions are available from the decision device. Then, the log-likelihood function takes on the form

$$L(r|\tilde{\theta},\tilde{\tau}) = \sum_i \text{Re}\left\{\hat{c}_i^* e^{-j\tilde{\theta}} \int_0^{T_0} r(t)g(t-iT-\tilde{\tau})dt\right\}$$

$$-\frac{1}{2}\sum_i\sum_m \text{Re}\{\hat{c}_i^*\hat{c}_m\}\int_0^{T_0} g(t-iT-\tilde{\tau})g(t-mT-\tilde{\tau})dt \quad (8.2.1)$$

We want to compute the location of the maximum of $L(r|\tilde{\theta},\tilde{\tau})$. This requires setting the following partial derivatives to zero:

$$\frac{\partial L(r|\tilde{\theta},\tilde{\tau})}{\partial \tilde{\theta}} = \sum_i \text{Im}\left\{\hat{c}_i^* e^{-j\tilde{\theta}} \int_0^{T_0} r(t)g(t-iT-\tilde{\tau})dt\right\} \quad (8.2.2)$$

$$\frac{\partial L(r|\tilde{\theta},\tilde{\tau})}{\partial \tilde{\tau}} = -\sum_i \text{Re}\left\{\hat{c}_i^* e^{-j\tilde{\theta}} \int_0^{T_0} r(t)g'(t-iT-\tilde{\tau})dt\right\}$$

$$+ \sum_i\sum_m \text{Re}\left\{\hat{c}_i^*\hat{c}_m \int_0^{T_0} g'(t-iT-\tilde{\tau})g(t-mT-\tilde{\tau})dt\right\} \quad (8.2.3)$$

where $g'(t)$ is the derivative of $g(t)$.

These equations can be written in a more suitable form by expanding the limits of the integrals to $\pm\infty$ while restricting the summations over index $i$ from 0 to $L_0 - 1$ with $L_0 \triangleq T_0/T$. Skipping the details this leads to

$$\frac{\partial L(r|\tilde{\theta},\tilde{\tau})}{\partial \tilde{\theta}} = \sum_{i=0}^{L_0-1} \text{Im}\{\hat{c}_i^* y(iT+\tilde{\tau})e^{-j\tilde{\theta}}\} \quad (8.2.4)$$

$$\frac{\partial L(r|\tilde{\theta},\tilde{\tau})}{\partial \tilde{\tau}} = \sum_{i=0}^{L_0-1} \text{Re}\left\{\hat{c}_i^*\left[y'(iT+\tilde{\tau})e^{-j\tilde{\theta}} - \sum_{m=i-D}^{i+D}\hat{c}_m h'[(i-m)T]\right]\right\} \quad (8.2.5)$$

where $y(t)$ is the matched-filter output, $y'(t)$ is its derivative, $h'(t)$ is the derivative of $h(t) \triangleq g(t) \otimes g(-t)$ and $D$ is an integer indicating the semi-duration of $h'(t)$ in symbol intervals. In other words, $h'(t)$ is vanishingly small for $|t| \geq DT$. Note that $y'(t)$ may also be viewed as the response of the *derivative matched-filter* $-g'(-t)$ to $r(t)$.

The simultaneous null of the derivatives is sought with the usual recursive method. In essence, the $k$-th terms in the sums in (8.2.4)-(8.2.5) are computed for $\tilde{\theta}$ and $\tilde{\tau}$ equal to the current phase and timing estimates and are then used as signal errors to update these estimates. This results in the following error signals [1]-[3]

$$e_P(k) = \text{Im}\left\{\hat{c}_k^* y(kT + \hat{\tau}_k) e^{-j\hat{\theta}_k}\right\} \quad (8.2.6)$$

$$e_T(k) = \text{Re}\left\{\hat{c}_k^* \left[ y'(kT + \hat{\tau}_k) e^{-j\hat{\theta}_k} - \sum_{m=k-D}^{k+D} \hat{c}_m h'[(k-m)T] \right]\right\} \quad (8.2.7)$$

and the corresponding updating equations

$$\hat{\theta}_{k+1} = \hat{\theta}_k + \gamma_P e_P(k) \quad (8.2.8)$$

$$\hat{\tau}_{k+1} = \hat{\tau}_k + \gamma_T e_T(k) \quad (8.2.9)$$

where $\gamma_P$ and $\gamma_T$ are step-size parameters.

Equations (8.2.8)-(8.2.9) describe the operation of two loops, a carrier-phase synchronizer and a timing synchronizer. The loops interact with each other, as is clear from the fact that $e_P(k)$ depends on $\hat{\tau}_k$ and, reciprocally,

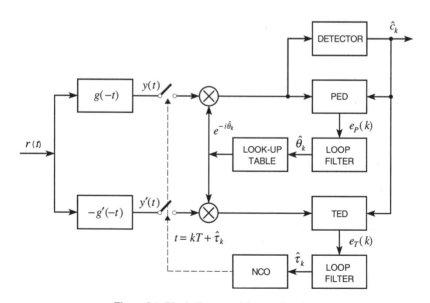

Figure 8.1. Block diagram of the synchronizer.

# Timing Recovery with Linear Modulations

$e_T(k)$ is a function of $\hat{\theta}_k$. We shall see later that these interactions may lead to synchronization failures.

A block diagram for the synchronizer is shown in Figure 8.1. The incoming waveform is fed to the matched filter $g(-t)$ and the derivative matched filter $-g'(-t)$ to produce $y(t)$ and $y'(t)$, respectively. Samples of $y(t)$ and $y'(t)$ are first derotated by $\hat{\theta}_k$ and then passed to the PED (phase error detector) and TED (timing error detector) to generate the error signals (8.2.6)-(8.2.7). Current phase and timing estimates are updated (in the loop filters) according to (8.2.8)-(8.2.9) and are fed back to the look-up table and the NCO.

## 8.2.2. Remark

At first sight the relation between (8.2.7) and the corresponding baseband algorithm derived in Section 7.4.1 may not be obvious. To reveal this relation let us rewrite here the expression of the baseband TED

$$e(k) = \hat{c}_k \left[ y'(kT + \hat{\tau}_k) - \sum_{m=k-D}^{k+D} \hat{c}_m h'[(k-m)T] \right] \qquad (8.2.10)$$

Now, recall that the quantities involved in (8.2.10) are all real-valued whereas those in (8.2.7) are generally complex. In particular so are the symbols and the derotated samples from the derivative matched filter. Then, expressing the complex variables in (8.2.7) in terms of their real and imaginary components yields

$$e_T(k) = \operatorname{Re}\{\hat{c}_k\} \operatorname{Re}\left\{ y'(kT + \hat{\tau}_k)e^{-j\hat{\theta}_k} - \sum_{m=k-D}^{k+D} \hat{c}_m h'[(k-m)T] \right\}$$

$$+ \operatorname{Im}\{\hat{c}_k\} \operatorname{Im}\left\{ y'(kT + \hat{\tau}_k)e^{-j\hat{\theta}_k} - \sum_{m=k-D}^{k+D} \hat{c}_m h'[(k-m)T] \right\} \qquad (8.2.11)$$

which appears as the sum of two baseband detectors of the type (8.2.10). In particular, for $D=0$ the sums in (8.2.11) vanish (since $h(t)$ is even and, therefore, $h'(0) = 0$) and the passband TED reduces to

$$e_T(k) = \operatorname{Re}\{\hat{c}_k\}\left[ \operatorname{Re}\left\{ y'(kT + \hat{\tau}_k)e^{-j\hat{\theta}_k} \right\} \right]$$

$$+ \operatorname{Im}\{\hat{c}_k\}\left[ \operatorname{Im}\left\{ y'(kT + \hat{\tau}_k)e^{-j\hat{\theta}_k} \right\} \right] \qquad (8.2.12)$$

The foregoing discussion may be summarized by saying that the passband timing algorithm can be derived from the corresponding baseband algorithm through three simple steps:

1. The output of the derivative matched filter is derotated by the carrier phase estimate $\hat{\theta}_k$.

2. Real and imaginary components of $y'(kT+\hat{\tau}_k)e^{-j\hat{\theta}_k}$ are separately used to form two timing errors, as would be done in a baseband transmission.

3. The baseband timing errors are added together to form the passband detector.

**Exercise 8.2.1.** Compute the S-curve for detector (8.2.7) assuming that: (*i*) phase estimates are ideal ($\hat{\theta}_k = \theta$); (*ii*) timing errors are small; (*iii*) data decisions are reliable ($\hat{c}_k \approx c_k$); (*iv*) symbols are uncorrelated.

*Solution.* The procedure strictly parallels the discussion in Sections 7.4.2 for baseband transmission. In particular, the waveform from the derivative matched filter is found to be

$$y'(t) = e^{j\theta}\sum_i c_i h'(t-iT-\tau) + n'(t) \qquad (8.2.13)$$

with

$$n'(t) \triangleq w(t) \otimes [-g'(-t)] \qquad (8.2.14)$$

Substituting into (8.2.7) for $\hat{\theta}_k = \theta$, $\hat{\tau}_k = \hat{\tau}$ and $\hat{c}_k = c_k$ yields

$$e_T(k) = \text{Re}\left\{\sum_i c_i c_k^* h'[(k-i)T+\hat{\tau}-\tau]\right.$$

$$\left. + c_k^* n'(kT+\hat{\tau})e^{-j\theta} - \sum_{m=k-D}^{k+D} c_m c_k^* h'[(k-m)T]\right\} \qquad (8.2.15)$$

As the symbols are assumed uncorrelated, we have

$$E\{c_{k+m}c_k^*\} = \begin{cases} C_2 & m=0 \\ 0 & m \neq 0 \end{cases} \qquad (8.2.16)$$

Hence, recalling that $h'(t)$ is an odd function and taking the expectation of (8.2.15) over data and noise gives the desired result

$$S(\tau-\hat{\tau}) = -C_2 h'(\tau-\hat{\tau}) \qquad (8.2.17)$$

# Timing Recovery with Linear Modulations

Comparing with the S-curve for the ML baseband algorithm, it appears that the two formulas coincide. It should be pointed out, however, that the constant $C_2$ is different in the two cases. For example, with 16-QAM modulation, the symbols are $c_k = a_k + j b_k$ with $a_k, b_k \in \{\pm 1, \pm 3\}$ and $C_2$ is 10. With quaternary baseband transmission, vice versa, we have $c_k \in \{\pm 1, \pm 3\}$ and $C_2$ is 5. Thus, the S-curve with 16-QAM is twice as large as with quaternary baseband transmission. This is intuitively clear in view of (8.2.11) which shows the timing error for pass-band transmission as the sum of two equivalent baseband timing errors.

**Exercise 8.2.2.** Compute the S-curve about the origin for the detector (8.2.7) assuming QAM modulation and phase errors $\phi_k \triangleq \theta - \hat{\theta}_k$ close to $m\pi/2$, with $m=0,1,2,3$.

*Solution.* With high SNR and $\phi_k \approx 0$ the decisions coincide (almost always) with the transmitted symbols and the S-curve is as indicated in (8.2.17). More generally, for $\phi \approx m\pi/2$, the decisions are given by

$$\hat{c}_k \approx c_k e^{jm\pi/2} \tag{8.2.18}$$

and reasoning as in Exercise 8.2.1 it is found that

$$S(\tau - \hat{\tau}) \approx -C_2 h'(\tau - \hat{\tau}) \tag{8.2.19}$$

Thus, the S-curve for the timing detector is insensitive to phase shifts by multiples of $\pi/2$.

## 8.2.3. Ad Hoc Timing Detectors

In the preceding chapter we have seen that other decision-directed TEDs (in addition to the ML-based detector) can be used for timing recovery in baseband transmission. In particular, the following detectors appear of prominent interest:

- the *early-late detector* (ELD) [4]:

$$e(k) = \hat{c}_k \left[ y(kT + T/2 + \hat{\tau}_k) - y(kT - T/2 + \hat{\tau}_{k-1}) \right] \tag{8.2.20}$$

- the *zero-crossing detector* (ZCD) [1]:

$$e(k) = (\hat{c}_{k-1} - \hat{c}_k) y(kT - T/2 + \hat{\tau}_{k-1}) \tag{8.2.21}$$

- the *Mueller and Mueller* detector (MMD) [5]

$$e(k) = \hat{c}_{k-1} y(kT + \hat{\tau}_k) - \hat{c}_k y[(k-1)T + \hat{\tau}_{k-1}] \qquad (8.2.22)$$

Paralleling the rules summarized at the end of Section 8.2.2, we may extend the above baseband algorithms to modulated carriers. For example, the ELD takes the form

$$e_T(k) = \text{Re}\{\hat{c}_k\} \text{Re}\left\{[y(kT + T/2 + \hat{\tau}_k) - y(kT - T/2 + \hat{\tau}_{k-1})] e^{-j\hat{\theta}_k}\right\}$$

$$+ \text{Im}\{\hat{c}_k\} \text{Im}\left\{[y(kT + T/2 + \hat{\tau}_k) - y(kT - T/2 + \hat{\tau}_{k-1})] e^{-j\hat{\theta}_k}\right\}$$

$$= \text{Re}\left\{\hat{c}_k^* e^{-j\hat{\theta}_k} [y(kT + T/2 + \hat{\tau}_k) - y(kT - T/2 + \hat{\tau}_{k-1})]\right\} \qquad (8.2.23)$$

Similarly, the ZCD becomes

$$e_T(k) = \text{Re}\left\{(\hat{c}_{k-1}^* - \hat{c}_k^*) y(kT - T/2 + \hat{\tau}_{k-1}) e^{-j\hat{\theta}_k}\right\} \qquad (8.2.24)$$

and the MMD

$$e_T(k) = \text{Re}\left\{\hat{c}_{k-1}^* y(kT + \hat{\tau}_k) e^{-j\hat{\theta}_k} - \hat{c}_k^* y[(k-1)T + \hat{\tau}_{k-1}] e^{-j\hat{\theta}_k}\right\} \qquad (8.2.25)$$

### 8.2.4. Equivalent Model of the Synchronizer

Several issues regarding the behavior of the scheme in Figure 8.1 are conveniently addressed by introducing an equivalent model for the synchronizer. In this section we derive this model taking detectors (8.2.6)-(8.2.7) as a reference but everything we shall say is equally valid for other phase/timing algorithms. In particular, this is true if the ML-based TED is replaced by either an ELD, or a ZCD or an MMD.

To begin, denote phase and timing errors with $\phi_k \triangleq \theta - \hat{\theta}_k$ and $\delta_k \triangleq \tau - \hat{\tau}_k$ and decompose both $e_P(k)$ and $e_T(k)$ into a deterministic part plus a zero-mean random component. For example, the deterministic part of the phase error is the expectation of $e_P(k)$ for fixed values of $\phi_k$ and $\delta_k$

$$S_P(\phi, \delta) \triangleq \text{E}\{e_P(k) | \phi_k = \phi, \delta_k = \delta\} \qquad (8.2.26)$$

Similarly, the deterministic part of $e_T(k)$ is defined as

# Timing Recovery with Linear Modulations

$$S_T(\phi,\delta) \triangleq E\{e_T(k)|\phi_k = \phi, \delta_k = \delta\} \tag{8.2.27}$$

These quantities may be viewed as two-dimensional S-curves and, accordingly, are referred to as *S-surfaces*. In particular, $S_P(\phi,\delta_{eq})$ represents the S-curve of the phase detector when the timing error has the fixed value $\delta_{eq}$ and, similarly, $S_T(\phi_{eq},\delta)$ is the S-curve of the timing detector when the phase error equals $\phi_{eq}$.

The random components of $e_P(k)$ and $e_T(k)$ are expressed as

$$n_P(k) \triangleq e_P(k) - S_P(\phi_k, \delta_k) \tag{8.2.28}$$

$$n_T(k) \triangleq e_T(k) - S_T(\phi_k, \delta_k) \tag{8.2.29}$$

Substituting into (8.2.8)-(8.2.9) produces the desired equivalent model [2], [6]

$$\hat{\theta}_{k+1} = \hat{\theta}_k + \gamma_P S_P(\phi_k, \delta_k) + \gamma_P n_P(k) \tag{8.2.30}$$

$$\hat{\tau}_{k+1} = \hat{\tau}_k + \gamma_T S_T(\phi_k, \delta_k) + \gamma_T n_T(k) \tag{8.2.31}$$

whose block diagram is illustrated in Figure 8.2.

One important issue about the synchronizer is concerned with its acquisition. In particular, one wonders whether the final equilibrium point will be $(\theta, \tau)$ (or something close to it). The answer requires knowledge of the S-surfaces $S_P(\phi,\delta)$ and $S_T(\phi,\delta)$ which can only be computed via intensive

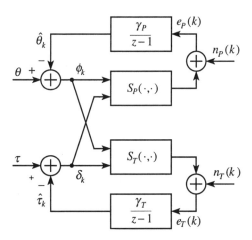

Figure 8.2. Equivalent model of the synchronizer.

simulations. The following remarks may be useful to carefully organize the simulations and save computer time.

For simplicity suppose that the disturbances $n_P(k)$ and $n_T(k)$ in (8.2.30)-(8.2.31) are negligible so that the equivalent model reduces to an autonomous system

$$\hat{\theta}_{k+1} = \hat{\theta}_k + \gamma_P S_P(\phi_k, \delta_k) \tag{8.2.32}$$

$$\hat{\tau}_{k+1} = \hat{\tau}_k + \gamma_T S_T(\phi_k, \delta_k) \tag{8.2.33}$$

Clearly, a necessary condition for an equilibrium point $(\hat{\theta}_{eq}, \hat{\tau}_{eq})$ is that the S-surfaces vanish at that point, i.e.,

$$S_P(\phi_{eq}, \delta_{eq}) = 0 \tag{8.2.34}$$

$$S_T(\phi_{eq}, \delta_{eq}) = 0 \tag{8.2.35}$$

Several interesting observations can be made about (8.2.34)-(8.2.35):

(*i*) The solutions to (8.2.34)-(8.2.35) are not necessarily all stable.

(*ii*) As the S-surfaces depend on the signal constellation, the signal-to-noise ratio and the pulse shape, so do the equilibrium points. In particular, number and location of these points may vary with SNR.

(*iii*) The ideal tracking point $\phi_{eq} = \delta_{eq} = 0$ need not be a solution to (8.2.34)-(8.2.35). With good channels, however, a stable point close to $\phi_{eq} = \delta_{eq} = 0$ is usually found.

(*iv*) As happens with any decision-directed algorithm, carrier phase is estimated within multiples of some angle $\alpha$, the symmetry angle of the signal constellation. In other words, if $(\phi_{eq}, \delta_{eq})$ is a stable point, so is $(\phi_{eq} + m\alpha, \delta_{eq})$, for any integer $m$. This phase ambiguity must be resolved by differential encoding/decoding or other methods.

(*v*) In addition to ambiguous points, *spurious points* are sometimes observed [7]-[8], especially at high SNR. Unfortunately they cannot be resolved by encoding/decoding. As the synchronizer will settle on that point (either spurious or ambiguous) which is closest to the initial conditions, the existence of spurious points is a serious hazard for correct synchronizer operation.

Returning to the solutions to (8.2.34)-(8.2.35), the question arises of identifying the stable points. A method to do so is now illustrated. Call $(\phi_{eq}, \delta_{eq})$ the generic solution. Linearizing the S-surfaces about $(\phi_{eq}, \delta_{eq})$ and denoting

## Timing Recovery with Linear Modulations

$$\begin{cases} d_P(k) \triangleq \phi_k - \phi_{eq} \\ d_T(k) \triangleq \delta_k - \delta_{eq} \end{cases} \quad (8.2.36)$$

yields

$$S_P(\phi_k, \delta_k) \approx A_{PP} d_P(k) + A_{PT} d_T(k) \quad (8.2.37)$$

$$S_T(\phi_k, \delta_k) \approx A_{TT} d_T(k) + A_{TP} d_P(k) \quad (8.2.38)$$

with

$$A_{PP} \triangleq \left[ \frac{\partial S_P(\phi, \delta)}{\partial \phi} \right]_{\phi = \phi_{eq}; \delta = \delta_{eq}} \quad (8.2.39)$$

$$A_{TT} \triangleq \left[ \frac{\partial S_T(\phi, \delta)}{\partial \delta} \right]_{\phi = \phi_{eq}; \delta = \delta_{eq}} \quad (8.2.40)$$

$$A_{PT} \triangleq \left[ \frac{\partial S_P(\phi, \delta)}{\partial \delta} \right]_{\phi = \phi_{eq}; \delta = \delta_{eq}} \quad (8.2.41)$$

$$A_{TP} \triangleq \left[ \frac{\partial S_T(\phi, \delta)}{\partial \phi} \right]_{\phi = \phi_{eq}; \delta = \delta_{eq}} \quad (8.2.42)$$

The parameters $A_{PP}$, $A_{TT}$, $A_{PT}$ and $A_{TP}$ have a simple interpretation. $A_{PP}$ represents the slope of the phase detector's S-curve at the equilibrium point. Similarly, $A_{TT}$ is the slope of the timing detector's S-curve, $S_T(\phi_{eq}, \delta)$. As for $A_{PT}$, it may be viewed as the sensitivity of $S_P(\phi, \delta_{eq})$ to changes in $\delta_{eq}$. An analogous interpretation applies to $A_{TP}$.

Substituting (8.2.37)-(8.2.38) into (8.2.32)-(8.2.33) and rearranging produces

$$d_P(k+1) = (1 - \gamma_P A_{PP}) d_P(k) - \gamma_P A_{PT} d_T(k) \quad (8.2.43)$$

$$d_T(k+1) = (1 - \gamma_T A_{TT}) d_T(k) - \gamma_T A_{TP} d_P(k) \quad (8.2.44)$$

The stability of a given solution $(\phi_{eq}, \delta_{eq})$ can now be assessed by standard techniques. In particular, $(\phi_{eq}, \delta_{eq})$ will be stable if the poles of the system (8.2.43)-(8.2.44) are all inside the unit circle. Assuming

$$0 < \gamma_P A_{PP} < 1 \quad (8.2.45)$$

$$0 < \gamma_T A_{TT} < 1 \qquad (8.2.46)$$

and working out the calculations, it turns out that this happens if and only if

$$A_{PP} > 0, \quad A_{TT} > 0 \qquad (8.2.47)$$

and

$$A_{PP} A_{TT} > A_{PT} A_{TP} \qquad (8.2.48)$$

In conclusion, while in *mono-dimensional* synchronizers stability is only guaranteed by the positive slope of the S-curves at the equilibrium point, in *two-dimensional* synchronizers the additional inequality (8.2.48) must be satisfied.

Figure 8.3 illustrates S-curves $S_P(\phi,0)$ for phase detector (8.2.6), as obtained by simulation with 64-QAM modulation and a Nyquist channel with 50% rolloff. Algorithm (8.2.25) has been chosen as the timing detector. Two potential spurious points at $\phi_{eq} = \pm 15°$ are pointed out. Further calculations (not included here) indicate that conditions (8.2.47)-(8.2.48) are satisfied at $E_s/N_0 = 23$ dB, which means that these points are stable and correspond to false locks. Simulations in Figure 8.4 indicate that timing errors will eventually settle to zero independently of the SNR. Vice versa, Figure 8.5 shows that phase errors may get stuck at about 15° when the SNR is sufficiently high. In

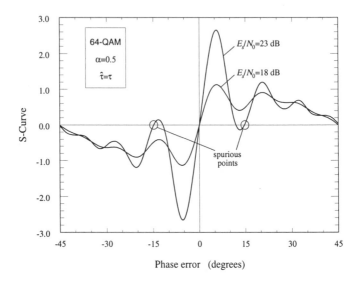

Figure 8.3. S-curves for the phase detector.

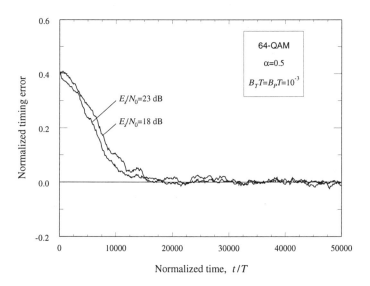

Figure 8.4. Timing acquisitions.

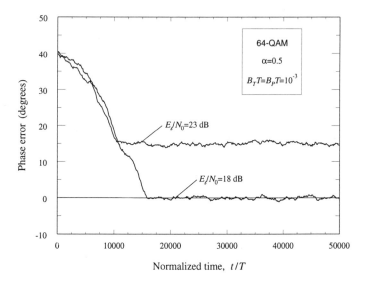

Figure 8.5. Phase acquisitions.

running these simulations the step sizes $\gamma_P$ and $\gamma_T$ have been computed by neglecting parameters $A_{PT}$ and $A_{TP}$ in (8.2.43)-(8.2.44). This turns the joint synchronizer into two separate tracking loops with bandwidths

$$B_P T \approx \frac{\gamma_P A_{PP}}{4} \qquad (8.2.49)$$

$$B_T T \approx \frac{\gamma_T A_{TT}}{4} \qquad (8.2.50)$$

The common value $10^{-3}$ has been assigned to these bandwidths.

**Exercise 8.2.3.** Consider a joint phase/timing synchronizer with the following detectors:

$$e_P(k) = \text{Im}\left\{\hat{c}_k^* y(kT + \hat{\tau}_k) e^{-j\hat{\theta}_k}\right\} \qquad (8.2.51)$$

$$e_T(k) = \text{Re}\left\{(\hat{c}_{k-1}^* - \hat{c}_k^*) y(kT - T/2 + \hat{\tau}_{k-1}) e^{-j\hat{\theta}_k}\right\} \qquad (8.2.52)$$

Show that: (i) a stable equilibrium point is located in the vicinity of the origin of the $(\phi, \delta)$ plane for a reasonable channel response; (ii) the parameters $A_{PT}$ and $A_{TP}$ vanish at that point.

*Solution.* To compute the S-surfaces about the origin we let $\hat{\theta}_k = \hat{\theta}$, $\hat{\tau}_k = \hat{\tau}_{k-1} = \hat{\tau}$, $\phi \triangleq \theta - \hat{\theta}$, $\delta \triangleq \tau - \hat{\tau}$ in (8.2.51)-(8.2.52). Then, performing straightforward calculations (under the assumption $\hat{c}_k \approx c_k$) leads to

$$S_P(\phi, \delta) = C_2 h(-\delta) \sin \phi \qquad (8.2.53)$$

$$S_T(\phi, \delta) = C_2 [h(T/2 - \delta) - h(-T/2 - \delta)] \cos \phi \qquad (8.2.54)$$

where $C_2$ is the expectation of $|c_i|^2$. Setting $S_P(\phi, \delta)$ and $S_T(\phi, \delta)$ to zero yields $\phi_{eq} = 0$ and $\delta_{eq}$ such that

$$h(T/2 - \delta_{eq}) = h(-T/2 - \delta_{eq}) \qquad (8.2.55)$$

Note that $\delta_{eq}$ is zero when $h(t)$ is even.

Finally, application of (8.2.39)-(8.2.42) results in

$$A_{PP} = C_2 h(-\delta_{eq}) \qquad (8.2.56)$$

$$A_{TT} = C_2 [h'(-T/2 - \delta_{eq}) - h'(T/2 - \delta_{eq})] \qquad (8.2.57)$$

$$A_{PT} = A_{TP} = 0 \qquad (8.2.58)$$

with $h'(t) \triangleq dh(t)/dt$.

# Timing Recovery with Linear Modulations

## 8.2.5. Tracking Performance

Tracking performance of the synchronizer is assessed by linearizing equations (8.2.30)-(8.2.31) about the operating point $(\phi_{eq}, \delta_{eq})$. The procedure for doing so is based on approximations (8.2.37)-(8.2.38) and leads to

$$d_P(k+1) = (1 - \gamma_P A_{PP})d_P(k) - \gamma_P A_{PT} d_T(k) - \gamma_P n_P(k) \quad (8.2.59)$$

$$d_T(k+1) = (1 - \gamma_T A_{TT})d_T(k) - \gamma_T A_{TP} d_P(k) - \gamma_T n_T(k) \quad (8.2.60)$$

where $d_P(k)$ and $d_T(k)$ are still as defined in (8.2.36). A block diagram illustrating the generation of $d_P(k)$ and $d_T(k)$ is displayed in Figure 8.6. As is seen, phase and timing loops are generally coupled. Coupling disappears only for $A_{PT} = A_{TP} = 0$, in which case each loop can be analyzed with the methods of Section 3.5.5 of Chapter 3. This is a crucial point as the coupling coefficients are zero for PSK/QAM modulations (see Exercise 8.2.3). It should be stressed that decoupling is a feature pertaining to steady-state operation conditions, not acquisition. Indeed, loop interactions take place during acquisition that may result in prolonged transients.

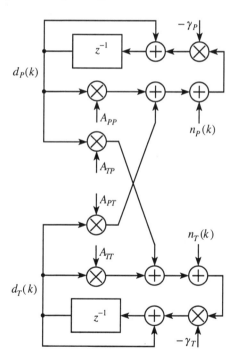

Figure 8.6. Linearized equivalent model of the synchronizer.

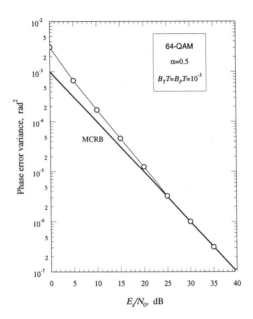

Figure 8.7. Phase error variance vs. $E_s / N_0$.

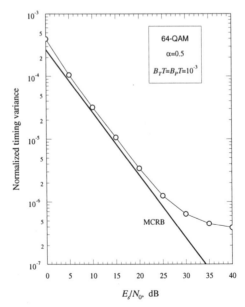

Figure 8.8. Normalized timing error variance vs. $E_s / N_0$.

# Timing Recovery with Linear Modulations

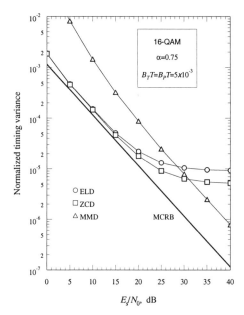

Figure 8.9. Normalized timing error variance vs. $E_s/N_0$ for $\alpha=0.75$.

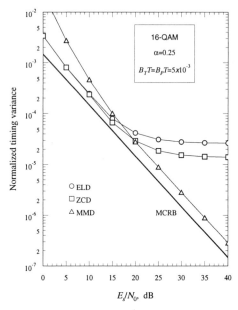

Figure 8.10. Normalized timing error variance vs. $E_s/N_0$ for $\alpha=0.25$.

Figures 8.7-8.8 illustrate phase and timing error variances for a synchronizer employing detectors (8.2.51)-(8.2.52). Modulation is 64-QAM and the channel is Nyquist with rolloff $\alpha = 0.5$. The loop bandwidths are both equal to $10^{-3}/T$. At SNR values of practical interest it is seen that the phase error variance achieves the MCRB and the timing errors (normalized to $T$) are within 1-2 dB of the MCRB.

Comparisons between timing detectors are shown in Figures 8.9-8.10. Modulation is 16-QAM and the channel is Nyquist with rolloff 0.75 or 0.25. Phase detector (8.2.51) has been used throughout. It is worth noting that, although the MMD has no self noise, its performance may be worse than that of other TEDs, especially with large rolloffs.

## 8.3. Non-Data-Aided Feedback Timing Recovery with Non-Offset Formats

So far timing has been recovered by exploiting carrier phase information. In fact, the preceding algorithms may be viewed as the combination of two *baseband* timing detectors operating on the in-phase and quadrature components of the demodulated signal. As long as the phase estimate is accurate, the I and Q signal components each provide some timing information that is eventually incorporated into a passband timing error.

Our next goal is to recover timing when no phase information is available. An obvious feature of this approach is that NDA methods are called for since data decisions cannot be relied upon unless the carrier phase is known. Two categories of NDA estimates may be envisaged, either feedback or feedforward. In this section we concentrate on feedback algorithms, starting with an ML-oriented approach.

### 8.3.1. ML-Oriented NDA Feedback Timing Recovery

Assume that any possible carrier frequency offset has been compensated so that $v$ can be set to zero in the signal model. In these circumstances the likelihood function takes the form

$$\Lambda(r|\tilde{\theta},\tilde{\tau},\tilde{c}) = \exp\left\{\frac{1}{N_0}\text{Re}\left\{\int_0^{T_0} r(t)\tilde{s}^*(t)dt\right\} - \frac{1}{2N_0}\int_0^{T_0}|\tilde{s}(t)|^2 dt\right\} \quad (8.3.1)$$

where $\tilde{s}(t)$ is the trial signal

$$\tilde{s}(t) \triangleq e^{j\tilde{\theta}}\sum_i \tilde{c}_i g(t - iT - \tilde{\tau}) \quad (8.3.2)$$

# Timing Recovery with Linear Modulations

Our aim is the location of the maximum of $\Lambda(r|\tilde{\tau})$, which is the average of $\Lambda(r|\tilde{\theta},\tilde{\tau},\tilde{c})$ with respect to $\tilde{\theta}$ and $\tilde{c}$. As usual, the problem proves to be intractable and we must resort to approximations. In this vein we drop the last integral in (8.3.1) and expand the exponential into a Taylor series. Keeping in mind that

$$\int_0^{T_0} r(t)\tilde{s}^*(t)dt \approx e^{-j\tilde{\theta}} \sum_{i=0}^{L_0-1} \tilde{c}_i^* y(iT + \tilde{\tau}) \tag{8.3.3}$$

and performing the expectation of $\Lambda(r|\tilde{\theta},\tilde{\tau},\tilde{c})$ with $\tilde{\theta}$ uniformly distributed over $[0,2\pi)$ produces (within immaterial constants)

$$\Lambda(r|\tilde{\tau}) \approx \sum_{i=0}^{L_0-1} |y(iT + \tilde{\tau})|^2 \tag{8.3.4}$$

Finally, the derivative of the $k$-th term in the right-hand side is computed for $\tilde{\tau} = \hat{\tau}_k$ and is used as an error signal to drive the function $\Lambda(r|\tilde{\tau})$ toward zero. This produces the timing error detector

$$e(k) = \text{Re}\{y^*(kT + \hat{\tau}_k)y'(kT + \hat{\tau}_k)\} \tag{8.3.5}$$

The calculation of $e(k)$ requires two separate filters to generate $y(t)$ and $y'(t)$, the matched filter and the derivative matched filter. As this complexity is often objectionable, it is of interest to dispense with the derivative matched filter by approximating the derivative $y'(t)$ with a finite difference

$$y'(kT + \hat{\tau}_k) \approx \frac{1}{T}[y(kT + T/2 + \hat{\tau}_k) - y(kT - T/2 + \hat{\tau}_{k-1})] \tag{8.3.6}$$

Note that this simplification does not come for free as it implies sampling $y(t)$ at twice the symbol rate. Anyway, substituting into (8.3.5) and dropping the factor $1/T$ yields

$$e(k) = \text{Re}\{y^*(kT + \hat{\tau}_k)[y(kT + T/2 + \hat{\tau}_k) - y(kT - T/2 + \hat{\tau}_{k-1})]\} \tag{8.3.7}$$

which looks like the ELD in (8.2.23), except that the rotated symbol $\hat{c}_k e^{j\hat{\theta}_k}$ is now replaced by the sample $y(kT + \hat{\tau}_k)$. For obvious reasons this detector will be referred to as NDA-ELD in the sequel.

**Exercise 8.3.1.** Compute the S-curve for the NDA-ELD.
*Solution.* Letting $\hat{\tau}_k = \hat{\tau}_{k-1} = \hat{\tau}$ in (8.3.7) and bearing in mind that

$$y(t) = e^{j\theta} \sum_i c_i h(t - iT - \tau) + n(t) \qquad (8.3.8)$$

yields after some algebra

$$S(\delta) = C_2 \sum_i h(iT - \delta)h(iT + T/2 - \delta) - C_2 \sum_i h(iT - \delta)h(iT - T/2 - \delta) \qquad (8.3.9)$$

As is apparent, the S-curve is independent of $\theta$, which means that the acquisition characteristics of the loop are not affected by the carrier phase.

A more succinct form for $S(\delta)$ is found by recalling the Poisson sum formula [9, p. 395]

$$\sum_i w(iT) = \frac{1}{T} \sum_m W\left(\frac{m}{T}\right) \qquad (8.3.10)$$

which applies to any finite-energy signal $w(t)$ with Fourier transform $W(f)$. Identifying $w(t)$ with $h(t - \delta)h(t + T/2 - \delta)$ and observing that the Fourier transform of $h_2(t) \triangleq h(t)h(t + T/2)$ is given by

$$H_2(f) \triangleq \int_{-\infty}^{\infty} H(v)H(f - v)e^{j\pi vT} dv \qquad (8.3.11)$$

produces

$$\sum_i h(iT - \delta)h(iT + T/2 - \delta) = \frac{1}{T} \sum_m H_2\left(\frac{m}{T}\right) e^{-j2\pi m\delta/T} \qquad (8.3.12)$$

Also, assuming a signal bandwidth less than $1/T$ (so that $H(f)=0$ for $|f| \geq 1/T$), it is recognized that $H_2(m/T)$ is zero except for the indexes $m=0$ and $m = \pm 1$. Hence, (8.3.12) reduces to

$$\sum_i h(iT - \delta)h(iT + T/2 - \delta) = \frac{1}{T} H_2(0) + \frac{2}{T} \text{Re}\left[H_2\left(\frac{1}{T}\right) e^{-j2\pi\delta/T}\right] \qquad (8.3.13)$$

A similar expression can be given to the second summation in (8.3.9):

$$\sum_i h(iT - \delta)h(iT - T/2 - \delta) = \frac{1}{T} H_2(0) - \frac{2}{T} \text{Re}\left[H_2\left(\frac{1}{T}\right) e^{-j2\pi\delta/T}\right] \qquad (8.3.14)$$

Then, substituting into (8.3.9) yields

$$S(\delta) = \frac{4C_2}{T} \text{Re}\left[H_2\left(\frac{1}{T}\right) e^{-j2\pi\delta/T}\right] \qquad (8.3.15)$$

which says that $S(\delta)$ is sinusoidal of period $T$.

Note that if $h(t)$ is an even function (as happens when the receive filter is matched to $g(t)$), then $H(f)$ is real-valued and $H_2(1/T)$ is an imaginary number. In fact, from (8.3.11) it can be shown that

$$H_2\left(\frac{1}{T}\right) = j\int_{-\infty}^{\infty} H\left(\frac{1}{2T}+f\right)H\left(\frac{1}{2T}-f\right)\cos(\pi fT)df \qquad (8.3.16)$$

Thus, inserting into (8.3.15) produces

$$S(\delta) = \frac{4C_2 K}{T}\sin(2\pi\delta/T) \qquad (8.3.17)$$

with

$$K \triangleq \int_{-\infty}^{\infty} H\left(\frac{1}{2T}+f\right)H\left(\frac{1}{2T}-f\right)\cos(\pi fT)df \qquad (8.3.18)$$

It is readily apparent from (8.3.18) that $K$ decreases with the bandwidth of $H(f)$ and, in particular, it vanishes as the bandwidth tends to $1/2T$. Thus, the NDA-ELD is not suitable for narrow-band signaling.

### 8.3.2. The Gardner Detector and Its Performance

The following alternative NDA detector has been proposed by Gardner [10]:

$$e(k) = \text{Re}\left\{[y(kT-T+\hat{\tau}_{k-1}) - y(kT+\hat{\tau}_k)]y^*(kT-T/2+\hat{\tau}_{k-1})\right\} \qquad (8.3.19)$$

It is identical to the GAD discussed in the context of baseband transmission except that the samples of $y(t)$ are now complex-valued. Its S-curve is found by paralleling the steps in Exercise 8.3.1 and is the same as with the NDA-ELD. Thus, as the latter is phase independent, so is GAD.

Figures 8.11-8.12 illustrate the tracking performance of NDA-ELD and GAD as obtained by simulation. Modulation is 16-QAM and the overall channel response is Nyquist with rolloffs of 75% and 25%. In both cases a loop bandwidth of $5 \cdot 10^{-3}/T$ is chosen. As is seen, performance is strongly dependent on the rolloff factor.

Self-noise elimination by means of prefiltering [11]-[12] can be pursued with the methods discussed in Section 6.5.3 in Chapter 6. It turns out that, even with modulated carriers, short prefilters (of 3-5 taps) are sufficient to practically achieve the goal. Figure 8.13 illustrates this fact by comparing the tracking

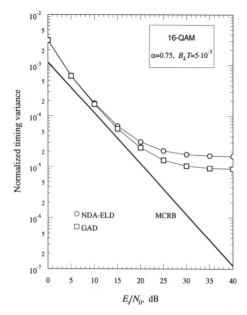

Figure 8.11. Normalized timing error variance vs. $E_s/N_0$ for $\alpha = 0.75$.

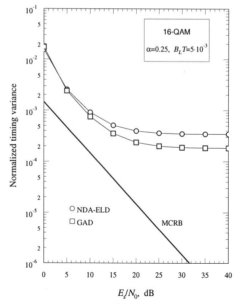

Figure 8.12. Normalized timing error variance vs. $E_s/N_0$ for $\alpha = 0.25$.

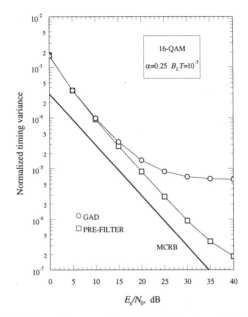

Figure 8.13. Tracking performance of GAD with/without pre-filtering for $\alpha = 0.25$.

performance of a plain GAD with that of a GAD with prefiltering. Modulation is 16-QAM with a 25% rolloff and a 5-tap prefilter is used.

## 8.4. Non-Data-Aided Feedforward Estimators with Non-Offset Formats

In the previous sections several feedback recovery schemes have been illustrated. Their tracking performance is generally good (with well-behaved channels) but their acquisitions may be too long for burst transmissions. Feedforward methods are preferable when short estimation times are needed. In the following we illustrate two NDA feedforward estimators for these applications. As they are simple extensions of the ML-based estimator and the Oerder and Meyr (O&M) method discussed in Chapter 7, the treatment will be rather concise as we can draw heavily for notations and approximations from the preceding chapter.

The derivation of the ML-based estimator strictly follows the approach in Section 7.6.1. In particular, the received waveform is first fed to an anti-alias filter (AAF) and then sampled at some rate $1/T_s$, a multiple $N$ of the symbol rate. An oversampling factor of 2 is generally sufficient but the ensuing formulas are developed for a generic $N$. We make the usual assumptions that: ($i$) the AAF has an ideal brick-wall transfer function with an edge frequency $B_{AAF}$ which is sufficiently large so as not to distort the signal components; ($ii$) the sampling rate equals $2B_{AAF}$. Then, denoting $x \triangleq \{x(kT_s)\}$ the samples from the AAF, the likelihood function $\Lambda(x|\tilde{\tau},\tilde{\theta},\tilde{c})$ is written as

$$\Lambda(x|\tilde{\tau},\tilde{\theta},\tilde{c}) = \exp\left\{\frac{T_s}{N_0}\sum_{k=0}^{NL_0-1}\text{Re}\left[x(kT_s)\tilde{s}^*(kT_s)\right] - \frac{T_s}{2N_0}\sum_{k=0}^{NL_0-1}|\tilde{s}(kT_s)|^2\right\} \quad (8.4.1)$$

where $\tilde{s}(t)$ is the trial signal

$$\tilde{s}(t) \triangleq e^{j\tilde{\theta}}\sum_i \tilde{c}_i g(t - iT - \tilde{\tau}) \quad (8.4.2)$$

Next, the second summation in (8.4.1) is dropped for simplicity and the exponential is expanded into a Taylor series truncated to the third term. Finally, the result is averaged with respect to data symbols and carrier phase. Skipping the details, this leads to

$$\Lambda(r|\tilde{\tau}) \approx \sum_{k_1=0}^{NL_0-1}\sum_{k_2=0}^{NL_0-1} x(k_1 T_s) x^*(k_2 T_s) F(k_1, k_2, \tilde{\tau}) \quad (8.4.3)$$

with

$$F(k_1, k_2, \tilde{\tau}) \triangleq \sum_i g(k_1 T_s - iT - \tilde{\tau})g(k_2 T_s - iT - \tilde{\tau}) \quad (8.4.4)$$

In Section 7.6.1 it is shown that the function $F(k_1, k_2, \tilde{\tau})$ is periodic over $\tilde{\tau}$ of period $T$ and, if $g(t)$ is bandlimited to $\pm 1/T$, it can be represented as

$$F(k_1, k_2, \tilde{\tau}) = F_0(k_1, k_2) + 2\text{Re}\left\{F_1(k_1, k_2)e^{j2\pi\tilde{\tau}/T}\right\} \quad (8.4.5)$$

with

$$F_m(k_1, k_2) = \frac{1}{T}\int_0^T F(k_1, k_2, \tilde{\tau})e^{-j2\pi m\tilde{\tau}/T}d\tilde{\tau} \quad (8.4.6)$$

Then, substituting into (8.4.3) and pruning off immaterial constants yields

$$\Lambda(r|\tilde{\tau}) \approx \text{Re}\left\{e^{j2\pi\tilde{\tau}/T}\sum_{k_1=0}^{NL_0-1}\sum_{k_2=0}^{NL_0-1} x(k_1 T_s) x^*(k_2 T_s) F_1(k_1, k_2)\right\} \quad (8.4.7)$$

from which the desired estimator is obtained by noting that the right-hand side in (8.4.7) achieves a maximum for

$$\hat{\tau} = -\frac{T}{2\pi} \arg\left\{ \sum_{k_1=0}^{NL_0-1} \sum_{k_2=0}^{NL_0-1} x(k_1 T_s) x^*(k_2 T_s) F_1(k_1, k_2) \right\} \quad (8.4.8)$$

This estimator may be written in an alternative form using the following formula from Appendix 7.A:

$$F_1(k_1, k_2) = \frac{1}{T} q[(k_1 - k_2) T_s] e^{-j\pi(k_1 + k_2)/N} \quad (8.4.9)$$

where $q(t)$ is an even function with a Fourier transform

$$Q(f) = G\left(f - \frac{1}{2T}\right) G^*\left(f + \frac{1}{2T}\right) \quad (8.4.10)$$

and $G(f)$ is the Fourier transform of $g(t)$. Inserting (8.4.9) into (8.4.8) and following the same steps as in Section 7.6.1 produces the estimator in its final form

$$\hat{\tau} = -\frac{T}{2\pi} \arg\left\{ \sum_{k=ND}^{N(L_0+D)-1} x[(k - ND) T_s] e^{-j\pi(k-ND)/N} z[(k - ND) T_s] \right\} \quad (8.4.11)$$

with

$$z(k T_s) \triangleq \left[ x^*(k_2 T_s) e^{-j\pi k_2 / N} \right] \otimes q(k T_s) \quad (8.4.12)$$

Here, the integer $D$ represents the semi-duration of $q(t)$ in symbol periods, i.e.,

$$q(t) \approx 0 \quad \text{for } |t| > DT \quad (8.4.13)$$

The estimator (8.4.11) is illustrated in Figure 8.14. As is seen, the samples from the anti-alias filter follow two parallel branches. In the upper branch $x(kT_s)$ is first complex-conjugated, then is multiplied by $e^{-j\pi k/N}$ and finally is filtered. In the lower branch it is multiplied by $e^{-j\pi k/N}$ and then is delayed. Branch outputs are multiplied together and the products are accumulated. The argument of the accumulator output gives the timing estimate within a factor of $-T/(2\pi)$.

The O&M estimator has the form [13]

$$\hat{\tau} = -\frac{T}{2\pi} \arg\left\{ \sum_{k=0}^{NL_0-1} |x(k T_s)|^2 e^{-j 2\pi k/N} \right\} \quad (8.4.14)$$

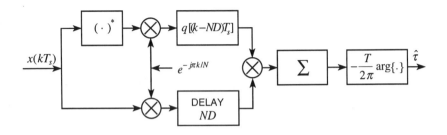

Figure 8.14. Block diagram of the ML-based estimator.

where $x(kT_s)$ are samples from the matched filter (not from the AAF, as happens in the ML-based scheme). As noted earlier with baseband transmission, the O&M algorithm is simpler to implement than the ML-based scheme as it does not involve any filtering of $x(kT_s)$. In fact it can be viewed as a limit case of the ML-based scheme when: (*i*) the AAF is replaced by a matched filter; (*ii*) the filter $q[(k-ND)T_s]$ and the delay block in Figure 8.14 are by-passed. The O&M estimator needs an oversampling of $N=4$, however, whereas $N=2$ is sufficient with the ML-based method.

Figures 8.15-8.16 compare the error variances of the two estimators in the case of 16-QAM modulation. The following assumptions are made with the ML-based algorithm:

(*i*) the anti-alias filter is an eight-pole Butterworth one with a −3 dB bandwidth of $1/T$;

(*ii*) the oversampling factor is $N=2$;

(*iii*) the pulse $g(t)$ corresponds to a root-raised-cosine rolloff function;

(*iv*) the impulse response $q[(k-ND)T_s]$ is derived by truncating $q(t)$ to $\pm 5T$;

(*v*) the delay $ND$ is set equal to 10.

With the O&M algorithm it is assumed that:

(*i*) the low-pass filter is matched to $g(t)$, which has the same shape as with the ML-based algorithm;

(*ii*) the oversampling factor is $N=4$.

It appears that the two methods are basically equivalent with large rolloffs whereas the ML-based scheme has significantly less self noise with small rolloffs.

# Timing Recovery with Linear Modulations

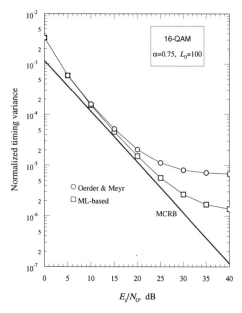

Figure 8.15. Performance comparisons with $\alpha=0.75$.

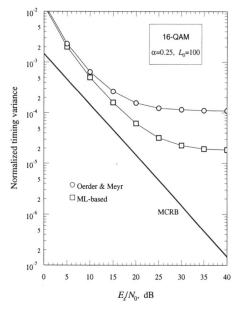

Figure 8.16. Performance comparisons with $\alpha=0.25$.

## 8.5. Timing Recovery with Frequency-Flat Fading Channels

In several mobile communication systems the transmission medium can be modeled as a frequency-flat fading channel [14]-[16] and the received signal component has the form

$$s(t) = a(t)\sum_i c_i g(t - iT - \tau) \qquad (8.5.1)$$

where the channel gain $a(t)$ is a complex-valued Gaussian process with a (Doppler) bandwidth $f_d$. Also, in many applications the frequency $f_d$ is small compared with the symbol rate $1/T$ so that $a(t)$ is nearly constant over the duration of the generic pulse in (8.5.1) and we can write

$$a(t)g(t - iT - \tau) \approx a(i)g(t - iT - \tau) \qquad (8.5.2)$$

where $a(i)$ is short for $a(iT+\tau)$. Correspondingly, equation (8.5.1) becomes

$$s(t) \approx \sum_i a(i) c_i g(t - iT - \tau) \qquad (8.5.3)$$

A warning about this approximation is useful. Filtering (8.5.3) into the matched filter $g(-t)$ produces

$$\sum_i a(i) c_i h(t - iT - \tau) \qquad (8.5.4)$$

with $h(t) \triangleq g(t) \otimes g(-t)$. So, if $h(t)$ is Nyquist, the samples of (8.5.4) taken at the correct decision times exhibit no intersymbol interference (ISI). By contrast, filtering (8.5.1) results in some amount of ISI which, of course, tends to vanish when the product $f_d T$ goes to zero [17]. In other words, approximation (8.5.3) implies *total* ISI elimination with Nyquist filtering. By contrast, filtering the true signal (8.5.1) will always result in some residual ISI. As we shall see, this residual ISI manifests itself as self noise even in synchronizers that would be self-noise free if (8.5.3) were exact.

For mathematical convenience in the following we investigate timing recovery making the approximation (8.5.3). In approaching this problem we shall distinguish between decision-directed (DD) and non-data-aided (NDA) methods. The first instance is motivated by the intuitive notion that timing accuracy can be improved by exploiting knowledge of the transmitted symbols. Furthermore, since data and channel are jointly estimated [18]-[21] in a coherent receiver, DD and channel-aided (CA) timing recovery becomes a practical option. This topic is investigated in the next sub-section.

# Timing Recovery with Linear Modulations

Interactions between data detection, channel estimation and timing recovery in DD-CA schemes might slow down the receiver start-up. As this subject has not yet been investigated in the literature, the range of application of DD-CA techniques is not clear at this time. Intensive computer simulations are required to understand the problem, even though some limitations are likely to arise with burst transmissions or digital voice services due to restrictions in acquisition times. For these applications a more secure route is to rely on NDA and non-channel-aided (NCA) methods. This subject is addressed in Sections 8.5.2 and 8.5.3.

## 8.5.1. DD-CA Timing Recovery

Assume that reliable estimates, $\{\hat{c}_i\}$ and $\{\hat{a}(i)\}$, of data symbols and channel gains are available. Then, the trial signal becomes

$$\tilde{s}(t) = \sum_i \hat{a}(i)\hat{c}_i g(t - iT - \tilde{\tau}) \tag{8.5.5}$$

Comparing (8.5.5) with the signal we would face with an AWGN channel and a variable carrier phase $\hat{\theta}_i$, i.e.,

$$\tilde{s}(t) = \sum_i e^{j\hat{\theta}_i} \hat{c}_i g(t - iT - \tilde{\tau}) \tag{8.5.6}$$

it appears that the two formulas coincide with the substitution

$$e^{j\hat{\theta}_i} \rightarrow a(i) \tag{8.5.7}$$

Hence, the same methods adopted in Section 8.2 can be used here with only marginal adjustments. Skipping the details, this leads to the following timing detectors:

(i) ML-based TED

$$e(k) = \text{Re}\left\{\hat{c}_k^* \hat{a}^*(k)\left[y'(kT + \hat{\tau}_k) - \sum_{m=k-D}^{k+D} \hat{c}_m \hat{a}(m) h'[(k-m)T]\right]\right\} \tag{8.5.8}$$

(ii) ELD

$$e(k) = \text{Re}\left\{\hat{c}_k^* \hat{a}^*(k)\left[y(kT + T/2 + \hat{\tau}_k) - y(kT - T/2 + \hat{\tau}_{k-1})\right]\right\} \tag{8.5.9}$$

(*iii*) ZCD

$$e(k) = \text{Re}\{(\hat{c}_{k-1}^* - \hat{c}_k^*)\hat{a}^*(k-1/2)y(kT - T/2 + \hat{\tau}_{k-1})\} \qquad (8.5.10)$$

(*iv*) MMD

$$e(k) = \text{Re}\{\hat{c}_{k-1}^*\hat{a}^*(k)y(kT + \hat{\tau}_k) - \hat{c}_k^*\hat{a}^*(k-1)y[(k-1)T + \hat{\tau}_{k-1}]\} \qquad (8.5.11)$$

where the notations $\hat{a}(k)$ and $\hat{a}(k-1/2)$ mean $\hat{a}(kT + \hat{\tau}_k)$ and $\hat{a}(kT - T/2 + \hat{\tau}_{k-1})$ respectively. For example, Figure 8.17 shows a block diagram for a symbol synchronizer endowed with an MMD.

The tracking behavior of timing loops endowed with these detectors can be assessed with the methods indicated in Section 8.2 for the AWGN channel. The only novelty is that, in the present context, expectations must also be performed with respect to the channel gains $\{a(i)\}$.

Figure 8.18 illustrates simulation results for the tracking performance of a synchronizer using an ML-based TED. Modulation is QPSK and the overall channel response is Nyquist with 50% rolloff. The simulated signal is modeled as indicated in (8.5.1) and the complex-valued gain $a(t)$ is generated by filtering two independent real-valued white Gaussian noise processes into fourth-order Butterworth filters. The 3 dB bandwidth of the filters is taken as the fading bandwidth $f_d$. For convenience, the fading power is normalized to unity so that the average received signal energy $E_s$ coincides with the transmitted energy. The detector parameter $D$ is set equal to 2. As perfect channel gain knowledge and data decisions are assumed, the indicated results represent lower bounds for practical systems employing channel estimation and data detection techniques (see [18]-[21]). It is seen that simulations stay close to the MCRB up to certain SNR values, depending on the fading rate. Then, self noise shows up as a consequence of fading-induced ISI as predicted earlier.

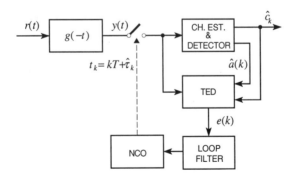

Figure 8.17. Block diagram of a symbol synchronizer.

# Timing Recovery with Linear Modulations

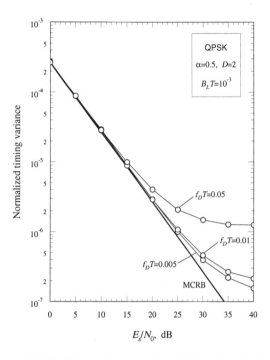

Figure 8.18. Tracking performance of an ML-based synchronizer.

**Exercise 8.5.1.** Compute the S-curve about the origin for an ELD assuming that: (*i*) timing errors are small and decisions are reliable ($\hat{c}_k \approx c_k$); (*ii*) symbols are uncorrelated, with zero-mean and variance $C_2$.

*Solution.* The matched-filter output has the form

$$y(t) = \sum_i a(i) c_i h(t - iT - \tau) + n(t) \qquad (8.5.12)$$

Inserting into (8.5.9) and letting $\hat{\tau}_{k-1} = \hat{\tau}_k = \hat{\tau}$ and $\hat{c}_k = c_k$ yields

$$e(k) = \mathrm{Re}\left\{\sum_i a(i)\hat{a}^*(k) c_i \hat{c}_k^* h[(k-i)T + T/2 - \delta]\right\}$$

$$- \mathrm{Re}\left\{\sum_i a(i)\hat{a}^*(k) c_i \hat{c}_k^* h[(k-i)T - T/2 - \delta]\right\}$$

$$+ \mathrm{Re}\left\{\hat{c}_k^* \hat{a}^*(k)[n(kT + T/2 + \hat{\tau}) - n(kT - T/2 + \hat{\tau})]\right\} \qquad (8.5.13)$$

Next, averaging over noise, data and fading produces

$$S(\delta) = R(0)C_2[h(T/2-\delta)-h(-T/2-\delta)] \qquad (8.5.14)$$

where $R(\tau) \triangleq \mathrm{E}\{a(t+\tau)a^*(t)\}$ is the fading autocorrelation function.

For $R(0) = 1$ equation (8.5.14) reduces to the same S-curve as with an AWGN channel (see Exercise 7.4.4). Also, as $h(t)$ is even, $S(\delta)$ has a null with a positive slope at the origin, which means that $\hat{\tau} = \tau$ is a stable tracking point.

### 8.5.2. NDA Timing Recovery

In Sections 8.3.1 and 8.3.2 two NDA detectors have been discussed for the AWGN channel, the NDA-ELD

$$e(k) = \mathrm{Re}\left\{y^*(kT+\hat{\tau}_k)[y(kT+T/2+\hat{\tau}_k)-y(kT-T/2+\hat{\tau}_{k-1})]\right\} \qquad (8.5.15)$$

and the GAD

$$e(k) = \mathrm{Re}\left\{[y(kT-T+\hat{\tau}_{k-1})-y(kT+\hat{\tau}_k)]y^*(kT-T/2+\hat{\tau}_{k-1})\right\} \qquad (8.5.16)$$

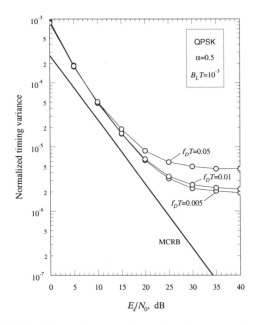

Figure 8.19. Tracking performance of a synchronizer endowed with GAD.

## Timing Recovery with Linear Modulations

Physical and mathematical considerations suggest that these detectors may also be useful with fading channels. Indeed, Exercise 8.5.2 shows that the NDA-ELD is a low-SNR approximation to the ML timing estimator. Furthermore, Exercise 8.5.3 indicates that the S-curve for GAD has essentially the same shape either with or without fading.

Figure 8.19 illustrates simulations for the tracking performance of a synchronizer endowed with GAD. Modulation is QPSK and the overall channel response is Nyquist with 50% rolloff. The channel gain is modeled as with the simulations in Figure 8.18. Again, the self-noise level increases with $f_d$ as a consequence of fading-induced ISI.

**Exercise 8.5.2.** Paralleling the arguments in Section 8.3.1, show that the NDA-ELD is a low-SNR approximation to the ML timing estimator [22].
*Solution.* Start with the likelihood function

$$\Lambda(r|\tilde{\tau},\tilde{a},\tilde{c}) = \exp\left\{\frac{1}{N_0}\text{Re}\left\{\int_0^{T_0} r(t)\tilde{s}^*(t)dt\right\} - \frac{1}{2N_0}\int_0^{T_0}|\tilde{s}(t)|^2 dt\right\} \quad (8.5.17)$$

with

$$\tilde{s}(t) = \sum_i \tilde{a}(i)\tilde{c}_i g(t - iT - \tilde{\tau}) \quad (8.5.18)$$

Neglect the last integral in (8.6.17) for simplicity. Then, letting $L_0 \triangleq T_0/T$ and writing

$$\int_0^{T_0} r(t)\tilde{s}^*(t)dt \approx \sum_{i=0}^{L_0-1} \tilde{a}^*(i)\tilde{c}_i^* y(iT + \tilde{\tau}) \quad (8.5.19)$$

yields

$$\Lambda(r|\tilde{\tau},\tilde{a},\tilde{c}) \approx \exp\left\{\frac{1}{N_0}\text{Re}\left\{\sum_{i=0}^{L_0-1} \tilde{a}^*(i)\tilde{c}_i^* y(iT + \tilde{\tau})\right\}\right\} \quad (8.5.20)$$

Assume a low SNR such that the exponential can be expanded into a Taylor series truncated to the third term. Then, averaging with respect to data symbols and channel gains leads to

$$\Lambda(r|\tilde{\tau}) \approx \sum_{i=1}^{L_0} |y(iT + \tilde{\tau})|^2 \quad (8.5.21)$$

where some irrelevant constants have been dropped.

Equation (8.5.21) is identical to (8.3.4) which applies for NDA-ML timing estimation with AWGN channels. On the other hand, the NDA-ELD has been brought out just from (8.3.4). Thus, the NDA-ELD is an approximation to the ML timing estimator even with fading channels.

**Exercise 8.5.3.** Compute the S-curve for the GAD in (8.5.16).
*Solution.* Letting $\hat{\tau}_{k-1} = \hat{\tau}_k = \hat{\tau}$ and $\delta = \tau - \hat{\tau}$ in (8.5.16) and using (8.5.12) yields

$$e(k) = \text{Re}\left\{\sum_{i_1}\sum_{i_2} a^*(i_1)a(i_2)c_{i_1}^*c_{i_2}h[(k-i_1)T - T/2 - \delta]h[(k-i_2-1)T - \delta]\right\}$$

$$-\text{Re}\left\{\sum_{i_1}\sum_{i_2} a^*(i_1)a(i_2)c_{i_1}^*c_{i_2}h[(k-i_1)T - T/2 - \delta]h[(k-i_2)T - \delta]\right\}$$

$$+N(k) \qquad (8.5.22)$$

where $N(k)$ is a zero-mean noise term. Next, averaging with respect to data, fading and thermal noise produces

$$S(\delta) = -R(0)C_2 \sum_i h(iT - \delta)h(iT - T/2 - \delta)$$

$$+ R(0)C_2 \sum_i h(iT - T/2 - \delta)h(iT - T - \delta) \qquad (8.5.23)$$

Finally, reasoning as in Exercise 8.3.1 gives

$$S(\delta) = \frac{4R(0)C_2 K}{T}\sin(2\pi\delta/T) \qquad (8.5.24)$$

where $K$ is defined as

$$K \triangleq \int_{-\infty}^{\infty} H\left(\frac{1}{2T} + f\right) H\left(\frac{1}{2T} - f\right) \cos(\pi f T) df \qquad (8.5.25)$$

and $H(f)$ is the Fourier transform of $h(t)$.

### 8.5.3. High SNR Approximation to the ML Estimator

In Exercise 8.5.2 it has been shown that the NDA-ELD is a low-SNR approximation to the ML timing estimator. Unfortunately, the low-SNR argument adopted there is questionable with transmissions over fading channels where

# Timing Recovery with Linear Modulations

$E_s/N_0$ values on the order of 20-30 dB are typical. Actually, a high-SNR approximation would be more useful. This route has been pursued by Ginesi in [23] under the assumption of Rayleigh fading. His method is now illustrated.

The starting point is still equation (8.5.17) but the subsequent steps involve somewhat milder approximations than in Exercise 8.5.2. In particular, the second integral in (8.5.17) is not discarded here. In fact, assuming an observation interval of many symbols ($L_0 \gg 1$), the integral is written as

$$\int_0^{T_0} |\tilde{s}(t)|^2 dt \approx \sum_{i_1=0}^{L_0-1} \sum_{i_2=0}^{L_0-1} \tilde{a}(i_1)\tilde{a}^*(i_2)\tilde{c}_{i_1}\tilde{c}_{i_2}^* h_{i_1,i_2} \qquad (8.5.26)$$

where

$$h_{i_1,i_2} \triangleq \int_0^{T_0} g(t - i_1 T - \tilde{\tau})g(t - i_2 T - \tilde{\tau})dt \qquad (8.5.27)$$

Thus, substituting (8.5.19) and (8.5.26) into (8.5.17) yields

$$\Lambda(r|\tilde{\tau},\tilde{a},\tilde{c}) = \exp\left\{\frac{1}{N_0}\sum_{i=0}^{L_0-1}\mathrm{Re}\left\{\tilde{a}^*(i)\tilde{c}_i^* y(iT + \tilde{\tau})\right\} \right.$$
$$\left. - \frac{1}{2N_0}\sum_{i_1=0}^{L_0-1}\sum_{i_2=0}^{L_0-1}\tilde{a}(i_1)\tilde{a}^*(i_2)\tilde{c}_{i_1}\tilde{c}_{i_2}^* h_{i_1,i_2}\right\} \qquad (8.5.28)$$

At this point $\Lambda(r|\tilde{\tau},\tilde{a},\tilde{c})$ is expressed in a more compact form by introducing the following matrix notations:

$$\tilde{a} \triangleq [\tilde{a}(0),\tilde{a}(1),\ldots,\tilde{a}(L_0-1)]^T \qquad (8.5.29)$$

$$\tilde{C} \triangleq \mathrm{diag}(\tilde{c}_0,\tilde{c}_1,\ldots,\tilde{c}_{L_0-1}) \qquad (8.5.30)$$

$$y \triangleq [y(\tilde{\tau}),y(T+\tilde{\tau}),\ldots,y[(L_0-1)T+\tilde{\tau}]]^T \qquad (8.5.31)$$

$$H \triangleq \{h_{i_1,i_2}\} \qquad (8.5.32)$$

$$\tilde{b} \triangleq \tilde{C}\tilde{a} \qquad (8.5.33)$$

where the superscript "T" means *transpose*. Inserting into (8.5.28) produces

$$\Lambda(r|\tilde{\tau},\tilde{a},\tilde{c}) = \exp\left\{\frac{1}{2N_0}\left[\tilde{b}^{*T}y + y^{*T}\tilde{b} - \tilde{b}^{*T}H\tilde{b}\right]\right\} \quad (8.5.34)$$

Note that the dependence of $\Lambda(r|\tilde{\tau},\tilde{a},\tilde{c})$ on $\tilde{\tau}$ takes place via $y$ and $H$.

The next step is to average the likelihood function over the channel gains (while keeping $\tilde{c}$ and $\tilde{\tau}$ fixed). To this end we observe that (8.5.34) depends on $\tilde{b} = \tilde{C}\tilde{a}$ rather than $\tilde{a}$ itself and, therefore, $\Lambda(r|\tilde{\tau},\tilde{a},\tilde{c})$ can be averaged with respect to $\tilde{b}$. On the other hand, for a given $\tilde{c}$ and with Rayleigh fading, $\tilde{b}$ is a zero-mean Gaussian vector with covariance matrix

$$R_{\tilde{b}} \triangleq E\{\tilde{b}\tilde{b}^{*T}\} = \tilde{C}R_{\tilde{a}}\tilde{C}^* \quad (8.5.35)$$

where $R_{\tilde{a}}$ is the covariance matrix of $\tilde{a}$. Thus, its probability density function has the form [9, p. 199]

$$f_{\tilde{b}}(\tilde{b}) = \frac{1}{\pi^{L_0}\det[R_{\tilde{b}}]}\exp\{-\tilde{b}^{*T}R_{\tilde{b}}^{-1}\tilde{b}\} \quad (8.5.36)$$

Then, multiplying $\Lambda(r|\tilde{\tau},\tilde{a},\tilde{c})$ by $f_{\tilde{b}}(\tilde{b})$ and integrating with respect to $\tilde{b}$ yields after some algebra

$$\Lambda(r|\tilde{\tau},\tilde{c}) = \frac{1}{\det\left[I + \frac{1}{2N_0}\tilde{C}R_{\tilde{a}}\tilde{C}^*H\right]}$$
$$\times \exp\left\{\left(\frac{1}{2N_0}\right)^2 y^{*T}\left[\tilde{C}R_{\tilde{a}}^{-1}\tilde{C}^* + \frac{1}{2N_0}H\right]^{-1}y\right\} \quad (8.5.37)$$

where $I$ is the $L_0 \times L_0$ identity matrix.

To go further we should average $\Lambda(r|\tilde{\tau},\tilde{c})$ with respect to the data. Unfortunately this appears a formidable task and, in consequence, we choose to make some approximations. In particular, we assume a high SNR and look for the asymptotic expression of $\Lambda(r|\tilde{\tau},\tilde{c})$ as $N_0$ tends to zero. As is argued in Appendix 8.A, decreasing $N_0$ makes $\Lambda(r|\tilde{\tau},\tilde{c})$ independent of $\tilde{C}$ and, in fact, the logarithm of $\Lambda(r|\tilde{\tau},\tilde{c})$ takes the limit form (within irrelevant constants)

$$\log \Lambda(r|\tilde{\tau},\tilde{c}) \approx y^{*T}H^{-1}y \quad (8.5.38)$$

Physically speaking this means that data symbols are useless for ML timing estimation when the SNR is large. An analogous property for AWGN channels has been pointed out in [8] and [24].

Next we concentrate on the location of the maximum of (8.5.38) or, equivalently, of the null of the derivative of $y^{*T}H^{-1}y$ over $\tilde{\tau}$. In Appendix 8.B it is shown that the derivative is given by

$$\frac{d}{d\tilde{\tau}}[y^{*T}H^{-1}y] = 2\operatorname{Re}\{z^{*T}[y' + Qz]\} \tag{8.5.39}$$

where $z \triangleq H^{-1}y$,

$$y' \triangleq \frac{dy}{d\tilde{\tau}} \tag{8.5.40}$$

and $Q$ is an $L_0 \times L_0$ matrix with entries

$$q_{i_1,i_2} \triangleq -\int_0^{T_0} g'(t - i_1 T - \tilde{\tau})g(t - i_2 T - \tilde{\tau})dt \tag{8.5.41}$$

$g'(t)$ being the derivative of $g(t)$.

Let us turn the right-hand side of (8.5.39) into scalar terms. Denoting by $z_k$ the elements of $z$ produces

$$\frac{d}{d\tilde{\tau}}[y^{*T}H^{-1}y] = 2\sum_{i=0}^{L_0-1}\operatorname{Re}\left\{z_i^*\left[y'(iT + \tilde{\tau}) + \sum_{m=0}^{L_0-1}z_m q_{i,m}\right]\right\} \tag{8.5.42}$$

The location of the null of (8.5.42) can now be found by the usual feedback method wherein the generic term in the sum over $i$ is used as an error signal to recursively adjust the timing estimate. Formally, we update $\hat{\tau}_k$ as

$$\hat{\tau}_{k+1} = \hat{\tau}_k + \gamma e(k) \tag{8.5.43}$$

where

$$e(k) = \operatorname{Re}\left\{z_k^*\left[y'(kT + \hat{\tau}_k) + \sum_{m=0}^{L_0-1}z_m q_{k,m}\right]\right\} \tag{8.5.44}$$

and $\gamma$ is a step-size parameter.

The error signal can be put in a simpler form under the following assumptions:

(*i*) the channel response $h(t)$ is Nyquist;
(*ii*) the time index $k$ is far from the extremes of the interval $(0, L_0 - 1)$.

The procedure can be divided into three steps. First, since $h(t) = g(t) \otimes g(-t)$, from assumption (*i*) and (8.5.27) we have

$$h_{i_1,i_2} = \begin{cases} 1 & i_1 = i_2 \\ 0 & i_1 \neq i_2 \end{cases} \qquad (8.5.45)$$

for indexes $i_1$ and $i_2$ away from 0 or $L_0-1$. Denoting by $\pm LT$ the range where $h(t)$ takes significant values, equation (8.5.45) holds as long as $i_1$ and $i_2$ are at least $L$ steps away from both 0 and $L_0-1$. Since $L \ll L_0$, this means that $\boldsymbol{H}$ is nearly an identity matrix and, in consequence, vectors $\boldsymbol{y}$ and $\boldsymbol{z}$ are approximately equal (since $\boldsymbol{z} = \boldsymbol{H}^{-1}\boldsymbol{y}$).

Second, from (*ii*) and (8.5.41) it is recognized that

$$q_{k,m} \approx -\int_{-\infty}^{\infty} g'(t - kT - \tilde{\tau})g(t - mT - \tilde{\tau})dt$$
$$= -h'[(k-m)T] \qquad (8.5.46)$$

where $h'(t)$ is the derivative of $h(t)$.

Third, as $h'[(i-m)T]$ decays to zero when the distance $|k-m|$ increases beyond some value $D$ and index $k$ is far from either 0 or $L_0-1$ by assumption, then the sum over $m$ in (8.5.44) can be extended over the interval $k - D \leq m \leq k + D$ without consequences.

Putting these facts together leads to the following expression of the NDA-ML detector:

$$e(k) = \text{Re}\left\{ y^*(kT + \hat{\tau}_k) \left[ y'(kT + \hat{\tau}_k) - \sum_{m=k-D}^{k+D} y(mT + \hat{\tau}_m) h'[(k-m)T] \right] \right\} \qquad (8.5.47)$$

which is strikingly similar to the error signal (8.2.7) for the DD joint phase and timing estimation. Indeed the two formulas coincide when $\hat{c}_k e^{j\hat{\theta}_k}$ in (8.2.7) is replaced by the sample $y(kT + \hat{\tau}_k)$ and the slowly varying nature of $\hat{\theta}_k$ is recognized.

Figure 8.20 illustrates simulations for the tracking performance of a synchronizer endowed with detector (8.5.47). Modulation is QPSK and the overall channel impulse response is Nyquist, with 50% rolloff. Parameter $D$ is set equal to 2. The fading distortion is modeled as described with the simulations in Figures 8.18 and 8.19. Comparing with those figures it appears that the NDA-ML detector loses 2-3 dB with the respect to the DD-CA-ML-detector in Figure 8.18 whereas it has better performance than the GAD in Figure 8.19.

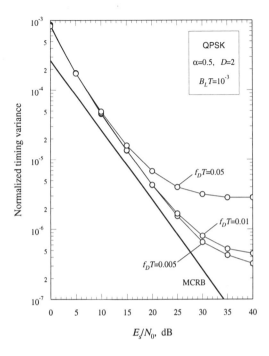

Figure 8.20. Tracking performance of the ML synchronizer.

### 8.5.4. Modified ML Algorithm

The NDA-ML-based algorithm (8.5.47) requires a derivative-matched filter to generate $y'(kT + \hat{\tau}_k)$. This represents an extra complexity that would be desirable to avoid. A simple method to achieve this goal is to make the approximation

$$y'(kT + \hat{\tau}_k) \approx \frac{1}{T}\left[y(kT + T/2 + \hat{\tau}_k) - y(kT - T/2 + \hat{\tau}_{k-1})\right] \quad (8.5.48)$$

as has been done in Section 8.3. It should be noted that (8.5.48) implies increasing the sampling rate to two samples per symbol. So, the elimination of the derivative-matched filter does not come entirely free. Also, it can be shown that replacing $y'(kT + \hat{\tau}_k)$ by (8.5.48) generates self noise even with very slow fading.

Ginesi [23] has pointed out that the self-noise degradation due to the approximation (8.5.48) can be eliminated by replacing the coefficients $h'[(k-m)T]$ in (8.5.47) by

$$h'[(k-m)T] \approx \frac{1}{T}\{h[(k-m)T+T/2]-h[(k-m)T-T/2]\} \quad (8.5.49)$$

or, in other words, by turning the algorithm (8.5.47) into

$$e(k) = \text{Re}\left\{y^*(kT+\hat{\tau}_k)\left[y(kT+T/2+\hat{\tau}_k)\right.\right.$$
$$\left.\left. - \sum_{m=k-D}^{k+D} y(mT+\hat{\tau}_m)h[(k-m)T+T/2]\right]\right\}$$
$$-\text{Re}\left\{y^*(kT+\hat{\tau}_k)\left[y(kT-T/2+\hat{\tau}_{k-1})\right.\right.$$
$$\left.\left. - \sum_{m=k-D}^{k+D} y(mT+\hat{\tau}_m)h[(k-m)T-T/2]\right]\right\} \quad (8.5.50)$$

A possible explanation for this favorable feature is as follows. As SNR grows large, the timing estimates become sufficiently accurate so that we can make the approximation

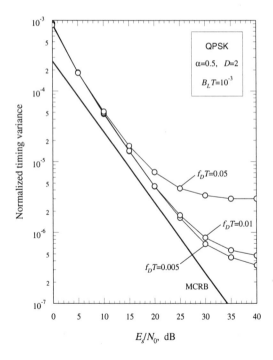

Figure 8.21. Tracking performance of the modified ML synchronizer.

# Timing Recovery with Linear Modulations

$$\hat{\tau}_k \approx \hat{\tau}_{k\pm 1/2} \approx \tau \qquad (8.5.51)$$

Furthermore, neglecting thermal noise we have

$$y(t) \approx \sum_i a(i) c_i h(t - iT - \tau) \qquad (8.5.52)$$

Then, substituting into (8.5.50) and letting $D \to \infty$ makes $e(k)$ vanish, which means that the error signal is self-noise free.

Figure 8.21 displays simulation results for the tracking performance of the modified ML algorithm (8.5.50). Modulation and channel model are as in Figure 8.20. Comparing the two figures it is seen that the original and modified ML detectors are essentially equivalent. Thus, we can obviate the derivative matched filter with negligible performance penalties. In retrospect, it is expected that this same method can be applied to the algorithms discussed in Section 8.3.

**Exercise 8.5.3.** Compute the S-curve for detector (8.5.47) when $D \to \infty$. Assume that $h(t)$ is Nyquist and is bandlimited to $\pm 1/T$.

*Solution.* The samples from the matched filter and the derivative matched filter have the form

$$y(kT + \hat{\tau}_k) = \sum_i a(i) c_i h[(k - i)T + \hat{\tau}_k - \tau] + n(kT + \hat{\tau}_k) \qquad (8.5.53)$$

$$y'(kT + \hat{\tau}_k) = \sum_i a(i) c_i h'[(k - i)T + \hat{\tau}_k - \tau] + n'(kT + \hat{\tau}_k) \qquad (8.5.54)$$

Letting $\hat{\tau}_k = \hat{\tau}$ and $\delta = \tau - \hat{\tau}$ in (8.5.47) yields

$$e(k) = \mathrm{Re}\left\{ \sum_{i_1} \sum_{i_2} a^*(i_1) a(i_2) c_{i_1}^* c_{i_2} h[(k - i_1)T - \delta] h'[(k - i_2)T - \delta] \right\}$$

$$- \sum_m h'[(k - m)T] \mathrm{Re}\left\{ \sum_{i_1} \sum_{i_2} a^*(i_1) a(i_2) c_{i_1}^* c_{i_2} \right.$$

$$\left. \times h[(k - i_1)T - \delta] h[(m - i_2)T - \delta] \right\} + N(k) \qquad (8.5.55)$$

where $N(k)$ represents zero-mean noise. Thus, averaging with respect to data, fading and thermal noise and rearranging produces

$$S(\delta) = R(0) C_2 \sum_i h(iT - \delta) h'(iT - \delta)$$

$$- R(0) C_2 \sum_m h'(mT) \left[ \sum_i h(iT - \delta) h[(i - m)T - \delta] \right] \qquad (8.5.56)$$

where $R(0)$ is the power of the fading distortion $a(t)$ and $C_2$ is the symbol variance.

Next, observe that the sum within square brackets is even with respect to $m$ whereas $h'(mT)$ is odd (since $h(t)$ is even). Thus the second line in (8.5.56) sums to zero and the S-curve reduces to

$$S(\delta) = R(0)C_2 \sum_i h(iT - \delta)h'(iT - \delta) \qquad (8.5.57)$$

Equation (8.5.57) can be put in a simpler form by application of the Poisson sum formula [9, p. 395] to $w(t) \triangleq h(t-\delta)h'(t-\delta)$. As a result we get

$$\sum_i h(iT - \delta)h'(iT - \delta) = \frac{1}{T}\sum_m W\left(\frac{m}{T}\right) \qquad (8.5.58)$$

with

$$W(f) \triangleq j2\pi e^{-j2\pi f\delta} \int_{-\infty}^{\infty} vH(v)H(f-v)dv \qquad (8.5.59)$$

As $H(f)$ is bandlimited to $\pm 1/T$ by assumption, it follows from (8.5.59) that $W(m/T)$ is zero unless $m = \pm 1$. Thus, collecting (8.5.57)-(8.5.59) yields the final result

$$S(\delta) = \frac{2R(0)C_2 K}{T}\sin(2\pi\delta/T) \qquad (8.5.60)$$

with

$$K \triangleq 2\pi \int_{-\infty}^{\infty} fH(f)H(1/T - f)df \qquad (8.5.61)$$

## 8.6. Decision-Directed Joint Phase and Timing Recovery with Offset Formats

We now return to the AWGN channel and address the problem of joint phase and timing estimation with OQPSK modulation. In doing so we follow the same steps as in Section 8.2 and, in particular, we assume that any possible carrier frequency offset has been perfectly compensated. In these conditions the received signal has the form

# Timing Recovery with Linear Modulations

$$s(t) = e^{j\theta} \left\{ \sum_i a_i g(t - iT - \tau) + j \sum_i b_i g(t - iT - T/2 - \tau) \right\} \quad (8.6.1)$$

where the symbols $a_i$ and $b_i$ take on the values $\pm 1$ independently and with the same probability. We first illustrate an estimation approach based on ML methods.

## 8.6.1. ML-Based Joint Phase and Timing Estimation

Assuming that reliable decisions from the data detector are available, the log-likelihood function is found to be

$$L(r|\tilde{\theta}, \tilde{\tau}) = \sum_i \text{Re} \left\{ \hat{a}_i e^{-j\tilde{\theta}} \int_0^{T_0} r(t) g(t - iT - \tilde{\tau}) dt \right\}$$

$$+ \sum_i \text{Im} \left\{ \hat{b}_i e^{-j\tilde{\theta}} \int_0^{T_0} r(t) g(t - iT - T/2 - \tilde{\tau}) dt \right\}$$

$$- \frac{1}{2} \sum_i \sum_m \hat{a}_i \hat{a}_m \int_0^{T_0} g(t - iT - \tilde{\tau}) g(t - mT - \tilde{\tau}) dt$$

$$- \frac{1}{2} \sum_i \sum_m \hat{b}_i \hat{b}_m \int_0^{T_0} g(t - iT - T/2 - \tilde{\tau}) g(t - mT - T/2 - \tilde{\tau}) dt \quad (8.6.2)$$

Maximizing this function over $\tilde{\theta}$ and $\tilde{\tau}$ requires looking for the nulls of the partial derivatives

$$\frac{\partial L(r|\tilde{\theta}, \tilde{\tau})}{\partial \tilde{\theta}} = \sum_i \text{Im} \left\{ \hat{a}_i e^{-j\tilde{\theta}} \int_0^{T_0} r(t) g(t - iT - \tilde{\tau}) dt \right\}$$

$$- \sum_i \text{Re} \left\{ \hat{b}_i e^{-j\tilde{\theta}} \int_0^{T_0} r(t) g(t - iT - T/2 - \tilde{\tau}) dt \right\} \quad (8.6.3)$$

$$\frac{\partial L(r|\tilde{\theta}, \tilde{\tau})}{\partial \tilde{\tau}} = -\sum_i \text{Re} \left\{ \hat{a}_i e^{-j\tilde{\theta}} \int_0^{T_0} r(t) g'(t - iT - \tilde{\tau}) dt \right\}$$

$$- \sum_i \text{Im} \left\{ \hat{b}_i e^{-j\tilde{\theta}} \int_0^{T_0} r(t) g'(t - iT - T/2 - \tilde{\tau}) dt \right\}$$

$$+ \sum_i \sum_m \hat{a}_i \hat{a}_m \int_0^{T_0} g'(t - iT - \tilde{\tau}) g(t - mT - \tilde{\tau}) dt$$

$$+ \sum_i \sum_m \hat{b}_i \hat{b}_m \int_0^{T_0} g'(t - iT - T/2 - \tilde{\tau}) g(t - mT - T/2 - \tilde{\tau}) dt \quad (8.6.4)$$

Equations (8.6.3)-(8.6.4) may be simplified by expanding the limits of the integrals to ±∞ while restricting the summations over $i$ from 0 to $L_0 - 1$. This results in

$$\frac{\partial L(r|\tilde{\theta},\tilde{\tau})}{\partial \tilde{\theta}} = \sum_{i=0}^{L_0-1} \text{Im}\left\{\hat{a}_i y(iT+\tilde{\tau})e^{-j\tilde{\theta}}\right\} - \sum_{i=0}^{L_0-1} \text{Re}\left\{\hat{b}_i y(iT+T/2+\tilde{\tau})e^{-j\tilde{\theta}}\right\} \quad (8.6.5)$$

$$\frac{\partial L(r|\tilde{\theta},\tilde{\tau})}{\partial \tilde{\tau}} = \sum_{i=0}^{L_0-1} \text{Re}\left\{\hat{a}_i \left[y'(iT+\tilde{\tau})e^{-j\tilde{\theta}} - \sum_{m=i-D}^{i+D}\hat{a}_m h'[(i-m)T]\right]\right\}$$

$$+ \sum_{i=0}^{L_0-1} \text{Im}\left\{\hat{b}_i \left[y'(iT+T/2+\tilde{\tau})e^{-j\tilde{\theta}} - j\sum_{m=i-D}^{i+D}\hat{b}_m h'[(i-m)T]\right]\right\} \quad (8.6.6)$$

where $y(t)$ is the matched-filter output and $y'(t)$ is its derivative. Also, $h(t)$ is the convolution $g(t) \otimes g(-t)$, $h'(t)$ is its derivative and $D$ is an integer such that $h'(mT) \approx 0$ for $|m| > D$.

The simultaneous nulls of the derivatives are sought resorting to the usual recursive method employing the following error signals:

$$e_P(k) = \text{Im}\left\{\hat{a}_k y(kT+\hat{\tau}_k)e^{-j\hat{\theta}_k}\right\} - \text{Re}\left\{\hat{b}_k y(kT+T/2+\hat{\tau}_k)e^{-j\hat{\theta}_k}\right\} \quad (8.6.7)$$

$$e_T(k) = \text{Re}\left\{\hat{a}_k \left[y'(kT+\hat{\tau}_k)e^{-j\hat{\theta}_k} - \sum_{m=k-D}^{k+D}\hat{a}_m h'[(k-m)T]\right]\right\}$$

$$+ \text{Im}\left\{\hat{b}_k \left[y'(kT+T/2+\hat{\tau}_k)e^{-j\hat{\theta}_k} - j\sum_{m=k-D}^{k+D}\hat{b}_m h'[(k-m)T]\right]\right\} \quad (8.6.8)$$

A more suitable form for $e_P(k)$ and $e_T(k)$ is obtained by introducing the in-phase (I) and quadrature (Q) signal components which are defined as

$$I(k) + jQ(k) \triangleq y(kT+\hat{\tau}_k)e^{-j\hat{\theta}_k} \quad (8.6.9)$$

$$I(k+1/2) + jQ(k+1/2) \triangleq y(kT+T/2+\hat{\tau}_k)e^{-j\hat{\theta}_k} \quad (8.6.10)$$

and their time derivatives

$$I'(k) + jQ'(k) \triangleq y'(kT+\hat{\tau}_k)e^{-j\hat{\theta}_k} \quad (8.6.11)$$

$$I'(k+1/2) + jQ'(k+1/2) \triangleq y'(kT+T/2+\hat{\tau}_k)e^{-j\hat{\theta}_k} \quad (8.6.12)$$

## Timing Recovery with Linear Modulations

Substituting into (8.6.7)-(8.6.8) yields

$$e_P(k) = \hat{a}_k Q(k) - \hat{b}_k I(k+1/2) \tag{8.6.13}$$

$$e_T(k) = \hat{a}_k \left[ I'(k) - \sum_{m=k-D}^{k+D} \hat{a}_m h'[(k-m)T] \right]$$

$$+ \hat{b}_k \left[ Q'(k+1/2) - \sum_{m=k-D}^{k+D} \hat{b}_m h'[(k-m)T] \right] \tag{8.6.14}$$

Equation (8.6.13) is an obvious extension of the phase detector obtained in Section 7.4 for clock-aided phase recovery. Equation (8.6.14), in turn, may be viewed as the sum of two timing measurements of the type (8.2.10) which are derived from the I and Q components. The disalignment by a half symbol between such components is reflected in the half index difference between the two terms in (8.6.14).

**Exercise 8.6.1.** Compute the S-surfaces about the origin in the $\phi - \delta$ plane for the detectors (8.6.13)-(8.6.14). Assume that $h(t)$ is Nyquist, i.e.,

$$h(t) = \frac{\sin(\pi t/T)}{\pi t/T} \frac{\cos(\alpha \pi t/T)}{1 - (2\alpha t/T)^2} \tag{8.6.15}$$

*Solution.* Letting $\hat{\theta}_k = \hat{\theta}$ and $\hat{\tau}_k = \hat{\tau}$, $\phi = \theta - \hat{\theta}$ and $\delta = \tau - \hat{\tau}$, the matched-filter output reads

$$y(t) = e^{j\theta} \left\{ \sum_i a_i h(t - iT - \tau) + j \sum_i b_i h(t - iT - T/2 - \tau) \right\} + n(t) \tag{8.6.16}$$

Inserting into (8.6.9)-(8.6.12) yields

$$I(k) = \cos\phi \sum_i a_i h[(k-i)T - \delta]$$

$$- \sin\phi \sum_i b_i h[(k-i)T - T/2 - \delta] + n_I(k) \tag{8.6.17}$$

$$Q(k) = \sin\phi \sum_i a_i h[(k-i)T - \delta]$$

$$+ \cos\phi \sum_i b_i h[(k-i)T - T/2 - \delta] + n_Q(k) \tag{8.6.18}$$

$$I(k+1/2) = \cos\phi \sum_i a_i h[(k-i)T + T/2 - \delta]$$
$$-\sin\phi \sum_i b_i h[(k-i)T - \delta] + n_I(k+1/2) \quad (8.6.19)$$

$$Q(k+1/2) = \sin\phi \sum_i a_i h[(k-i)T + T/2 - \delta]$$
$$+\cos\phi \sum_i b_i h[(k-i)T - \delta] + n_Q(k+1/2) \quad (8.6.20)$$

$$I'(k) = \cos\phi \sum_i a_i h'[(k-i)T - \delta]$$
$$-\sin\phi \sum_i b_i h'[(k-i)T - T/2 - \delta] + n_I'(k) \quad (8.6.21)$$

$$Q'(k+1/2) = \sin\phi \sum_i a_i h'[(k-i)T + T/2 - \delta]$$
$$+\cos\phi \sum_i b_i h'[(k-i)T - \delta] + n_Q'(k+1/2) \quad (8.6.22)$$

where terms of the type $n_I(k)$, $n_Q(k)$ represent thermal noise contributions.

Next, keeping in mind that the I and Q detectors make decisions according to

$$\hat{a}_k = \text{sgn}[I(k)] \quad (8.6.23)$$

$$\hat{b}_k = \text{sgn}[Q(k+1/2)] \quad (8.6.24)$$

and the synchronization errors $\phi$ and $\delta$ are small by assumption, from (8.6.17) and (8.6.20) it is recognized that the decisions are correct (apart from noise-induced errors), i.e.,

$$\hat{a}_k \approx a_k, \quad \hat{b}_k \approx b_k \quad (8.6.25)$$

Thus, substituting into the expressions for $e_P(k)$ and $e_T(k)$ and averaging with respect to data and noise gives the desired result

$$S_P(\phi,\delta) = 2h(\delta)\sin\phi \quad (8.6.26)$$

$$S_T(\phi,\delta) = -2h'(\delta)\cos\phi \quad (8.6.27)$$

# Timing Recovery with Linear Modulations

**Exercise 8.6.2.** Under the same assumptions as in the previous exercise show that the following tracking points are stable:

$$\delta_{eq} = 0, \quad \phi_{eq} = 0, \pi \qquad (8.6.28)$$

$$\delta_{eq} = T/2, \quad \phi_{eq} = \pm \pi/2 \qquad (8.6.29)$$

*Solution.* The goal is to demonstrate that, at each of the above points, the S-surfaces are zero and the synchronizer's parameters $A_{PP}$, $A_{TT}$, $A_{PT}$ and $A_{TP}$ satisfy conditions (8.2.47)-(8.2.48).

Consider $\delta_{eq} = 0, \phi_{eq} = 0$, for example. From the previous exercise we know that in the vicinity of $\delta_{eq} = 0, \phi_{eq} = 0$ the S-surfaces are expressed by (8.6.26)-(8.6.27). By inspection it is seen that $S_P(0,0) = S_T(0,0) = 0$, which means that $\delta_{eq} = 0, \phi_{eq} = 0$ is an equilibrium point. Furthermore, application of (8.2.39)-(8.2.42) yields

$$A_{PP} = 2 \qquad (8.6.30)$$

$$A_{TT} = \frac{2\pi^2}{T^2}\left[\frac{1}{3} + \alpha^2(1 - \frac{8}{\pi^2})\right] \qquad (8.6.31)$$

$$A_{PT} = A_{TP} = 0 \qquad (8.6.32)$$

which satisfy the stability conditions (8.2.47)-(8.2.48).

A similar procedure can be followed in the other cases. For example, in the vicinity of $\delta_{eq} = T/2, \phi_{eq} = \pi/2$ the S-surfaces are found to be

$$S_P(\phi, \delta) = -2h(T/2 - \delta)\cos\phi \qquad (8.6.33)$$

$$S_T(\phi, \delta) = 2h'(T/2 - \delta)\sin\phi \qquad (8.6.34)$$

By inspection it is seen that $S_P(\pi/2, T/2) = S_T(\pi/2, T/2) = 0$. Also, from (8.2.39)-(8.2.42) we get for $A_{PP}, A_{TT}, A_{PT}$ and $A_{TP}$ the same values as indicated in (8.6.30)-(8.6.32). Thus, point $\delta_{eq} = T/2, \phi_{eq} = \pi/2$ is stable.

It is worth noting that the parameters $A_{PP}, A_{TT}, A_{PT}$ and $A_{TP}$ are the same at any tracking point. In particular, as $A_{PT}$ and $A_{TP}$ are both zero, phase and timing loops are decoupled.

**Exercise 8.6.3.** Neglecting thermal noise and retaining the assumptions of the previous exercise, compute the decisions from the I and Q detectors for each tracking condition (8.6.28)-(8.6.29). As will be seen, each detector produces one of the four sequences $\{a_i\}$, $\{-a_i\}$, $\{b_i\}$, $\{-b_i\}$, depending on the actual tracking point. Indicate a coding scheme to cope with this ambiguity.

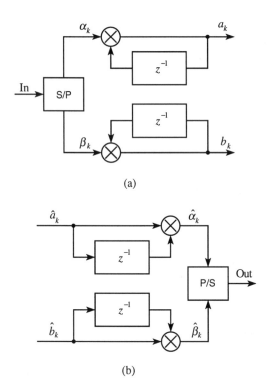

Figure 8.22. Differential encoding and decoding schemes.

*Solution.* Substituting (8.6.17) and (8.6.20) into (8.6.23)-(8.6.24) results in

$$\delta_{eq} = 0, \phi_{eq} = 0 \quad \Rightarrow \quad \begin{cases} \hat{a}_k = a_k \\ \hat{b}_k = b_k \end{cases} \quad (8.6.35)$$

$$\delta_{eq} = 0, \phi_{eq} = \pi \quad \Rightarrow \quad \begin{cases} \hat{a}_k = -a_k \\ \hat{b}_k = -b_k \end{cases} \quad (8.6.36)$$

$$\delta_{eq} = T/2, \phi_{eq} = \pi/2 \quad \Rightarrow \quad \begin{cases} \hat{a}_k = -b_{k-1} \\ \hat{b}_k = a_k \end{cases} \quad (8.6.37)$$

$$\delta_{eq} = T/2, \phi_{eq} = -\pi/2 \quad \Rightarrow \quad \begin{cases} \hat{a}_k = b_{k-1} \\ \hat{b}_k = -a_k \end{cases} \quad (8.6.38)$$

As is intuitively clear, the I detector releases $\{a_i\}$ and the Q detector releases $\{b_i\}$ only if $\delta_{eq} = 0, \phi_{eq} = 0$. Sign inversions and/or sequence exchanges between I and Q rails occur in all of the other cases. As the actual equilibrium point is unpredictable (it depends on the receiver's initial conditions), something must be done to avoid an invalid output when the tracking point is other than $\delta_{eq} = 0, \phi_{eq} = 0$.

A simple solution is to encode the data on the I/Q rails as indicated in Figure 8.22(a). Here, the information sequence $\ldots,\alpha_{k-1},\beta_{k-1},\alpha_k,\beta_k$ is split into two paths by the serial-to-parallel (S/P) converter and the resulting streams $\{\alpha_k\}$ and $\{\beta_k\}$ are differentially encoded as

$$a_k = \alpha_k a_{k-1} \quad b_k = \beta_k b_{k-1} \tag{8.6.39}$$

At the receiver side, differential decoding is performed as indicated in Figure 8.22(b). It is readily verified that, apart from decision errors, the sequence from the parallel to serial converter (P/S) is again $\ldots,\alpha_{k-1},\beta_{k-1},\alpha_k,\beta_k$, independently of the synchronizer's tracking point.

## 8.6.2. Other Timing Detectors

Two NDA schemes have been suggested by Gardner in [1, Chapter 11] as an alternative to the ML-based detector (8.6.14). One is obtained from (8.6.14) setting $D$ to zero and replacing the decisions $\hat{a}_k$ and $\hat{b}_k$ by the corresponding samples $I(k)$ and $Q(k+1/2)$. This results in

$$e_T(k) = I(k)I'(k) + Q(k+1/2)Q'(k+1/2) \tag{8.6.40}$$

The second is even more interesting as it does not require derivative matched filters to generate $I'(k)$ and $Q'(k+1/2)$. It consists of the sum of two baseband Gardner algorithms as derived from the I and Q signal components, i.e.,

$$e_T(k) = -I(k-1/2)[I(k) - I(k-1)] - Q(k)[Q(k+1/2) - Q(k-1/2)] \tag{8.6.41}$$

For obvious reasons this algorithm is denoted by I/Q-GAD in the sequel.

The tracking performance of the ML-based detector and the I/Q-GAD have been assessed by simulation and some results are displayed in Figures 8.23-8.24. The phase detector (8.6.13) has been used in both cases. Parameter $D$ in the ML-based algorithm has been chosen equal to 2. A Nyquist channel response with either 75% or 25% rolloff is assumed. Phase and timing loops have a noise bandwidth of $5 \cdot 10^{-3}/T$. At high rolloffs the two detectors appear nearly equivalent whereas the ML-based is definitely superior as $\alpha$ decreases.

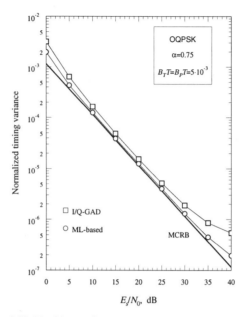

Figure 8.23. Tracking performance comparisons with 75% rolloff.

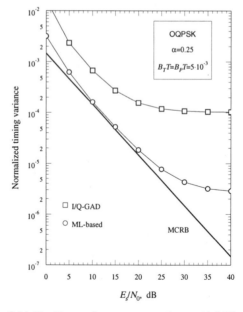

Figure 8.24. Tracking performance comparisons with 25% rolloff.

## Timing Recovery with Linear Modulations

**Exercise 8.6.4.** Compute $S_T(\phi,\delta)$ for the I/Q-GAD assuming a Nyquist channel.

*Solution.* Substituting (8.6.17)-(8.6.20) into (8.6.41) and averaging with respect to data and noise yields, after some manipulations,

$$S_T(\phi,\delta) = 2\cos(2\phi)\sum_i h(iT - \delta)[h(iT + T/2 - \delta) - h(iT - T/2 - \delta)] \quad (8.6.42)$$

Then, proceeding as in Exercise 8.3.1 produces

$$S_T(\phi,\delta) = \frac{8K}{T}\cos(2\phi)\sin(2\pi\delta/T) \quad (8.6.43)$$

with

$$K \triangleq \int_{-\infty}^{\infty} H\left(\frac{1}{2T} + f\right) H\left(\frac{1}{2T} - f\right) \cos(\pi f T) df \quad (8.6.44)$$

It is easily seen that $K$ decreases with the bandwidth of $H(f)$. Correspondingly, function $S_T(\phi,\delta)$ decreases and the synchronizer's performance worsens. In the limit, if $H(f)=0$ for $|f|>1/2T$, the S-surface vanishes and the timing recovery method fails.

**Exercise 8.6.5.** Consider a synchronizer using the phase detector (8.6.7) and an I/Q-GAD algorithm. Assuming a Nyquist channel, show that there are the following stable points: $\delta_{eq} = 0, \phi_{eq} = 0,\pi$, and $\delta_{eq} = T/2, \phi_{eq} = 0,\pm\pi/2$. Compute the parameters $A_{PP}$, $A_{TT}$, $A_{PT}$ and $A_{TP}$ at these points.

*Solution.* The phase detector is as in Exercise 8.6.2 and, in consequence, function $S_P(\phi,\delta)$ remains unchanged. Function $S_T(\phi,\delta)$, vice versa, is given by (8.6.43). It is readily verified that the S-surfaces vanish at $\delta_{eq} = 0$, $\phi_{eq} = 0, \pi$ and $\delta_{eq} = T/2, \phi_{eq} = 0,\pm\pi/2$. In addition, parameters $A_{PP}$, $A_{TT}$, $A_{PT}$ and $A_{TP}$ take on the values

$$A_{PP} = 2 \quad (8.6.45)$$

$$A_{TT} = 16\pi\frac{K}{T^2} \quad (8.6.46)$$

$$A_{PT} = A_{TP} = 0 \quad (8.6.47)$$

at any equilibrium point. As they satisfy conditions (8.2.47)-(8.2.48), it is concluded that $\delta_{eq} = 0$, $\phi_{eq} = 0,\pi$ and $\delta_{eq} = T/2, \phi_{eq} = 0,\pm\pi/2$ are all stable.

## 8.7. NDA Feedforward Joint Phase and Timing Recovery with Offset Formats

In the preceding section we have investigated *feedback* recovery schemes. Although they have good tracking performance, they are not suitable for applications requiring short acquisitions. Here we describe a *feedforward* joint phase and timing ML-based estimator developed in [25].

### 8.7.1 Computation of the Likelihood Function

The received waveform is first passed through an anti-aliasing filter (AAF) and then is sampled at rate $1/T_s$, a multiple $N$ of $1/T$. Although an oversampling factor of 2 is sufficient with signals bandlimited to $1/T$, the ensuing formulas are developed for a generic $N$. Also, the assumption is made that: (*i*) the AAF has a bandwidth $B_{AAF}$ sufficiently large so as not to distort the signal components; (*ii*) the sampling rate equals $2B_{AAF}$; (*iii*) the noise samples are independent. In particular, independence occurs if the AAF transfer function is rectangular but other (more practical) filter transfer functions are conceivable.

Denoting by $x \triangleq \{x(kT_s)\}$ the samples from the AAF, the likelihood function reads

$$\Lambda(x|\tilde{\tau},\tilde{\theta},\tilde{a},\tilde{b}) = \exp\left\{\frac{T_s}{N_0}\sum_{k=0}^{NL_0-1}\operatorname{Re}\left[x(kT_s)\tilde{s}^*(kT_s)\right] - \frac{T_s}{2N_0}\sum_{k=0}^{NL_0-1}|\tilde{s}(kT_s)|^2\right\} \quad (8.7.1)$$

where $\tilde{a}\triangleq\{\tilde{a}_i\}$, $\tilde{b}\triangleq\{\tilde{b}_i\}$ are the data sequences and $\tilde{s}(t)$ is the trial signal:

$$\tilde{s}(t) = e^{j\tilde{\theta}}\left\{\sum_i \tilde{a}_i g(t-iT-\tilde{\tau}) + j\sum_i \tilde{b}_i g(t-iT-T/2-\tilde{\tau})\right\} \quad (8.7.2)$$

It is a simple matter to show that the first sum in (8.7.1) may be written as

$$\sum_{k=0}^{NL_0-1}\operatorname{Re}\left[x(kT_s)\tilde{s}^*(kT_s)\right] = Z_e + Z_o \quad (8.7.3)$$

with

$$Z_e \triangleq \sum_i \tilde{a}_i \operatorname{Re}\left\{e^{-j\tilde{\theta}}z(2i)\right\} \quad (8.7.4)$$

# Timing Recovery with Linear Modulations

$$Z_o \triangleq \sum_i \tilde{b}_i \operatorname{Im}\left\{e^{-j\tilde{\theta}} z(2i+1)\right\} \qquad (8.7.5)$$

$$z(i) \triangleq \sum_{k=0}^{NL_0-1} x(kT_s) g(kT_s - iT/2 - \tilde{\tau}) \qquad (8.7.6)$$

Thus, dropping the second sum in (8.7.1) for simplicity and assuming a low SNR such that the exponential can be approximated by the first three terms of its Taylor expansion produces (within immaterial constants)

$$\Lambda(x|\tilde{\tau},\tilde{\theta},\tilde{a},\tilde{b}) \approx Z_e + Z_o + \frac{T_s}{2N_0}(Z_e + Z_o)^2 \qquad (8.7.7)$$

As we look for an NDA estimator, we need the marginal likelihood function $\Lambda(x|\tilde{\tau},\tilde{\theta})$, which is the expectation of (8.7.7) over $\tilde{a}$ and $\tilde{b}$. The calculation of $\Lambda(x|\tilde{\tau},\tilde{\theta})$ is carried out in Appendix 8.C where it is shown that

$$\mathrm{E}_{\tilde{a},\tilde{b}}\{Z_e + Z_o\} = 0 \qquad (8.7.8)$$

and

$$\mathrm{E}_{\tilde{a},\tilde{b}}\{(Z_e + Z_o)^2\} = \frac{1}{T}\operatorname{Re}\left\{e^{-j2\tilde{\theta}}\left(Xe^{j2\pi\tilde{\tau}/T} + Ye^{-j2\pi\tilde{\tau}/T}\right)\right\} + C \qquad (8.7.9)$$

where the following notations have been used:

$$X \triangleq \sum_{k=0}^{NL_0-1}\left[x(kT_s)e^{-j\pi k/N}\right]u(kT_s) \qquad (8.7.10)$$

$$Y \triangleq \sum_{k=0}^{NL_0-1}\left[x(kT_s)e^{j\pi k/N}\right]v(kT_s) \qquad (8.7.11)$$

$$u(kT_s) \triangleq \sum_{n=0}^{NL_0-1}\left[x(nT_s)e^{-j\pi n/N}\right]q[(k-n)T_s] \qquad (8.7.12)$$

$$v(kT_s) \triangleq \sum_{n=0}^{NL_0-1}\left[x(nT_s)e^{j\pi n/N}\right]q^*[(k-n)T_s] \qquad (8.7.13)$$

$$C \triangleq \frac{1}{2}\sum_i |z(i)|^2 \qquad (8.7.14)$$

In the above equations function $q(t)$ has Fourier transform

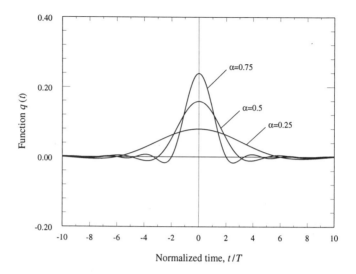

Figure 8.25. Function $q(t)$.

$$Q(f) \triangleq G\left(f - \frac{1}{2T}\right) G^*\left(f + \frac{1}{2T}\right) \tag{8.7.15}$$

where $G(f)$ is the Fourier transform of $g(t)$. From (8.7.15) and the property $G(-f) = G^*(f)$ (which holds because $g(t)$ is real-valued) it can be shown that $q(t)$ is even. Figure 8.25 illustrates the shape of $q(t)$ for a root-raised-cosine rolloff function $G(f)$ with rolloff $\alpha$.

It is worth noticing that $X$ and $Y$ are both independent of $\tilde{\theta}$ and $\tilde{\tau}$. In Appendix 8.C it is shown that this is also true for the sum in (8.7.14). Putting these facts together it is concluded that maximizing $\Lambda(x|\tilde{\tau},\tilde{\theta})$ is equivalent to maximizing

$$\lambda(\tilde{\theta},\tilde{\tau}) \triangleq \text{Re}\left\{e^{-j2\tilde{\theta}}\left(Xe^{j2\pi\tilde{\tau}/T} + Ye^{-j2\pi\tilde{\tau}/T}\right)\right\} \tag{8.7.16}$$

In the next section this result is exploited to estimate $\theta$ and $\tau$.

### 8.7.2 ML-Based Estimator

To locate the maximum of $\lambda(\tilde{\theta},\tilde{\tau})$ we first compute the partial derivatives

$$\frac{\partial \lambda(\tilde{\theta},\tilde{\tau})}{\partial \tilde{\theta}} = 2\,\text{Im}\left\{e^{-j2\tilde{\theta}}\left(Xe^{j2\pi\tilde{\tau}/T} + Ye^{-j2\pi\tilde{\tau}/T}\right)\right\} \tag{8.7.17}$$

# Timing Recovery with Linear Modulations

$$\frac{\partial \lambda(\tilde{\theta}, \tilde{\tau})}{\partial \tilde{\tau}} = \frac{2\pi}{T} \mathrm{Im}\left\{e^{-j2\tilde{\theta}}\left(-Xe^{j2\pi\tilde{\tau}/T} + Ye^{-j2\pi\tilde{\tau}/T}\right)\right\} \quad (8.7.18)$$

Then, setting them to zero and solving for $\tilde{\theta}$ and $\tilde{\tau}$ yields

$$\hat{\theta} = \frac{1}{4}[\arg(X) + \arg(Y)] + m_\theta \frac{\pi}{4} \quad (8.7.19)$$

$$\hat{\tau} = \frac{T}{4\pi}[-\arg(X) + \arg(Y)] + m_\tau \frac{T}{4} \quad (8.7.20)$$

where $m_\theta$ and $m_\tau$ are arbitrary integers and the arguments of $X$ and $Y$ are restricted to the interval $(-\pi, \pi]$. It should be noted that the pairs $(\hat{\theta}, \hat{\tau})$ from (8.7.19)-(8.7.20) do not all correspond to maxima for $\lambda(\tilde{\tau}, \tilde{\theta})$. In fact, inserting (8.7.19)-(8.7.20) into (8.7.16) and rearranging yields

$$\lambda(\tilde{\tau}, \tilde{\theta}) = \left[|X| + (-1)^{m_\tau}|Y|\right] \mathrm{Re}\left\{e^{j(m_\tau - m_\theta)\pi/2}\right\} \quad (8.7.21)$$

which indicates that the maxima occur only when $(m_\tau - m_\theta)/2$ and $m_\tau$ are even numbers, say $2l$ and $2m$. In other words, the locations for the maxima are expressed by

$$\hat{\theta} = \frac{1}{4}[\arg(X) + \arg(Y)] + m\frac{\pi}{2} - l\pi \quad (8.7.22)$$

$$\hat{\tau} = \frac{T}{4\pi}[-\arg(X) + \arg(Y)] + m\frac{T}{2} \quad (8.7.23)$$

for arbitrary integers $l$ and $m$.

The following remarks are of interest.

(i) From (8.7.23) it appears that the timing estimates are ambiguous by multiples of $T/2$. This should be expected in view of the $T/2$ delay between I and Q signal components.

(ii) For a given timing estimate (i.e., for a fixed $m$), equation (8.7.22) tells us that the phase estimates are ambiguous by multiples of $\pi$. Again, this is intuitively obvious since an OQPSK signal may be written as either

$$s(t) = e^{j\theta}\left\{\sum_i a_i g(t - iT - \tau) + j\sum_i b_i g(t - iT - T/2 - \tau)\right\} \quad (8.7.24)$$

or

$$s(t) = e^{j(\theta \pm \pi)} \left\{ \sum_i (-a_i) g(t - iT - \tau) + j \sum_i (-b_i) g(t - iT - T/2 - \tau) \right\} \quad (8.7.25)$$

which suggests that either $\{\theta, a, b\}$ or $\{\theta \pm \pi, -a, -b\}$ may be viewed as unknown signal parameters.

(*iii*) As differential encoding/decoding can be used to resolve the above ambiguities (see Exercise 8.6.3), in the sequel we arbitrarily set $l = m = 0$ for simplicity.

The quantities $X$ and $Y$ appearing in (8.7.19)-(8.7.20) are now written in a more convenient form following the same arguments as in Section 6.6.1 of Chapter 6. The passages may be divided into three major steps as follows.

First, since the function $q(t)$ has a duration of a few symbol intervals about the origin (see Figure 8.25) and the estimation interval is usually much longer than $T$, equations (8.7.12)-(8.7.13) are only marginally affected if the summations are extended from $-\infty$ to $+\infty$. This turns the right-hand sides into convolutions so that $u(kT_s)$ and $v(kT_s)$ may be viewed as outputs of two (non-causal) filters, $q(kT_s)$ and $q^*(kT_s)$, driven by $x(kT_s)e^{-j\pi k/N}$ and $x(kT_s)e^{j\pi k/N}$, respectively.

Second, the filters can be made causal by shifting their impulse responses rightward by $ND$ samples. The shift must be sufficient to make the tails on the negative axis negligible. In doing so the filter outputs are delayed $ND$ steps and become

$$u[(k-ND)T_s] = [x(kT_s)e^{-j\pi k/N}] \otimes q[(k-ND)T_s] \quad (8.7.26)$$

$$v[(k-ND)T_s] = [x(kT_s)e^{j\pi k/N}] \otimes q^*[(k-ND)T_s] \quad (8.7.27)$$

Finally, $u[(k-ND)T_s]$ and $v[(k-ND)T_s]$ are inserted into (8.7.10)-(8.7.11) to produce the desired expressions of $X$ and $Y$:

$$X = \sum_{k=ND}^{N(L_0+D)-1} \left[ x[(k-ND)T_s] e^{-j\pi(k-ND)/N} \right] u[(k-ND)T_s] \quad (8.7.28)$$

$$Y = \sum_{k=ND}^{N(L_0+D)-1} \left[ x[(k-ND)T_s] e^{j\pi(k-ND)/N} \right] v[(k-ND)T_s] \quad (8.7.29)$$

A block diagram for the estimator is depicted in Figure 8.26. Some observations can be made about this scheme. First, when $G(f)$ is real-valued it follows from (8.7.15) that $Q^*(f) = Q(-f)$. This implies $q^*(t) = q(t)$ which means that the filters in the estimator are real. Second, a considerable simplification

# Timing Recovery with Linear Modulations

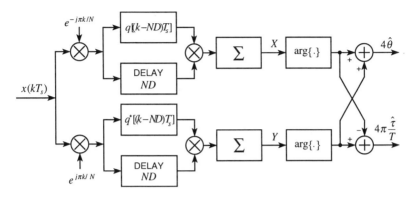

Figure 8.26. Block diagram of the joint phase and timing estimator.

occurs when the oversampling factor $N$ equals 2, as occurs with rolloff values less than unity. In these circumstances, in fact, the exponentials $e^{j\pi k/N}$ and $e^{-j\pi k/N}$ reduce to $(j)^k$ and $(-1)^k(j)^k$, respectively, and the multiplications by $e^{\pm j\pi k/N}$ involve only sign inversions and/or exchanges between real and imaginary parts of $x(kT_s)$. Thus, significant savings in computational load are possible. Third, it should be stressed that phase and timing are estimated in parallel, *not sequentially*. In particular, the timing estimator remains the same even if no phase information is required (as happens with differential detection).

**Exercise 8.7.1.** The approach described above differs from that followed in Section 8.4 with non-offset formats insofar as the carrier phase is no longer viewed as a nuisance parameter and, as such, is not averaged out from the likelihood function. What happens if the averaging procedure is pursued with OQPSK signals?

*Solution.* The starting point is equation (8.7.16) which is an (approximate) expression for the likelihood function in the parameters $(\tilde{\theta}, \tilde{\tau})$. We should average $\lambda(\tilde{\theta}, \tilde{\tau})$ over $\tilde{\theta}$, taking $\tilde{\theta}$ uniformly distributed over $[0, 2\pi)$. As is readily seen, however, the average is zero and searching for its maximum is meaningless.

This failure is clearly due to the rough approximations leading to (8.7.16). Still, it is somewhat surprising that the obstacle disappears if we look for a joint estimator of $(\theta, \tau)$ rather than a carrier-independent estimator of $\tau$.

## 8.7.3 Estimator Performance

An analysis of the joint estimator (8.7.19)-(8.7.20) is discussed in [25]. In the sequel we only report on simulations which have been run under the following conditions:

(*i*) The AAF is implemented as an 8-pole Butterworth FIR filter with bandwidth $1/T$.

(*ii*) Sampling is performed at twice the symbol rate ($N=2$).

(*iii*) Function $g(t)$ has a root-raised-cosine Fourier transform with rolloff $\alpha$.

(*iv*) Parameter $D$ in equations (8.7.28)-(8.7.29) is a function of the rolloff and is indicated in the figures. In general $D$ is larger with small rolloffs.

(*v*) An observation interval of $L_0=100$ symbols has been adopted.

Figure 8.27 illustrates the phase error variance as a function of $E_s/N_0$ with rolloff as a parameter. The reader may want to compare these results with those shown in Section 5.10 which correspond to a different clockless phase recovery method. It is seen that the present method is slightly superior. Other differences are that the algorithm considered in Chapter 5 is simpler to implement but requires a sampling rate of four samples per symbol, as opposed to two samples per symbol in the present scheme.

Figures 8.28-8.29 display timing error variance versus $E_s/N_0$. With $\alpha = 0.75$ it appears that the performance of the timing estimator is within 1 dB of the MCRB.

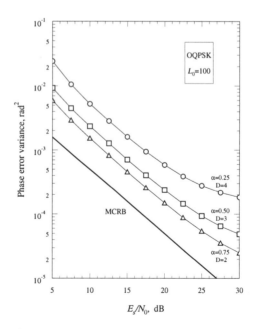

Figure 8.27. Performance of the phase estimator.

# Timing Recovery with Linear Modulations

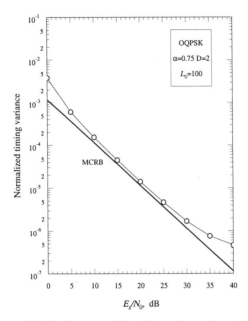

Figure 8.28. Performance of the timing estimator for $\alpha=0.75$.

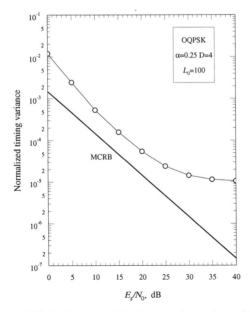

Figure 8.29. Performance of the timing estimator for $\alpha=0.25$.

## 8.8. Key Points of the Chapter

- Several methods are available for joint phase and timing recovery with non-offset formats. They can be divided into two categories: either feedback or feedforward.

- Feedback schemes consist of two distinct loops, one for phase tracking and the other for timing tracking. Interactions between loops may lead to acquisition failures (spurious locks). The occurrence of spurious lock points is related to the signal constellation and the operating SNR. Roughly speaking, large signal constellations and high SNR increase the chances of spurious locks.

- With PSK/QAM modulations, loop interactions are negligible in the steady state. Thus, the tracking performance of a joint phase/timing synchronizer can be established with the methods applicable to isolated loops.

- Feedback schemes have typically good tracking performance but have poor acquisition properties. Feedforward methods are preferable with burst mode transmissions. Two NDA feedforward schemes have been indicated. They allow some trade-off between performance and implementation complexity.

- Timing recovery for transmissions over frequency-flat fading channels may be pursued using either DD and channel-aided methods or NDA techniques. The former route is only viable with coherent receivers where data and channel estimates are generated in the Viterbi detector.

- With noncoherent detectors, vice versa, NDA synchronizers are indispensable. A close approximation to the NDA-ML timing estimator has been indicated.

- Feedback methods for joint phase and timing recovery are readily available for OQPSK transmissions over the AWGN channel.

- Joint phase and timing recovery with OQPSK may also be realized in a feedforward fashion. One method is described in the last section of the chapter. Alternately, the carrier phase can first be estimated in a clockless manner, as indicated in Chapter 4. Then, timing can be extracted (still, in a feedforward fashion) from the resulting I and Q signal components.

## Appendix 8.A

In this Appendix we show that, as $N_0$ decreases, the logarithm of the likelihood function

# Timing Recovery with Linear Modulations

$$\Lambda(r|\tilde{\tau},\tilde{c}) = \frac{1}{\det\{I + [1/(2N_0)]\tilde{C}R_{\tilde{a}}\tilde{C}^*H\}} \quad (8.A.1)$$

$$\times \exp\left\{[1/(2N_0)]^2 y^{*T}\left[\tilde{C}R_{\tilde{a}}^{-1}\tilde{C}^* + [1/(2N_0)]H\right]^{-1} y\right\}$$

takes on the form

$$\ln \Lambda(r|\tilde{\tau},\tilde{c}) \approx -L_0 \ln[1/(2N_0)] + [1/(2N_0)]y^{*T}H^{-1}y \quad (8.A.2)$$

As a first step in this direction we write the logarithm of (8.A.1) as

$$\ln \Lambda(r|\tilde{\tau},\tilde{c}) = \Gamma_1(r|\tilde{\tau},\tilde{c}) + \Gamma_2(r|\tilde{\tau},\tilde{c}) \quad (8.A.3)$$

with

$$\Gamma_1(r|\tilde{\tau},\tilde{c}) \triangleq -\ln\left\{\det\left[I + [1/(2N_0)]\tilde{C}R_{\tilde{a}}\tilde{C}^*H\right]\right\} \quad (8.A.4)$$

$$\Gamma_2(r|\tilde{\tau},\tilde{c}) \triangleq [1/(2N_0)]^2 y^{*T}\left\{\tilde{C}R_{\tilde{a}}^{-1}\tilde{C}^* + [1/(2N_0)]H\right\}^{-1} y \quad (8.A.5)$$

Let us first concentrate on $\Gamma_1(r|\tilde{\tau},\tilde{c})$. For $N_0$ sufficiently small it is obvious that

$$\det\{I + [1/(2N_0)]\tilde{C}R_{\tilde{a}}\tilde{C}^*H\} \approx \det\{[1/(2N_0)]\tilde{C}R_{\tilde{a}}\tilde{C}^*H\}$$
$$= [1/(2N_0)]^{L_0} \det[\tilde{C}R_{\tilde{a}}\tilde{C}^*H] \quad (8.A.6)$$

and substituting into (8.A.4) yields

$$\Gamma_1(r|\tilde{\tau},\tilde{c}) \approx -L_0 \ln[1/(2N_0)] - \ln\{\det[\tilde{C}R_{\tilde{a}}\tilde{C}^*H]\} \quad (8.A.7)$$

Next we turn our attention to $\Gamma_2(r|\tilde{\tau},\tilde{c})$. Rewrite (8.A.5) in the form

$$\Gamma_2(r|\tilde{\tau},\tilde{c}) \triangleq [1/(2N_0)] y^{*T}\left\{2N_0\tilde{C}R_{\tilde{a}}^{-1}\tilde{C}^* + H\right\}^{-1} y \quad (8.A.8)$$

For $N_0$ sufficiently small this equation becomes

$$\Gamma_2(r|\tilde{\tau},\tilde{c}) \approx [1/(2N_0)]y^{*T}H^{-1}y \quad (8.A.9)$$

Then, substituting (8.A.7) and (8.A.9) into (8.A.3) and letting $N_0 \to 0$ gives the desired result (8.A.2).

## Appendix 8.B

In this Appendix we show that the derivative of $y^{*T}H^{-1}y$ with respect to $\tilde{\tau}$ may be written in the form

$$\frac{d}{d\tilde{\tau}}(y^{*T}H^{-1}y) = 2\operatorname{Re}\{z^{*T}(y' + Qz)\} \tag{8.B.1}$$

where $y' \triangleq dy/d\tilde{\tau}$, $z \triangleq H^{-1}y$ and $Q$ is an $L_0 \times L_0$ matrix with entries

$$q_{i_1,i_2} \triangleq -\int_0^{T_0} g'(t - i_1 T - \tilde{\tau})g(t - i_2 T - \tilde{\tau})dt \tag{8.B.2}$$

$g'(t)$ being the derivative of $g(t)$.

The derivative of $y^{*T}H^{-1}y$ is given by

$$\frac{d}{d\tilde{\tau}}(y^{*T}H^{-1}y) = y'^{*T}H^{-1}y + y^{*T}H^{-1}y' + y^{*T}\left(\frac{d}{d\tilde{\tau}}H^{-1}\right)y \tag{8.B.3}$$

Next observe that

$$y'^{*T}H^{-1}y = (y^{*T}H^{-1}y')^* \tag{8.B.4}$$

$$y^{*T}\left(\frac{d}{d\tilde{\tau}}H^{-1}\right)y = \left[y^{*T}\left(\frac{d}{d\tilde{\tau}}H^{-1}\right)y\right]^* \tag{8.B.5}$$

$$\frac{d}{d\tilde{\tau}}H^{-1} = -H^{-1}\left(\frac{d}{d\tilde{\tau}}H\right)H^{-1} \tag{8.B.6}$$

Equations (8.B.4)-(8.B.5) follow from the fact that $H$, $H^{-1}$ and $dH^{-1}/d\tilde{\tau}$ are all real-valued and symmetric matrixes while equation (8.B.6) is obtained by differentiating the identity $HH^{-1} = I$. Inserting into (8.B.3) and recalling the definition $z \triangleq H^{-1}y$ yields

$$\frac{d}{d\tilde{\tau}}(y^{*T}H^{-1}y) = 2\operatorname{Re}\{z^{*T}y'\} - z^{*T}\left(\frac{d}{d\tilde{\tau}}H\right)z \tag{8.B.7}$$

Now, keep in mind that $H$ has the entries

$$h_{i_1,i_2} \triangleq \int_0^{T_0} g(t - i_1 T - \tilde{\tau})g(t - i_2 T - \tilde{\tau})dt \tag{8.B.8}$$

Thus

# Timing Recovery with Linear Modulations

$$\frac{d}{d\tilde{\tau}}H = -(Q + Q^T) \tag{8.B.9}$$

where the elements of $Q$ are defined in (8.B.2). On the other hand, since $z^T Q z^*$ is a scalar, it equals its transpose $z^{*T} Q^T z$, i.e.,

$$z^{*T} Q^T z = z^T Q z^* = (z^{*T} Q z)^* \tag{8.B.10}$$

Inserting (8.B.9)-(8.B.10) into (8.B.7) yields the desired result (8.B.1).

## Appendix 8.C

In this Appendix we compute the expectations of $Z_e + Z_o$ and $(Z_e + Z_o)^2$ over the data $\tilde{a} \triangleq \{\tilde{a}_i\}$, $\tilde{b} \triangleq \{\tilde{b}_i\}$.

Start with the definition of $Z_e + Z_o$:

$$Z_e + Z_o = \sum_i \tilde{a}_i \operatorname{Re}\{e^{-j\tilde{\theta}} z(2i)\} + \sum_i \tilde{b}_i \operatorname{Im}\{e^{-j\tilde{\theta}} z(2i+1)\} \tag{8.C.1}$$

where

$$z(i) = \sum_{k=0}^{NL_0 - 1} x(kT_s) g(kT_s - iT/2 - \tilde{\tau}) \tag{8.C.2}$$

As the data are zero mean, it is clear that

$$E_{\tilde{a},\tilde{b}}\{Z_e + Z_o\} = 0 \tag{8.C.3}$$

Next we concentrate on $(Z_e + Z_o)^2$. From (8.C.1) and the statistics of the sequences $\{\tilde{a}_i\}$ and $\{\tilde{b}_i\}$ (recall that they are independent and each consists of independent and zero-mean symbols) it is readily shown that

$$E_{\tilde{a},\tilde{b}}\{(Z_e + Z_o)^2\} = \sum_i \left[\operatorname{Re}\{e^{-j\tilde{\theta}} z(2i)\}\right]^2 + \sum_i \left[\operatorname{Im}\{e^{-j\tilde{\theta}} z(2i+1)\}\right]^2 \tag{8.C.4}$$

Next, writing $[\operatorname{Re}\{x\}]^2$ and $[\operatorname{Im}\{x\}]^2$ in the form

$$[\operatorname{Re}\{x\}]^2 = \frac{1}{2}\operatorname{Re}\{x^2\} + \frac{1}{2}|x|^2 \tag{8.C.5}$$

$$[\operatorname{Im}\{x\}]^2 = -\frac{1}{2}\operatorname{Re}\{x^2\} + \frac{1}{2}|x|^2 \tag{8.C.6}$$

and substituting into (8.C.4) yields

$$E_{\tilde{a},\tilde{b}}\{(Z_e + Z_o)^2\} = \frac{1}{2}\text{Re}\left\{e^{-j2\tilde{\theta}}\sum_i (-1)^i z^2(i)\right\} + C \qquad (8.\text{C}.7)$$

with

$$C \triangleq \frac{1}{2}\sum_i |z(i)|^2 \qquad (8.\text{C}.8)$$

Assuming that $g(t)$ in (8.C.2) is bandlimited to $\pm 1/T$, i.e.,

$$G(f) = 0, \quad |f| \geq 1/T \qquad (8.\text{C}.9)$$

we now prove that $C$ is independent of $\tilde{\tau}$ and the sum in (8.C.7) may be written as

$$\sum_i (-1)^i z^2(i) = \frac{2}{T}\left[Xe^{j2\pi\tilde{\tau}/T} + Ye^{-j2\pi\tilde{\tau}/T}\right] \qquad (8.\text{C}.10)$$

where

$$X \triangleq \sum_{k_1=0}^{NL_0-1}\sum_{k_2=0}^{NL_0-1} x(k_1 T_s)x(k_2 T_s)q[(k_1-k_2)T_s]e^{-j\pi(k_1+k_2)/N} \qquad (8.\text{C}.11)$$

$$Y \triangleq \sum_{k_1=0}^{NL_0-1}\sum_{k_2=0}^{NL_0-1} x(k_1 T_s)x(k_2 T_s)q^*[(k_1-k_2)T_s]e^{j\pi(k_1+k_2)/N} \qquad (8.\text{C}.12)$$

$$q(t) \triangleq \int_{-\infty}^{\infty} G^*\left(f + \frac{1}{2T}\right)G\left(f - \frac{1}{2T}\right)e^{j2\pi ft}df \qquad (8.\text{C}.13)$$

Note that function $q(t)$ is even. This is seen making the change of variable $f' = -f$ in (8.C.13) and using the relation $G(-f) = G^*(f)$.

The proof relies on the following version of the Poisson sum formula:

$$T\sum_i g_1(t_1 - iT)g_2(t_2 - iT)$$

$$= \sum_n e^{j\pi n(t_1+t_2)/T}\int_{-\infty}^{\infty} G_1^*\left(f - \frac{n}{2T}\right)G_2\left(f + \frac{n}{2T}\right)e^{-j2\pi f(t_1-t_2)}df \qquad (8.\text{C}.14)$$

where $g_1(t)$ and $g_2(t)$ are arbitrary functions and $G_1(f)$, $G_2(f)$ are their Fourier transforms.

Let us first concentrate on $C$. Inserting (8.C.2) into (8.C.8) yields

# Timing Recovery with Linear Modulations

$$C = \frac{1}{2} \sum_{k_1=0}^{NL_0-1} \sum_{k_2=0}^{NL_0-1} x(k_1 T_s) x^*(k_2 T_s)$$
$$\times \left[ \sum_i g(k_1 T_s - iT/2 - \tilde{\tau}) g(k_2 T_s - iT/2 - \tilde{\tau}) \right] \quad (8.C.15)$$

On the other hand, application of (8.C.14) under the bandlimiting condition (8.C.9) produces

$$\sum_i g(k_1 T_s - iT/2 - \tilde{\tau}) g(k_2 T_s - iT/2 - \tilde{\tau}) = \frac{2}{T} \int_{-\infty}^{\infty} |G(f)|^2 e^{-j2\pi(k_1-k_2)fT_s} df \quad (8.C.16)$$

which indicates that $C$ is independent of $\tilde{\tau}$.

Relation (8.C.10) is derived as follows. Using (8.C.2) yields

$$\sum_i (-1)^i z^2(i) = \sum_{k_1=0}^{NL_0-1} \sum_{k_2=0}^{NL_0-1} x(k_1 T_s) x(k_2 T_s)$$
$$\times \left[ \sum_i (-1)^i g(k_1 T_s - iT/2 - \tilde{\tau}) g(k_2 T_s - iT/2 - \tilde{\tau}) \right] \quad (8.C.17)$$

Next write

$$\sum_i (-1)^i g(k_1 T_s - iT/2 - \tilde{\tau}) g(k_2 T_s - iT/2 - \tilde{\tau})$$
$$= \sum_i g(k_1 T_s - iT - \tilde{\tau}) g(k_2 T_s - iT - \tilde{\tau})$$
$$- \sum_i g(k_1 T_s - iT - T/2 - \tilde{\tau}) g(k_2 T_s - iT - T/2 - \tilde{\tau}) \quad (8.C.18)$$

Application of (8.C.14) to each sum in the right-hand side of (8.C.18) leads eventually to (8.C.10).

## References

[1] F.M.Gardner, *Demodulator Reference Recovery Techniques Suited for Digital Implementation*, European Space Agency, Final Report, ESTEC Contract No. 6847/86/NL/DG, August, 1988.

[2] H.Kobayashi, Simultaneous Adaptive Estimation and Decision Algorithm for Carrier Modulated Data Transmission Systems, *IEEE Trans. Commun.*, **COM-19**, 268-280, June 1971.

[3] M.H.Meyers and L.E.Franks, Joint Carrier Phase and Symbol Timing Recovery for PAM Systems, *IEEE Trans. Commun.*, **COM-28**, 1121-1129, Aug. 1980.

[4] W.C.Lindsey and M.K.Simon, *Telecommunication Systems Engineering*, Englewood Cliffs, NJ: Prentice Hall, 1972.
[5] K.H.Mueller and M.Mueller, Timing Recovery in Digital Synchronous Data Receivers, *IEEE Trans. Commun.*, **COM-24**, 516-531, May 1976.
[6] U.Mengali, Joint Phase and Timing Acquisition in Data Transmission, *IEEE Trans. Commun.*, **COM-25**, 1174-1185, Oct. 1977.
[7] M.K.Simon and J.G.Smith, Carrier Synchronization and Detection of QASK Signal Sets, *IEEE Trans. Commun.*, **COM-22**, 98-106, Feb. 1974.
[8] D.D.Falconer and J.Salz, Optimal Reception of Digital Data Over the Gaussian Channel with Unknown Delay and Phase Jitter, IEEE Trans. Inf. Theory, **IT-23**, 117-126, Jan. 1977.
[9] A.Papoulis, *Probability, Random Variables, and Stochastic Processes*, New York: McGraw-Hill, 1991.
[10] F.M.Gardner, A BPSK/QPSK Timing-Error Detector for Sampled Receivers, *IEEE Trans. Commun.*, **COM-34**, 423-429, May 1986.
[11] A.N.D'Andrea and M.Luise, Design and Analysis of Jitter-Free Clock Recovery Scheme for QAM Systems, *IEEE Trans. Commun.*, **COM-41**, 1296-1299, Sept. 1993.
[12] A.N.D'Andrea and M.Luise, Optimization of Symbol Timing Recovery for QAM Data Demodulators, *IEEE Trans. Commun.*, **COM-44**, 399-406, March 1996.
[13] M.Oerder and H.Meyr, Digital Filter and Square Timing Recovery, *IEEE Trans. Commun.*, **COM-36**, 605-611, May 1988.
[14] P.A.Bello, Characterization of Randomly Time-Variant Linear Channels, *IEEE Trans. Commun. Systems*, 360-393, Dec. 1965.
[15] J.G.Proakis, *Digital Communications*, New York: McGraw-Hill, 1983.
[16] S.Stein, Fading Issues in System Engineering, *IEEE J. Selec. Areas Commun.*, **SAC-5**, 68-89, Feb. 1987.
[17] K.Goethals and M.Moeneclaey, PSK Symbol Synchronization Performance of ML-Oriented Data-Aided Algorithms for Nonselective Fading Channels, *IEEE Trans. Commun.*, **COM-43**, 767-772, Feb./March/April 1995.
[18] G.T.Irvine and P.J.McLane, Symbol-Aided plus Decision-Directed Reception for PSK/TCM Modulation on Shadowed Mobile Satellite Fading Channels, *IEEE J. Select. Areas Commun.*, **SAC-10**, 1289-1299, Oct. 1992.
[19] A.Aghamohammadi, H.Meyr and G.Asheid, A New Method for Phase Synchronization and Automatic Gain Control of Linearly Modulated Signals on Frequency-Flat Fading Channels, *IEEE Trans. Commun.*, **COM-39**, 25-29, Jan. 1991.
[20] A.N.D'Andrea, A.Diglio and U.Mengali, Symbol-Aided Channel Estimation with Nonselective Rayleigh Fading Channels, *IEEE Trans. Vehic. Technol.*, **VT-44**, 41-49, Feb. 1995.
[21] G.M.Vitetta and D.Taylor, Maximum Likelihood Decoding of Uncoded and Coded PSK Signals Sequences Transmitted over Rayleigh Flat-Fading Channels, *IEEE Trans. Commun.*, **COM-43**, 2750-2758, Nov. 1995.
[22] K.Goethals and M.Moeneclaey, Tracking Performance of ML-Oriented NDA Symbol Synchronizers for Nonselective Fading Channels, *IEEE Trans. Commun.*, **COM-43**, 1179-1184, Feb./March/April 1995.
[23] A.Ginesi, Timing Synchronization for Transmissions over Frequency-Flat Fading Channels, Ph. D. Thesis (in Italian), Department of Information Engineering, University of Pisa, November 1997.
[24] M.Moeneclaey, Synchronization Problems in PAM Systems, *IEEE Trans. Commun.*, **COM-28**, 1130-1136, Aug. 1980.
[25] A.D'Amico, A.N.D'Andrea and U.Mengali, Feedforward Joint Phase and Timing Estimation with OQPSK Modulation, Submitted to IEEE Trans. Vehic. Technol., Oct. 1996.

# 9

# Timing Recovery with CPM Modulations

## 9.1. Introduction

The last topic in this book is timing recovery with CPM signals. The chapter has the same profile as the previous one, except that we shall only be concerned with transmissions over AWGN channels since CPM timing recovery with multipath channels has received scarce attention thus far. Although it is widely believed that conventional clock synchronizers can be used even with fading channels, a closer look at the question will be worthwhile as it is likely that better methods can be discovered by taking the fading channel statistics into account. This problem is left as a subject of further study.

For convenience, feedback algorithms are investigated first (Sections 9.2 to 9.4) while feedforward methods are left to the second part of the chapter.

As happens with PAM modulations, carrier phase plays an important role in timing recovery. In a coherent receiver, phase and timing parameters are required in the detection process and, accordingly, Section 9.2 investigates DD joint phase and timing recovery. The resulting algorithms have excellent tracking performance but their acquisitions are comparatively slow and prone to spurious locks. In particular, spurious locks occur with partial response formats and multilevel alphabets. Thus, the methods proposed in Section 9.2 are only useful with either full response formats or binary (possibly partial response) modulations. Alternative schemes are needed otherwise.

One alternative is to use phase-independent timing algorithms so as to eliminate interactions between timing and phase recovery. This is the only sensible approach in differential receivers (where no phase information is required). Phase-independent algorithms are addressed in Section 9.3 making use of ML methods. As we shall see, the resulting schemes have good accuracy with full response signaling; with partial response formats, vice versa, their performance is poor, especially with long frequency pulses.

In summary, multilevel partial response systems cannot be synchronized either by the methods of Section 9.2 (because of spurious locks) or by those in Section 9.3 (because of poor performance). What can be done then? A solution is proposed in Section 9.3.4 where a lock detector is employed with the joint phase and timing estimator of Section 9.2. In this way the joint estimator can be used even with multilevel partial response formats.

Section 9.4 concludes our discussion on feedback algorithms and describes a simple *ad hoc* detector for MSK-type signals.

Sections 9.5 and 9.6 deal with feedforward timing synchronizers. In particular, Section 9.5 derives a feedforward version of the timing loop proposed in Section 9.3. Its estimation accuracy is comparable with that of the feedback synchronizer, assuming that the usual relation

$$L_0 T = \frac{1}{2B_T} \qquad (9.1.1)$$

holds between the length of the observation interval and the bandwidth of the feedback loop.

Section 9.6 discusses two *ad hoc* timing synchronizers for MSK. They are simple to implement and have fairly good accuracy. One of them can be extended to GMSK signaling with only minor adjustments.

## 9.2. Decision-Directed Joint Phase and Timing Recovery

### 9.2.1. ML Formulation

Before addressing the estimation problem it is useful to refresh some notations about CPM signals and their ML detection [1]-[2]. To begin, the received signal is modeled as

$$s(t) = e^{j(2\pi v t + \theta)} \sqrt{\frac{2E_s}{T}} e^{j\psi(t-\tau,\alpha)} \qquad (9.2.1)$$

where the phase $\psi(t,\alpha)$ has the form

$$\psi(t,\alpha) \triangleq \eta(t, C_k, \alpha_k) + \Phi_k, \qquad kT \le t < (k+1)T \qquad (9.2.2)$$

with

$$\eta(t,C_k,\alpha_k) \triangleq 2\pi h \sum_{i=k-L+1}^{k} \alpha_i q(t-iT) \qquad (9.2.3)$$

$$h = \frac{K}{P} \qquad (9.2.4)$$

$$C_k \triangleq (\alpha_{k-L+1},\ldots,\alpha_{k-2},\alpha_{k-1}) \qquad (9.2.5)$$

$$\Phi_k \triangleq \pi h \sum_{i=0}^{k-L} \alpha_i \mod 2\pi \qquad (9.2.6)$$

In the above equations $\alpha \triangleq \{\alpha_i\}$ are the data symbols, $C_k$ is the *correlative state* and $\Phi_k$ is the *phase state* of the modulator. Note that there are $M^{L-1}$ correlative states and $P$ phase states. The symbols belong to the $M$-ary alphabet $\{\pm 1, \pm 3, \ldots, \pm(M-1)\}$ and are equally likely and independent. The modulation index is denoted by $h$. Finally, the phase response of the modulator is normalized so that

$$q(t) = \begin{cases} 0 & t \leq 0 \\ 1/2 & t \geq LT \end{cases} \qquad (9.2.7)$$

where $L$ is a parameter called the *correlation length*.

Figure 9.1 illustrates a block diagram for the ML receiver under the assumption that the synchronization parameters $\{v,\theta,\tau\}$ are all known. The filter bank is made of an array of $M^L$ filters with impulse responses

$$h^{(l)}(t) \triangleq \begin{cases} e^{-j\eta_l(T-t,C_0^{(l)},\alpha_0^{(l)})} & 0 \leq t \leq T \\ 0 & \text{elsewhere} \end{cases} \qquad (9.2.8)$$

with $(l=1,2,\ldots,M^L)$. In this equation $(C_0^{(l)},\alpha_0^{(l)}) = (\alpha_{-L+1}^{(l)},\ldots,\alpha_{-1}^{(l)},\alpha_0^{(l)})$ is a generic realization of $(\alpha_{-L+1},\ldots,\alpha_{-1},\alpha_0)$ and $\eta_l(t,C_0^{(l)},\alpha_0^{(l)})$ is given by

$$\eta_l(t,C_0^{(l)},\alpha_0^{(l)}) = 2\pi h \sum_{i=-L+1}^{0} \alpha_i^{(l)} q(t-iT) \qquad (9.2.9)$$

The filters are all driven by the common voltage $r(t)e^{-j2\pi v t}$ and their outputs are sampled at $(k+1)T+\tau$ to produce the statistics

$$Z_k(C_k,\alpha_k,\tau) \triangleq \int_{\tau+kT}^{\tau+(k+1)T} r(t)e^{-j2\pi v t} e^{-j\eta(t-\tau,C_k,\alpha_k)} dt \qquad (9.2.10)$$

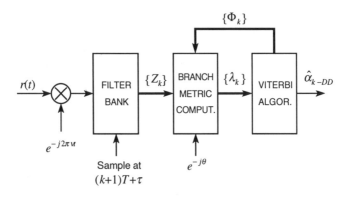

Figure 9.1. Block diagram of the ML receiver.

which are needed for the computation of the branch metrics. In fact, the Viterbi algorithm operates on a trellis of $PM^{L-1}$ nodes, the generic of which (at the $k$-th step) is denoted $S_k \triangleq (C_k, \Phi_k)$. There are $M$ branches stemming from a given node $\overline{S}_k = (\overline{C}_k, \overline{\Phi}_k)$, one for each possible transmitted symbol $\alpha_k$ and the metric for the branch associated with $\tilde{\alpha}_k$ is given by

$$\lambda_k(\overline{S}_k, \tilde{\alpha}_k) \triangleq \mathrm{Re}\left\{ Z_k(\overline{C}_k, \tilde{\alpha}_k, \tau) e^{-j(\theta + \overline{\Phi}_k)} \right\} \qquad (9.2.11)$$

Decisions are delivered by the Viterbi decoder with a delay $DD$.

With these definitions at hand we now turn our attention to the estimation problem. In doing so we assume that the carrier frequency offset and the data symbols are known, with the understanding that the latter will eventually be replaced by the decisions from the Viterbi algorithm. In these conditions the only unknown parameters are $\theta$ and $\tau$ and the associated likelihood function becomes

$$\Lambda(r|\tilde{\theta}, \tilde{\tau}) = \exp\left\{ \frac{1}{N_0} \sqrt{\frac{2E_s}{T}} \sum_{k=0}^{L_0-1} \mathrm{Re}\left\{ Z_k(C_k, \alpha_k, \tilde{\tau}) e^{-j(\tilde{\theta} + \Phi_k)} \right\} \right\} \qquad (9.2.12)$$

where, as usual, $L_0$ is the observation length. Clearly, the maximum of $\Lambda(r|\tilde{\theta}, \tilde{\tau})$ corresponds to the maximum of the sum in (9.2.12). So, setting to zero the partial derivatives of the sum yields

$$\sum_{k=0}^{L_0-1} \mathrm{Im}\left\{ Z_k(C_k, \alpha_k, \tilde{\tau}) e^{-j(\tilde{\theta} + \Phi_k)} \right\} = 0 \qquad (9.2.13)$$

# Timing Recovery with CPM Modulations

$$\sum_{k=0}^{L_0-1} \text{Re}\left\{Y_k(C_k, \alpha_k, \tilde{\tau})e^{-j(\tilde{\theta}+\Phi_k)}\right\} = 0 \qquad (9.2.14)$$

where $Y_k$ is the derivative of $Z_k$ with respect to $\tilde{\tau}$.

It is worth noting that $\{Y_k\}$ can be derived in two ways: either by differentiating the output of the filters $\{h^{(l)}(t)\}$ (and sampling at $(k+1)T+\tau$) or by feeding $r(t)e^{-j2\pi vt}$ into a bank of *derivative filters* with impulse responses $\{dh^{(l)}(t)/dt\}$. The first approach seems difficult as it is known that an exact derivative is hard to obtain by digital means. The other looks awkward as it requires building a second bank of filters, an expensive task in many practical cases. In spite of this apparent deadlock, we shall see in Section 9.2.3 that the question can be settled in a rather simple manner.

To solve equations (9.2.13)-(9.2.14) we use the $k$-th term in each sum as an error signal to update the current estimates of $\theta$ and $\tau$. In doing so two questions arise. The first is to replace the true data sequence $\{\ldots, \alpha_{k-2}, \alpha_{k-1}, \alpha_k\}$ involved in the calculation of $Z_k$, $Y_k$ and $\Phi_k$ by some reliable estimate thereof. As we have argued in other similar circumstances, a reasonable procedure is to exploit the decisions from the best survivor $\{\ldots, \alpha_{k-3}^{(b)}, \alpha_{k-2}^{(b)}, \alpha_{k-1}^{(b)}\}$ in the Viterbi algorithm. The second question is that, as the elements of $\{\ldots, \alpha_{k-3}^{(b)}, \alpha_{k-2}^{(b)}, \alpha_{k-1}^{(b)}\}$ become more and more reliable as the sequence is traced backward, it seems sensible to use only those elements at some distance $D$ or greater from the current time $k$. Putting these facts together leads to the following updating equations:

$$\hat{\theta}_{k+1} = \hat{\theta}_k + \gamma_P e_P(k-D) \qquad (9.2.15)$$

$$\hat{\tau}_{k+1} = \hat{\tau}_k + \gamma_T e_T(k-D) \qquad (9.2.16)$$

with

$$e_P(k-D) \triangleq \text{Im}\left\{Z_{k-D}(C_{k-D}^{(b)}, \alpha_{k-D}^{(b)}, \hat{\tau}_{k-D})e^{-j[\hat{\theta}_{k-D}+\Phi_{k-D}^{(b)}]}\right\} \qquad (9.2.17)$$

$$e_T(k-D) \triangleq \text{Re}\left\{Y_{k-D}(C_{k-D}^{(b)}, \alpha_{k-D}^{(b)}, \hat{\tau}_{k-D})e^{-j[\hat{\theta}_{k-D}+\Phi_{k-D}^{(b)}]}\right\} \qquad (9.2.18)$$

where $D$ is a delay parameter. Note that a large $D$ entails more reliable decisions but also large delays in the loops. Thus, some trade-off is called for between opposite requirements. Simulations indicate that a good compromise is achieved with $D=1$.

**Exercise 9.2.1.** The *S-surfaces* of a joint phase and timing synchronizer have been defined as

$$S_P(\phi, \delta) \triangleq \text{E}\{e_P(k)|\phi_k = \phi, \delta_k = \delta\} \qquad (9.2.19)$$

$$S_T(\phi,\delta) \triangleq \mathrm{E}\{e_T(k)|\phi_k = \phi, \delta_k = \delta\} \tag{9.2.20}$$

where $\phi_k \triangleq \theta - \hat{\theta}_k$ and $\delta_k \triangleq \tau - \hat{\tau}_k$ are phase and timing errors. Compute $S_P(\phi,0)$ assuming ideal decisions from the Viterbi detector (i.e., replace $\{\ldots,\alpha_{k-3}^{(b)},\alpha_{k-2}^{(b)},\alpha_{k-1}^{(b)}\}$ by the transmitted sequence $\{\ldots,\alpha_{k-3},\alpha_{k-2},\alpha_{k-1}\}$).

*Solution.* For our purposes we can set $D$ to zero in (9.2.17) as it turns out that the S-surfaces do not depend on time. Thus, letting $\hat{\tau}_k = \tau$ (which follows from the assumption $\delta = 0$) and replacing $\{\ldots,\alpha_{k-3}^{(b)},\alpha_{k-2}^{(b)},\alpha_{k-1}^{(b)}\}$ by $\{\ldots,\alpha_{k-3},\alpha_{k-2},\alpha_{k-1}\}$ in (9.2.17) yields

$$e_P(k) = \mathrm{Im}\{Z_k(C_k,\alpha_k,\tau)e^{-j[\theta-\phi_k+\Phi_k]}\} \tag{9.2.21}$$

where $C_k$ and $\Phi_k$ are the "true" correlative and phase states. On the other hand we have by definition

$$Z_k(C_k,\alpha_k,\tau) = \int_{\tau+kT}^{\tau+(k+1)T} r(t)e^{-j2\pi\nu t}e^{-j\eta(t-\tau,C_k,\alpha_k)}dt \tag{9.2.22}$$

and

$$r(t) = e^{j(2\pi\nu t+\theta)}\sqrt{\frac{2E_s}{T}}e^{j[\eta(t-\tau,C_k,\alpha_k)+\Phi_k]} + w(t) \tag{9.2.23}$$

Thus, substituting into (9.2.21) yields

$$e_P(k) = \sqrt{2E_sT}\sin\phi_k + n(k) \tag{9.2.24}$$

where $n(k)$ is a zero-mean noise term. Hence, setting $\phi_k = \phi$ and taking the expectation of (9.2.24) over $n(k)$ produces the desired result

$$S_P(\phi,0) = \sqrt{2E_sT}\sin\phi \tag{9.2.25}$$

**Exercise 9.2.2.** Compute the slope of the S-curve $S_T(0,\delta)$ at $\delta = 0$ assuming ideal decisions from the Viterbi detector.

*Solution.* Letting $D = 0$, $\theta(k) = \theta$ and replacing $\{\ldots,\alpha_{k-3}^{(b)},\alpha_{k-2}^{(b)},\alpha_{k-1}^{(b)}\}$ by $\{\ldots,\alpha_{k-3},\alpha_{k-2},\alpha_{k-1}\}$ in (9.2.18) yields

$$e_T(k) = \mathrm{Re}\{Y_k(C_k,\alpha_k,\hat{\tau}_k)e^{-j(\theta+\Phi_k)}\} \tag{9.2.26}$$

with

# Timing Recovery with CPM Modulations

$$Y_k = \frac{\partial Z_k}{\partial \hat{\tau}_k} \tag{9.2.27}$$

and

$$Z_k(C_k, \alpha_k, \hat{\tau}_k) = \int_{\hat{\tau}_k + kT}^{\hat{\tau}_k + (k+1)T} r(t) e^{-j2\pi vt} e^{-j\eta(t - \hat{\tau}_k, C_k, \alpha_k)} dt \tag{9.2.28}$$

Next, recalling that

$$S_T(0, \delta) = \mathrm{E}\{e_T(k) | \phi = 0, \hat{\tau}_k = \tau - \delta\} \tag{9.2.29}$$

the slope of $S_T(0, \delta)$ at $\delta = 0$ may be written as

$$A_{TT} = -\mathrm{Re}\left\{ \mathrm{E}\left[ \frac{\partial^2 Z_k(C_k, \alpha_k, \hat{\tau}_k)}{\partial \hat{\tau}_k^2} \right]_{\hat{\tau}_k = \tau} e^{-j(\theta + \Phi_k)} \right\} \tag{9.2.30}$$

On the other hand, from (9.2.28) it is found (after some manipulations) that

$$\left[ \frac{\partial^2 Z_k(C_k, \alpha_k, \hat{\tau}_k)}{\partial \hat{\tau}_k^2} \right]_{\hat{\tau}_k = \tau} =$$
$$-\sqrt{\frac{2E_s}{T}} e^{j(\theta + \Phi_k)} \int_{kT}^{(k+1)T} \left[ \eta'^2(t, C_k, \alpha_k) - j\eta''(t, C_k, \alpha_k) \right] dt + n(k) \tag{9.2.31}$$

where $\eta'(t, C_k, \alpha_k)$ and $\eta''(t, C_k, \alpha_k)$ are the first and second derivatives of $\eta(t, C_k, \alpha_k)$ with respect to time and $n(k)$ is a zero-mean noise term. Substituting into (9.2.30) and performing some further calculations yields the desired result

$$A_{TT} = \mathrm{E}\{\alpha_i^2\} \sqrt{\frac{2E_s}{T}} (2\pi h)^2 \int_0^{LT} g^2(t) dt \tag{9.2.32}$$

## 9.2.2. Approximate Digital Differentiation

In the previous section we have pointed out the necessity of approximating the statistics $\{Y_k\}$ in some manner. As is now explained, a solution to this problem is readily found when the receiver is implemented in digital form. In these circumstances the waveform $r(t)e^{-j2\pi vt}$ is first fed to an anti-aliasing filter (AAF) and then is sampled at some rate $1/T_s = N/T$. The sampling times $\{t_m\}$ are related to the timing estimates $\hat{\tau}_k$ by the relation

$$t_m = kT + nT_s + \hat{\tau}_k \tag{9.2.33}$$

with

$$k \triangleq \text{int}\left(\frac{m}{N}\right) \tag{9.2.34}$$

$$n \triangleq m \mod N \tag{9.2.35}$$

and $\text{int}(m/N)$ is the integer part of $m/N$. The meaning of the three indexes $m$, $k$ and $n$ is as follows: $m$ counts samples, $k$ counts symbol intervals and $n$ counts samples within a symbol interval.

The samples from the AAF are fed to a filter bank, as indicated in Figure 9.2. The impulse response $h^{(l)}(mT_s)$ of the generic filter in the bank is the sampled version of its analog counterpart $h^{(l)}(t)$ in Figure 9.1. Actually, the sequence $Z^{(l)}(m)$ from $h^{(l)}(mT_s)$ closely approximates the samples of $Z^{(l)}(t)$ from $h^{(l)}(t)$, provided that the oversampling factor $N$ is adequate.

Our task is to approximate the statistics $\{Y_k\}$. As mentioned earlier, they may be thought of as the samples of the waveforms $Z'^{(l)}(t) \triangleq dZ^{(l)}(t)/dt$ ($l=1,2,...,M^L$) taken at the instants $t = t_{(k+1)N}$. Formally

$$Y_k^{(l)} = Z'^{(l)}(t_{(k+1)N}) \tag{9.2.36}$$

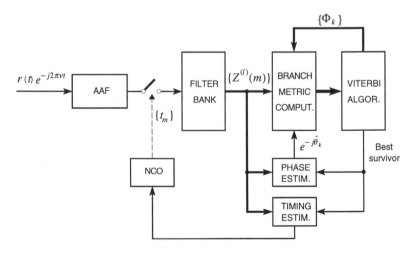

Figure 9.2. Digital implementation of the ML receiver.

Therefore, a simple approximation to $Z'^{(l)}(t_{(k+1)N})$ is provided by the two-point differentiation formula

$$Z'^{(l)}(t_m) \approx \frac{1}{2T_s}\left[Z^{(l)}(m+1) - Z^{(l)}(m-1)\right] \qquad (9.2.37)$$

Note that in this manner the outputs of the filter bank $\{h^{(l)}(mT_s)\}$ are sufficient to compute both $\{Z_k^{(l)}\}$ and $\{Y_k^{(l)}\}$ and, ultimately, the error signals $e_P(k-D)$ and $e_T(k-D)$, as well as the branch metrics

$$\lambda_k(\overline{S}_k, \tilde{\alpha}_k) \triangleq \mathrm{Re}\left\{Z_k(\overline{C}_k, \tilde{\alpha}_k, \hat{\tau}_k) e^{-j(\hat{\theta}_k + \overline{\Phi}_k)}\right\} \qquad (9.2.38)$$

### 9.2.3. Tracking Performance and Spurious Locks

Figures 9.3-9.4 show simulations illustrating the tracking performance of the synchronizer under the following conditions: (*i*) the symbols are binary and the frequency pulses are 3RC; (*ii*) the modulation index equals 1/2; (*iii*) the noise equivalent bandwidths of the loops are $B_T = 5 \cdot 10^{-3}/T$ and $B_P = 10^{-2}/T$; (*iv*) the oversampling factor $N$ is chosen equal to 4; (*v*) the approximation (9.2.37) is adopted; (*vi*) phase errors are expressed in radians, timing errors in cycles (meaning that they are normalized to the symbol period $T$). The modified Cramer-Rao bounds are also shown as benchmarks. We see that the error variances are very close to the MCRBs for $E_s/N_0$ values of practical interest.

Similar results are obtained with full response formats (either binary or multilevel) and with binary partial response schemes. In all these cases the synchronization errors may be made so small as to achieve ideal receiver performance. When it comes to partial response *and* multilevel schemes, however, the situation is more complex insofar as spurious locks take place (see also [3]). A spurious lock is illustrated in the simulation shown in Figure 9.5. Here, 3RC pulses are used as in Figures 9.3-9.4 but the alphabet is quaternary, not binary. As is seen, while the phase errors wander around zero, the timing errors fluctuate around 0.35. In a practical situation this would result in a very poor receiver performance.

In summary, the joint synchronizer discussed so far is only suitable with full response schemes (for example, with continuous-phase frequency shift keying) or with binary modulations. We shall see later that it can also be used with partial response multilevel schemes provided that a *lock detector* is employed. Before describing the lock detector, however, it is useful to discuss NDA timing estimation.

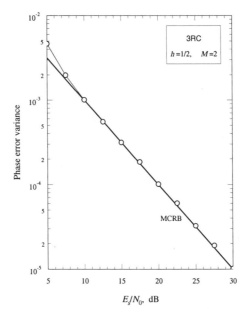

Figure 9.3. Simulation results for phase error variance.

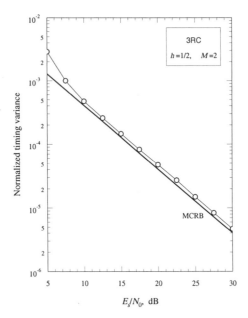

Figure 9.4. Simulation results for timing variance.

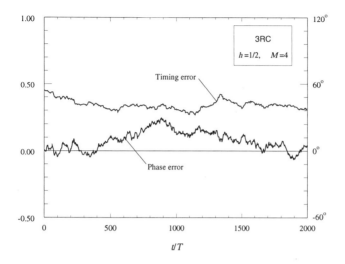

Figure 9.5. Simulation results showing a spurious lock.

## 9.3. NDA Feedback Timing Recovery

### 9.3.1. Approximate Expression for the Likelihood Function

In this section we investigate an ML-oriented timing error detector (TED) for use with general CPM formats. The algorithm is non-data-aided (NDA) and is independent of the carrier phase. Although it is noisier than the DD detector discussed previously, it does not suffer from spurious locks and, as we shall see, it can be conveniently exploited as a false lock detector in conjunction with the DD synchronizer in Section 9.2.

The starting point in our discussion is the likelihood function for the data $\boldsymbol{\alpha} \triangleq \{\alpha_i\}$ and the synchronization parameters $\theta$ and $\tau$ (carrier frequency is assumed known). Denoting $0 \le t \le L_0 T$ the observation interval, this function is found to be

$$\Lambda(r|\tilde{\boldsymbol{\alpha}},\tilde{\theta},\tilde{\tau}) = \exp\left\{\frac{1}{N_0}\sqrt{\frac{2E_s}{T}}\operatorname{Re}\left[e^{-j\tilde{\theta}}\int_0^{L_0 T} r(t) e^{-j[2\pi\nu t + \psi(t-\tilde{\tau},\tilde{\boldsymbol{\alpha}})]} dt\right]\right\} \quad (9.3.1)$$

with

$$r(t) = e^{j\theta}\sqrt{\frac{2E_s}{T}} e^{j[2\pi\nu t + \psi(t-\tau,\boldsymbol{\alpha})]} + w(t) \quad (9.3.2)$$

$$\psi(t,\boldsymbol{\alpha}) = 2\pi h \sum_{i=-\infty}^{\infty} \alpha_i q(t-iT) \tag{9.3.3}$$

Letting

$$\int_0^{L_0 T} r(t) e^{-j[2\pi v t + \psi(t-\tilde{\tau},\tilde{\boldsymbol{\alpha}})]} dt \triangleq |Y(\tilde{\boldsymbol{\alpha}},\tilde{\tau})| e^{j\gamma(\tilde{\boldsymbol{\alpha}},\tilde{\tau})} \tag{9.3.4}$$

equation (9.3.1) may be rewritten in the form

$$\Lambda(r|\tilde{\boldsymbol{\alpha}},\tilde{\theta},\tilde{\tau}) = \exp\left\{\frac{1}{N_0}\sqrt{\frac{2E_s}{T}}|Y(\tilde{\boldsymbol{\alpha}},\tilde{\tau})|\cos\left[\gamma(\tilde{\boldsymbol{\alpha}},\tilde{\tau}) - \tilde{\theta}\right]\right\} \tag{9.3.5}$$

Our task is to compute the marginal likelihood function $\Lambda(r|\tilde{\tau})$. As a first step in this direction we average (9.3.5) over the carrier phase, taking $\tilde{\theta}$ uniformly distributed over $[0,2\pi)$. This produces

$$\Lambda(r|\tilde{\boldsymbol{\alpha}},\tilde{\tau}) = I_0\left(\frac{1}{N_0}\sqrt{\frac{2E_s}{T}}|Y(\tilde{\boldsymbol{\alpha}},\tilde{\tau})|\right) \tag{9.3.6}$$

where $I_0(x)$ is the zero-order modified Bessel function. Next, assuming a low $E_s/N_0$ so that we can make the approximation

$$I_0(x) \approx 1 + \frac{x^2}{4} \tag{9.3.7}$$

from (9.3.6) we get

$$\Lambda(r|\tilde{\boldsymbol{\alpha}},\tilde{\tau}) \approx |Y(\tilde{\boldsymbol{\alpha}},\tilde{\tau})|^2 \tag{9.3.8}$$

where immaterial constants, independent of $\tilde{\boldsymbol{\alpha}}$ and $\tilde{\tau}$, have been dropped for simplicity. Making use of (9.3.4), equation (9.3.8) becomes

$$\Lambda(r|\tilde{\boldsymbol{\alpha}},\tilde{\tau}) \approx \int_0^{L_0 T}\int_0^{L_0 T} r(t_1)r^*(t_2) e^{j[2\pi v(t_2-t_1)+\psi(t_2-\tilde{\tau},\tilde{\boldsymbol{\alpha}})-\psi(t_1-\tilde{\tau},\tilde{\boldsymbol{\alpha}})]} dt_1 dt_2 \tag{9.3.9}$$

Thus, defining

$$F(\Delta t, t) \triangleq E_{\tilde{\boldsymbol{\alpha}}}\left\{e^{j[\psi(t,\tilde{\boldsymbol{\alpha}})-\psi(t-\Delta t,\tilde{\boldsymbol{\alpha}})]}\right\} \tag{9.3.10}$$

and averaging $\Lambda(r|\tilde{\boldsymbol{\alpha}},\tilde{\tau})$ over the data yields

## Timing Recovery with CPM Modulations

$$\Lambda(r|\tilde{\tau}) \approx \int_0^{L_0T}\int_0^{L_0T} r(t_1)r^*(t_2)e^{j2\pi v(t_2-t_1)}F(t_2-t_1, t_2-\tilde{\tau})dt_1dt_2 \quad (9.3.11)$$

The expectation (9.3.10) has been computed in Appendix 4.B and reads

$$F(\Delta t, t) = \prod_{i=-\infty}^{\infty} \frac{1}{M} \frac{\sin[2\pi M p(t-iT, \Delta t)]}{\sin[2\pi p(t-iT, \Delta t)]} \quad (9.3.12)$$

where $M$ is the alphabet size and $p(t, \Delta t)$ is defined as

$$p(t, \Delta t) \triangleq q(t) - q(t - \Delta t) \quad (9.3.13)$$

Our ultimate goal is to compute the argument $\tilde{\tau}$ that maximizes $\Lambda(r|\tilde{\tau})$. It is clear from (9.3.11), however, that this is a formidable task because of the cumbersome form of $F(\Delta t, t)$. In an attempt to sidestep the obstacle we observe that $F(\Delta t, t)$ is periodic with respect to $t$ of period $T$ and, as such, it can be expanded into a Fourier series. Our hope is that only a few terms in the series are significant. To explore this point we write $F(t_2 - t_1, t_2 - \tilde{\tau})$ in the form

$$F(t_2 - t_1, t_2 - \tilde{\tau}) = \sum_{m=-\infty}^{\infty} C_m(t_1, t_2) e^{j2\pi m\tilde{\tau}/T} \quad (9.3.14)$$

and we substitute it into (9.3.11) to produce

$$\Lambda(r|\tilde{\tau}) \approx \sum_{m=-\infty}^{\infty} A(m) e^{j2\pi m\tilde{\tau}/T} \quad (9.3.15)$$

with

$$A(m) \triangleq \int_0^{L_0T}\int_0^{L_0T} r(t_1)r^*(t_2)e^{j2\pi v(t_2-t_1)} C_m(t_1, t_2) dt_1 dt_2 \quad (9.3.16)$$

The coefficients $C_m(t_1, t_2)$ are computed in Appendix 9.A and are found to be

$$C_m(t_1, t_2) = h_m(t_1 - t_2) e^{-j\pi m(t_1+t_2)/T} \quad (9.3.17)$$

where $h_m(t)$ is a real-valued function

$$h_m(t) \triangleq e^{j\pi mt/T} \frac{1}{T}\int_0^T F(-t, u) e^{j2\pi mu/T} du \quad (9.3.18)$$

which satisfies the relation

$$h_m(t) = h_{-m}(t) \tag{9.3.19}$$

Inserting these results into (9.3.15)-(9.3.16) and performing some ordinary manipulations gives

$$\Lambda(r|\tilde{\tau}) \approx \text{Re}\left[\sum_{m=1}^{\infty} A(m) e^{j2\pi m\tilde{\tau}/T}\right] \tag{9.3.20}$$

with

$$A(m) = \int_0^{L_0 T} \left[r(t) e^{-j2\pi vt} e^{-j\pi mt/T}\right] y_m^*(t) dt \tag{9.3.21}$$

and

$$y_m(t) \triangleq \int_0^{L_0 T} \left[r(u) e^{-j2\pi vu} e^{j\pi mu/T}\right] h_m(t-u) du \tag{9.3.22}$$

At this point the question arises of how many terms in the sum (9.3.20) must be retained. An indication comes from Figure 9.6, which illustrates the shape of $h_m(t)$ for 1REC pulses, binary symbols and $h = 1/2$. As is seen, $h_1(t)$ is by far the largest pulse. This indicates that $y_1^*(t)$ is the dominant function in

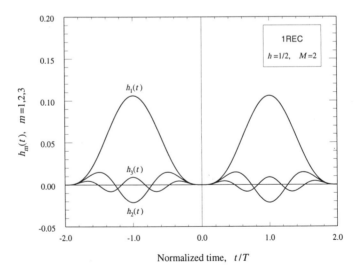

Figure 9.6. Shape of $h_m(t)$ for 1REC pulses, binary symbols and $h=1/2$.

Table 9.1. Ratio $E_m/E_1$ for some LREC formats.

| M | h | L | $E_2/E_1$ | $E_3/E_1$ | $E_4/E_1$ |
|---|---|---|---|---|---|
| 2 | 1/2 | 1 | $3.40 \cdot 10^{-2}$ | $6.05 \cdot 10^{-3}$ | $1.85 \cdot 10^{-3}$ |
| 2 | 1/2 | 2 | $4.33 \cdot 10^{-2}$ | $8.09 \cdot 10^{-3}$ | $2.51 \cdot 10^{-3}$ |
| 4 | 1/4 | 1 | $3.00 \cdot 10^{-2}$ | $5.32 \cdot 10^{-3}$ | $1.61 \cdot 10^{-3}$ |
| 8 | 1/8 | 1 | $3.06 \cdot 10^{-2}$ | $5.31 \cdot 10^{-3}$ | $1.61 \cdot 10^{-3}$ |

Table 9.2. Ratio $E_m/E_1$ for some LRC formats.

| M | h | L | $E_2/E_1$ | $E_3/E_1$ | $E_4/E_1$ |
|---|---|---|---|---|---|
| 2 | 1/2 | 1 | $1.80 \cdot 10^{-2}$ | $1.08 \cdot 10^{-4}$ | $4.33 \cdot 10^{-6}$ |
| 2 | 1/2 | 2 | $3.55 \cdot 10^{-4}$ | $2.76 \cdot 10^{-6}$ | $2.01 \cdot 10^{-7}$ |
| 4 | 1/4 | 1 | $1.85 \cdot 10^{-2}$ | $2.27 \cdot 10^{-4}$ | $5.79 \cdot 10^{-6}$ |
| 8 | 1/8 | 1 | $1.88 \cdot 10^{-2}$ | $2.76 \cdot 10^{-4}$ | $6.25 \cdot 10^{-6}$ |

the set $\{y_m^*(t)\}$ and, ultimately, the term with $m=1$ in (9.3.20) is the most significant. It turns out that this conclusion applies to most modulation formats. For example, Tables 9.1 and 9.2 show the ratio of the energy of $h_m(t)$, $E_m$, to the energy of $h_1(t)$ for a few LREC and LRC formats.

The foregoing considerations indicate that we can limit the summation in (9.3.20) to the first term, which implies that we can concentrate on the simple formula

$$\Lambda(r|\tilde{\tau}) \approx \text{Re}\left[A(1)e^{j2\pi \tilde{\tau}/T}\right] \quad (9.3.23)$$

This approximation is now used to derive a timing error detector.

### 9.3.2. Timing Error Detector

The following procedure may be applied to locate the maximum of $\Lambda(r|\tilde{\tau})$. Rewrite $A(1)$ as

$$A(1) = \sum_{k=0}^{L_0-1} A_k(1) \quad (9.3.24)$$

with

$$A_k(1) \stackrel{\Delta}{=} \int_{kT}^{(k+1)T} \left[ r(t)e^{-j2\pi vt}e^{-j\pi t/T} \right] y_1^*(t)dt \qquad (9.3.25)$$

Also, assume that the duration of $h_1(t)$ is much shorter than the observation interval so that $y_1(t)$ in (9.3.22) can be approximated as the output of a filter $h_1(t)$ driven by $r(t)e^{-j2\pi vt}e^{j\pi t/T}$, i.e.,

$$y_1(t) \approx \int_{-\infty}^{\infty} \left[ r(u)e^{-j2\pi vu}e^{j\pi u/T} \right] h_1(t-u)du \qquad (9.3.26)$$

Then, equation (9.3.23) becomes

$$\Lambda(r|\tilde{\tau}) \approx \sum_{k=0}^{L_0-1} \text{Re}\left[ A_k(1)e^{j2\pi \tilde{\tau}/T} \right] \qquad (9.3.27)$$

and the maximum of $\Lambda(r|\tilde{\tau})$ can be found through the usual recursive method wherein the derivative of the generic term in the sum (9.3.27) is used as an error signal. Simple calculations show that this signal is given by

$$e(k) = -\text{Im}\left[ A_k(1)e^{j2\pi \hat{\tau}_k/T} \right] \qquad (9.3.28)$$

and the recursive equation becomes

$$\hat{\tau}_{k+1} = \hat{\tau}_k + \gamma e(k) \qquad (9.3.29)$$

Figure 9.7 depicts a block diagram for the computation of the error signal. In drawing this figure we have made some minor changes motivated by practical considerations. First, the filter $h_1(t)$ has been made causal by rightward

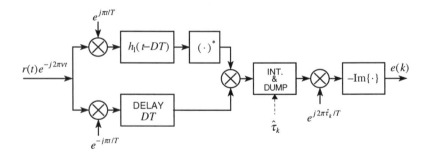

Figure 9.7. Timing error generator in analog form.

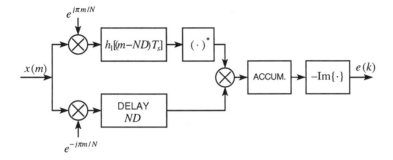

Figure 9.8. Digital implementation of the timing error detector.

shifting its impulse response by $DT$ seconds, the semi-duration of $h_1(t)$ (for example, $D$ may be chosen equal to 2 in the case of Figure 9.6). Second, to compensate for this delay, an identical delay has been introduced in the lower branch of the diagram. Third, the extremes of integration in the integrate-and-dump circuit have been changed as follows:

$$A_{k-D}(1) = \int_{(k-D)T+\hat{\tau}_k}^{(k+1-D)T+\hat{\tau}_k} [r(t)e^{-j2\pi vt}e^{-j\pi t/T}]y_1^*(t)dt \qquad (9.3.30)$$

In a digital implementation the incoming waveform is first passed through an anti-aliasing filter and then is sampled with some oversampling factor $N$. Denoting $T_s = T/N$ the sampling period and $x(m)$ the sample at $t_m = mT_s + \hat{\tau}_k$ from the AAF, the block diagram of the detector becomes as indicated in Figure 9.8 where the accumulator computes the sum

$$A_{k-D}(1) = \sum_{m=kN+DN}^{(k+1)N+DN-1} [x(m-ND)e^{-j\pi(m-ND)/N}]y_1^*(m-ND) \qquad (9.3.31)$$

### 9.3.3. Performance

The acquisition capability of the tracking system (9.3.29) is established by the S-curve of the timing detector. The calculation of this curve is a boring task, however, and is not pursued here. A detailed derivation is provided in [4] and the outcome is

$$S(\delta) = \frac{2E_s NH}{T}\sin\left(\frac{2\pi\delta}{T}\right) \qquad (9.3.32)$$

where $\delta \triangleq \tau - \hat{\tau}$ is the timing error and $H$ is the energy of $\{h_1(mT_s)\}$, i.e.,

$$H \triangleq \sum_{m=-\infty}^{\infty} h_1^2(mT_s) \tag{9.3.33}$$

Some remarks about (9.3.32) are useful. First, the curve has a sinusoidal shape of period $T$, with a zero upcrossing at the origin. This says that the only stable point of the loop is $\delta = 0$. Second, as the S-curve is independent of the carrier phase, there cannot be interactions between timing and phase recovery and, in consequence, no false locks can occur. This is in contrast with the behavior of the joint phase and timing recovery scheme discussed in Section 9.2.3. Third, the amplitude of the S-curve depends on the modulation parameters through the coefficient $H$. As this coefficient decreases when the length $L$ of the frequency pulses increases, the S-curve may eventually become too small for proper loop operation. Figure 9.9 shows $H$ as a function of $L$ for LREC pulses with $h=1/2$ and binary modulation. As is seen, $H$ decreases very rapidly. For example, $H$ is reduced to one-tenth in passing from $L=1$ to $L=2$.

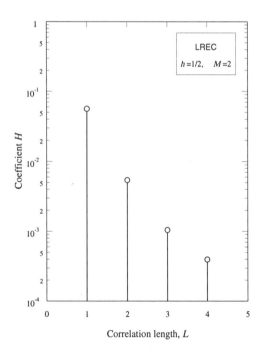

Figure 9.9. Coefficient $H$ versus correlation length for LREC pulses.

# Timing Recovery with CPM Modulations

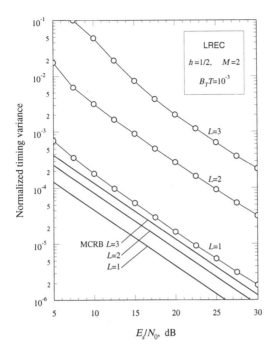

Figure 9.10. Tracking performance with LREC pulses.

Figure 9.10 illustrates simulation results for the loop performance with LREC pulses, binary modulation and $h=1/2$. The loop bandwidth is $B_T = 10^{-3}/T$ and the anti-aliasing filter is implemented as an 8-th order Butterworth type of bandwidth $2/T$. The oversampling factor is $N=4$. As expected, the tracking accuracy worsens dramatically as $L$ increases as a consequence of the diminishing amplitude of the S-curve.

## 9.3.4. False Lock Detection

As pointed out earlier, the DD detector in the previous section (say, DD-TED) has good tracking performance even with long correlation lengths but suffers from false locks when operating with multilevel signaling. On the other hand, the NDA detector in Section 9.3.2 (say, NDA-TED) has poor tracking performance with long frequency pulses but has no false locks. Thus, the idea arises of putting together the two detectors so as to make up for their individual weaknesses. A possible way to achieve this goal is now explained with the help of Figure 9.11.

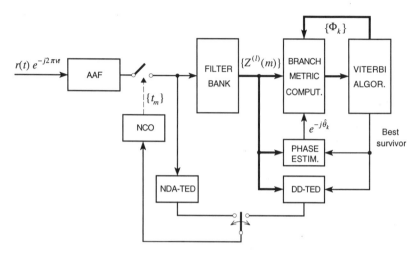

Figure 9.11. Receiver block diagram.

At any given time the timing loop is driven by either NDA-TED or DD-TED. In particular, the former is in command during acquisition, the latter in the tracking mode. When the NDA-TED is in stand-by, its error signal $e_T^{(NDA)}(k)$ is fed to a first-order IIR filter (not shown in the diagram) to produce the statistic $u(k)$ given by

$$u(k+1) = (1-\alpha)u(k) + \alpha e_T^{(NDA)}(k) \qquad (9.3.34)$$

from which a lock signal is derived in the form $|u(k)|$. The rationale behind this procedure is that, if the synchronizer is in a false lock, then $e_T^{(NDA)}(k)$ exhibits some DC component which can be exploited to switch the loop from DD-TED to NDA-TED mode. For this purpose the signal $e_T^{(NDA)}(k)$ is filtered (to smooth out as much noise as possible) and the result, $u(k)$, is monitored. An out-of-lock is declared as soon as $|u(k)|$ overcomes some fixed threshold $\lambda$. When this happens the DD-TED is switched off and the NDA-TED is activated in its place. Furthermore, the control signal $u(k)$ is set to zero and the NDA-TED is put in command for the next $K$ symbols, with $K$ on the order of the inverse of the loop bandwidth. This allows the loop to reach a steady-state condition. As soon as this occurs, the DD-TED is switched on again and the tracking operation restarts.

A simulation study illustrating an acquisition from a false lock is shown in Figure 9.12. Modulation is 3RC with quaternary symbols and a modulation index of 1/2. A bandwidth of $B_P = 10^{-2}/T$ is chosen for the phase loop whereas the timing loop bandwidth is either $B_T = 5 \cdot 10^{-3}/T$ or $B_T = 5 \cdot 10^{-4}/T$,

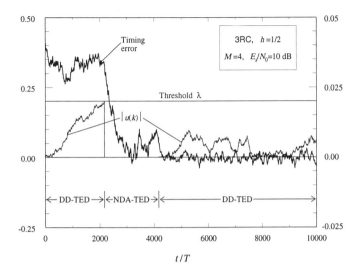

Figure 9.12. Timing acquisition with quaternary 3RC and $h=1/2$.

depending on whether the DD-TED or the NDA-TED is used. This is so because the latter is noisier and we have tried to keep the same steady-state errors in both cases. The parameter $K$ equals 2000 and the step size $\alpha$ in (9.3.34) is set to $5 \cdot 10^{-4}$. The threshold $\lambda$ is chosen equal to 0.02, a bit lower than the amplitude of the NDA-TED S-curve at the false-lock point (which is about 0.025).

As indicated in Figure 9.12, the experiment begins with the lock indicator $|u(k)|$ equal to zero. At the start the DD-TED is on and the normalized timing errors fluctuate around 0.3. The system is clearly in a false lock and this makes the lock indicator output increase. When $|u(k)|$ reaches the threshold, the NDA-TED takes control and keeps operating for 2000 symbols. Then, the DD-TED is turned on again and the synchronizer proceeds in the tracking mode. During the first part of the acquisition, carrier phase errors (not shown) wander in a random manner. They converge to zero, however, as soon as the timing errors decrease.

## 9.4. Ad Hoc Feedback Schemes for MSK-Type Modulations

We now restrict our attention to MSK-type modulations. In this context two viable routes to timing recovery have been identified so far, either the DD method described in Section 9.2 or the NDA approach in Section 9.3. The choice between the two methods depends on the application. For example, the DD method has superior tracking accuracy but involves a phase recovery section which would be useless in a differential receiver. In the following we consider again carrier-phase-independent timing detectors as in Section 9.3 but we proceed on a heuristic ground as indicated in [5]. This will lead us to a new scheme which is considerably simpler to implement than those encountered previously.

The TED proposed in [5] may be described as follows. After compensation for the frequency offset, the incoming waveform is passed through an AAF and is sampled at the instants

$$t_m = kT + nT_s + \hat{\tau}_k \quad (9.4.1)$$

with

$$k \triangleq \text{int}\left(\frac{m}{N}\right) \quad (9.4.2)$$

$$n \triangleq m \mod N \quad (9.4.3)$$

As usual, the integer $N$ represents the oversampling factor and $T_s = T/N$ is the sampling period. Of particular interest to us are the samples $x(kT + T_s + \hat{\tau}_k)$ and $x(kT - T_s + \hat{\tau}_{k-1})$ taken one step after and before the start of the generic symbol interval, $kT + \hat{\tau}_k$. They are used to form the following TED:

$$e(k) = (-1)^{D+1} \text{Re}\left\{x^2(kT - T_s + \hat{\tau}_{k-1})x^{*2}[(k-D)T - T_s + \hat{\tau}_{k-D-1}]\right\}$$
$$- (-1)^{D+1} \text{Re}\left\{x^2(kT + T_s + \hat{\tau}_k)x^{*2}[(k-D)T + T_s + \hat{\tau}_{k-D}]\right\} \quad (9.4.4)$$

where $D$ is a design parameter taking integer and positive values. This parameter is set to unity in [5]. Further simulations have shown that $D=1$ is a good choice with MSK whereas $D=2$ is preferable with GMSK.

An intuitive motivation for (9.4.4) is not available and the simplest way to explain the detector operation is to show that its S-curve has a regular form, with a unique up-crossing at the origin. In the following we derive this curve making two simplifying assumptions:

(*i*) The AAF bandwidth is sufficiently large to pass the signal components undistorted.

(*ii*) Thermal noise is negligible.

In a practical situation the noise may not be negligible but, as indicated in [5], its effects on the calculation of the S-curve can be ignored.

## Timing Recovery with CPM Modulations

To begin we observe that (as a consequence of assumptions ($i$)-($ii$)) the output $x(t)$ from the AAF coincides with the received signal. So, we have

$$x(t) = e^{j\theta}\sqrt{\frac{2E_s}{T}}e^{j\psi(t-\tau,\alpha)} \tag{9.4.5}$$

with

$$\psi(t,\alpha) = 2\pi h \sum_{i=-\infty}^{\infty} \alpha_i q(t-iT) \tag{9.4.6}$$

Next, recalling the definition

$$S(\delta) \triangleq E\{e(k)|\tau - \hat{\tau}_k = \delta\} \tag{9.4.7}$$

from (9.4.4) we get

$$S(\delta) = (-1)^{D+1}\left(\frac{2E_s}{T}\right)^2 \text{Re}\{c(kT-T_s-\delta)-c(kT+T_s-\delta)\} \tag{9.4.8}$$

with

$$c(t) \triangleq E_\alpha\{e^{j2[\psi(t,\alpha)-\psi(t-DT,\alpha)]}\} \tag{9.4.9}$$

Note that the carrier phase $\theta$ does not appear in (9.4.8), which means that the detector is phase-insensitive.

A general expression for the expectation in (9.4.9) is computed in Appendix 4.B for an arbitrary modulation. In particular, with a binary alphabet and a modulation index of 1/2 this expression becomes

$$c(t) = \prod_{i=-\infty}^{\infty} \cos[2\pi p(t-iT,DT)] \tag{9.4.10}$$

with

$$p(t,DT) \triangleq q(t) - q(t-DT) \tag{9.4.11}$$

Collecting (9.4.8) and (9.4.10) yields $S(\delta)$ for any $q(t)$.

Unfortunately, the actual shape of the S-curve is not readily recognizable from the above equations. The only exception occurs with MSK and $D=1$, in which case the following simple result is found in Exercise 9.4.1:

$$S(\delta) = \left(\frac{2E_s}{T}\right)^2 \sin\left(\frac{2\pi}{N}\right)\sin\left(\frac{2\pi\delta}{T}\right) \qquad (9.4.12)$$

Note that the maximum amplitude of the sinusoid is achieved for $N=4$ whereas $S(\delta)$ vanishes for $N=2$. Thus, the synchronizer would not work with a sampling rate of $2/T$.

Figure 9.13 shows S-curves for MSK ($D=1$) and GMSK ($D=2$). In drawing these curves the factor $(2E_s/T)^2$ in (9.4.8) has been set to unity for simplicity. Also, the AAF is implemented as an eighth-order Butterworth filter of bandwidth $1/T$ and the pre-modulation filter with GMSK has a normalized bandwidth $BT = 0.3$. The oversampling factor is set to $N=4$ and the ratio $E_s/N_0$ equals 10 dB. Solid lines represent the theoretical curves as given by (9.4.8) while circles indicate simulation results. As is seen, simulation and theory agree fairly well. The small discrepancy with MSK is due to signal distortions in the anti-aliasing filter (GMSK has a smaller bandwidth and is less affected by the AAF).

Figure 9.14 compares the *ad hoc* algorithm (9.4.4) with the ML-based NDA detector in (9.3.28). The parameter $D$ equals 1 with MSK whereas it is set to 2 with GMSK. As in Figure 9.13, the bandwidth of the pre-modulation filter is $BT = 0.3$ and a loop bandwidth of $B_T T = 10^{-3}$ is chosen. It appears that *ad hoc* and ML-based detectors have comparable performance with MSK but the latter is considerably worse with GMSK. This fact should not sound paradoxical as the derivation of the ML-based TED has involved a number of

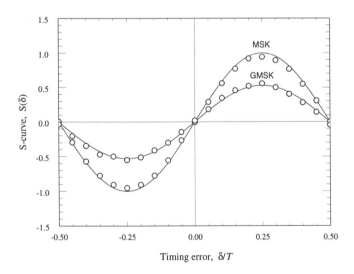

Figure 9.13. S-curves for MSK and GMSK.

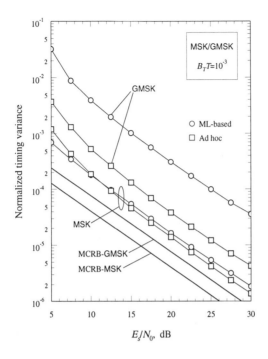

Figure 9.14. Comparison between algorithms.

approximations and deviations from the true ML method. At any rate, it should be stressed that the ML-based TED is useful with general CPM modulation whereas the *ad hoc* algorithm works only with MSK-type signals.

**Exercise 9.4.1.** Derive equation (9.4.12) assuming MSK modulation and $D=1$.

*Solution.* The phase response $q(t)$ with MSK is given by

$$q(t) = \begin{cases} 0 & t < 0 \\ t/(2T) & 0 \leq t \leq T \\ 1/2 & t > T \end{cases} \qquad (9.4.13)$$

Thus, (9.4.11) becomes

$$p(t,T) = \begin{cases} t/(2T) & 0 \leq t \leq T \\ 1 - t/(2T) & T < t \leq 2T \\ 0 & \text{elsewhere} \end{cases} \qquad (9.4.14)$$

Next we concentrate on the function $c(t)$ in (9.4.10). Clearly $c(t)$ is periodic of period $T$ and, as such, it is completely defined by the values it takes on the interval $(0,T)$. On the other hand, as $p(t,T)$ is zero outside the interval $(0,2T)$, it follows that $\cos[2\pi p(t-iT,T)]$ is unity over $(0,T)$ except for $i = 0$ and $i = -1$. Thus, for $0 \le t \le T$ equation (9.4.10) becomes

$$c(t) = \cos[2\pi p(t,T)]\cos[2\pi p(t+T,T)] \qquad (9.4.15)$$

and taking (9.4.14) into account, we get

$$c(t) = -\frac{1}{2}\left[1 + \cos\left(\frac{2\pi t}{T}\right)\right] \qquad (9.4.16)$$

Substituting this result into (9.4.8) with $D=1$ and rearranging yields (9.4.12).

## 9.5. NDA Feedforward Timing Estimation

Henceforth we concentrate on feedforward estimation methods. We begin with an ML-oriented scheme which is based on the same ideas used in Section 9.3 in the context of feedback synchronization. In effect, as the present derivation has many points in common with that discussion, our account will be rather condensed for we can draw from previous results. A more comprehensive description of this estimator may be found in [6].

A general CPM modulation format is assumed as in Section 9.3 but the signal samples are no longer synchronized to the transmitter clock. In fact they are derived from a free-running oscillator at some rate $1/T_s$. As usual, we suppose that the signal components at the output of the anti-aliasing filter are undistorted and the noise samples are independent. Thus, denoting by $L_0$ the length of the observation interval and choosing the ratio $T/T_s$ as an integer $N$ for convenience, the likelihood function for the data $\boldsymbol{\alpha} \triangleq \{\alpha_i\}$ and the synchronization parameters $\theta$ and $\tau$ becomes (carrier frequency is assumed known)

$$\Lambda(\boldsymbol{x}|\tilde{\boldsymbol{\alpha}},\tilde{\theta},\tilde{\tau}) = \exp\left\{-\frac{1}{2\sigma_n^2}\sum_{k=0}^{NL_0-1}\left|x(kT_s) - \sqrt{\frac{2E_s}{T}}e^{j[\tilde{\theta}+\psi(kT_s-\tilde{\tau},\tilde{\boldsymbol{\alpha}})]}\right|^2\right\} \qquad (9.5.1)$$

Here, $x(t)$ represents the AAF output, $kT_s$ ($k = 0,1,2...$) are the sampling times and $2\sigma_n^2$ is the variance of the noise samples. Noting that

# Timing Recovery with CPM Modulations

$$\left| x(kT_s) - \sqrt{\frac{2E_s}{T}} e^{j[\tilde{\theta}+\psi(kT_s-\tilde{\tau},\tilde{\alpha})]} \right|^2 =$$

$$|x(kT_s)|^2 + \frac{2E_s}{T} - 2\sqrt{\frac{2E_s}{T}} \operatorname{Re}\left\{ x(kT_s) e^{-j[\tilde{\theta}+\psi(kT_s-\tilde{\tau},\tilde{\alpha})]} \right\} \quad (9.5.2)$$

and eliminating immaterial constants, equation (9.5.1) becomes

$$\Lambda(x|\tilde{\alpha},\tilde{\theta},\tilde{\tau}) = \exp\left\{ \frac{1}{\sigma_n^2} \sqrt{\frac{2E_s}{T}} \operatorname{Re}\left[ e^{-j\tilde{\theta}} \sum_{k=0}^{NL_0-1} x(kT_s) e^{-j\psi(kT_s-\tilde{\tau},\tilde{\alpha})} \right] \right\} \quad (9.5.3)$$

The computation of the marginal likelihood $\Lambda(x|\tilde{\tau})$ can now be performed following the same arguments used in Section 9.3. Skipping the details, the final result is very similar to that in (9.3.23) and has the form

$$\Lambda(x|\tilde{\tau}) \approx \operatorname{Re}\left[ A(1) e^{j2\pi\tilde{\tau}/T} \right] \quad (9.5.4)$$

with

$$A(1) \triangleq \sum_{k=0}^{NL_0-1} \left[ x(kT_s) e^{-j\pi k/N} \right] y^*(kT_s) \quad (9.5.5)$$

$$y(kT_s) \triangleq \sum_{n=0}^{NL_0-1} \left[ x(nT_s) e^{j\pi n/N} \right] h_1\left[ (k-n)T_s \right] \quad (9.5.6)$$

where $h_1(t)$ is as given in (9.3.18).

As the observation interval is usually much longer than the duration of $h_1(t)$, equation (9.5.5) can be rearranged as follows. First, the summation in (9.5.6) is extended from $-\infty$ to $+\infty$ and, correspondingly, $y(kT_s)$ can be viewed as the output of a (non-causal) filter $h_1(kT_s)$ driven by $x(kT_s)e^{j\pi k/N}$, i.e.,

$$y(kT_s) \approx \left[ x(kT_s) e^{j\pi k/N} \right] \otimes h_1(kT_s) \quad (9.5.7)$$

Second, the filter $h_1(kT_s)$ is made causal by rightward shifting its impulse response by some steps, say $ND$. In doing so the filter output becomes

$$y\left[ (k-ND)T_s \right] \approx \left[ x(kT_s) e^{j\pi k/N} \right] \otimes h_1\left[ (k-ND)T_s \right] \quad (9.5.8)$$

and equation (9.5.5) may be rewritten as

$$A(1) = \sum_{k=ND}^{N(L_0+D)-1} x[(k-ND)T_s] e^{-j\pi(k-ND)/N} y^*[(k-ND)T_s] \quad (9.5.9)$$

At this point, substituting into (9.5.4) and recognizing that $\text{Re}\{A(1)e^{j2\pi \tilde{\tau}/T}\}$ achieves a maximum when the argument of $A(1)$ equals $-2\pi\tilde{\tau}/T$ leads to the estimation rule

$$\hat{\tau} = -\frac{T}{2\pi} \arg\left\{ \sum_{k=ND}^{N(L_0+D)-1} x[(k-ND)T_s] e^{-j\pi(k-ND)/N} y^*[(k-ND)T_s] \right\} \quad (9.5.10)$$

whose block diagram is shown in Figure 9.15.

A performance analysis of this estimator is rather lengthy and is provided in [6]. Roughly speaking, the estimation accuracy is the same as that of the feedback scheme in Section 9.3 provided that the observation length $L_0$ is related to the bandwidth $B_T$ of the feedback loop by

$$L_0 T = \frac{1}{2B_T} \quad (9.5.11)$$

Thus, the merits and weaknesses of the present method are the same as those pointed out in Section 9.3. On the bright side, the algorithm is carrier-phase-insensitive and can be used with multilevel signaling. Its performance is poor with long frequency pulses however. Figure 9.16 illustrates the variance of the normalized estimates $\hat{\tau}/T$ for quaternary 1REC and 1RC pulses with a modulation index $h = 1/2$. The observation length is $L_0 = 100$ and the oversampling factor equals 4. The bandwidth of the anti-alias filter is $2/T$. We see that 1REC format is much more difficult to synchronize than 1RC.

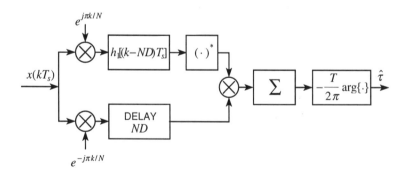

Figure 9.15. Block diagram of the estimator.

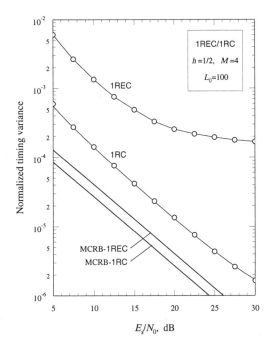

Figure 9.16. Timing error variance with $M=4$, 1RC pulses and $h=1/2$.

## 9.6. Ad Hoc Feedforward Schemes for MSK Modulation

We conclude this chapter with two more feedforward timing schemes for MSK modulation. Both of them are based on heuristic arguments and are considerably simpler to implement than the scheme in the previous section. We first report on the method by Mehlan, Chen and Meyr (MCM) [7] and then on that by Lambrette and Meyr (LM) [8]. It turns out that the former can be extended to GMSK modulation with only minor adjustments.

### 9.6.1. MCM Scheme

The incoming signal is fed to an anti-aliasing filter whose output is denoted $x(t)$. For the time being we assume a negligible noise level so that we can write

$$x(t) = e^{j(2\pi v t + \theta)} \sqrt{\frac{2E_s}{T}} e^{j\psi(t-\tau,\alpha)} \qquad (9.6.1)$$

with

$$\psi(t,\alpha) = \pi \sum_{i=-\infty}^{\infty} \alpha_i q(t - iT) \qquad (9.6.2)$$

$$q(t) = \begin{cases} 0 & t < 0 \\ t/(2T) & 0 \le t \le T \\ 1/2 & t > T \end{cases} \qquad (9.6.3)$$

and $\alpha_i = \pm 1$. Note that a non-zero carrier frequency offset has been assumed in (9.6.1). This is in keeping with the signal model adopted in [7] and is necessary to arrive at the same estimator developed in the original paper.

The rationale behind the MCM algorithm is best understood focusing on the process

$$z(t) \triangleq \left[ x(t) x^*(t - T) \right]^2 \qquad (9.6.4)$$

As we shall see, its statistics depend on the timing parameter $\tau$ and therefore we can estimate $\tau$ by processing $z(t)$ in some manner. The following considerations are useful to understand how this can be done.

Collecting (9.6.1) and (9.6.4) yields

$$z(t) = \left( \frac{2E_s}{T} \right)^2 e^{j4\pi vT} e^{j2[\psi(t-\tau,\alpha)-\psi(t-T-\tau,\alpha)]} \qquad (9.6.5)$$

which indicates that $z(t)$ depends not only on $\tau$ but also on $v$ and the data. To eliminate the dependence on the data we average $z(t)$ over $\alpha$. Using the relation (see Exercise 9.4.1)

$$E_\alpha \left\{ e^{j2[\psi(t,\alpha)-\psi(t-T,\alpha)]} \right\} = -\frac{1}{2} \left[ 1 + \cos\left( \frac{2\pi t}{T} \right) \right] \qquad (9.6.6)$$

and denoting by $\bar{z}(t)$ the expectation $E_\alpha\{z(t)\}$, it is found that

$$\bar{z}(t) = -\frac{1}{2} \left( \frac{2E_s}{T} \right)^2 e^{j4\pi vT} \left[ 1 + \cos\left[ \frac{2\pi(t-\tau)}{T} \right] \right] \qquad (9.6.7)$$

Next, the residual dependence on $v$ is removed by taking the absolute value of $\bar{z}(t)$

$$|\bar{z}(t)| = K \left[ 1 + \cos\left[ \frac{2\pi(t-\tau)}{T} \right] \right] \qquad (9.6.8)$$

# Timing Recovery with CPM Modulations

where $K$ is a positive constant.

Equation (9.6.8) establishes a link between $\tau$ and $\bar{z}(t)$. An explicit expression of $\tau$ is now obtained observing that $\bar{z}(t)$ is periodic and computing the coefficient at frequency $f = 1/T$ in its Fourier series

$$\frac{1}{T}\int_0^T |\bar{z}(t)| e^{-j2\pi t/T} dt = \frac{K}{2} e^{-j2\pi \tau/T} \qquad (9.6.9)$$

Solving for $\tau$ yields

$$\tau = -\frac{T}{2\pi} \arg\left\{ \frac{1}{T}\int_0^T |\bar{z}(t)| e^{-j2\pi t/T} dt \right\} \qquad (9.6.10)$$

This result is important as it indicates a procedure to derive $\tau$ from $\bar{z}(t)$ (and ultimately from $x(t)$) in the absence of noise. In practice the noise can seldom be neglected and the right-hand side of (9.6.10) gives only an estimate of $\tau$.

The estimator (9.6.10) can be implemented in digital form as follows. As a first step, the integral in (9.6.10) is expressed as a function of the samples of $\bar{z}(t)$ taken at some rate $1/T_s$. Choosing $T_s = T/N$ produces

$$\frac{1}{T}\int_0^T |\bar{z}(t)| e^{-j2\pi t/T} dt \approx \frac{1}{N} \sum_{k=0}^{N-1} |\bar{z}(kT_s)| e^{-j2\pi k/N} \qquad (9.6.11)$$

for $N$ sufficiently large. The second step is to compute the expectation $\bar{z}(kT_s)$ as a function of the samples of $z(t)$. As $z(t)$ is a cyclostationary process of period $T$, its samples have the same statistics at multiples of $T$, and a reasonable approximation to $\bar{z}(kT_s)$ is obtained by averaging over $T$-spaced samples of $z(t)$:

$$\bar{z}(kT_s) \approx \frac{1}{L_0} \sum_{i=0}^{L_0-1} z[(k+iN)T_s] \qquad (9.6.12)$$

Incorporating these approximations into (9.6.10) produces the MCM estimator

$$\hat{\tau} = -\frac{T}{2\pi} \arg\left\{ \sum_{k=0}^{N-1} \left| \sum_{i=0}^{L_0-1} z[(k+iN)T_s] \right| e^{-j2\pi k/N} \right\} \qquad (9.6.13)$$

with

$$z(kT_s) = \left[ x(kT_s) x^*[(k-N)T_s] \right]^2 \qquad (9.6.14)$$

The performance of the MCM algorithm is investigated in [7] and is now discussed making use of some simulations. Before doing so, however, a few words about the applicability of (9.6.13) are useful. It is clear from the previous derivation that the MCM algorithm is specifically tailored for MSK. Thus, it is not obvious that it also will work with other modulation formats. In fact its performance with GMSK is found to be poor. It turns out, however, that significant improvements are achieved by replacing $x^*[(k-N)T_s]$ by $x^*[(k-2N)T_s]$ in (9.6.14), i.e., choosing

$$z(kT_s) = \left[x(kT_s)x^*[(k-2N)T_s]\right]^2 \qquad (9.6.15)$$

As is seen, (9.6.15) involves samples of $x(t)$ at a distance $2T$ instead of $T$. In the simulations to follow, the variant (9.6.15) is always used with GMSK.

Simulations with MSK and GMSK are shown in Figures 9.17-9.18. The AAF is implemented as an eighth-order Butterworth filter of bandwidth $1/T$ and the oversampling factor equals 4. Figure 9.17 deals with MSK and compares the timing error variance with MCM to that of the feedback scheme

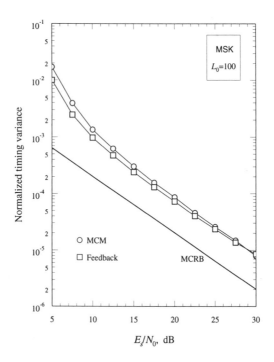

Figure 9.17. Timing error variance with MSK.

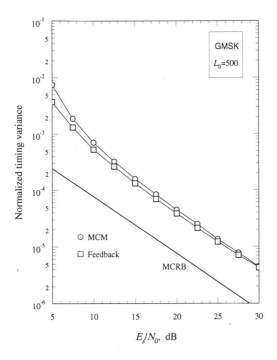

Figure 9.18. Timing error variance with GMSK.

described in Section 9.4. The loop bandwidth in the latter is $B_T T = 5 \cdot 10^{-3}$, which corresponds to $L_0 = 100$. It appears that the two methods give equivalent results.

Figure 9.18 illustrates similar results for GMSK (with a pre-modulation bandwidth of $BT = 0.3$). Here the loop bandwidth of the feedback scheme is set to $10^{-3}/T$, which corresponds to an observation length of $L_0 = 500$. Again, the MCM method and the feedback scheme exhibit comparable performance.

### 9.6.2. LM Scheme

We assume MSK modulation and make the same approximations as in the previous section, with the only exception that we now take $v = 0$ for further simplicity. As happens with the MCM algorithm, the incoming signal is first passed through an AAF to produce $x(t)$. Neglecting thermal noise, we have

$$x(t) = e^{j\theta} \sqrt{\frac{2E_s}{T}} e^{j\psi(t-\tau,\alpha)} \qquad (9.6.16)$$

The motivation for the LM scheme can be explained by considering the following process:

$$z(t) \triangleq \left| \arg\{x(t)x^*(t-T)\} \right| \qquad (9.6.17)$$

where the argument function is taken in the interval $(-\pi,\pi]$. To proceed, denote by $[x]_{-\pi}^{\pi}$ the value of $x$ reduced to the interval $(-\pi,\pi]$. Then, from (9.6.16) we have

$$z(t) = \left| [\psi(t-\tau,\boldsymbol{\alpha}) - \psi(t-T-\tau,\boldsymbol{\alpha})]_{-\pi}^{\pi} \right| \qquad (9.6.18)$$

or, making use of (9.6.2),

$$z(t) = \left| \left[ \pi \sum_i \alpha_i p(t-iT-\tau,T) \right]_{-\pi}^{\pi} \right| \qquad (9.6.19)$$

with

$$p(t,T) \triangleq q(t) - q(t-T) \qquad (9.6.20)$$

Figure 9.19(a) shows the shape of $p(t,T)$ for MSK signaling. It is clear from the figure that the sum

$$\pi \sum_i \alpha_i p(t-iT-\tau,T) \qquad (9.6.21)$$

takes values between $\pm\pi/2$ and, in consequence, (9.6.19) reduces to

$$z(t) = \left| \pi \sum_i \alpha_i p(t-iT-\tau,T) \right| \qquad (9.6.22)$$

Figure 9.19(b) depicts $z(t)$ for the symbol pattern $\ldots,+1,+1,-1,+1,+1,\ldots$ (the reader may check that this same $z(t)$ would also be generated by the opposite pattern $\ldots,-1,-1,+1,-1,-1,\ldots$). We see that $z(t)$ has a constant value $\pi/2$ as long as there is no pattern variation. When a variation occurs, $z(t)$ may be viewed as the sum of $\pi/2$ and a couple of triangular pulses as indicated in Figure 9.19(c). Extending these results to a general data pattern, it is realized that $z(t)$ can be written as

$$z(t) = \frac{\pi}{2}\left[ 1 - \frac{1}{2}\sum_i (1-\alpha_i\alpha_{i-1})g(t-iT-\tau) \right] \qquad (9.6.23)$$

# Timing Recovery with CPM Modulations

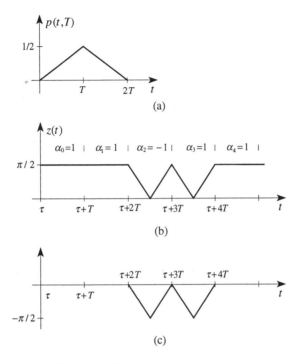

Figure 9.19. Illustrating the shape of $z(t)$.

where $g(t)$ is a triangular pulse given by

$$g(t) = \begin{cases} 2t/T & 0 \le t \le T/2 \\ 2(1-t/T) & T/2 \le t \le T \\ 0 & \text{elsewhere} \end{cases} \quad (9.6.24)$$

From (9.6.23) it appears that $z(t)$ depends not only on the timing parameter but also on the data. The latter can be removed by averaging over the possible symbol sequences. Denoting by $\bar{z}(t)$ the resulting average yields

$$\bar{z}(t) = \frac{\pi}{2}\left[1 - \frac{1}{2}\sum_i g(t - iT - \tau)\right] \quad (9.6.25)$$

The next step is to establish an explicit expression of $\tau$ as a function of $\bar{z}(t)$. To this end we observe that $\bar{z}(t)$ is periodic of period $T$ and, as such, it can be expanded into a Fourier series. Computing the coefficient at frequency $f = 1/T$ in the series gives

$$\frac{1}{T}\int_0^T \bar{z}(t)e^{-j2\pi t/T}dt = \frac{1}{2\pi}e^{-j2\pi\tau/T} \qquad (9.6.26)$$

from which the following expression of $\tau$ is obtained:

$$\tau = -\frac{T}{2\pi}\arg\left\{\frac{1}{T}\int_0^T \bar{z}(t)e^{-j2\pi t/T}dt\right\} \qquad (9.6.27)$$

At this point we face the same situation encountered with (9.6.10). In particular we have a procedure to derive $\tau$ from the observed waveform $x(t)$ and we want to implement this procedure in digital form. Using the same method adopted with the MCM leads to the LM estimator

$$\hat{\tau} = -\frac{T}{2\pi}\arg\left\{\sum_{k=0}^{N-1}\sum_{i=0}^{L_0-1} z[(k+iN)T_s]e^{-j2\pi k/N}\right\} \qquad (9.6.28)$$

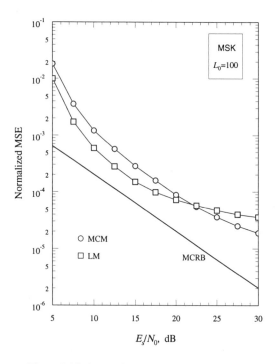

Figure 9.20. Comparison between LM and MCM.

with

$$z(t) = \left|\arg\left\{x(kT_s)x^*[(k-N)T_s]\right\}\right| \quad (9.6.29)$$

The performance of the LM estimator is investigated in [8] and is compared in Figure 9.20 with that of the MCM. The AAF bandwidth equals $1/T$ and the oversampling factor is set to $N = 4$ with both algorithms. An observation interval of 100 symbols is considered. As the LM exhibits some bias (in the range $\pm 1\%$ of $T$, depending on the value of $\tau$), the algorithm accuracy is expressed in terms of mean square error rather than variance with both LM and MCM. The parameter $\tau$ is taken uniformly distributed in the range $(0,T)$ in the simulations. It appears that LM is slightly superior at SNR values of practical interest.

## 9.7. Key Points of the Chapter

- Several timing synchronizers are available for CPM modulations. For convenience they have been divided into two categories, feedback and feedforward schemes. Feedforward schemes are better suited for burst mode data communications.

- The most accurate feedback method is described in Section 9.2 and is based on ML estimation criteria. The synchronizer is decision-directed and performs joint carrier phase and timing estimation. Its performance is close to the modified CRB at SNR values of practical interest. Although it is applicable to general CPM modulation, the synchronizer has a rather complex structure and is associated with an ML coherent receiver. Moreover, with multilevel and partial response formats it needs a lock detector to resolve false locks.

- A simpler feedback method, still based on ML estimation concepts, is described in Section 9.3. It applies to general CPM modulation and operates in a non-data-aided and carrier-independent manner. This makes it useful with less complex detection schemes and, in particular, with differential receivers. Its performance depends on the modulation parameters and worsens as the length of the frequency pulse increases.

- An even simpler method is reported in Section 9.4. This is based on *ad hoc* reasoning and, like the scheme in Section 9.3, operates in a non-data-aided and carrier-phase-independent manner. With a loop bandwidth of $B_T T = 10^{-3}$ the tracking accuracy is fairly good with MSK modulation (about 5 dB from the MCRB), a little worse with GMSK (about 8 dB from the MCRB).

- In summary, efficient feedback timing recovery appears to be viable only with short frequency pulses. With long pulses the only possible route is to resort to decision-directed estimation which, however, is more complex to implement.
- Feedforward timing methods are reported in Sections 9.5 and 9.6. In particular, in Section 9.5 a non-data-aided scheme is discussed which is useful with general CPM formats and has a performance comparable with that of the feedback synchronizer in Section 9.3.
- Section 9.6 describes two *ad hoc* synchronizers for MSK signaling. They have comparable performance and are very easy to implement.

## Appendix 9.A

Letting

$$F(\Delta t, t) = \prod_{i=-\infty}^{\infty} \frac{1}{M} \frac{\sin[2h\pi M p(t - iT, \Delta t)]}{\sin[2h\pi p(t - iT, \Delta t)]} \quad (9.A.1)$$

in this Appendix we investigate the properties of the coefficients $C_m(t_1, t_2)$ in the following Fourier expansion:

$$F(t_2 - t_1, t_2 - \tilde{\tau}) = \sum_{m=-\infty}^{\infty} C_m(t_1, t_2) e^{j 2\pi m \tilde{\tau}/T} \quad (9.A.2)$$

In particular, we show that the coefficients can be written as

$$C_m(t_1, t_2) = h_m(t_1 - t_2) e^{-j\pi m(t_1 + t_2)/T} \quad (9.A.3)$$

where $h_m(t)$ is defined by

$$h_m(t) \triangleq e^{j\pi m t/T} \frac{1}{T} \int_0^T F(-t, u) e^{j 2\pi m u/T} du \quad (9.A.4)$$

and satisfies the relation

$$h_m(t) = h_{-m}(t) \quad (9.A.5)$$

As a first step we prove that $F(\Delta t, t)$ satisfies the relations

$$F(-\Delta t, t) = F(\Delta t, t + \Delta t) \quad (9.A.6)$$

$$F(-\Delta t, t) = F(\Delta t, -t) \qquad (9.A.7)$$

provided that the condition

$$q(t) = \frac{1}{2} - q(LT - t) \qquad (9.A.8)$$

is met. Bearing in mind that

$$q(t) = \begin{cases} 0 & t \leq 0 \\ 1/2 & t \geq LT \end{cases} \qquad (9.A.9)$$

the meaning of (9.A.8) is that $q(t)$ must be anti-symmetric around $t=LT/2$, which is true in most practical cases.

To prove (9.A.6) consider the definition

$$p(t, \Delta t) = q(t) - q(t - \Delta t) \qquad (9.A.10)$$

It is readily seen that $p(t, -\Delta t) = -p(t + \Delta t, \Delta t)$. Then, substituting into (9.A.1) yields (9.A.6).

To arrive at (9.A.7) we write (9.A.10) in the form

$$p(t, -\Delta t) = q(t) - q(t + \Delta t) \qquad (9.A.11)$$

Then, combining with (9.A.8) results in

$$p(t, -\Delta t) = -p(LT - t, \Delta t) \qquad (9.A.12)$$

from which (9.A.7) follows, bearing in mind that $F(\Delta t, LT - t) = F(\Delta t, -t)$ (since $F(\Delta t, t)$ is periodic of period $T$ with respect to $t$).

Next, we turn our attention to the coefficients $C_m(t_1, t_2)$ in (9.A.2). They are expressed by

$$C_m(t_1, t_2) = \frac{1}{T} \int_0^T F(t_2 - t_1, t_2 - \tilde{\tau}) e^{-j2\pi m\tilde{\tau}/T} d\tilde{\tau} \qquad (9.A.13)$$

Letting $\tilde{\tau} = t_2 - t$ and integrating yields

$$C_m(t_1, t_2) = c_m(t_2 - t_1) e^{-j2\pi m t_2/T} \qquad (9.A.14)$$

with

$$c_m(t) \triangleq \frac{1}{T} \int_0^T F(t, x) e^{j2\pi mx/T} dx \qquad (9.A.15)$$

As $F(\Delta t, t)$ is real valued, from (9.A.15) it follows that

$$c_{-m}(t) = c_m^*(t) \tag{9.A.16}$$

Also, substituting (9.A.6)-(9.A.7) into (9.A.15) yields $c_m(-t) = c_m(t)e^{-j2\pi mt/T}$ and $c_m(-t) = c_m^*(t)$. Putting these facts together it follows that the function

$$h_m(t) = c_m(-t)e^{j\pi mt/T} \tag{9.A.17}$$

is real valued and satisfies the relations

$$h_{-m}(t) = h_m(t), \quad h_m(-t) = h_m(t) \tag{9.A.18}$$

Hence, combining (9.A.14) with (9.A.17) leads to (9.A.3) whereas (9.A.4) follows from (9.A.15) and (9.A.17).

# References

[1] J.B.Anderson, T.Aulin and C.-E.Sundberg, *Digital Phase Modulation*, New York: Plenum, 1986.
[2] J.B.Anderson and C-E.Sundberg, Advances in Constant Envelope Coded Modulation, *IEEE Communications Magazine*, **29**, No. 12, 36-45, Dec. 1991.
[3] J. Huber and W. Liu, Data-Aided Synchronization of Coherent CPM Receivers, *IEEE Trans. Commun.*, **COM-40**, 178-189, Jan. 1992.
[4] M.Morelli, U.Mengali and G.M.Vitetta, Joint Phase and Timing Recovery with CPM Signals, to appear in *IEEE Trans. Commun.*
[5] A.N.D'Andrea, U.Mengali and R.Reggiannini, A Digital Approach to Clock Recovery in Generalized Minimum Shift Keying, *IEEE Trans. Veh. Tech.*, **VT-39**, 227-234, Aug. 1990.
[6] A.N.D'Andrea, U.Mengali and M.Morelli, Symbol Timing Estimation with CPM Modulation, *IEEE Trans. Commun*, **COM-44**, 1362-1372, Oct. 1996.
[7] R.Mehlan, Y-E.Chen and H.Meyr, A Fully Digital Feedforward MSK Demodulator with Joint Frequency Offset and Symbol Timing Estimation for Burst Mode Mobile Radio, *IEEE Trans. Veh. Tech.*, **VT-42**, 434-443, Nov. 1993.
[8] U.Lambrette and H.Meyr, Two Timing Recovery Algorithms for MSK, Proc. IEEE ICC'94, New Orleans, Louisiana, May 1994, pp. 1155-1159.

# Index

Acquisition
  time, 127
  of frequency, 120
  of phase, 206
  of timing, 419
Ambiguity resolution, 204, 282, 369
AWGN (additive white Gaussian noise), 10, 17

Bandpass signal: *see* Signal
Baseband signal: *see* Signal
Basepoint index, 364
Bound
  Cramer-Rao (CRB), 54
  modified Cramer-Rao (MCRB), 55
    in frequency estimation, 58, 60
    in phase estimation, 61
    in timing estimation, 64, 67

Channel
  AWGN, 10, 17
  estimation, 251, 252, 259, 329, 334, 338-339
  Rayleigh fading, 246
  Ricean fading, 246
Closed-loop: *see* Feedback
Coherent receiver: *see* Receiver
Complex envelope
  of CPM signals, 28
  of linerly modulated signals, 16
  of bandpass noise, 18
Correlative state, 29
Costas loop, 206
CPM (continuous-phase modulation)
  ML sequence estimation with, 29, 320-321, 479
  simplified receiver for, 31
CRB (Cramer-Rao bound), 54
Cycle slipping, 210

Damping factor, 222
Degradations
  due to frequency errors, 27, 198, 217
  due to phase errors, 20, 24-25, 36
  due to timing errors, 12, 31
Delay-and-multiply method, 97, 106, 133, 169, 176, 506
Demodulation, 17
Differentially coherent detection, 204, 310, 458
Differential detection, 26, 34, 257
Doppler bandwidth, 247
Dual filter detector, 133

Energy per symbol, 71
Equilibrium point
  spurious, 212, 387, 420, 485, 495
  stable, 208, 420
  unstable, 208, 420
Equivalent model, 122, 207, 225, 239, 419
Error detector
  frequency: *see* FED
  phase: *see* PED
  timing: *see* TED
Estimation
  bounds in parameter, 53
  frequency, 79-188
  maximum likelihood, 38, 47
  phase, 189-352
  timing, 353-516

Eye opening, 12

Fading
  fast, 333, 336
  frequency-flat, 245, 327, 438
  linearly time-selective, 336
  Rayleigh, 245
  Ricean, 245
  slow, 248
False lock detection, 495
FED (frequency error detector)
  for CPM, 158, 167, 174, 175
  for linear modulation, 102, 105, 118, 130
Feedback
  frequency estimation
    with CPM, 158, 166, 175
    with linear modulation, 100, 104, 119
  phase estimation
    with CPM, 316, 323, 481
    with linear modulation, 203, 206, 230, 240, 273, 275, 289, 414, 455
  timing estimation
    with baseband transmission, 354, 374, 384, 385, 392, 393
    with CPM, 481, 484, 492, 496, 498
    with linear modulation, 414, 417, 429, 431, 440, 442, 448, 454, 459
Feedforward
  frequency estimation
    with CPM, 155, 156, 169, 177, 180
    with linear modulation, 83, 84, 105, 115, 133, 140
  phase estimation
    with CPM, 312, 341, 345, 349
    with linear modulation, 192, 195, 277, 281, 291, 295, 298, 301, 467
  timing estimation
    with baseband transmission, 354, 402, 403
    with CPM, 504, 507, 512
    with linear modulation, 436, 467
Filter
  derivative matched, 117
  matched, 10
Fisher information matrix, 57
Fractional interval, 364

Frequency offset, 17
Frequency estimation
  with CPM
    data-aided (DA), 154, 156
    clock-aided, 154, 156, 176
    clockless, 157, 169
    ML-based, 157
    non-data-aided (NDA), 157, 169, 176
  with linear modulation
    data-aided (DA), 80
    decision-directed (DD), 97
    clock-aided, 80, 97, 100
    clockless, 108, 133
    Fitz method, 88
    Kay method, 86
    Luise and Reggiannini method, 89
    ML, 80
    non-data-aided (NDA), 100, 108, 133
Frequency offset, 17
Fractional interval, 364
Full response, 28

Gardner detector: *see* TED
GMSK (Gaussian MSK), 29, 149

Hangup, 209

Interleaving, 249
Interpolator
  ideal, 360
  polinomial, 362

Joint phase and timing estimation, 412, 452, 462, 478

Laurent expansion, 148
Likelihood function, 43
Limiter-discriminator receiver, 32
Linear modulation, 15
Lodge and Moher receiver, 333
Locking range, 220
Loop bandwidth: *see* Noise equivalent bandwidth
Loop equivalent model: *see* Equivalent model
LRC (raised-cosine pulse of length $L$), 149
LREC (rectangular pulse of length $L$), 149

# Index

Matched-filter: *see* Filter
MCRB (modified Cramer-Rao bound), 55
  with baseband transmission for timing, 64
  with CPM
    for carrier frequency, 60
    for carrier phase, 61
    for timing, 67
  with linear modulation
    for carrier frequency, 58
    for carrier phase, 61
    for timing, 64
Mean-time-to-slip, 244
ML (maximum likelihood)
  estimation criterion, 38
  frequency estimation, 80
  parameter estimation, 38
  phase estimation
    with CPM, 311
    with linear modulation, 190, 194
  sequence detection, 30, 248, 320, 479
  timing estimation, 371
Modulation
  CPM, 28
  Gaussian MSK (GMSK), 29
  linear, 15
  MSK, 28
  MSK-type, 33, 148, 308
M-power method, 279
Mueller and Mueller timing detector: *see* TED
Multipath channel: *see* Channel
Multiple synchronizers, 240

Natural frequency, 222
NCO (number controlled oscillator), 356
Noise equivalent bandwidth, 126, 214, 326
Nyquist criterion, 11

Oerder and Meyr algorithm, 402, 435
Open-loop: *see* Feedforward
OQPSK (offset QPSK), 16

PAM (pulse amplitude modulation), 10
Partial response, 28
PED (phase error detector)
  for CPM, 315, 323, 481
  for linear modulation, 206, 229, 240, 271, 273, 274, 289, 414, 454

Per-survivor-processing, 240, 259, 328
Phase ambiguity, 204
Phase error, 20
Phase estimation
  with CPM
    clock-aided, 311, 314, 319, 339
    clockless, 346
    data-aided (DA), 311
    decision-directed (DD), 314, 319, 481
    non-data-aided (NDA), 339, 346
  with linear modulation
    data-aided (DA), 190
    clock-aided, 190, 201, 228, 236, 248, 266, 286
    clockless, 297, 299
    decision-directed (DD), 201, 228, 236, 245, 412, 452
    maximum-likelihood, 190, 194
    non-data-aided (NDA), 266, 286, 297, 299
Phase noise, 224
Phase response, 28, 148
Phase state, 29
Pilot-symbol, 252
Pilot-tone, 251
Poisson sum formula, 393
Power spectral density,
  with baseband transmission, 71
  with linear modulation, 72
Pseudo-symbols, 33
PSK (phase shift keying), 16

QAM, 16
QPSK, 20
Q-function, 23
Quadricorrelator, 131

Raised-cosine rolloff function, 11
Rayleigh distribution, 246
Rayleigh fading (*see* Channel)
Receiver,
  coherent
    with CPM, 30, 35, 310, 316, 322
    with linear modulation, 19, 206, 230
  differential
    with linear modulation, 26
    with MSK-type modulation, 34
  limiter-discriminator, 32
  Lodge and Moher, 333

maximum likelihood
  with CPM, 322
  with linear modulation, 248-249
  MSK-type, 310
Ricean distribution, 246
Ricean fading: *see* Channel
Rolloff factor, 11
Root-raised-cosine rolloff function, 12

Sampling
  non-synchronous, 359
  synchronous, 355
Second-order loop, 220
Self noise, 130
Self-noise elimination, 396
S-curve, 120, 206, 211, 232, 375, 386-387
Signal
  baseband, 10
  CPM, 28
  linearly modulated, 15
Simplified CPM receiver, 31
Slip: *see* Cycle slipping
Static phase error, 219
Step-size, 118
S-surface, 419, 481
Survivor, 236, 259, 328
Synchronization
  bounds in, 53
  function
    frequency recovery, 18
    phase recovery, 18
    timing recovery, 10

TCM (trellis-coded modulation), 29, 236
TED (timing error detector)
  for baseband transmission
    early-late, 384
    Gardner, 393
    ML-based, 371, 491
    Mueller and Mueller, 385
    zero-crossing, 385,
  for linear modulation

  early-late, 429, 439
  Gardner, 431, 442, 459
  ML-based, 414, 429, 439, 448, 450, 454
  Mueller and Mueller, 418, 440
  zero-crossing, 417, 440
  for CPM
    ML-based, 481, 492
    ad hoc, 498
Tentative decision, 236
Timing adjustment
  with non-synchronous sampling, 366
  with synchronous sampling, 357
Timing jitter, 13
Timing estimation
  with baseband transmission,
    decision-directed (DD), 371
    non-data-aided (NDA), 391, 398
  with CPM,
    decision-directed (DD), 478
    non-data-aided (NDA), 487
  with linear modulation,
    decision-directed (DD), 413, 439, 452
    non-data-aided (NDA), 428, 433, 442, 448, 450, 467
Tracking
  frequency, 124
  phase, 206, 213, 232, 481
  timing, 377, 395, 414, 417, 425, 429, 431, 439-440, 442, 448, 450, 454, 459, 481, 492

Unstable equilibrium point, 208
Unique word, 282
Unwrapping algorithm
  for phase recovery, 284
  for timing recovery, 369

Viterbi detector, 29, 253, 322
Viterbi and Viterbi method, 280
VCO (voltage controlled oscillator), 100